*Soil Fertility
and Fertilizers*

SAMUEL L. TISDALE
LATE PRESIDENT, THE SULPHUR INSTITUTE

WERNER L. NELSON
LATE SENIOR VICE PRESIDENT, POTASH & PHOSPHATE INSTITUTE

JAMES D. BEATON
POTASH & PHOSPHATE INSTITUTE OF CANADA

JOHN L. HAVLIN
KANSAS STATE UNIVERSITY

Soil Fertility and Fertilizers

FIFTH EDITION

MACMILLAN PUBLISHING COMPANY
NEW YORK
Maxwell Macmillan Canada
TORONTO
Maxwell Macmillan International
NEW YORK   OXFORD   SINGAPORE   SYDNEY

Editor: Paul F. Corey
Production Supervisor: Margaret Comaskey
Production Manager: Paul Smolenski
Cover Designer: Cathleen Norz
Cover illustration: photograph © Mark Gibson

This book was set in Baskerville by Crane Typesetting Service, Inc., printed and bound by Hamilton Printing Company. The cover was printed by Lehigh Press.

Macmillan Publishing Company
866 Third Avenue, New York, New York 10022

Macmillan Publishing Company is part
of the Maxwell Communication Group of Companies.

Maxwell Macmillan Canada, Inc.
1200 Eglinton Avenue East
Suite 200
Don Mills, Ontario M3C 3N1

LIBRARY OF CONGRESS CATALOGING IN PUBLICATION DATA
Soil fertility and fertilizers / Samuel L. Tisdale . . . [et al.]. —
    5th ed.
        p.   cm.
    Rev. ed. of: Soil fertility and fertilizers / Samuel L. Tisdale,
Werner L. Nelson, James D. Beaton. 4th ed. c1985.
    Includes bibliographical references and index.
    ISBN 0-02-420835-3
    1. Fertilizers.   2. Soil fertility.   3. Crops—Nutrition.
I. Tisdale, Samuel L.   II. Tisdale, Samuel L.   Soil fertility and
fertilizers.
S633.S715   1993
631.4'22—dc20
                                                           92-32122
                                                           CIP

Printing: 2 3 4 5 6 7 8   Year: 4 5 6 7 8 9 0 1 2

Dedication

Soil Fertility and Fertilizers was first published in 1956 and it is our honor to dedicate this fifth edition to the memory of Dr. Samuel L. Tisdale (1918–1989) and Dr. Werner L. Nelson (1914–1992). To adequately describe and summarize their remarkable accomplishments and numerous contributions to agricultural science would require a lengthy chapter for each. Knowing their intense interest in and dedication to agricultural education, we know they would want us to devote all of this edition to soil fertility, plant nutrition, and nutrient management subject matter. However, it is important for those who did not have the opportunity to know them that we briefly describe their impressive careers.

Dr. Tisdale was born in Tampa, Florida in 1918, finished high school in 1936 (Hattiesburg, Mississippi), and earned his B.S. in agricultural science from Auburn University in 1942. After serving in World War II, he enrolled at Purdue University, where he completed his Ph.D. in soil fertility and plant nutrition in 1949. His scientific and leadership skills were demonstrated and recognized early in his career. After two years as assistant professor at North Carolina State University and two years as agronomist with the Tennessee Valley Authority, he returned to North Carolina in 1953 as associate professor of soils and was promoted to full professor and director of the state's soil testing laboratory in 1955, succeeding Dr. Nelson in that position. His leadership greatly contributed to North Carolina State University's strong international and national reputation in soil testing, soil fertility, and plant nutrition. After serving as southeastern regional director of the National Plant Food Institute (1958–1960), he joined The Sulphur Institute as director of agricultural research, where he retired as president in 1984. For two and a half decades he tirelessly promoted and communicated the importance of sulfur in agriculture throughout the world. His international contributions as a scientist and industry leader are remembered through numerous publications, awards, and honors, but more importantly, by many agriculturalists who were touched by his ideas, kindness, and humor.

Dr. Nelson was born in Sheffield, Illinois, in 1914 and received his B.S. in agronomy (1937) and M.S. in soil chemistry (1938) from the University of Illinois and his Ph.D. in soil physics (1940) from Ohio State University. After two years as faculty member at the University of Idaho (1940–1941), he spent thirteen years (1941–1954) in research and teaching at North Carolina State University, where he was promoted to professor in charge of soil fertility research and director of the state's soil testing laboratory. In 1954 he joined The Potash and Phosphate Institute, where he rapidly rose to senior vice-president and served in that capacity until his retirement in 1986. Those of us who knew him know that he never retired, as he continued writing and contributing to the fifth edition of this text until his death in 1992. His contributions to efficient and profitable agriculture in the United States and to solving food stress problems in underdeveloped nations are evidenced through his leadership in the United Nations Food and Agriculture Organization, Council on Agricultural Science and Technology, the American Society of Agronomy, the Soil Science Society of America, and many other national and regional organizations. His immense accomplishments and contributions have been recognized through numerous university, industry, and society honors and awards. Three awards have been established in his name: the National Fertilizer Solutions Association Werner L. Nelson Award, the Indiana Plant Food and Agricultural Chemicals Association Werner L. Nelson Award, and the American Society of Agronomy Werner L. Nelson Award for Diagnosis of Yield Limiting Factors. Although his leadership in agricultural science, industry, and education will be greatly missed, the way he touched people through his intelligence, inquisitiveness, intensity, caring, and encouragement is the greatest loss.

We dedicate the fifth edition of *Soil Fertility and Fertilizers* to Samuel L. Tisdale and Werner L. Nelson. No matter how future editions may be changed, their contributions will always remain the strength of this text.

John L. Havlin
James D. Beaton

Preface

The importance of soil fertility and plant nutrition to the health and survival of all life can not be understated. As human populations continue to increase, human disturbance of earth's ecosystem to produce food and fiber will place greater demand on soils to supply essential nutrients. Therefore, it is critical that we increase our understanding of the chemical, biological, and physical properties and relationships in the soil–plant–atmosphere continuum that control nutrient availability.

The evidence is clear that the soil's native ability to supply sufficient nutrients has decreased with the higher plant productivity levels associated with increased human demand for food. One of the greatest challenges of our generation will be to develop and implement soil, crop, and nutrient management technologies that enhance the quality of the soil, water, and air. If we do not improve and/or sustain the productive capacity of our fragile soils, we can not continue to support the food and fiber demand of our growing population.

To the Student

The goal of this book is to establish a thorough understanding of plant nutrition, soil fertility, and nutrient management in the student so that she or he can (1) describe the influence of numerous soil biological, physical, and chemical properties on nutrient availability to crops, (2) identify plant nutrition–soil fertility problems and recommend proper corrective action, and (3) identify soil and nutrient management practices that maximize productivity and profitability, while maintaining or enhancing the productive capacity of the soil and the quality of the environment.

The specific objectives are to (1) describe how plants take up or absorb plant nutrients and how the soil system supplies these nutrients, (2) identify and describe plant nutrient deficiency symptoms and methods used to quantify nutrient problems, (3) describe how soil organic matter, cation exchange capacity, soil pH, parent material, climate, and human activities affect nutrient availability, (4) evaluate nutrient and soil amendment materials on the basis of content, use, and effects on the soil and the crop, (5) quantify, using basic chemical principles, application rates of nutrients and amendments needed

vii

to correct plant nutrition problems in the field, (6) describe nutrient response patterns, fertilizer use efficiency, and the economics involved in fertilizer use, and (7) describe and evaluate soil and nutrient management practices that either impair or sustain soil productivity and environmental quality.

To the Teacher

Motivate your students to learn by showing them how the knowledge and skills gained through the study of soil fertility will be essential for success in their careers. Use teaching methodologies that enhance their critical thinking and problem solving skills. In addition to understanding the qualitative soil fertility and plant nutrition relationships, students must know how to quantitatively evaluate nutrient availability. Environmental protection demands that nutrients be added in quantities and by methods that maximize crop productivity and recovery of the added nutrients.

Since some of the examples used in this text may not be representative of your specific region, frequently integrate additional field examples from your region to illustrate the qualitative and quantitative principles. Strongly reinforce the reality that production agriculture, sustainability, and environmental quality are compatible provided soil, crop, and nutrient management technologies are used properly. Develop in your students the desire and discipline to expand beyond this text through reading and self-learning. Demand of your students what will be demanded of them after they graduate—to think, communicate, cooperate, and solve problems from an interdisciplinary perspective.

We hope your students find the text a valuable resource throughout their career. Please feel free to provide suggestions for enhancing the effectiveness of the text as a teaching and learning aid.

Acknowledgments

We wish to thank the following reviewers for their helpful comments on the manuscript:

Mark M. Alley, Virginia Polytechnic Institute and State University; James R. Brown, University of Missouri at Columbia; Richard H. Fox, Pennsylvania State University; Jay W. Johnson, Ohio State University at Columbus; and H. M. Reisenauer, University of California at Davis.

John L. Havlin
James D. Beaton

Contents

CHAPTER 1

Soil Fertility—Past and Present   *1*

Ancient Records   2
Soil Fertility During the First Eighteen Centuries
     A.D.   4
Progress During the Nineteenth Century   6
The Development of Soil Fertility in the United
     States   9
Increasing Crop Yields   10
Looking to the Twenty-first Century   11
Selected References   13

CHAPTER 2

Growth and the Factors Affecting It   *14*

Factors Affecting Plant Growth   14
Growth and Growth Expressions   33
Modeling Applications in Soil Fertility   41
Summary   42
Questions   42
Selected References   44

CHAPTER 3

Elements Required in Plant Nutrition   *45*

Essentiality of Elements in Plant Nutrition   45
Elements Required in Plant Nutrition   46
Function of Nutrients in Plants: Macronutrients   48
Function of Nutrients in Plants: Micronutrients   66
Summary   78

Questions   78
Selected References   79

CHAPTER 4

Basic Soil–Plant Relationships   *80*

Ion Exchange in Soils   81
Movement of Ions from Soils to Roots   96
Ion Absorption by Plants   102
Summary   106
Questions   107
Selected References   108

CHAPTER 5

Soil and Fertilizer Nitrogen   *109*

The N Cycle   109
N Fixation by *Rhizobium* and Other Symbiotic
    Bacteria   110
N Fixation by Nonsymbiotic Soil
    Microorganisms   117
N Addition from the Atmosphere   118
Industrial Fixation of N   119
Forms of Soil N   119
Forms of N Absorbed by Plants   120
N Transformations in Soils   122
Fertilizer N Materials   153
Summary   172
Questions   173
Selected References   175

CHAPTER 6

Soil and Fertilizer Phosphorus   *176*

The P Cycle   176
P Fertilizers   205
Summary   227
Questions   228
Selected References   229

CHAPTER 7

Soil and Fertilizer Potassium   *230*

K Content of Soils   230
Forms of Soil K   231

Soil Factors Affecting K Availability   242
Plant Factors Affecting K Availability   251
K Soil Testing   256
K Fertilizers   258
Agronomic Value of K Fertilizers   261
Summary   262
Questions   263
Selected References   265

CHAPTER 8
Soil and Fertilizer Sulfur, Calcium,
and Magnesium   *266*

Sulfur   266
Calcium   289
Magnesium   296
Summary—S   300
Summary—Ca and Mg   301
Questions—S   302
Questions—Ca and Mg   303
Selected References   303

CHAPTER 9
Micronutrients and Other Beneficial Elements
in Soils and Fertilizers   *304*

General Relationships of Micronutrients
    in Soils   304
Iron   304
Zinc   319
Copper   327
Manganese   332
Boron   337
Chloride   342
Molybdenum   346
Cobalt   350
Sodium   351
Silicon   353
Selenium   356
Summary   359
Questions   361
Selected References   363

CHAPTER 10
Soil Acidity and Basicity

364

General Concepts   364
Soil Acidity   367
The Soil as a Buffer   373
Determination of Active and Potential Acidity
    in Soils   374
Soil pH for Crop Production   377
Liming Materials   378
Use of Lime in Agriculture   384
Application of Liming Materials   388
Factors Determining the Selection of a Liming
    Program   392
Acidulating the Soil   393
Saline, Sodic, and Saline/Sodic Soils   396
Summary   401
Questions   402
Selected References   404

CHAPTER 11
Soil Fertility Evaluation

405

Nutrient-Deficiency Symptoms of Plants   405
Plant Analyses   410
Biological Tests   426
Soil Testing   428
Remote Sensing   455
Summary   456
Questions   460
Selected References   461

CHAPTER 12
Fundamentals of Fertilizer Application

462

Crop Characteristics   462
Soil Characteristics   469
Fertilizer Placement   474
General Considerations   483
Utilization of Nutrients from the Subsoil   502
Fluid Fertilizers   505
Summary   507
Questions   509
Selected References   510

CHAPTER 13

Fertilizers, Water Use, and Other Interactions *511*

Nutrient–Water Interactions   513
Other Interactions   528
Summary   536
Questions   538
Selected References   539

CHAPTER 14

Economics of Plant-Nutrient Use *540*

Fertilizer Use and Prices   540
Maximum Economic Yield   542
Yield Level and Unit Cost of Production   543
Returns per Dollar Spent or Profit per Acre   546
What Is the Most Profitable Rate of Plant
    Nutrients?   547
Price of Fertilizer in Relation to the Value
    of the Crop   548
Price per Pound of Nutrients   553
Liming   554
Animal Wastes   555
Plant Nutrients as Part of Increasing Land
    Value   555
Additional Benefits from Maximum Economic
    Yields   556
Summary   558
Questions   559
Selected References   560

CHAPTER 15

Cropping Systems and Soil Management *561*

Soil and Crop Productivity   563
OM in the Soil   568
Tillage   572
Legumes in the Rotation   583
Rotations Versus Continuous Cropping   588
Winter Cover/Green Manure Crops   592
Animal Manure   594
Sewage   601
Summary   605
Questions   607
Selected References   607

Common Conversions and Constants   *609*

Index   *613*

CHAPTER 1

Soil Fertility—Past and Present

During most of our existence on earth, food has been procured by hunting and gathering. The various systems of agriculture in the past and the estimated global population each system was capable of sustaining are summarized in Table 1.1. All of these systems exist in various parts of the globe today since the entire world has not advanced in unison from one system to the next. Presently there is starvation in large areas where the systems for producing and/or transporting food are inadequate. In contrast, developed countries utilizing improved cultural practices and varieties, fertilizers, and pesticides are generally self-sufficient in food production.

By 1975 the world's population reached 4 billion, close to the value of 4.2 billion appearing in Table 1.1. With the global population anticipated to exceed 6 billion by the year 2000 and to stabilize at 11 billion in 2050, it is obvious that large and sustained increases in food production will be needed. Since cultivated land areas are expected to expand by only about 20%, it will be necessary to intensify agricultural production, wherein fertilizers will play a vital role.

Because of the importance of soil fertility in food production, it is fitting to trace the highlights in our understanding and development of sound soil fertility practices. The period during which humans began cultivation of plants marks the dawn of agriculture. The exact time is not known, but it was certainly several thousand years before the birth of Christ.

TABLE 1.1   Capability of Agricultural Systems to Produce Food and Support Population

Agricultural System	Cultural Stage or Time	Cereal Yield (t/ha)	World Population (millions)	Hectares per Person
Hunting and gathering	Paleolithic		7	
Shifting agriculture	Neolithic (10,000 years ago)	1	35	40.0
Medieval rotation	500–1450	1	900	1.5
Livestock farming	Late 1700s	2	1800	0.7
Fertilizers/pesticides	Twentieth century	4	4200	0.3

SOURCE: McCloud, *Agron. J.,* **67:**1 (1975).

Ancient Records

As man became less of a wanderer and more of a settler, families, clans, and villages developed, and with them came the development of agriculture. One area in the world that shows evidence of early civilization is Mesopotamia, situated between the Tigris and Euphrates rivers in what is now Iraq. Writings dating back to 2500 B.C. mention the fertility of the land. It is recorded that the yield of barley was 86-fold and even 300-fold in some areas, which means that for every seed planted, 86 to 300 seeds were harvested.

Herodotus, the Greek historian, reporting on his travels through Mesopotamia some 2,000 years later, mentioned the phenomenal yields obtained by these people. The high production was probably the result of a well-developed irrigation system and soil of high fertility, attributable in part to annual flooding by the river. Theophrastus, writing around 300 B.C., referred to the richness of the Tigris alluvium and stated that the water was allowed to remain on the land as long as possible so that a large amount of silt might be deposited.

In time man learned that certain soils failed to produce satisfactory yields when cropped continuously. The practice of adding animal and vegetable manures to the soil to restore fertility probably developed from such observations, but how or when fertilization actually began is not known.

Around 800 B.C., Homer mentioned in the Greek poem The Odyssey the manuring of vineyards by the father of Odysseus. References to a "manure heap" also suggest its systematic collection and storage. These writings suggest that manuring was an agricultural practice in Greece nine centuries before the birth of Christ.

Theophrastus (372–287 B.C.) recommended manuring of thin soils but suggested that rich soils be manured sparingly. He also endorsed a practice considered good today—the use of bedding in the stall, which conserved urine and bulk and increased the humus value of the manure. Theophrastus also suggested that plants with high nutrient requirements also had a high water requirement.

The truck gardens and olive groves around Athens were enriched by city sewage sold to farmers. A canal system was used, and there is evidence of a device for regulating the flow. The ancients also fertilized vineyards and groves with water that contained dissolved manure.

Manures were classified according to their concentration. Theophrastus, for example, listed them in the following order of decreasing value: human, swine, goat, sheep, cow, oxen, and horse. Later, Varro, an early writer on Roman agriculture, developed a similar list but rated bird and fowl manure as superior to human excrement. Columella suggested the feeding of clover to cattle because it enriched the excrement.

Not only did the ancients recognize the merits of manure, but they also observed the effect that dead bodies had on increasing crop growth. Archilochus made such an observation around 700 B.C., and the Old Testament records are even earlier. In Deuteronomy it is directed that animals' blood should be poured on the ground. The increased fertility of land that has received the bodies of the dead has been acknowledged through the years, probably most poetically by Omar Khayyam, the astronomer-poet of Persia, who around the end of the eleventh century wrote:

> I sometime think that never blows so red
> The rose as where some buried Caesar bled;
> That every hyacinth the garden wears
> Dropt in her lap from some once lovely head.
>
> And this delightful herb whose tender green
> Fledges the rivers lip on which we lean—
> Ah, lean upon it lightly! for who knows
> From what once lovely lip it springs unseen.

The value of green-manure crops, particularly legumes, was also recognized. Theophrastus noted that when farmers plowed under residues from a good bean crop (*Vicia faba*), the soil was enriched. Virgil (70–19 B.C.) advocated the application of legumes, as indicated in the following passage:

> Or, changing the season, you will sow there yellow wheat, whence before you have taken up the joyful pulse, with rustling pods, or the vetch's slender offspring and the bitter lupine's brittle stalks, and rustling grove.

The use of mineral fertilizers or soil amendments was not entirely unknown to the ancients. Theophrastus suggested mixing different soils to "remedy defects and add heart to the soil." This practice may have been beneficial from several standpoints. The addition of fertile soil to infertile soil could lead to increased fertility and may have provided better inoculation of legume seed on some fields. Also, the mixing of coarse-textured soils with fine-textured soils may have improved water and air relations in the soils.

The value of liming was also recognized. Early dwellers of Aegina dug up marl and applied it to their land. The Romans, who learned this practice from the Greeks, classified the various liming materials and recommended that one type be applied to grain and another to meadow. Pliny (A.D. 62–113) stated that lime should be spread thinly on the ground and that one treatment was "sufficient for many years, though not 50." Columella also recommended spreading marl on a gravelly soil and mixing gravel with a dense calcareous soil.

The Bible records the value of wood ashes in its reference to the burning of briars and bushes by the Jews, and Xenophon and Virgil both reported the burning of stubble to clear fields and destroy weeds. Cato advised the vine keeper to burn prunings on the spot and to plow in the ashes to enrich the soil. Pliny stated that lime from lime kilns was excellent for olive groves, and some farmers burned manure and applied the ashes to their fields. Columella also suggested the spreading of ashes or lime on lowland soils to reduce acidity.

Saltpeter, or potassium nitrate, was mentioned by both Theophrastus and Pliny as useful for fertilizing plants and is referred to in the Bible in the book of Luke. Brine was mentioned by Theophrastus. Apparently recognizing that palm trees required large quantities of salt, early farmers poured brine around the roots of their trees.

Virgil wrote on the soil characteristic known today as *bulk density*. His advice on determining this property was as follows:

> ... first, you shall mark out a place with your eye, and order a pit to be sunk deep in solid ground, and again return all the mold into its place, and level with your feet the sands at top. If they prove deficient, the soil is loose and more fit for cattle and bounteous vines; but if they deny the possibility of returning to their places, and there be an over plus of mold after the pit is filled up, it is a dense soil; expect reluctant clods and stiff ridges, and give the first plowing to the land with sturdy bullocks.

Virgil and Columella suggested a taste test to measure the degree of acidity and salinity of soils, and Pliny stated that the bitterness of soils might be detected by the presence of black and undergound herbs.

Pliny wrote that "among the proofs of the goodness of soil is the comparative thickness of the stem in corn," and Columella stated that the best test for the suitability of land for a given crop was whether it would grow.

Many of the early writers believed that the color of the soil was a criterion of its fertility in that black soils were fertile and light or gray soils infertile. Columella disagreed, pointing to the infertility of the black marshland soils and the high fertility of the light-colored soils of Libya. He felt that such factors as structure, texture, and acidity were far better estimates of soil fertility.

The age of the Greeks from perhaps 800 to 200 B.C. was indeed a Golden Age. Many of the people of this period displayed genius that was unequaled for centuries to come. Their writings, culture, and agriculture were copied by the Romans, and the philosophy of many of the Greeks of this period dominated human thinking for more than 2,000 years.

Soil Fertility During the First Eighteen Centuries A.D.

After the decline of Rome there were few advances in agriculture until the publication of *Opus ruralium commodorum*, a collection of local agricultural practices, by Pietro de Crescenzi (1230–1307). De Crescenzi is considered the founder of modern agronomy, but his manuscript is confined to the work of writers from the time of Homer. His contribution consisted largely of summarizing the material; however, he did suggest increasing manure application rates.

After the appearance of de Crescenzi's work, little was added to agricultural knowledge for many years, although Palissy in 1563 is credited with the observation that the ash content of plants represented the material they had removed from the soil.

Francis Bacon (1561–1624) suggested that the principal nourishment of plants was water. He believed that the main purpose of the soil was to keep plants erect and to protect them from heat and cold, and he stated that each plant drew from the soil a substance unique for its own particular nourishment. Bacon maintained further that continued production of the same type of plant would impoverish the soil for that particular species.

During this same period, Jan Baptiste van Helmont (1577–1644), a Flemish physician/chemist, reported experimental results that he believed proved that water was the sole nutrient of plants. He placed 200 lb of soil in an earthen container, moistened the soil, and planted a willow shoot weighing 5 lb. He carefully shielded the soil in the crock from dust, and only rain or distilled water was added. After five years, van Helmont terminated the experiment. The tree weighed 169 lb and about 3 oz. He could account for all but about 2 oz of the 200 lb of soil originally used. Because he had added only water, his conclusion was that water was the sole nutrient of the plant, for he attributed the 2 oz soil loss to experimental error. Van Helmont's conclusion, while wrong, stimulated later investigations, which further contributed to the understanding of plant nutrition.

The work of van Helmont was repeated several years later by Robert Boyle (1627–1691) of England. Boyle is probably best known for expressing the relation of the volume of a gas to its pressure. He was also interested in biology and believed in the experimental approach to solving problems in science. Boyle confirmed the findings of van Helmont, but he concluded from chemical analyses of plant samples that plants contained salts, spirits, earth, and oil, which were formed from water.

About this same time, J. R. Glauber (1604–1668), a German chemist, suggested that saltpeter (KNO_3), and not water, was the "principle of vegetation." He collected the salt from soil under cattle pens and argued that it must have come from animals' droppings. He further stated that because the animals ate forage, the saltpeter must have come originally from the plants. When he applied this salt to plants and observed the large increases in growth, he was convinced that soil fertility and the value of manure were due to saltpeter.

John Mayow (1643–1679), an English chemist, estimated the quantities of nitrate in the soil at various times during the year and found it in its greatest concentration in the spring. Failing to find any during the summer, he concluded that the saltpeter had been absorbed by the plant during its period of rapid growth.

About 1700, a study was made that represented a considerable advance in agricultural science. Acquainted with the work of Boyle and van Helmont, John Woodward grew spearmint in rainwater, river water, sewage water, and sewage water plus garden mold. He carefully measured the quantity of water transpired by the plants and recorded plant weight at the beginning and end of the experiment. He found that the growth of the spearmint was proportional to the amount of impurities in the water and concluded that terrestrial matter, or earth, rather than water, was the principle of vegetation. Although his conclusion is not correct in its entirety, his experimental technique was considerably better than any that had been used before.

There was much understandable ignorance concerning plant nutrition during this period. Many quaint ideas came into being and then disappeared. Not the least of these ideas was introduced by another enterprising Englishman, Jethro Tull (1674–1741). Tull, educated at Oxford, appears to have been interested in politics, but ill health forced his retirement to the farm. There he carried out numerous experiments, most of which dealt with cultural practices. He believed that the soil should be finely pulverized to provide the "proper pabulum" for the growing plant. According to Tull, the soil particles were actually ingested through openings in the plant roots. The pressure caused by the swelling of the growing roots was thought to force this finely divided soil into "the lacteal mouths of the roots," after which it entered the "circulatory system" of the plant.

Although Tull's ideas about plant nutrition were a bit odd, his experiments led to the development of two valuable pieces of farm equipment, the grain drill and the horse-drawn cultivator. His book *Horse Hoeing Husbandry* was considered an authoritative text in English agricultural circles.

Around 1762, John Wynn Baker, a Tull adherent, established an experimental farm in England to exhibit the results of experiments in agriculture. Baker's work was praised later by Arthur Young (1741–1820), an English agriculturist, who admonished his readers to beware of giving too much credit to calculations based on the results of only a few years' work, an admonition that is as timely today as it was then. Young also experimented with substances that would improve crop yields. He grew barley in sand to which he added such materials as charcoal, train oil, poultry dung, spirits of wine, niter, gunpowder, pitch, oyster shells, and numerous other materials. Some of the materials produced plant growth; others did not. Young published a work entitled *Annals of Agriculture*, in 46 volumes, which was highly regarded and made a considerable impact on English agriculture.

Many of the agricultural writings of the seventeenth and eighteenth centuries reflected the idea that plants were composed of one substance, and most of the workers were searching for this *principle of vegetation*. Around 1775, however, Francis Home stated that there was not one principle but probably many, among which he included air, water, earth, salts, oil, and fire. Home felt that the problems of agriculture were essentially related to plant nutrition. He carried out pot experiments to measure the effects of different substances on plant growth and made chemical analyses of plant materials.

The discovery of oxygen (O_2) by Priestley was the key to a number of other discoveries that went far toward unlocking the mystery of plant life. Jan Ingenhousz (1730–1799) showed that air was purified in the presence of light. Coupled with this discovery was the statement by Jean Senebier (1742–1809), a Swiss natural philosopher and historian, that the increase in the weight of van Helmont's willow tree was the result of air!

Progress During the Nineteenth Century

These discoveries stimulated the thinking of Theodore de Saussure. He worked on two problems—the effect of air on plants and the origin of salts in plants. As a result, de Saussure demonstrated that plants absorbed O_2 and liberated carbon dioxide (CO_2), the central process of respiration. In addition, he found that plants absorb CO_2 and release O_2 in the presence of light; however, plants died when kept in a CO_2-free environment.

De Saussure concluded that the soil furnishes only a small fraction of the nutrients needed by plants, but he demonstrated that it does supply both ash (calcium [Ca], magnesium [Mg], potassium [K], and other minerals) and nitrogen (N). He dispelled the idea that plants generate potash spontaneously and that the plant root does not behave as a filter. Rather, the membranes are selectively permeable, allowing for a more rapid uptake of water than of salts. He also showed the differential absorption of salts and the inconstancy of plant composition, which varies with soil, plant age, and plant part.

De Saussure's conclusion that the carbon (C) contained by plants was derived from the air did not meet with immediate acceptance by his colleagues. No less a figure than Sir Humphry Davy, who published *The Elements of Agricultural Chemistry* about 1813, stated that although plants received some carbon from the air, most was taken in through the roots. Davy was so enthusiastic in this belief that he recommended the use of oil as a fertilizer because of its C and hydrogen [H] content.

During the late nineteenth to the early twentieth century, much progress was made in the understanding of plant nutrition and crop fertilization. Among the men of this period whose contributions loom large was Jean Baptiste Boussingault (1802–1882), a French chemist, who established a farm in Alsace on which he carried out field-plot experiments and who is considered the father of field-plot technique. Boussingault employed the techniques of de Saussure in weighing and analyzing the manures he added to his plots and the crops he harvested. He maintained a balance sheet that showed how much of the various plant-nutrient elements came from rain, soil, and air; analyzed the composition of his crops during various stages of growth; and determined that the best rotation was the one that produced the largest amount of organic matter in addition to that added in the manure.

Justus von Liebig (1803–1873), a German chemist, effectively destroyed the humus myth. The presentation of his paper at a prominent scientific meeting jarred the conservative thinkers of the day to such an extent that only a few scientists since that time have dared to suggest that the C in plants comes from any source other than CO_2. Liebig made the following statements:

1. Most of the carbon in plants comes from the CO_2 of the atmosphere.
2. Hydrogen and O_2 come from water.
3. The alkaline metals are needed for the neutralization of acids formed by plants as a result of their metabolic activities.
4. Phosphorus (P) is necessary for seed formation.
5. Plants absorb everything indiscriminately from the soil but excrete from their roots those materials that are nonessential.

Not all of Liebig's ideas, of course, were correct. He thought that acetic acid was excreted by the roots. He also believed that the ammonium (NH_4^+) form of N was the one absorbed and that plants might obtain this compound from soil, manure, or air.

Liebig firmly believed that by analyzing the plant and studying the elements it contained, one could formulate a set of fertilizer recommendations based on these analyses. It was also his opinion that plant growth was proportional to the amount of mineral substances available in the fertilizer.

The *law of the minimum*, stated by Liebig in 1862, is a simple but logical guide for predicting crop response to fertilization. This law states that

> every field contains a maximum of one or more and a minimum of one or more nutrients. With this minimum, be it lime, potash, nitrogen, phosphoric acid,

magnesia or any other nutrient, the yields stand in direct relation. It is the factor that governs and controls . . . yields. Should this minimum be lime . . . yield . . . will remain the same and be no greater even though the amount of potash, silica, phosphoric acid, etc. . . . be increased a hundred fold.

Liebig's law of the minimum dominated the thinking of agricultural workers for a long time, and it has been of universal importance in soil fertility management. The contributions that Liebig made to the advancement of agriculture were monumental, and he is recognized as the father of agricultural chemistry.

Following Liebig's now famous paper was the establishment in 1843 of an agricultural experiment station at Rothamsted, England, by J. B. Lawes and J. H. Gilbert. Many of the experiments were similar to those carried out in France by Boussingault. Lawes and Gilbert did not believe that all of the maxims set down by Liebig were correct. Twelve years after the station was founded, they settled the following points:

1. Crops require both P and K, but the composition of the plant ash is no measure of the amounts of these constituents required by the plant.
2. Nonlegume crops require N. Without this element, no growth will be obtained, regardless of the quantities of P and K present. The amount of ammonia (NH_3) contributed by the atmosphere is insufficient for the needs of crops.
3. Soil fertility can be maintained for some years by chemical fertilizers.
4. The beneficial effect of fallow lies in the increase in available N compounds in the soil.

The problem of soil and plant N remained unsolved. Several workers had observed the unusual behavior of legumes. In some instances they grew well without added N, whereas in others no growth was obtained. Nonlegumes always failed to grow when there was insufficient soil N.

In 1878 some light was thrown on the situation by two French bacteriologists, Theodore Schloessing and Alfred Müntz. These scientists purified sewage water by passing it through a sand and limestone filter. They analyzed the filtrate periodically, and for 28 days only NH_3-N was detected. At the end of this time nitrates (NO_3^-) began to appear in the filtrate. Schloessing and Müntz found that the production of NO_3^- could be stopped by adding chloroform and that it could be started again by adding a little fresh sewage water. They concluded that nitrification was the result of bacterial action.

These experimental results were applied to soils by Robert Warrington of England. He showed that nitrification could be stopped by carbon disulfide and chloroform and that it could be started again by adding a small amount of unsterilized soil. He also demonstrated that the reaction was a two-step phenomenon, the NH_3 first being converted to nitrites and the nitrites subsequently to nitrates. Although Warrington conducted numerous experiments, it was S. Winogradsky who is credited with isolating the organisms responsible for nitrification.

In 1886 two German scientists, Hellriegel and Wilfarth, concluded that bacteria in the nodules attached to legume roots assimilated gaseous N from the atmosphere and converted it to a form used by higher plants. This was the first specific information regarding N fixation by legumes. They did not, however, isolate the responsible organisms. It was M. W. Beijerinck, who isolated *Bacillus radicicola*, the organism responsible for N fixation.

The Development of Soil Fertility in the United States

Although most of the agricultural advances during the eighteenth century were accomplished on the Continent, a few early American contributions were significant. Squanto, an Indian chief, taught the Pilgrims in the 1620s to bury a fish under hills of corn to improve yields. In 1733, James E. Oglethorpe established an experimental garden on the bluffs of the Savannah River, which is the present site of the city of Savannah, Georgia. This garden was devoted to the production of exotic food crops and is said to have been a place of beauty while it was maintained. Interest in it was lost, however, and it soon ceased to exist, but because it was largely the result of British interests, it probably cannot be truly considered an American undertaking.

Benjamin Franklin applied gypsum to a prominent hillside in a pattern that outlined the words "This land has been plastered." The increased growth of pasture to which the gypsum had been applied effectively demonstrated its fertilizer value.

In 1785 a society was formed in South Carolina that had among its objectives the setting up of an experimental farm. Eleven years later President Washington, in his annual message to the Congress, pleaded for the establishment of a national board of agriculture. Some of the most important contributions to early American agriculture were made by Edmond Ruffin of Virginia from about 1825 to 1845. He may have been the first to use lime on humid-region soils to replace nutrients lost by crop removal and leaching. Although his use of lime to bolster crop yields was known to the ancients, it was apparently a new experience in America.

In 1862 the Department of Agriculture was established, and, in the same year, the Morrill Act provided for state colleges of agriculture and mechanical arts. The first organized agricultural experiment station, set up in 1875 at Middletown, Connecticut, was supported by state funds. In 1877 North Carolina established a similar station, followed closely by New Jersey, New York, Ohio, and Massachusetts. In 1888 the Hatch Act established state experiment stations to be operated in conjunction with the land-grant colleges, and an annual grant of $15,000 was made available to each state for their support. Although much of the early experimental work consisted largely of demonstrations, a scientific approach to agricultural problems was gradually developed.

Two early workers who contributed to the development of soil fertility in the United States were Milton Whitney and C. G. Hopkins. Around the beginning of the twentieth century, these two men engaged in a controversy that attracted national attention. Whitney maintained that the total supply of nutrients in soils was inexhaustible and that the important factor from the standpoint of plant nutrition was the rate at which these nutrients entered the soil solution. Hopkins, on the other hand, felt that this philosophy would lead to soil depletion and a serious decline in crop production. He made a survey of

Illinois soils and reduced soil fertility to a system of bookkeeping. As a result of these exhaustive studies, he concluded that Illinois soils required only lime and phosphate. So effectively did he preach this doctrine that the use of lime and rock phosphate in a corn, oats, and clover rotation was a continuous practice in this state for many years. Whitney's ideas were shown to be at least partly incorrect, but the argument did stimulate scientific thought.

In the early twentieth century, most experiment stations established field plots that showed the remarkable benefits of fertilization. It was shown, for example, that there was a widespread need for P fertilizers, that K was generally lacking in the coastal plains regions, and that N was particularly deficient in the soils of the South. The soils east of the Mississippi River were generally acid and needed lime, whereas those west of the river were as a rule fairly well supplied with Ca. Even though the general fertility status of soils in the United States had been fairly well defined, it was soon apparent that blanket fertilizer recommendations based on such knowledge could not be made. Each field required individual attention, and the interest in soil testing increased.

During the last 40 years, much headway has been made toward understanding soil fertility problems. To enumerate the men whose contributions have advanced our knowledge would require far more space than is available here. These advances have been the work of scientists from many countries.

Increasing Crop Yields

The fruits of past studies are apparent, for agricultural production is greater today than it has ever been, and the world is generally better fed, clothed, and housed than at any time in the past. This could not be so if the production of crops today were at the level of Europe in the Early Middle Ages, when the average yield of grain was 6 to 10 bu/a.

Many developing countries have achieved self-sufficiency in some of the basic food and fiber crops. For example, traditionally food deficit countries like India, China, Indonesia, Pakistan, and Brazil are either self-sufficient or net exporters of wheat, rice, maize, soybeans, sugarcane, and/or cotton.

The annual growth rate in dietary energy supplies (DES) climbed sharply between 1961 and 1983. As a result, 75 of the 112 developing countries showed a DES growth rate greater than the growth of their population. This is the result of development of high-yielding varieties and the use of other inputs such as fertilizer, water, and pesticides.

Some appreciation of the progress made in increasing the long-term average yields of major U.S. crops can be obtained from Table 1.2. The principal

TABLE 1.2 Increasing Yields of Major Crops in United States (1950–1987)

	Corn, bu/a	Wheat, bu/a	Soybeans, bu/a	Alfalfa, ton/a
1950	37.6	14.3	21.7	2.1
1964	62.1	26.2	22.8	2.4
1972	96.9	32.7	28.0	2.9
1982	113.2	35.5	31.5	3.4
1987	119.9	39.4	34.2	3.4

factors contributing to higher corn yields include the introduction of hybrids in the 1940s, the availability and use of relatively inexpensive fertilizers in the early 1950s, more effective pest management, and improved cultural practices.

Corn yields exceeding 200 bu/a now occur frequently in the United States, while yields reported by some researchers and farmers have exceeded 300 bu/a. It is apparent that an adequate supply of plant nutrients is an important component in high-yield crop production (Table 1.3).

Soybean yields have not shown the marked increases that corn has (Table 1.2). This crop did not gain prominence until after 1940, and it is only recently that the benefits of agronomic research in breeding, fertilization, and management have begun to be realized, as can be seen in the higher yields of the late 1970s and 1980s. Yields in excess of 100 bu/a have been obtained. Improvements effective for corn were also effective for soybeans.

The steady improvement in wheat yields since 1940 (Table 1.2) is attributed to the introduction of new types, such as the semi-dwarf varieties, and the increased use of fertilizers and pesticides. Yields well over 100 bu/a are common, with some reported yields of over 200 bu/a.

Alfalfa yields have increased steadily (Table 1.2). Improved varieties, cutting management, lime and fertilizer, and pest control have been key to this increase. Yields of 10 ton/a without irrigation have been obtained. Yields of 15 to 20^+ ton/a have been obtained with irrigation and more frequent cutting in areas with longer growing seasons.

Looking to the Twenty-first Century

Continued improvements in crop yields through improved genetics, cultural practices, and pest management will be needed because increasing world population will increase the demand for food and fiber. Although new lands can still be brought into production, the most productive soils are already under cultivation. Therefore, increasing the food supply will require higher yields per unit land area, which will increase the demand placed on the soil to provide adequate nutrients. With greater nutrient demand, increased nutrient

TABLE 1.3 Average Management Practices by Region Associated with 200+ bu/a Corn Yields, Throughout the United States, 1975–1976

		Region*				
	Overall	West	Great Plains	Midwest	Northeast	Southeast
Yield (bu/a)	218	220	219	216	217	221
Planting date	Apr. 22	Apr. 17	Apr. 27	Apr. 29	May 6	Mar. 28
Harvest population	25,500	27,350	26,300	24,400	23,950	25,550
Fertilizer rate (N-P_2O_5-K_2O)	220-95-109	227-65-61	234-59-29	204-120-144	184-90-93	258-113-172
Number of growers	549	99	102	275	15	58

*West: CA, OR, WA, UT; Great Plains: CO, KS, OK, NE, NM, TX; Midwest: IL, IN, IA, KY, MI, MN, MO, OH, WI; Northeast: DE, MD, NJ, NY, PA, VA; Southeast: AL, FL, GA, LA, NC, SC, TN, MS.

SOURCE: Dibb and Walker, *Better Crops Plant Food*, **62**:17 (Winter 1978–1979).

inputs will be required. As nutrient use increases, the potential for surface and groundwater contamination with applied nutrients grows. Development and use of nutrient management technologies that increase the recovery of applied nutrients by the crop will reduce the potential for environmental degradation.

Research is showing that conservation tillage can decrease soil erosion while increasing water-use efficiency and crop yields, which contribute to agricultural sustainability. Conservation tillage can have a considerable effect on nutrient requirements, and further research is needed to develop soil fertility practices suitable for use with conservation technologies. Remote sensing, which is infrared photography from high altitudes, is used to determine crop conditions. Problems arising from soil, irrigation, or pest conditions can often be detected and corrected in time to prevent serious depressions in yield.

In the western regions of the United States and Canada, crop productivity has greatly increased with the development of irrigation systems. Because of the elimination of moisture as a limiting factor, greater fertilizer-use efficiency can be obtained and unit production costs lowered.

Groundwater supplies, however, are being depleted by widespread irrigation. This depletion has created the need for improved efficiency of water use. More efficient irrigation systems, in addition to technologies in water capture and more efficient use of moisture for dryland crop production, are being developed. Crop production systems involving these modern methods of moisture management, in conjunction with other high-yield factors, such as fertilization, hybrid or variety, and seeding date, need to be examined.

Soil and plant analyses as a means of determining fertilizer and lime requirements for crops have been used for many years. Additional research and information are needed to improve the accuracy of these tests as guides to crop fertilization and liming.

Developments will continue in more effective fertilizers; however, continued evaluation of fertilizer effectiveness through short- and long-term experimentation is needed for continued increases in crop production efficiency. Higher crop yields impose different nutrient requirements. Fertilizer rates that gave satisfactory responses with corn yields of 150 bu/a will not be adequate at yield levels of 200+ bu/a. Agronomic research with a multidisciplinary approach will enhance positive yield interactions. It is essential that rates of inputs and treatments be compared in experiments so that the sources of responses can be identified and economics can be calculated. Long-term crop production studies are needed to determine the effects of alternative practices and/or new crop management technologies on the nutrient supplying capacity of soils.

New developments in biotechnology may have profound impacts on agricultural production. Through the technique of gene transplants, desirable qualities of one genus or species may be transferred to another. It is conceivable that greater photosynthetic efficiency, higher protein and vitamin content, better disease and insect resistance, and other factors can be introduced into otherwise desirable crop species. Such genetic alterations could have a marked impact on yield and nutrient requirements.

Progress in agriculture depends on high-caliber research. For every problem solved by the scientist today, many more are raised. Agricultural scientists must answer fundamental questions that deal more with the *why* of things than with the *what.*

It is not the purpose of this chapter to cover all of the significant events in the development of the science of soil fertility. Certainly, the advances made toward the end of the nineteenth and the early twentieth centuries have been largely responsible for the present state of our learning. Although brief, it is hoped that this sketch will give the student some idea of the time, effort, and thought that have been devoted in the last 4,500 years to accumulating what is still insufficient knowledge.

Selected References

RUSSEL, D. A., and G. G. WILLIAMS. 1977. History of chemical fertilizer development. *SSSAJ*, **41**:260–265.

VIETS, F. G. 1977. A perspective on two centuries of progress in soil fertility and plant nutrition. *SSSAJ*, **41**:242–249.

CHAPTER 2

Growth and the Factors Affecting It

Obtaining the maximum production potential of a particular crop depends on the environment and the skill of the farmer in identifying and eliminating those factors that reduce the production potential. This chapter will discuss those factors that influence crop response to plant nutrients and consider how crop growth and response can be quantified.

Factors Affecting Plant Growth

Over 50 factors affect crop growth and yield; a partial list appears in Table 2.1. Many of these factors can be controlled by the grower, but for high yields these factors must operate in unison, as many of them are interrelated. Growers are unable to manage all of the climate factors, except for rainfall through irrigation and wind through windbreaks. Growers can, however, modify almost all of the crop factors and many of the soil factors.

TABLE 2.1 Factors Affecting Crop Yield Potential

Climate Factors	Soil Factors	Crop Factors
Precipitation	Organic matter	Crop species/variety
Quantity	Texture	Planting date
Distribution	Structure	Seeding rate and geometry
Air temperature	Cation exchange capacity	Row spacing
Relative humidity	Base saturation	Seed quality
Light	Slope and topography	Evapotranspiration
Quantity	Soil temperature	Water availability
Intensity	Soil management factors	Nutrition
Duration	Tillage	Pests
Altitude/latitude	Drainage	Insects
Wind	Others	Diseases
Velocity	Depth (root zone)	Weeds
Distribution		Harvest efficiency
CO_2 concentration		

Two major factors establishing the upper limit of the potential yield of crops are (1) the amount of moisture available during the growing season and (2) the length of the growing season. To produce the maximum yield possible, crop plants must utilize a high percentage of the available solar energy.

The potential maximum yield of most crops far exceeds current yield levels. For example, maximum yields of corn and soybeans are on the order of 490 to 580 bu/a and 140 to 225 bu/a, respectively. Wheat genotypes presently available have the potential for producing 178 to 223 bu/a of grain (12 to 15 tons/ha).

Factors involved in crop growth can be classified as genetic or environmental. A review of the major factors follows.

Genetic Factors

The importance of genetics in the growth of agricultural crops is illustrated by the large increases in yield resulting from the introduction of new corn hybrids and improved wheat varieties (Fig. 2.1). Yields obtained with corn hybrids of the recent past were 38 to 61% greater under good and poor conditions, respectively, than they were with hybrids developed in the 1930s. Since 1930, hard red winter wheat grain yields increased 16.4 kg/ha/yr, or about 0.7%/yr, and were directly attributable to genetic improvement.

VARIETY AND PLANT NUTRIENT NEEDS The high crop yields produced with modern hybrids and varieties will require more plant nutrients. This

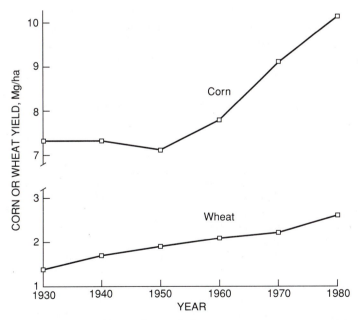

FIGURE 2.1   Effects of genetic improvements on corn and hard red winter wheat yields from 1930 to 1980. *Corn: Carlone and Russel,* CropSci., *24:465, 1987; wheat: Cox et al.,* Crop Sci., *28:756, 1988.*

important fact, which has often been overlooked in shifts to higher-yielding varieties, is evident in Table 2.2 for several soybean varieties. Under low-fertility conditions a new high-yielding variety cannot develop its full yield potential. In fertile soils the same new variety will deplete the soil more rapidly, and eventually yields will decline if supplemental nutrients are not provided.

An example of the importance of genetics as a yield-limiting factor is shown in Figure 2.2. Notice the inability of the Arksoy variety to produce the yields obtained with the other varieties in the moderate- and high-fertility soils. One of the first steps in a successful farming enterprise is the selection of hybrids or varieties that are genetically capable of producing high yields and of utilizing the supply of plant-available nutrients. In the development of new hybrids or varieties, pest control must be used whenever necessary. If such pests are not controlled, many otherwise promising high-yielding materials will be discarded in a research program or fail to perform adequately on the farm. This is an example of the limiting factor concept and of the importance of optimum management of all factors influencing crop yield.

VARIETY—FERTILITY INTERACTIONS   In general, varieties that have a narrow range of adaptation tend to show significant variety—fertilizer interactions, whereas those with a wide range of adaptation do not. As early as 1922, several states recommended corn varieties on the basis of the fertility level of the soil. The varieties selected for poor soils were different from those suggested for more fertile soils.

In most developed agricultural countries it is customary to supply adequate plant nutrients on low-fertility soils. Thus recommendations for varieties or hybrids generally do not need to be made on the basis of the fertility level of the soil but rather on their ability to withstand insects, diseases, or unfavorable moisture or temperature conditions. However, one strategy for managing land areas with low iron (Fe) availability or an excess of aluminum (Al) or soluble salts is to select a cultivar or species capable of tolerating the specific condition. Although soil fertility need no longer be a limiting factor in most instances, knowledge of nutrient uptake mechanisms and their cultivar specificity is needed.

TABLE 2.2   Nutrient Removal in Seed and Nutrient Concentrations in Leaves of Various Varieties and Lines of Soybeans

Soybean Variety or PI Line	Yield of Seed (kg/ha)	Nutrient Removal in Seed (kg/ha)			Nutrient Concentration in Leaves (%)		
		N	P	K	N	P	K
PI92561	2914	196	20	58	4.9	0.38	1.1
Ford	2903	195	17	57	4.7	0.34	0.8
Adams	2882	191	17	59	4.9	0.39	1.0
Seneca	2840	183	16	56	4.3	0.38	1.4
PI80536	1082	75	8	21	4.8	0.39	1.4
PI200479	1024	73	6	19	3.6	0.29	0.8

SOURCE: DeMooy et al., in B.E. Caldwell, ed., *Soybeans: Improvement, Production, and Uses,* pp. 267–352. *ACS Agron. Monog. 16.* Madison, Wis.: American Society of Agronomy, 1973.

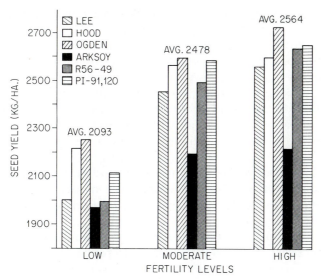

FIGURE 2.2   Average yields of six genetic lines of
soybeans when grown at three levels of soil fertility for 6
years.   *Caviness and Hardy*, Agron. J., *62:236, 1970.*

IMPORTANCE OF PROGRESSIVE RESEARCH IN GENETICS   The genetic constitution of a given plant species limits the extent to which that plant may
develop. No environmental conditions, no matter how favorable, can further
extend these limits. It is imperative that there be a program to produce
new varieties or hybrids capable of achieving greater yields under specified
conditions.

Environmental Factors

Environment is defined as the *aggregate of all the external conditions and influences
affecting the life and development of an organism.* Among the environmental factors known to influence plant growth, the following are probably most important:

1. Temperature.
2. Moisture supply.
3. Radiant energy.
4. Composition of the atmosphere.
5. Soil structure and composition of soil air.
6. Soil reaction.
7. Biotic factors.
8. Supply of mineral nutrient elements.
9. Absence of growth-restricting substances.

Many environmental factors do not behave independently. An example is
the inverse relationship between soil air and soil moisture or between the O_2
and CO_2 content in the soil atmosphere. As soil moisture increases, soil air

decreases, and as the CO_2 content of the soil air increases, the O_2 content decreases.

The effects of environmental factors acting individually and collectively were used to estimate the potential corn yields shown in Table 2.3. A higher yield potential exists in the eastern parts of these selected states because of more favorable growing conditions, including cooler temperatures, higher rainfall, lower altitude, less wind velocity, and better physical conditions of the soil. An appreciation of the impact of frost-free days or the length of the growing season can be gained by comparing yields in a north-to-south direction.

These examples illustrate the dependence on other factors of so-called independent variables of plant environment. Even though these factors are not independent of one another, they will be treated separately.

TEMPERATURE   The limit of survival of living organisms has generally been reported to be between $-35°$ and $+75°C$. The range of growth for most agricultural plants, however, is usually much narrower—perhaps between 15° and 40°C. At temperatures much below or above these limits, growth decreases rapidly.

Optimum temperatures for plant growth change with the species and varieties, duration of exposure, plant age, stage of development, and the particular growth criterion used to evaluate performance. Temperature directly affects photosynthesis, respiration, cell-wall permeability, absorption of water and nutrients, transpiration, enzyme activity, and protein coagulation. This influence is reflected in the growth of the plant. A plant's capacity for growth of new photosynthetic area can substantially influence total photosynthesis and plant productivity. Therefore, the initiation and expansion rate of new leaves and the duration of the various phases of plant development contribute greatly to crop productivity. Figure 2.3 shows the effect of temperature on development of corn leaves.

Respiration and transpiration are directly affected by temperature, decreasing as temperature declines and increasing as temperature rises. At very high temperatures the rate of respiration is initially great, but after a few hours respiration rates for some plants drop off rapidly.

For many crop plants the optimum temperature for photosynthesis is lower than that for respiration. This has been suggested as one reason for the higher yield of starchy crops, such as corn and potatoes, in cool climates compared with the yield of these crops in warmer regions. It is possible that under conditions of prolonged temperatures above the optimum, a plant may liter-

TABLE 2.3   Relationship of Climatic Factors to Potential Yields of Corn in Three Great Plains States (bu/a)

	Nonirrigated			Irrigated
	S. Dakota	Nebraska	Kansas	Kansas
Western	42	54	59	270
Central	89	132	178	319
Eastern	188	209	217	327

SOURCE: Strauss, *Fert. Solut.*, **22**(1):68 (1978).

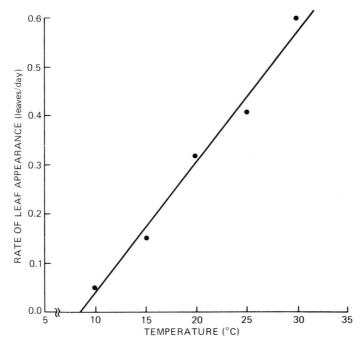

FIGURE 2.3   Influence of temperature on the rate of
corn leaf appearance.   *Tollenaar et al.,* Crop Sci., *19:363,
1979.*

ally suffer from starvation simply because respiration is taking place more
rapidly than photosynthesis.

Under high temperatures, water loss by excessive transpiration may exceed
water intake, and wilting soon follows. The absorption of water by plant roots
is affected by temperature, where absorption increases with temperature of
the rooting medium from 0°C to about 60° or 70°C. Low soil temperature may
adversely affect the growth of plants by its effect on the absorption of water.
If soil temperatures are low, yet excessive transpiration is taking place, the
plant may be injured because of tissue dehydration. The moisture supply of
the soil also may be influenced to some extent by temperature, for unusually
warm weather produces more rapid evaporation of water from the soil surface.

Temperature also affects nutrient absorption. In many plant species the
absorption of solutes by roots is retarded at low soil temperatures. This may be
caused by lower respiratory activity or by reduced cell membrane permeability,
both of which could affect uptake, as well as the rate and extent of root
permeation in the soil.

The effect of temperature on the uptake of nutrients by the potato plant is
illustrated by the data shown graphically in Figure 2.4. For example, the P
content in both tops and roots increased with an increase in temperature, but
the opposite effect was observed with K in the roots.

Temperature exerts its influence on plant growth indirectly by its effect on
the microbial population of the soil. The activity of the nitrobacteria, as well

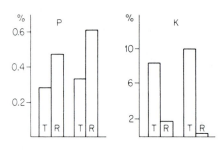

FIGURE 2.4   Effect of soil temperature on P and K in potato tops (T) and roots (R). *Epstein,* Agron. J., *63:664, 1971.*

as that of most heterotrophic organisms, increases with temperature. Soil pH may change with temperature, which, in turn, may affect plant growth. It has been observed that soil pH increases in winter and decreases in summer. This is generally considered to be related to the activities of microorganisms, since microbial activity is accompanied by the release of CO_2, which combines with water to form carbonic acid (H_2CO_3). In soils that are only slightly acid, this small change in pH may influence the availability of micronutrients such as manganese (Mn), zinc (Zn), or Fe.

Numerous studies on the direct relationship between yield or dry-matter production and temperature have been made. Figure 2.5 shows how heating of field soil in northern Ohio improved the dry-matter production and grain yield of corn. These data illustrate the importance of the limiting factor concept stated earlier. Knowledge of temperature–plant growth relationships

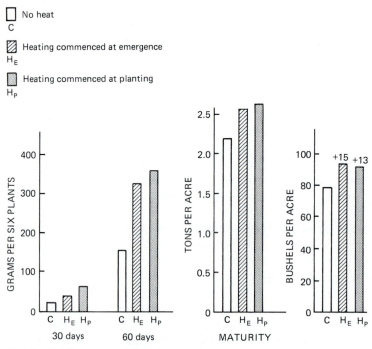

FIGURE 2.5   Dry weight of corn plants at several stages of development and corn yield as affected by soil heating. *Mederski and Jones,* SSSA J., *27:188, 1963.*

is important because planting a crop or variety not adapted to prevailing temperatures will result in reduced yield potential, and reseeding may be required.

Temperature may also alter the composition of the soil air, which is the result of increases or decreases in microbial activity. When the activity of the soil microorganisms is great, there will be higher partial pressure of the CO_2 of the soil atmosphere as the O_2 content decreases. Under conditions restricting the diffusion of gases into and out of the soil, a decrease in the O_2 pressure might influence the rate of respiration of the plant roots, and hence their ability to absorb nutrients.

MOISTURE SUPPLY   The growth of plants is proportional to the amount of water present. Water is required for the manufacture of carbohydrates, to maintain hydration of protoplasm, and as a vehicle for the translocation of carbohydrates and nutrients. Internal moisture stress causes reduction in both cell division and cell elongation, and hence in growth.

Water stress results when the extractable water in the root zone is insufficient to meet the plant's transpirational demands. Variations in soil water deficits are mainly responsible for year-to-year fluctuations in crop yields. Various physiological processes in plants are affected differently by water stress. For example, leaf elongation is more sensitive to soil water deficits than the other processes and will cease before all the extractable soil water is consumed. Roots grow best when soils are well supplied with moisture, but as demonstrated in Figure 2.6, growth can occur even in relatively dry soils. When water deficits limit root development, the uptake of nutrients and water will be curtailed.

The effects of irrigated and dryland moisture regimes and of increasing N rates on the yield of rapeseed are shown in Figure 2.7. Increasing N rates up

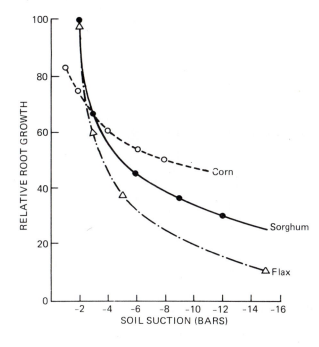

FIGURE 2.6   The relative growth of roots of maize, sorghum, and flax as influenced by soil water potential (suction in bars).   *Hurd and Spratt,* Physiological Aspects of Dryland Farming, *p. 183. New Delhi: Oxford & IBH, 1975.*

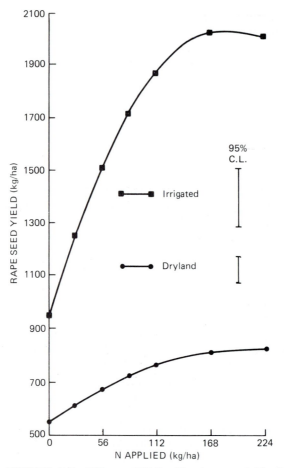

FIGURE 2.7   Effects of N fertilization on yield of
rapeseed under irrigated and dryland conditions.   *Henry
and MacDonald, Can. J. Soil Sci., 58:305, 1978.*

to 168 kg/ha under both sets of moisture conditions increased yields; however, optimizing available soil water with irrigation greatly increased the yield response to N compared to dryland conditions. This illustrates clearly the limiting effect that moisture stress can have on plants' responses to applied fertilizer. It should also be noted that it is necessary to provide sufficient fertilizer to make the greatest use of available water.

Yield is not the only plant property affected by soil moisture. Higher percentages of protein are generally associated with low soil moisture, as demonstrated for barley in Figure 2.8.

Soil moisture level also has a pronounced effect on the uptake of plant nutrients. Low levels of extractable water in the root zone retard nutrient availability by impairing each of the three major processes involved in nutrient uptake. These processes are (1) diffusion, (2) mass flow, and (3) root interception. As a general rule, there is an increase in nutrient uptake when extractable water is high rather than low. The effect of decreasing soil moisture stress on P uptake by soybean tops is apparent in Table 2.4. Flooding of soil pores by

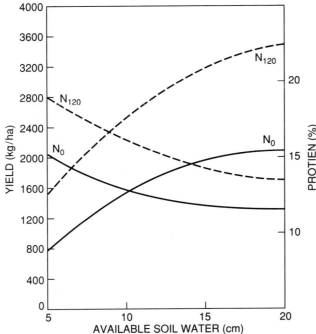

FIGURE 2.8   Increase in yield and decrease in protein content of barley as a function of available soil water at two N levels (kg/ha), with an average (117 mm) June–July precipitation. Lines descending from left to right are protein; those ascending from left to right are yield.   *Bole and Pittman*, Proc. 1978 Sask. Soil Crops Workshop, *February 8–9, 1978.*

TABLE 2.4   Influence of Soil Moisture Stress on P Uptake by Soybeans at Two Levels of Soil P

| | P Content of Tops (mg/plant) | | | |
| *Soil Moisture Status* | | *Derived from:* | | |
(tension in bars)	*Total*	*Fertilizer*	*Soil*	*P (%)*
	Soil P, 24 ppm			
Continuously < 2	15.3	3.6	11.7	0.254
Depletion from 0.03 to 15	13.0	3.1	9.9	0.230
Maintained near 15	9.5	2.0	7.5	0.164
	Soil P, 243 ppm			
Continuously < 2	30.3	20.9	9.4	0.214
Depletion from 0.03 to 15	24.5	15.9	8.6	0.170
Maintained near 15	10.1	6.7	3.4	0.105

For the 10 days preceding harvest.
SOURCE: Marais and Wiersma, *Agron. J.*, **67:**778 (1975).

excessive amounts of moisture is detrimental since the resultant lack of O_2 restricts root respiration and ion absorption.

The tendency for plant nutrients to be taken up more readily as soil moisture increases has a favorable effect on the water-use efficiency (WUE) of plants. WUE is the amount of dry matter produced from a given quantity of water, usually expressed as grams of dry weight per hectare-centimeter (acre-inch) of water. Table 2.5 summarizes the effects of two fertilizer application programs and three water management systems on the WUE of corn grain production. Repeated small applications of fertilizer throughout the growing season, coupled with the tensiometer-scheduled irrigation system, produced the highest WUE on this sandy soil.

Placement of fertilizer nutrients is an important consideration in situations where upper portions of the root zone are subject to rapid and prolonged drying. Fertilizers placed at depths in the root zone where soil is moist will be more effective. In arid and semiarid regions, where leaching is not a problem, improved distribution of fertilizer nutrients in the root zone can result from occasional heavy surface dressings. Fertilizer placement is discussed at length in Chapter 12.

Soil moisture level also influences plant growth indirectly by its effect on the behavior of soil microorganisms. At extremely low or extremely high moisture levels, the activity of organisms responsible for the transformation of nutrients into plant-available forms is inhibited.

RADIANT ENERGY   Radiant energy is a significant factor in plant growth and development. The quality, intensity, and duration of light are all important. Clear-day radiation is a useful indicator of the amount of solar energy available for physiological processes within plants. The values for the periods between tassel initiation (anthesis) and maturity in corn suggest that the highest production potential for this crop in the United States exists near the 40° latitude, a zone running east to west from the southern border of Pennsylvania, through the middle of Ohio, Indiana, and Illinois, across northern Missouri, and astride the Nebraska and Kansas borders. Much of the Corn Belt's top production occurs between 38° and 43° latitude.

Studies of the effect of light quality on plant growth suggest that the full

TABLE 2.5   Yield of Corn and WUE as Affected by Water Management and Fertilizer Application Method

Fertilizer Treatment	*Natural Rainfall: 49 cm*		*Daily Irrigation (0.64 cm): 49 + 50 cm*		*Tensiometer Scheduled: 49 + 26 cm*	
	Yield (kg/ha)	*WUE*	*Yield (kg/ha)*	*WUE*	*Yield (kg/ha)*	*WUE*
Conventional†	2860	59	3380	34	4760	64
Improved program‡	2710	56	4950	50	6600	80

WUE = kg/(ha · cm).
† 112, 98, and 280 kg/ha of N, P, K, broadcast preplant plus two later topdressings each of 112 kg/ha of N.
‡ 336, 98, and 280 kg/ha of N, P, K, broadcast in increments of 5%, 5%, 10%, 20%, 20%, 20%, and 20% during the growing season.
SOURCE: Rhoads et al., *Agron J.*, **70**:306 (1978).

spectrum of sunlight is generally satisfactory for plant growth. Even though light quality is known to affect plant growth, it is not likely that in the foreseeable future this factor can be controlled on a large-scale field basis.

It has been shown that most plants are generally able to achieve good growth at light intensities of less than full daylight. However, plants do differ in their response to light of varying intensity, as illustrated in Figure 2.9. Note that the two forest species, oak and maple, were the least responsive to increasing radiation intensities. Light intensity in their usual forest habitats is considerably less than that of full sunlight, especially where maple is present in the understory.

For corn, which continues to respond to increasing insolation, the interception of solar radiation by its canopy and the associated efficiency of photosynthesis are principal determinants of growth and yield. The net photosynthesis rate in corn is proportional to radiation interception, provided that soil moisture is adequate.

Changes in light intensity caused by shading can exert considerable influence on crop growth. With high plant populations, light penetration to lower positions in the plant canopy may be inadequate for bottom leaves to carry on photosynthesis. Table 2.6 shows how corn yields were increased by providing artificial light. The importance of light distribution within crop canopies is further illustrated in Figure 2.10, where corn grain yields are compared for hybrids with diverse leaf orientation grown at different plant densities. An advantage for leaf erectness is evident in the yield response.

Shading of crops can also occur when two different species are grown in a mixture, such as a grass-clover pasture. Balanced growth between the grass

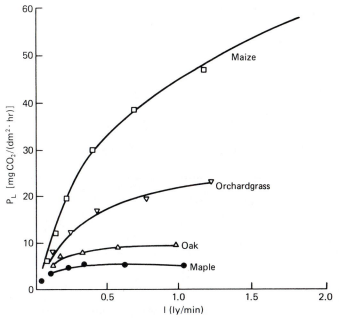

FIGURE 2.9   Relation of CO_2 assimilation (P_L) to light intensity (I) among species.   *Hesketh and Baker*, Crop Sci., *7:286, 1967.*

TABLE 2.6   Effect of Supplemental Light on Corn
Yields

Row from Lights	Corn Yield (bu/a)	Yield Increase (%)
1	346	70
2	288	41
3	242	19
4	230	13
5	257	26
6	223	9
7	219	7
8–9	204	0

SOURCE: Welch, L.F. High yield corn panel on physiology,
biochemistry and chemistry associated with maximum yield corn.
Proceedings of a research rountable sponsored by FAR and PPI
198–199 (1985).

and clover is an important problem in good pasture management. The loss
of clover from clover-grass pastures heavily fertilized with N is frequently
observed. This has been attributed largely to competition for moisture and
nutrients, although reduced light density may also be an important factor that
is explained by these factors:

1. Increased rates of N application give increased yields of grass.
2. Increased yields of grass give higher leaf areas of grass above the clover-
   leaf canopy.
3. Higher leaf areas above the clover reduce light density at the clover-leaf
   canopy.
4. Reduced light density at the clover-leaf canopy causes reduced growth of
   clover.

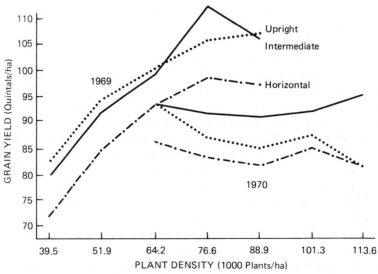

FIGURE 2.10   Yield response of hybrids grouped by leaf
angle. *Hicks and Stucker, Agron. J., 64:486, 1972.*

Studies by Japanese workers, who used wheat as the test plant indicate that the absorption of NH_4^+, sulfate (SO_4^{2-}), and water was increased with increasing light intensity but that absorption of Ca and Mg was little affected. Light intensity had marked effects on the uptake of P and K. It was also observed that O_2 uptake by the roots increased with increasing light intensity.

PHOTOPERIODISM   The behavior of the plant in relation to day length is termed *photoperiodism*. On the basis of their reaction to photoperiod, plants have been classed as short-day, long-day, or indeterminate. Short-day plants are those that will flower only when the photoperiod is as short as or shorter than some critical period of time. If the time of exposure to light is longer than this critical period, the plants will develop vegetatively without completing their reproductive cycle. Corn, sorghum, and rice are examples. Long-day plants are those that will bloom only if the period of time during which they are exposed to light is as long as or longer than some critical period. If the plants are exposed to light for periods shorter than this critical time, they will develop only vegetatively. Grains and clovers are members of this group. Plants that flower and complete their reproductive cycle over a wide range of day lengths are classed as indeterminate. Cotton and buckwheat are representative.

COMPOSITION OF THE ATMOSPHERE   C is required for plant growth and, except for water, is the most abundant material within plants. The principal source of C for plants is CO_2 gas in the atmosphere. It is taken into their leaves and, through photosynthetic activity, is chemically bound in organic molecules.

CO_2 is continually being returned to the atmosphere as a product of respiration of animals and plants and microbial decomposition of organic residues. Although the normal value of atmospheric CO_2 is approximately 0.03%, the concentration may range from one-half to several times this figure. Within a thick plant canopy on a still day, the CO_2 concentration may become measurably less during daylight hours when there is a high rate of photosynthesis. Similarly, in a dense forest, the content may drop considerably.

Many economic plants respond to elevated levels of CO_2 by increased growth and productivity (Table 2.7). Plants that fix CO_2 through the C_3 enzyme route (ribulose 1,5-diphosphate carboxylase) have a greater potential response to increased atmospheric levels of CO_2 than do C_4 plants, which utilize this source of C via phosphoenolpyruvate carboxylase.

Plants will grow more rapidly if they are supplied with supplementary CO_2; about 0.1% (1000 ppm) is generally thought to be ideal. Depressed productivity has been observed with concentrations in the range 0.3 to 0.5%.

Supplementary CO_2 has greater potential in a greenhouse than in open-field situations, where wind action makes it difficult to maintain elevated CO_2 concentrations. However, in densely planted crops that restrict air movement it may be possible to raise the CO_2 concentration through microbial breakdown of fresh, active sources of organic matter like manures or crop residues.

Increasing atmospheric temperature and CO_2 concentrations, described as the *greenhouse effect*, may have considerable effects on crop growth and yield. Some projections indicate that crop yields should increase, while others predict

TABLE 2.7   Differences in Leaf Photosynthesis Among Plants and Percentage
Increases in Growth or Yield from Elevated Levels of Carbon Dioxide

Plant	CO_2 Fixation $[mg/(dm^2 \cdot hr)]$ (Normal CO_2 Levels)	Increase in Growth or Yield at Elevated CO_2 Levels (%)
Corn, grain sorghum, and sugarcane	60–75	100
Rice	40–75	135
Sunflower	50–65	130
Cotton	40–50	100
Soybean, sugar beet	30–40	56
Oats, wheat, barley	30–35	66
Tobacco	20–25	67
Tomato, cucumber, lettuce	20–25	50
Tree species, grapes and ornamentals, citrus	10–20	40

SOURCE: Wittwer, *Proc. Agr. Res. Inst. 21st Annu. Meet.*, pp. 69–86 (1973).

decreasing yields. Although the productivity of some crops may increase with
increasing CO_2, higher temperatures may increase drought stress and reduce
yield. Most scientists conclude that although the greenhouse effect may cause
changes in cropping patterns, crop productivity will continue to increase since
farming practices will most likely adapt to a changing environment.

TOXIC ATMOSPHERIC SUBSTANCES   The quality of the atmosphere surrounding aboveground plant parts may influence growth. Certain gases, such
as sulfur dioxide (SO_2), carbon monoxide (CO), and hydrofluoric acid (HF),
when released into the air in sufficient quantities, are toxic to plants. Although
the exception rather than the rule, isolated cases of injury from these gases
have been reported.

Strong acids such as sulfuric (H_2SO_4), nitric (HNO_3), and hydrochloric
(HCl) acids have lowered the pH of rain and snow falling on much of northern
Europe and the eastern sections of the United States and Canada to between
4 and 5. Values between pH 2.1 and 3.0 have been observed during storms
at various locations. Acid rain is often due mainly to relatively high concentrations of SO_2 and $SO_4{}^{2-}$. Some of the effects that acid rain can have on plants
and soil include increased leaching of inorganic nutrients and organic substances from foliage; accelerated cuticular erosion of leaves; leaf damage when
pH values fall below 3.5; altered response to associated pathogens, symbiants,
and saprophytes; lowered germination and establishment of conifers; reduced
availability of soil N; decreased respiration; and increased leaching of nutrient
ions from soils.

Injury to vegetation by fluorine released during the manufacture of metallic
Al and the production of phosphatic fertilizers has been reported. Damage to
crops, however, may not be so important as the toxicity to grazing livestock.

The release of chlorofluorohydrocarbons (CFC) and other gases into the
atmosphere has been related to degradation of the ozone layer, which filters
out harmful radiation. Although this environmental disturbance has been
primarily observed at the North and South poles, continued ozone degradation may result in long-term animal and plant health problems.

Considerable progress has been made by industry over the last decade in reducing the amount of SO_2, CFC, and other gases discharged into the atmosphere. The environmental benefits of the regulatory legislation limiting atmospheric emissions should be realized in the next decade.

SOIL STRUCTURE AND COMPOSITION OF SOIL AIR   Soil structure and texture determine the bulk density of a soil. As a rule, the higher the bulk density, the more compact the soil, the more poorly defined the structure, and the smaller the amount of pore space. Such conditions are frequently reflected in restricted plant growth.

High bulk densities inhibit the emergence of seedlings and offer increased mechanical resistance to root penetration. They reduce the rate of O_2 diffusion into the soil pores, and root respiration is directly related to a continuing and adequate supply of this gas. The effect of increasing the percentage of O_2 on top and root growth is shown in Figures 2.11 and 2.12, respectively.

Under field conditions, O_2 diffusion into the soil is determined largely by the moisture level of the soil if bulk density is not a limiting factor. On well-drained soils with good structure, O_2 content is not likely to retard plant growth except during periods of flooding, when reduced O_2 supply may restrict ion uptake. The increasing importance of O_2 with decreasing moisture tension is illustrated in Figure 2.13. Note that raising the O_2 supply at low moisture tensions results in a continued increase in rubidium (Rb) ion uptake to an O_2 percentage of 8 to 10. Rb is similar to K in plant uptake.

The O_2 supply at the root-absorbing surface is critical. Hence not only is

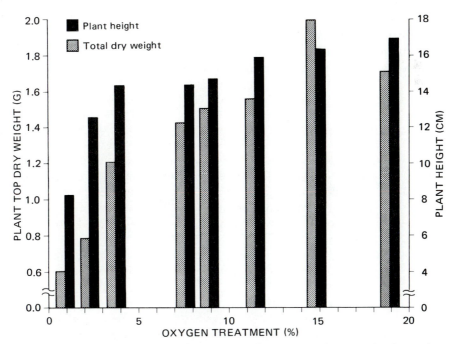

FIGURE 2.11   Effect of O_2 concentration of soil air on the growth of snapdragon plants. *Letey et al., SSSA Proc., 25:184, 1961.*

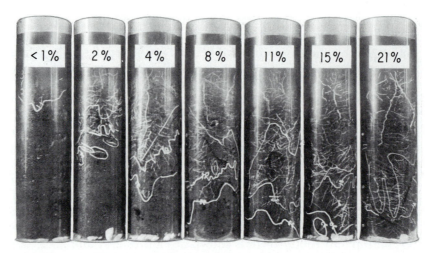

FIGURE 2.12   Effect of O_2 concentration maintained at the soil surface on root development.   *Stolzy et al. SSSA Proc., 25:464, 1961.*

the gross O_2 level of the soil air important, but also the rate at which O_2 diffuses through the soil to maintain an adequate partial pressure at the root surface. The influence of various rates of O_2 diffusion on the growth of pea plants on soils of three different fertility levels is illustrated in Figure 2.14. At low rates of O_2 diffusion, small increases in diffusion rate have a much greater impact on corn grown on soils with medium or high fertility levels than on soils with a low fertility level. Adequate fertility could benefit crops under high soil moisture conditions.

SOIL REACTION   Soil reaction or pH may affect plant development by influencing the availability of certain plant nutrients. Examples are the reduced availability of phosphates in acid soils high in Fe and Al and of Mn in high

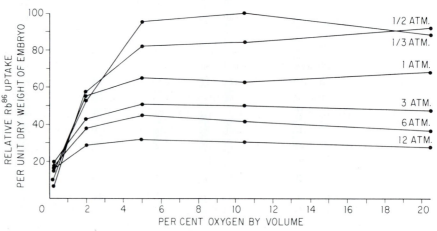

FIGURE 2.13   Effect of O_2 level and soil moisture tension on ^{86}Rb uptake by corn seedlings.   *Danielson et al., SSSA Proc., 21:5, 1957.*

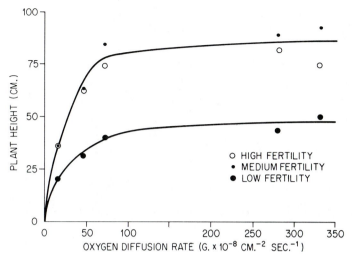

FIGURE 2.14   Effect of O_2 diffusion rate and fertility level on the growth of pea plants.   *Cline and Erickson,* SSSA Proc., **23**:*334, 1959.*

organic matter soils with high pH values. A decline in the availability of molybdenum (Mo) results from a decrease in soil pH. Acid mineral soils are frequently high in soluble Al and Mn, and excessive amounts of these elements are toxic to plants.

When NH_4-N fertilizers are left on the surface of soil with pH values greater than 7, NH_3 may be lost by volatilization. Soil pH values lower than 5.0 and above 7.5 will favor the conversion of water-soluble fertilizer P into forms of lower availability to crops. Certain soilborne diseases are influenced by soil pH. Scab of Irish potatoes, pox of sweet potatoes, and black root rot of tobacco are favored by neutral to alkaline conditions.

The importance of soil acidity to crop growth and the availability of plant nutrients are treated in detail in Chapter 10. Soil acidity is a property of the greatest importance to the grower and one that is easily altered.

BIOTIC FACTORS   Many biotic factors can limit plant growth and reduce crop yields. Heavy fertilization may encourage greater vegetative growth and better environmental conditions for certain disease organisms. The imbalance of nutrients may also be a reason for the increased incidence of disease. The data in Table 2.8 show how a proper balance of nutrients, especially K in combination with N, reduced stem rot of rice and increased yields.

Certain pests may impose an added fertilizer requirement. Viruses and nematodes, for example, attack the roots of certain crops and reduce absorption, requiring a greater supply of nutrients. The beneficial effect on soybean yield of using both K and a nematicide on a low K soil is shown in Figure 2.15.

Insect infestation also may seriously limit plant growth. Heavy fertilization may encourage certain insects, such as the cotton boll weevil, by greater vegetative growth. Definite advances have been made in breeding insect-resistant strains of certain crops and in developing insecticides.

TABLE 2.8   Stem Rot Incidence in Rice and Grain Yield
Index as Affected by Fertilizer

Treatment (lb/a)				
N	P₂O₅	K₂O	Percent Diseased	Yield Index
0	0	0	47.0	100
120	0	60	7.2	163
120	60	60	4.4	187
120	60	0	69.2	66

SOURCE: Ismunadji, *Proc. 12th Colloq. Int. Potash Inst.*, pp. 47–60
(1976).

To reduce pest effects on crop productivity, proper crop rotation and integrated pest management strategies should be practiced, followed by adequate
but not excessive fertilization to keep plants in a vigorous, healthy condition
to maximize competitiveness.

Weeds are another serious deterrent to efficient crop production, for they
compete for moisture, nutrients, and in many instances light. In addition to
these competitive effects of weeds, crop growth may also be suppressed by
biochemical interference or allelopathy. Some weeds are known to produce
and release harmful substances into the root environment.

SUPPLY OF MINERAL NUTRIENT ELEMENTS   From 5 to 10% of the dry
weight of plants is composed of the essential plant nutrients, as well as other
elements derived from the soil that are beneficial for plant growth. Various
aspects of supply, availability, and uptake of these elements will be discussed
in subsequent chapters.

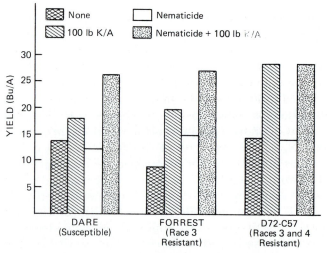

FIGURE 2.15   Yield of soybean varieties as affected by
nematicide treatment and K fertility.   *Shannon et al.,*
Better Crops Plant Food, *61(1):14, 1977.*

ABSENCE OF GROWTH-RESTRICTING SUBSTANCES   Normal development of plants can be restricted or stopped completely by toxic substances. Almost all soil elements, even those essential and beneficial to plant growth, will become toxic to plants when they are present in abnormally high concentrations in the root zone. It is not definitely known if the toxicity of these elements is a direct result of their presence in excessive quantities or if it is due to their interference with uptake of plant nutrients with similar chemical properties.

Probably one of the most widespread toxicity problems involves excessive levels of soluble Al. Other elements that are potentially toxic are nickel (Ni), lead (Pb), mercury (Hg), cadmium (Cd), chromium (Cr), Mn, Cu, Zn, selenium (Se), arsenic (As), molybdenum (Mo), chloride (Cl), boron (B), and fluorine (F). Fortunately, toxicity problems associated with these various elements are infrequent in most agricultural soils. They are most likely to occur in situations involving disposal of waste materials such as those from mines and metallurgical operations, municipal systems, and pulp and paper mills.

Organic compounds, including phenol, cresols, hydrocarbons, substituted ureas, and chlorinated hydrocarbon insecticides, can be toxic if they are present in high concentrations. Such chemicals are not usually a problem at low concentrations since soil microorganisms can acquire the ability to decompose them.

Growth and Growth Expressions

Growth is defined as the *progressive development of an organism*, and there are several ways in which this development can be expressed. Growth may be expressed in terms of dry weight, height, or diameter. Scientists have used various mathematical models to describe or define plant growth. These models can be useful in predicting the crop response to plant nutrients and other growth factors.

Whether growth is expressed as the increase in dry weight or plant height, there is a fairly constant relationship between growth and time. The general pattern is one of initially small increases in size, followed by large increases, and then by a period during which the size of the plant increases slowly or not at all (Fig. 2.16). Thus, the rate of plant growth changes with time, and the maximum growth rate occurs at a point on the curve where the slope is a maximum. Although the general shape of the curve is determined by the genetic constitution of each plant species and by the environment, numerous growth factors can alter the shape of the growth curve.

Although growth curves are helpful in understanding the general pattern of plant development, they indicate nothing, about the factors affecting growth, such as the supply of nutrients, light, CO_2, and water. The plant is a product of both its genetic constitution and its environment. The genetic pattern is a fixed quantity for a given plant and determines its potential for growth in a favorable environment. Plant growth is a function of various growth factors, which may be expressed as

$$G = f(x_1, x_2, x_3, \ldots, x_n) \tag{1}$$

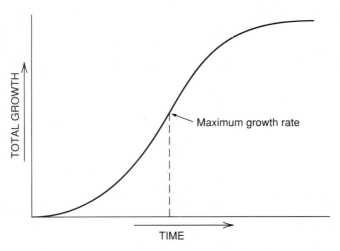

FIGURE 2.16   Generalized curve illustrating the growth
pattern of an annual plant.

where

$$G = \text{some measure of plant growth}$$

and

$$x_1, x_2, x_3, \ldots, x_n = \text{the various growth factors}$$

If all but one of the growth factors are present in adequate amounts, an
increase in the quantity of this limiting factor will generally increase plant
growth. This relationship was named the *law of the minimum* by Liebig and was
discussed in Chapter 1. This, however, is not a simple linear relationship.
Although linear responses occur over small portions of the yield response
curve, the addition of each successive increment of a growth factor results in
a progressively smaller increase in growth (Figure 2.17).

   In 1909, E. A. Mitscherlich of Germany was among the first to quantify the
relationship between plant growth response and the addition of a growth
factor. He stated that "yield can be increased by each single growth factor
even when it is not present in the minimum as long as it is not present in the
optimum" and that an "increase in yield of a crop as a result of increasing a
single growth factor is proportional to the decrement from the maximum
yield obtainable by increasing the particular growth factor."

Mitscherlich's Equation

The growth equation defined in the preceding section is a generalized expression relating growth to all the factors involved. Mitscherlich developed an
equation that related growth to the supply of plant nutrients. He observed

that when plants were supplied with adequate amounts of all but one nutrient, their growth was proportional to the amount of this one limiting element that was supplied to the soil. Plant growth increased as more of this element was added, but the increase in growth was progressively smaller with each successive addition of the element (Figure 2.17). Mitscherlich expressed this mathematically as

$$dy/dx = (A - y)c \tag{2}$$

where dy is the increase in yield resulting from an increment dx of the growth factor x, A is the maximum possible yield obtained by supplying all growth factors in optimum amounts, y is the yield obtained after any given quantity of the factor x has been applied, and c is a proportionality constant that might be considered as an efficiency factor.

The Mitscherlich equation could be reduced to

$$y = A(1 - 10^{-cx}) \tag{3}$$

None of these expressions is conveniently handled as written, but they may also be stated as the integral of Eq. (2) using common logarithms:

$$\log (A - y) = \log A - c(x) \tag{4}$$

The symbols used are the same as those in Eq. (2). If the function is graphed, the curve obtained appears as shown in Figure 2.17.

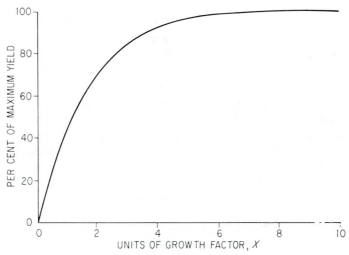

FIGURE 2.17   Percentage of maximum yield as a function of increasing additions of a growth factor, x.

CALCULATION OF THE VALUE OF THE PROPORTIONALITY FACTOR c The constant c in equation (4) becomes 0.301 when yields are expressed on a relative basis of $A = 100$ and x is a quantity of a growth factor. This is shown by first rewriting Eq. (4) as follows:

$$\log A - \log (A - y) = cx$$

or

$$\log \frac{A}{A - y} = cx$$

When the nutrient supply (x) is increased to produce 50% of the maximum yield,

$$\frac{A}{A - y} = \frac{100}{50} = 2$$

Thus $\log 2 = c(1)$ and $c = 0.301$.

The value of c varies with the particular growth factor. Mitscherlich found that the value of c was 0.122 for N, 0.60 for P, and 0.40 for K. He claimed that it was constant for each fertilizer nutrient, independent of the crop, the soil, or other conditions. The average value for c in British experiments conducted before 1940 was 1.1 for N, 0.80 for P, and 0.80 for K. In numerous other investigations, it has been observed that c is not a constant term and that it varies rather widely for different crops grown under different conditions.

The significance of the c term is that it gives an indication of whether the maximum yield level can be achieved by a relatively low or high quantity of the specific growth factor. When the value of c is small, a large quantity is needed, and vice versa.

CALCULATION OF RELATIVE YIELDS FROM ADDITION OF INCREASING AMOUNTS OF A GROWTH FACTOR   If A, the maximum yield, is considered to be 100%, equation (4) reduces to

$$\log (100 - y) = \log 100 - 0.301(x) \tag{5}$$

It is possible to determine the relative yield expected from the addition of a given number of units of x. It will be helpful if the student observes how these calculations are made.

If none of the growth factor is available, that is, $x = 0$, then $y = 0$; but suppose that 1 unit of x is present. Then

$$\log (100 - y) = \log 100 - 0.301(1)$$
$$\log (100 - y) = 2 - 0.301$$
$$\log (100 - y) = 1.699$$
$$100 - y = 50$$
$$y = 50$$

and the addition of 1 unit of the growth factor x results in a yield that is 50% of the maximum.

Assume, however, that 2 units of the growth factor were present. In this instance

$$
\begin{aligned}
\log (100 - y) &= \log 100 - 0.301(2) \\
\log (100 - y) &= 2.000 - 0.602 \\
\log (100 - y) &= 1.398 \\
100 - y &= 25 \\
y &= 75
\end{aligned}
$$

The same operation may be repeated until 10 units of the growth factor have been added. The result of such a series of calculations is given in Table 2.9.

It is obvious that the successive increases of a growth factor result in a yield increase that is 50% of that resulting from addition of the preceding unit until a point is reached at which further increases are of no consequence. Again, this relationship is shown in Figure 2.17.

Plant growth as a function of nutrient inputs *is* logarithmic and generally follows a pattern of diminishing increases, as expressed in the Mitscherlich equation (Fig. 2.17). The growth of annual plants *does* tend to reach a maximum with increasing inputs of nutrients under a particular set of environmental conditions, and often the plants that produce the highest yield of dry matter have the lowest percentage of N in their tissues. However, it remains for posterity to determine whether a single expression can be developed that will universally predict the amount of growth that can be produced from the input of a given quantity of plant nutrients when environmental and genetic growth factors are adequately described.

TABLE 2.9 Example of Diminishing Returns to Increasing Addition of Factor x as Described by the Mitscherlich Equation

Units of Growth Factor, x	Yield (%)	Increase in Yield (%)
0	0	
1	50	50
2	75	25
3	87.5	12.5
4	93.75	6.25
5	96.88	3.125
6	98.44	1.562
7	99.22	0.781
8	99.61	0.390
9	99.80	0.195
10	99.90	0.098

Factorial Experiments and Regression Equations

Much fertilizer research has made use of the factorial experiment. In such studies the effect on crop yield of several levels of different input factors is evaluated simultaneously. For example, the effect of several levels of N and P on the yield of a crop can be studied simultaneously in one experimental. If three levels of N and three levels of P are examined, the experiment is said to be a 3^2 *factorial* and requires nine treatments to evaluate the effect on yield of all combinations of N and P. A regression equation is developed in which the yield is functionally related to the inputs of the fertilizer variables.

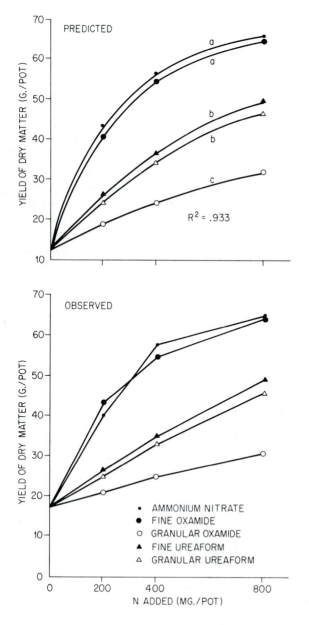

FIGURE 2.18 Observed and predicted corn forage yield as affected by rate, source, and granule size of fertilizer nitrogen. *Engelstad and Khasawneh, Agron. J., **61:473, 1969.***

When climate, soil type, plant population, fertilizer placement, and other factors are uniform and constant, such studies help to predict fertilizer requirements, but the equations developed from them are not universally applicable. The models generated by the data from these studies are commonly of the diminishing-return type, similar to those of the Mitscherlich and related expressions, and can be used to calculate the maximum economic yield or fertilizer rate needed for maximum profit (see Chapter 14).

It should be noted that regression techniques can only organize data already collected under a particular set of conditions and have little predictive value for other situations. Regression techniques are commonly used for making fertilizer recommendations.

Nutrient Interactions

Interactions among plant nutrients are often overlooked even though they can have considerable influence on plant growth. The interplay of plant nutrients is best studied in factorial experiments that test each nutrient at three or more rates.

Two or more growth factors are said to interact when their influence individually is modified by the presence of one or more of the others. An interaction takes place when the response of two or more inputs used in combination is unequal to the sum of their individual responses. There can be both positive and negative interactions in soil fertility studies (Fig. 2.19). In addition, there can be circumstances where there is no interaction, with the action of factors being only additive.

In negative interactions, the two nutrients combined increase yields less than when they are applied separately. This kind of interaction can be the result of substitution for and/or interference of one treatment with the other. Lime × P, lime × Mo, Mo × P, and Na × K are common negative interactions involving apparent substitution effects. Changes in soil pH will result in numerous interactions where one ion or nutrient interferes with or competes with the uptake and utilization of other nutrients by plants.

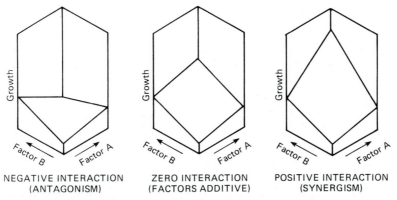

FIGURE 2.19   Influence of interactions between two nutrient factors on the growth of a crop.   *Sumner and Farina, in J. K. Syers, Ed.,* Phosphorus in Agricultural Systems, *Elsevier, New York, 1983.*

Positive interactions are in accordance with Liebig's law of the minimum. If two factors are limiting, or nearly so, addition of one will have little effect on growth, whereas provision of both together will have a much greater influence. In severe deficiencies of two or more nutrients, all fertilizer responses will result in strong positive interactions.

Yield increases from an application of one nutrient can reduce the concentration of a second nutrient, but the higher yields result in greater uptake of the second nutrient. This is a dilution effect, which should be distinguished from an antagonistic effect.

In addition to interactions between two or more nutrients, there are numerous opportunities for other kinds of interactions: for example, nutrients and disease, nutrients and cultural practice, nutrients and crop species, nutrients and hybrid or variety, nutrients and seeding date, nutrients and plant population or spacing, and nutrients and environmental conditions. Many of these types of interactions are discussed in Chapter 13.

Bray's Nutrient Mobility Concept

A modification of the Mitscherlich concept was proposed by R. Bray and his co-workers at the University of Illinois. In brief, crop yields obey the percentage sufficiency law of Mitscherlich for such elements as P and K, which are relatively immobile in the soil. This concept, in turn, is based on Bray's nutrient mobility concept, which states that

> as the mobility of a nutrient in the soil decreases, the amount of that nutrient needed in the soil to produce a maximum yield (the soil nutrient requirement) increases from a value determined by the magnitude of the yield and the optimum percentage composition of the crop, to a constant value.

The magnitude of this constant is independent of the amount of crop yield, provided that the kind of plant, planting pattern and rate, and fertility pattern remain constant and that similar soil and seasonal conditions prevail. Bray further states that for a mobile element such as NO_3-N, Liebig's law of the minimum best expresses the growth of a crop. Bray has modified the Mitscherlich equation to

$$\log (A - Y) = \log A - C_1 b - Cx \qquad (6)$$

where A, Y, and x have the connotations already given; C_1 is a constant representing the efficiency of b for yields in which b represents the amount of an immobile but available form of nutrient, such as P or K, measurable by some suitable soil test; and C represents the efficiency factor for x, which is the added fertilizer form of the nutrient b.

Bray showed that the values for C_1 and C are specific and fairly constant over a wide area, regardless of yield and season, for each of the following crops: corn, wheat, and soybeans. The factors that will alter the values, however, are wide differences in soil series, plant population and planting patterns, and the form and distribution in the soil of the immobile nutrient under study. Hence, as management practices and fertilizer placement methods are changed to obtain higher yields, the values change and must be reexamined.

Limited Applications of Growth Expressions

Numerous equations or functions have been used to describe the relation between plant growth and nutrient input. Steenbjerg and Jakobsen of Denmark, in commenting on the variability among growth response curves, point out that "the constants in formulas are not constants because the variables in the formulas are not independent variables." Factors other than nutrient interactions obviously affect the shape of yield curves. They include other environmental factors, which were discussed in the preceding sections. The change in the shape and position of yield–plant nutrient input curves with changes in environmental conditions is of the greatest importance to the practical agriculturalist. Understanding the interactions between these crop growth factors is essential to identifying soil and crop management practices needed for profitable crop production.

The term *limiting growth factor,* used frequently throughout this book, has been clearly illustrated by the variable nature of the response curves and surfaces previously discussed. If, for example, a crop has inadequate moisture, the application of a given amount of fertilizer will provide a lower yield than if moisture were adequate. Another example, and an important one, is the application of fertilizer to a crop growing on a soil that is too acid for maximum growth, regardless of the amount of fertilizer added. If lime is not applied, acidity becomes the limiting factor that reduces both the yield response to fertilizer and the farmer's return on the investment. The importance to practical farm operations of the concept of a growth pattern and how it may be altered by various limiting factors cannot be overstated.

Modeling Applications in Soil Fertility

The use of computers in agriculture is essential for analyzing management decisions before going to the field. Models have been developed that provide information on crop growth and development in response to various alternatives. Practices such as tillage; planting dates; fertilizer rates, timing, and placement; irrigation scheduling; and so on can be explored with the computer models, and outcomes from various scenarios can be tested before field work begins.

Researchers use models to test treatment alternatives to help ensure that the range of treatments is adequate to test the hypothesis. Farmers test alternatives to help reduce risk and maximize profit. Models are actually only a computerized representation of what we know about the alternatives being studied. Thus the computer does no more than ensure that we consider many of the factors affecting the outcome, and it analyzes the possible responses faster than could be done without computer help.

Every person involved in decision making uses a model, whether he or she thinks of it as using a computer or not. Knowledge, experience, and interpretations of the crop–soil–environment system make up the model. As the person learns more, experiences more, and makes more decisions, the model is refined and the ability to predict the response improves. Putting this model into a computer forces a more organized approach and more thoroughness in analysis, but the model is still limited by that person's knowledge of the subject.

When a model is used to make predictions, or *simulations*, its reliability is only as good as the model and the data being used. The more accurate and complete the information, the more reliable the simulation. In the process of running the simulation, the researcher or farmer may discover weaknesses in the model or the data and may be able to avoid costly mistakes before going to the field.

A new technology goes a step further with the computer, allowing the computer to draw conclusions and make interpretations. Expert systems, a form of artificial intelligence, draws upon a set of facts, data, models, and opinions of various experts regarding the relationships being studied. The expert system takes this information and develops a set of decision-making rules and determines the probabilities associated with different alternatives.

While expert systems are relatively new in agriculture, they have been used in the medical profession and in engineering for several years. In some cases, the technology has proved to be superior to decisions made by trained professionals. Again, the computer is limited by the accuracy of the facts it is given.

Modeling is a technology that needs further testing to become widely accepted. But there have been many successes, and it is an important tool for developing new crop management technologies. Computers are better at handling details and thoroughly analyzing alternatives. They are important management tools for the researcher and farmer, never replacing the individual's thinking process but supplementing it so that more options can be considered in making a decision.

Summary

1. Plant growth as a function of time, genetic makeup of the plant, and environmental factors was discussed. The importance of selecting crops that are genetically capable of making maximum use of the supply of available plant nutrients was pointed out.

2. The environmental factors were considered in relation to their effect on plant growth, as well as their impact on limiting the crop response to applied plant nutrients. The concept of limiting growth factors was discussed, and the need to recognize the importance of this concept in practical farming operations was pointed out.

3. Growth of annual plants follows a well-defined pattern. Plant responses to environmental conditions, including the supply of plant nutrients, also follow a set pattern. When growth is plotted as a function of increasing amounts of applied nutrients, successive increments of fertilizer give successively smaller increases in plant growth. Such curves, known as *response curves*, have been studied by numerous investigators, and various mathematical formulas have been developed to describe them.

Questions

1. Equations that express plant growth as a function of inputs of a plant nutrient differ from one another to varying degrees. To what do you attribute these differences?

2. Despite the differences among the

various equations referred to in Question 1, they all seem to have one thing in common. What is it?

3. Crop yields have been increasing over the years because of improvements in tillage, varieties, pest control, fertilization, and so on. From a *theoretical* standpoint, what do you consider to be the factor that will *ultimately* limit further increases in plant growth? From a *practical* standpoint, what do you think this factor will be?

4. Plant yields in terms of forage, grain, or fruit are the criteria of the effectiveness of various fertilizer inputs. If one is interested in the effect of the imposed treatment on the plants' ability to convert radiant energy to a usable form, are these yield figures the best criteria of treatment effect? Why? (*Hint*: Compare the highest yields of soybeans with the highest yields of corn.) What criterion would you use?

5. What growth factor in the past has been frequently overlooked by plant breeders in developing new crop varieties?

6. Among the environmental factors limiting crop response to nutrients, which is probably the most easily and inexpensively changed?

7. What are some of the environmental factors affecting plant response to applied nutrients that are more easily controlled in the greenhouse than in the field? If you were planning to control these factors in a greenhouse operation set up for the commercial production of crops, what factors would you consider before instituting control?

8. In your opinion, could the limited or partial control of the CO_2 content of the aboveground atmosphere under field conditions be done successfully? Under what conditions would you expect such control to be successful?

9. Of *all* the growth factors limiting crop production, which is most easily corrected in commercial farming operations?

10. Study the various growth curves discussed in this chapter. Should a commercial grower attempt to produce maximum yields? Why? At what point along the yield–fertilizer input curve do you feel the grower should operate? Elaborate.

11. Why is soil structure so important in influencing crop responses to applied plant nutrients? Be specific. What can be done about soil structure? On which type of soil is structure more important—loamy sands or silty clay loams? Why?

12. Light or radiant energy was listed as an environmental factor affecting growth and response to plant nutrients. In what ways does light influence growth under field conditions? Under controlled conditions? What can be done about the effect of light on growth?

13. Although not listed as such, human beings comprise the biotic factor that influences crop growth to the greatest extent. Why is this statement made, especially in relation to commercial farming?

14. In the strictest sense, is it correct to refer to the various environmental factors as independent variables? Why?

15. Which plant physiological process (or processes) is (or are) most affected by soil water deficits?

16. List the principal atmospheric pollutants that might affect plant growth. Are any of them beneficial for plants?

17. Describe the various ways that weeds compete with crops.

18. Are there differences in the ability of crop species to compete with a given weed species?

19. Identify one major growth factor that may strongly influence successful corn production in the U.S. Corn Belt.

20. Why is temperature an important factor affecting crop growth?
21. Are current yields of corn and soybeans approaching their potential maximum yields?
22. High-yield potential is an important plant characteristic. What are other important properties?
23. What plant growth factors can be largely controlled by humans? Which ones are beyond our control?
24. Is fertilizer response related to soil moisture supply?
25. Does soil moisture supply influence plant nutrient availability and uptake? How does it affect nutrient availability?
26. Does photoperiod affect crop growth? How?
27. What is an interaction?
28. How might modeling be helpful in soil fertility?

Selected References

BRAY, R. H. 1963. Confirmation of the nutrient mobility concept of soil–plant relationships. *Soil Sci.*, **95:**124.

COX, T. S., J. P. SHROYER, L. BEN-HUI, R. G. SEARS, and T. J. MARTIN. 1988. Genetic improvement in agronomic traits of hard red winter wheat cultivars from 1919 to 1987. *Crop Sci.* **28:**756.

ENGELSTAD, O. P., and F. E. KHASAWNEH. 1969. Use of a concurrent Mitscherlich model in fertilizer evaluation. *Agron. J.*, **61:**473.

LEVITT, J. 1980. *Responses of Plants to Environmental Stresses.* Vol. I. *Chilling, Freezing, and High Temperature Stresses.* Academic Press, New York.

LEVITT, J. 1980. *Responses of Plants to Environmental Stresses.* Vol. II. *Water, Radiation, Salt, and Other Stresses.* Academic Press, New York.

MILTHORPE, F. L., and J. MOORBY. 1974. *An Introduction to Crop Physiology.* Cambridge University Press, London.

PEARSON, C. J. 1984. *Control of Crop Productivity.* Academic Press, New York.

STEENBJERG, F., and S. T. JAKOBSEN. 1963. Plant nutrition and yield curves. *Soil Sci.*, **95:**69.

TOLLENAAR, M. 1985. What is the current upper limit of corn productivity? Physiology, biochemistry and chemistry associated with maximum yield corn. *Found. Agric. Res. Pot. Phosphate Inst.*, 167–173.

WAGGONER, P. E. 1983. Agriculture and a climate changed by more carbon dioxide. *Changing Climate.* National Academy Press, Washington, D.C., pp. 383–418.

WILKS, D. S. 1988. Estimating the consequences of CO_2-induced climatic change on North American grain agriculture using general circulation model information. *Climatic Change*, **13:**19.

WITTWER, S. H. 1974. Maximum production capacity of food crops. *Bioscience*, **24**(4):216.

CHAPTER 3

Elements Required in Plant Nutrition

Essentiality of Elements in Plant Nutrition

A mineral element is considered essential to plant growth and development if the element is involved in plant metabolic functions and the plant cannot complete its life cycle without the element. Usually the plant exhibits a visual symptom indicating a deficiency in a specific nutrient, which normally can be corrected or prevented by supplying that nutrient. Visual nutrient deficiency symptoms can be caused by many other plant stress factors; therefore, caution should be exercised when diagnosing deficiency symptoms. These interactions are discussed in Chapter 11. The following terms are commonly used to describe levels of nutrients in plants:

Deficient: When the concentration of an essential element is low enough to limit yield severely and distinct deficiency symptoms are visible. Extreme deficiencies can result in plant death. With moderate or slight deficiencies, symptoms may not be visible, but yields will still be reduced.

Critical range: The nutrient concentration in the plant below which a yield response to added nutrient occurs. Critical levels or ranges vary among plants and nutrients, but occur somewhere in the transition between nutrient deficiency and sufficiency.

Sufficient: The nutrient concentration range in which added nutrient will not increase yield but can increase nutrient concentration. The term *luxury consumption* is often used to describe nutrient absorption by the plant that does not influence yield.

Excessive or *toxic:* When the concentration of essential or other elements is high enough to reduce plant growth and yield. Excessive nutrient concentration can cause an imbalance in other essential nutrients, which also can reduce yield.

The general relationship between nutrient concentration in plant tissue and plant yield is shown in Figure 3.1. Yield is severely affected when a nutrient is deficient, and when the nutrient deficiency is corrected, growth increases more rapidly than nutrient concentration. Under severe deficiency, rapid

45

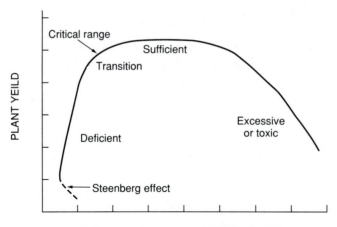

FIGURE 3.1   Relationship between essential plant nutrient concentration and plant growth or yield. As plant nutrient concentration increases toward the critical level, plant yield increases. Above the critical level the plant contains sufficient levels for normal growth and can continue to absorb the nutrient without increasing yield (luxury consumption). Excessive absorption of a nutrient or element can be toxic to the plant and reduce yield or cause plant death.

increases in yield with added nutrient can cause a small decrease in nutrient concentration. This is called the *Steenberg effect* and results from dilution of the nutrient in the plant by the rapid plant growth. When the concentration reaches the critical range, plant yield is generally maximized. Nutrient sufficiency occurs over a wide concentration range, wherein yield is unaffected. Increases in nutrient concentration above the critical range indicate that the plant is absorbing nutrients above that needed for maximum yield. This luxury consumption is common in most plants. Elements absorbed in excessive quantities can reduce plant yield directly through toxicity or indirectly by reducing concentrations of other nutrients below their critical ranges.

Elements Required in Plant Nutrition

Sixteen elements are considered essential to plant growth; their relative concentrations in plants are listed in Table 3.1. Carbon (C), hydrogen (H), and oxygen (O) are the most abundant elements in plants. The photosynthetic process in green leaves converts CO_2 and H_2O into simple carbohydrates from which amino acids, sugars, proteins, nucleic acid, and other organic compounds are synthesized. Carbon, H, and O are not considered mineral nutrients. The supply of CO_2 is relatively constant (see p. 27). The supply of H_2O rarely limits photosynthesis directly but does indirectly though the various effects resulting from moisture stress (see p. 21).

The remaining 13 essential elements are classified as macronutrients and micronutrients, and the classification is based on their relative abundance in plants (Table 3.1). The macronutrients are nitrogen (N), phosphorus (P), potassium (K), sulfur (S), calcium (Ca), and magnesium (Mg). Compared to the macronutrients, the concentrations of the seven micronutrients—iron (Fe), zinc (Zn), manganese (Mn), copper (Cu), boron (B), chlorine (Cl), and molybdenum (Mo)—are very small. Five additional elements—sodium (Na), cobalt (Co), vanadium (Va), nickel (Ni), and silicon (Si)—have been established as essential micronutrients in *some* plants. Micronutrients are often referred to as *minor* elements, but this does not mean that they are less important than macronutrients. Micronutrient deficiency or toxicity can reduce plant yield similar to macronutrient deficiency or toxicity.

Although aluminum (Al) is not an essential plant nutrient, its concentration in plants can be high when soils contain relatively large amounts of Al in soil solution. In fact, plants absorb many nonessential elements, and over 60 elements have been identified in plant materials. When plant material is burned, the remaining *plant ash* contains all the essential and nonessential mineral elements except C, H, O, N, and S, which are burned off as gases.

The plant content of mineral elements is affected by many factors, and their concentration in crops varies considerably. Shown in Table 12.1 (Chapter 12) are the contents of some of the mineral elements in common crop plants. Although the latest available data have been used, the figures should be regarded only as averages. Soil, climate, crop variety, and management factors exert considerable influence on plant composition and in individual cases may cause appreciable variation from the values in the table.

Plant nutrient concentration data are valuable to successful fertilizer management programs and can be used to help establish fertilizer recommendations. Because many biological and chemical reactions occur with fertilizers in soils, the quantity of nutrients absorbed by plants does not equal the quantity

TABLE 3.1 Relative and Average Plant Nutrient Concentrations

Plant Nutrient	Relative Concentration	Average Concentration*
H	60,000,000	6.0 %
O	30,000,000	45.0 %
C	30,000,000	45.0 %
N	1,000,000	1.5 %
K	400,000	1.0 %
Ca	200,000	0.5 %
Mg	100,000	0.2 %
P	30,000	0.2 %
S	30,000	0.1 %
Cl	3,000	100 ppm (0.01%)
Fe	2,000	100 ppm
B	2,000	20 ppm
Mn	1,000	50 ppm
Zn	300	20 ppm
Cu	100	6 ppm
Mo	1	0.1 ppm

*Concentration expressed by weight on a dry matter basis.

applied as a fertilizer. Proper fertilizer management can maximize the proportion of fertilizer nutrient absorbed by the plant. As plants absorb nutrients from the soil, complete their life cycle, and die, the nutrients in the plant residue are returned to the soil. These plant nutrients are subject to the same biological and chemical reactions as fertilizer nutrients. Although this cycle varies somewhat among nutrients, understanding nutrient dynamics in the soil–plant–atmosphere system is essential to successful fertilizer management.

Function of Nutrients in Plants: Macronutrients

Nitrogen

Nitrogen (N) is a vitally important plant nutrient and is the most frequently deficient of all nutrients. Plants normally contain between 1 and 5% N by weight. It is absorbed by plants as nitrate (NO_3^-) and ammonium (NH_4^+) ions and as urea. In moist, warm, well-aerated soils the NO_3^- form is dominant.

Before NO_3^- can be used in the plant, it must be reduced to NH_4^+ or NH_3. Nitrate reduction involves two enzyme-catalyzed reactions that occur in roots and/or leaves, depending on plant species. Both reactions occur in series so that nitrite (NO_2^-) does not accumulate.

	Reduction Reaction	Enzyme	Reaction Site
Step 1.	$NO_3 \rightarrow NO_2$	Nitrate reductase	Cytoplasm
Step 2.	$NO_2 \rightarrow NH_3$	Nitrite reductase	Chloroplast

The NH_3 produced in these reactions is assimilated into numerous amino acids that are subsequently incorporated into proteins and nucleic acids. Proteins provide the framework for chloroplasts, mitochondria, and other structures in which most biochemical reactions occur. The type of protein formed is controlled by a specific genetic code found in nucleic acids, which determines the quantity and arrangement of amino acids in each protein. One of these nucleic acids, deoxyribonucleic acid (DNA), present in the nucleus and mitochondria of the cell, duplicates the genetic information in the chromosomes of the parent cell to the daughter cell. Ribonucleic acid (RNA), present in the nucleus and cytoplasm of the cell, executes the instructions coded within the DNA molecules. Most of the enzymes controlling these metabolic processes are also proteins. These functional proteins are not stable entities, for they are continually being degraded and resynthesized.

In addition to its role in the formation of proteins, N is an integral part of chlorophyll, which is the primary absorber of light energy needed for photosynthesis. The basic unit of chlorophyll's structure is the porphyrin ring system, composed of four pyrrole rings, each containing one N and four C atoms (Fig. 3.2). A single Mg atom is bonded in the center of each porphyrin ring.

An adequate supply of N is associated with high photosynthetic activity, vigorous vegetative growth, and a dark green color. An excess of N in relation to other nutrients, such as P, K, and S, can delay crop maturity. Stimulation of heavy vegetative growth early in the growing season can be a disadvantage

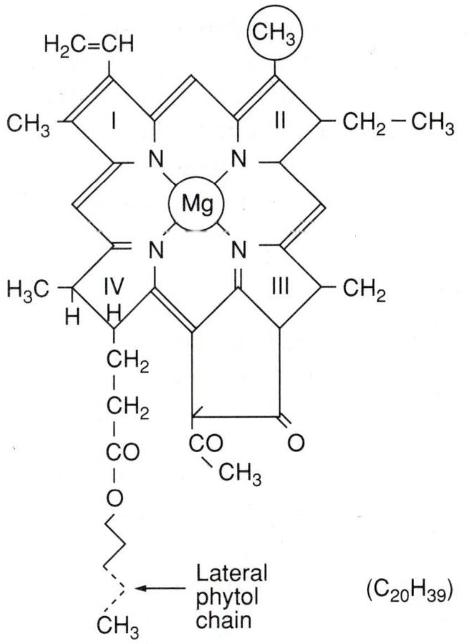

FIGURE 3.2   A simplified representation of a chlorophyll molecule.

in regions where soil moisture limits plant growth. Early-season depletion of soil moisture without adequate replenishment prior to the grain-filling period can depress yields.

If N is used properly in conjunction with other needed soil fertility inputs, it can speed the maturity of crops such as corn and small grains (Table 3.2). Applications of up to 120 lb N/a lower the percentage of water in corn grain at harvest. This favorable influence saves energy required to dry grain to 15.5% moisture content and/or permits an earlier harvest.

The supply of N is related to carbohydrate utilization. When N supplies are insufficient, carbohydrates will be deposited in vegetative cells, causing them to thicken. When N supplies are adequate and conditions are favorable for growth, proteins are formed from the manufactured carbohydrates. Less carbohydrate is thus deposited in the vegetative portion, more protoplasm is formed, and, because protoplasm is highly hydrated, a more succulent plant results.

TABLE 3.2   Effect of N on the Moisture Content and Yield of Corn Grain (Averages for the Years 1967–1977)

N (lb/a)	Yield (bu/a)	Moisture in Grain at Harvest (%)
0	66	36.1
60	101	30.0
120	135	27.9
180	158	26.9
240	167	28.2
300	168	27.2

SOURCE: Ohio State Univ., *17th Annu. Agron. Demonstration, Farm Sci. Rev.* (1979).

Excessive succulence in some crops may have a harmful effect. With cotton, a weakening of the fiber may result, and with grain crops, lodging may occur, particularly with a low K supply or with varieties not adapted to high levels of N. In some cases, excessive succulence may make a plant more subsceptible to disease or insect attack. Crop plants, such as wheat and rice, have been modified for growth at higher densities and at higher levels of N fertilization. Shorter plant height and improved lodging resistance have been bred into the plants, which respond in yield to much higher rates of N than in the past.

When plants are deficient in N, they become stunted and yellow in appearance. The loss of protein N from chloroplasts in older leaves produces the yellowing or *chlorosis* indicative of N deficiency. Chlorosis usually appears first on the lower leaves; the upper leaves remain green, while under severe N deficiency lower leaves will turn brown and die. This necrosis begins at the leaf tip and progresses along the midrib until the entire leaf is dead. The appearance of normal and N-deficient corn plants is shown in Figure 3.3.

The tendency of the young upper leaves to remain green as the lower leaves yellow or die is an indication of the mobility of N in the plant. When the roots are unable to absorb sufficient N to meet their growing requirement, protein in the older plant parts is converted to soluble N, translocated to the active meristematic tissues, and reused in the synthesis of new protein.

FIGURE 3.3   Corn plants receiving adequate and inadequate N. Notice that the N-deficient plants in the center rows have a thin, spindly appearance and light color compared to the vigorous growth and dark color of the rows to the right and far left.

Phosphorus

Phosphorus (P) occurs in most plants in concentrations between 0.1 and 0.4%, considerably lower than those typically found for N and K. Plants absorb either $H_2PO_4^-$ or HPO_4^{2-} orthophosphate ions. Absorption of $H_2PO_4^-$ is greatest at low pH values, whereas uptake of HPO_4^{2-} is greatest at higher values of soil pH.

Plants may also absorb certain soluble organic phosphates. Nucleic acid and phytin occur as degradation products of the decomposition of soil organic matter and can be taken up by growing plants. Because of the instability of many organic P compounds in the presence of an active microbial population, their importance as sources of P for higher plants is limited.

The most essential function of P in plants is in energy storage and transfer. Adenosine di- and triphosphates (ADP and ATP) act as "energy currency" within plants (Fig. 3.4). When the terminal phosphate molecule from either ATP or ADP is split off, a relatively large amount of energy (12,000 cal/mol) is liberated. Energy obtained from photosynthesis and metabolism of carbohydrates is stored in phosphate compounds for subsequent use in growth and reproductive processes.

Donation or transfer of the energy-rich phosphate molecules from ATP to energy-requiring substances in the plant is known as *phosphorylation*. In this reaction ATP is converted back to ADP. The compounds ADP and ATP are formed and regenerated in the presence of sufficient P.

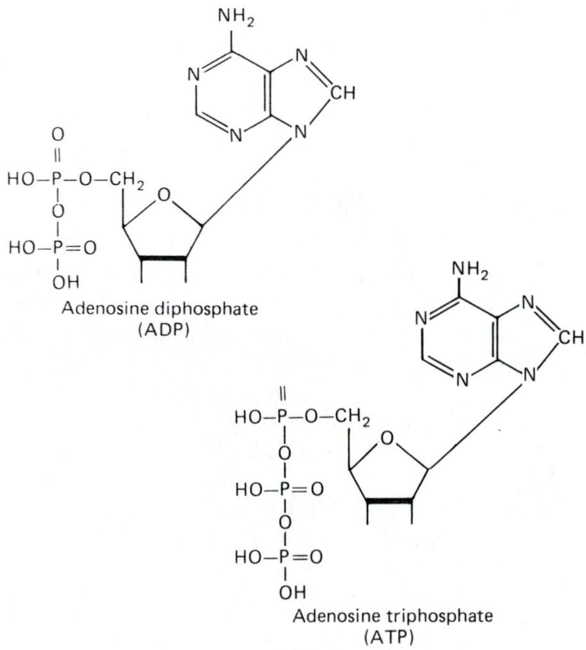

FIGURE 3.4   Structure of ADP and ATP.   *Wallingford, in* Phosphorus for Agriculture: A Situation Analysis, *p. 7. Atlanta, Ga.: Potash & Phosphate Institute, 1978.*

ATP is the source of energy that powers practically every energy-requiring biological process in plants. Almost every metabolic reaction of any significance proceeds via phosphate derivatives. Table 3.3 lists some of the more important metabolic processes that involve ATP or its equivalent. P also is an important structural component of a wide variety of biochemicals, including nucleic acids, coenzymes, nucleotides, phosphoproteins, phospholipids, and sugar phosphates. An adequate supply of P early in the life of a plant is important in the development of its reproductive parts. Large quantities of P are found in seed and fruit, and it is considered essential for seed formation.

A good supply of P is associated with increased root growth. When soluble phosphate compounds are applied in a band, plant roots proliferate extensively in that area of treated soil. Similar observations are made with both NO_3^- and NH_4^+ applied in a band near roots. Figure 3.5 demonstrates how the exposure of parts of the main seminal roots of barley to zones of high P concentration causes initiation and prolific development of both first- and second-order laterals. This results in considerable modification of the root form but has little influence on the extension of the main seminal roots. The greatly increased root proliferation should encourage extensive exploitation of the treated soil areas for nutrients and moisture.

Several other effects on plant growth are attributed to P fertilization. P is associated with early maturity of crops, particularly grain crops. The effect of ample P nutrition on reducing the time required for grain ripening is illustrated in Figure 3.6.

An adequate supply of P is associated with greater straw strength in cereals. The quality of certain fruit, forage, vegetable, and grain crops is improved and disease resistance increased when these crops have satisfactory P nutrition. The effect of P on raising the tolerance of small grains to root-rot diseases is particularly noteworthy. Also, the risk of winter damage to small grains can be substantially lowered by applications of P, particularly on low-P soils and with unfavorable growing conditions.

P is mobile in plants, and when a deficiency occurs, it is translocated from older tissues to the active meristematic regions. Because of the marked effect of P deficiency on retarding overall growth, the striking foliar symptoms that are evident with N or K deficiency are seldom observed. In corn and some other grass species, P deficiency symptoms also are expressed by purple discoloration of the leaves or leaf edges (Fig. 3.7).

TABLE 3.3   Processes or Pathways Involving ATP

Membrane transport	Generation of membrane electrical potentials
Cytoplasmic streaming	Respiration
Photosynthesis	Biosynthesis of cellulose, pectins, hemicellulose, and lignin
Protein biosynthesis	
Phospholipid biosynthesis	Lipid biosynthesis
Nucleic acid synthesis	Isoprenoid biosynthesis → steroids and gibberellins

SOURCE: Glass et al., *Proc. Western Canada Phosphate Symp.*, p. 358 (1980).

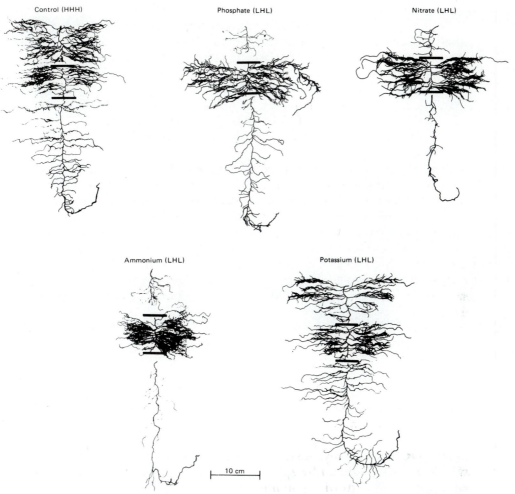

FIGURE 3.5   Effect of a localized supply of phosphate, nitrate, ammonium, and potassium on root form. Control plants (HHH) received the complete nutrient solution to all parts of the root system. The other roots (LHL) received the complete nutrient solution only in the middle zone, the top and bottom being supplied with a solution deficient in the specified nutrient.   *Drew*, New Phytol., **75:486, 1975.**

Potassium

The potassium ion (K^+) is actively taken up from soil solution by plant roots. The concentration of K^+ in vegetative tissue usually ranges from 1 to 4% on a dry matter basis. Thus, plant requirements for available K are quite high. K, unlike N, P, and most other nutrients, forms no coordinated compounds in the plant. Instead it exists solely as the K^+ ion, either in solution or bound to negative charges on such organic radicals as the acid radical :R-C-O⁻ . As
                                                                    ‖
                                                                    O

FIGURE 3.6   Effect of P fertilization on the maturity of small grains. Notice the more advanced maturity of the small grains receiving the P *(left)* in contrast to those that received no P *(right).*   *Courtesy of O. H. Long, Univ. of Tennessee.*

FIGURE 3.7   P deficiency symptoms in corn. Note the purpling along the leaf edges.

a result of its strictly ionic nature, K^+ has functions particularly related to the ionic strength of solutions within plant cells.

1. *Enzyme activation.* Enzymes are involved in many important plant physiological processes, and over 80 plant enzymes require K for their activation. Enzyme activation is regarded as the single most important function of K. These enzymes tend to be most abundant in meristematic tissue at the growing points, both above and below ground level, where cell division takes place rapidly and where primary tissues are formed.

Starch synthetase is an enzyme involved in the conversion of soluble sugars into starch, which is a vital step in the grain-filling process. Nitrogenase is the enzyme responsible for reducing atmospheric N_2 to NH_3 in the cells of *Rhizobium* bacteria. The NH_3 formed is released into cells of the host legume plant, where it is used for synthesis of amino acids. The intensity of the N_2 reduction process depends on the supply of carbohydrates. K enhances carbohydrate transport to nodules and utilization for synthesis of amino acids.

2. *Water relations.* The predominance of K over other cations in plants makes its role in osmotic regulation particularly important. K provides much of the osmotic "pull" that draws water into plant roots. Plants that are K deficient are less able to withstand water stress, mostly because of their inability to make full use of available water.

Maintenance of plant turgor is essential to the proper functioning of photosynthetic and metabolic processes. The opening of stomata occurs when there is an increase of turgor pressure in the guard cells surrounding each stoma, which is brought about by an influx of K. Malfunctioning of stomata due to a deficiency of this nutrient has been related to lower rates of photosynthesis and less efficient use of water.

Transpiration is the loss of water through stomata, and it accounts for the major portion of a plant's water use. K can affect the rate of transpiration and water uptake through regulation of stomatal opening. An example of how improved K nutrition reduced the transpiration rate of peas by more complete closing of the stomata is shown in Figure 3.8.

3. *Energy relations.* Plants require K for the production of high-energy phosphate molecules (ATP), which are produced in both photosynthesis and respiration. The amount of CO_2 that is assimilated into sugars during photosynthesis increases sharply with increasing K (Fig. 3.9).

4. *Translocation of assimilates.* Once CO_2 is assimilated into sugars during photosynthesis, the sugars are transported to plant organs, where they are stored or used for growth. Translocation of sugars requires energy in the form of ATP—which requires K for its synthesis.

The translocation of sugar from leaves is greatly reduced in K-deficient plants. For example, normal translocation in sugarcane leaves is approximately 2.5 cm/min; however, the rate is reduced by half in K-deficient plants.

5. *N uptake and protein synthesis.* Total N uptake and protein synthesis are reduced in K-deficient plants, as indicated by a buildup of amino acids. Again, the involvement of K is through the need for ATP for both processes.

The beneficial effect of K on increased yields of winter wheat is due mainly to higher kernel weight (Table 3.4). This striking effect of K on both grain yield and test weight results from increasing either the photosynthetic capacity

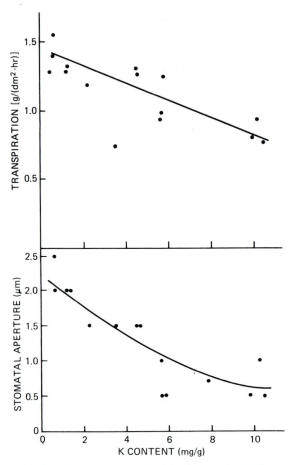

FIGURE 3.8   Improved K nutrition reduced the
transpiration rate of peas due to smaller stomatal
opertures.   *Brag,* Physiol. Plant., *26:254, 1972.*

or the productive life of flag leaves, which accounts for up to 80% of grain
filling.

When K is limiting, characteristic deficiency symptoms appear in the plant.
Typical K deficiency symptoms in alfalfa consist of white spots on the leaf
edges, whereas chlorosis and necrosis of the leaf edges are observed with corn
and other grasses (Fig. 3.10).

Since K is mobile in the plant, visual deficiency symptoms usually appear
first in the lower leaves, progressing toward the top as the severity of the
deficiency increases. Potassium deficiency also can occur in young leaves at
the top of high-yielding, fast-maturing crops like cotton and wheat.

Another symptom of insufficient K is weakening of straw in grain crops,
which causes lodging in small grains and stalk breakage in corn (Fig. 3.11)
and sorghum. The results shown in Table 3.5 are indicative of how seriously
stalk breakage can affect production through impaired yields and harvesting
losses. K deficiencies greatly reduce crop yields. In fact, serious yield reduc-

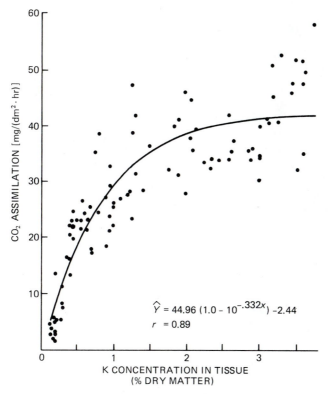

FIGURE 3.9   Adequate K in corn leaves increases
photosynthesis as measured by CO_2 fixation. *Smid and
Peaslee, Agron. J., 68:907, 1976.*

tions may occur without the appearance of deficiency symptoms. This phenomenon has been termed *hidden hunger* and is not restricted to K.

K stress can increase the degree of crop damage by bacterial and fungal
diseases, insect and mite infestation, and nematode and virus infection. Soybeans are highly susceptible to pod and stem blight caused by the fungus
Diaporthe sojae L. The gray, moldy seed resulting from infection by this pathogen means not only lower yield but also lower seed quality (Fig. 3.12). The
relationship between percentage of soybean seed infected by *D. sojae* L. and

TABLE 3.4   Effect of K Nutrition on Growth Factors of Winter Wheat
(Average of Five Varieties)

	Relative Responses to Increasing K Nutrition			
Growth Factor	Deficient	Mildly Deficient	Adequate	High
Grain yield	100	143	199	225
Single kernel wt.	100	132	181	192
Kernels per head	100	109	111	112
Heads per plant	100	101	99	105

SOURCE: Forster, *Proc. 8th Inter. Fert. Congr.,* Vol. 1; p. 41 (1976).

(a)

(b)

FIGURE 3.10   K-deficient alfalfa (a) and corn (b).   *Western Potash News Lett.,*
W-29 (April 1963)

FIGURE 3.11   Response of corn to K on a low-K soil. Note the poor growth and lodged condition of the crop on the right.   *Courtesy of the Potash & Phosphate Institute, Atlanta, Ga.*

TABLE 3.5   Effect of N and K on Yields and Stalk Breakage of Corn

K_2O Applied (lb/a)	Nitrogen Applied (lb/a)		
	0	80	160
	Yield (bu/a)		
0	48	33	38
80	73	116	119
160	59	122	129
	Stalk Breakage (%)		
0	9	57	59
80	4	3	8
160	4	4	4

SOURCE: Schulte, *Proc. Wisconsin Fert. and Aglime Conf.*, p. 58 (1975).

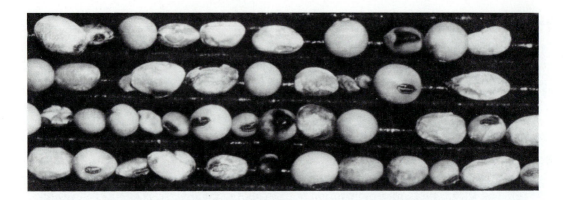

FIGURE 3.12   Low K causes moldy, diseased, low-quality soybeans.   *Courtesy of the Potash & Phosphate Institute, Atlanta, Ga.*

K treatment is illustrated in Table 3.6. Higher rates of K as either KCl or K_2SO_4 markedly decreased the incidence of disease in each variety. Lack of K in wetland rice greatly increases the severity of foliar diseases such as stem rot, sheath blight, and brown leaf spot.

TABLE 3.6   Effect of K on Soybean Seed Yield and Disease

KCl or K_2SO_4 *(g/cylinder)*	*Seeds per Plant*		*Diseased Seed (%)**	
	Var. A	*Var. B*	*Var. A*	*Var. B*
Control	254	200	87	62
2	262	207	65	58
10	275	209	21	33
30 + 10 sidedress	264	200	13	14
	NS†		LSD = 6.0	

*Percent gray, moldy seed (*D. sojae* infected).
† Not significant.
SOURCE: Crittenden and Svec, *Agron. J.,* **66:**697 (1974).

Calcium

Calcium (Ca) is absorbed by plants as Ca^{2+}, and its concentration ranges from 0.2 to 1.0%. Ca has an important role in the structure and permeability of cell membranes. Lack of Ca^{2+} produces a general breakdown of membrane structures, with resultant loss in retention of cellular diffusible compounds. Ca enhances uptake of NO_3-N and therefore is interrelated with N metabolism. The presence of Ca^{2+} also provides some regulation of cation uptake. For example, studies have shown that K^+ and Na^+ uptake are about equal in the absence of Ca^{2+}, but in its presence K^+ uptake greatly exceeds Na^+ uptake.

Ca is essential for cell elongation and division, and Ca^{2+} deficiency manifests itself in the failure of terminal buds of shoot and apical tips of roots to develop, which causes plant growth to cease.

In corn, Ca^{2+} deficiency prevents the emergence and unfolding of new leaves, the tips of which are almost colorless and are covered with a sticky gelatinous material that causes them to adhere to one another.

In fruits and vegetables, the most frequent indicator of Ca^{2+} deficiency consists of disorders in the storage tissues. Examples of Ca^{2+} disorders are blossom-end rot in tomato and bitter pit of apples. The unsatisfactory condition of apples affected by bitter pit is evident in Figure 3.13.

Finally, Ca^{2+} is generally considered to be an immobile element in the plant. There is very little translocation of Ca^{2+} in the phloem, and for this reason there is often a poor supply of Ca^{2+} to fruits and storage organs. Downward translocation of Ca^{2+} is also limited in roots, which usually prevents them from entering low-Ca soils.

Magnesium

Magnesium (Mg), is absorbed as Mg^{2+}, and its concentration in crops varies between 0.1 and 0.4%. The importance of Mg^{2+} is obvious, since it is a primary constituent of the chlorophyll molecule, and without chlorophyll the autotrophic green plant would fail to carry on photosynthesis (Fig. 3.2). Chlorophyll usually accounts for about 15 to 20% of the total Mg^{2+} content of plants.

Mg also serves as a structural component in ribosomes, stabilizing them in the configuration necessary for protein synthesis. As a consequence of Mg^{2+} deficiency, the proportion of protein N decreases and that of nonprotein N generally increases in plants.

Mg is involved in a number of physiological and biochemical functions. It is associated with transfer reactions involving phosphate-reactive groups. Mg is required for maximal activity of almost every phosphorylating enzyme in carbohydrate metabolism. Most reactions involving phosphate transfer from ATP require Mg^{2+}. Since the fundamental process of energy transfer occurs in photosynthesis, glycolysis, the tricarboxylic acid cycle (citric acid or Krebs cycle), and respiration, Mg^{2+} is important throughout plant metabolism.

Because of the mobility of a substantial portion of the plant Mg^{2+} and its ready translocation from older to younger plant parts, deficiency symptoms often appear first on the lower leaves. In many species, shortage of Mg^{2+} results in interveinal chlorosis of the leaf, in which only the veins remain green. In more advanced stages the leaf tissue becomes uniformly pale yellow,

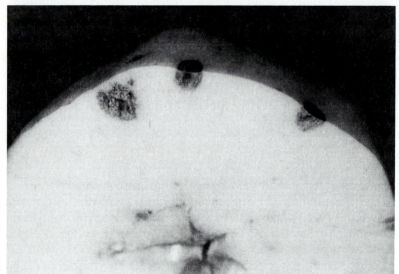

FIGURE 3.13   *Top:* Bitter pit development in a Golden
Delicious apple. *Bottom:* Cross-sectional view of bitter pit
development in a Golden Delicious apple.   *Courtesy of*
B. Fleming and Dr. G. Neilsen, Agriculture Canada,
Summerland, B.C.

then brown and necrotic. In other species, notably cotton, the lower leaves
may develop a reddish-purple cast, gradually turning brown and finally necrotic. Mg deficiency of corn is illustrated in Figure 3.14.

Sulfur

Sulfur (S) is absorbed by plant roots almost exclusively as the sulfate ion,
SO_4^{2-}. Small quantities of SO_2 can be absorbed through plant leaves and
utilized within plants, but high concentrations are toxic. Elemental S dusted

FIGURE 3.14   Mg-deficiency symptoms in corn.   *Courtesy of Dr. G.R. Hagstrom, Duval Sales Corp., Houston, Tex.*

on fruit tree leaves finds its way in small amounts to the internal plant system relatively soon after application. The mechanism by which this water-insoluble source of S penetrates the plant is not known.

Typical concentrations of S in plants range between 0.1 and 0.4%. S is present in equal or lesser amounts than P in such plants as wheat, corn, beans, and potatoes but in larger amounts in alfalfa, cabbage, and turnips. Among the families of crop plants, the requirement increases in the order Gramineae < Leguminosae < Cruciferae and is also reflected in the corresponding differences in the S content of their seeds: 0.18–0.19%, 0.25–0.3%, and 1.1–1.7%, respectively. Much of the SO_4^{2-} is reduced in the plant to —S—S and —SH forms, although SO_4^{2-} occurs in plant tissues and cell sap.

S deficiency which has a pronounced retarding effect on plant growth and is characterized by uniformly chlorotic plants—stunted, thin-stemmed, and spindly. In many plants these symptoms resemble those of N deficiency and have undoubtedly led to many incorrect diagnoses. Unlike N, however, S does not appear to be easily translocated from older to younger plant parts; therefore, deficiency symptoms occur first in younger leaves. Typical S deficiency symptoms in winter wheat are shown in Figure 3.15.

S-deficient cruciferous crops such as cabbage and canola/rapeseed will initially develop a reddish color on the underside of the leaves. In canola/rapeseed the leaves are also cupped inward. As the deficiency progresses in cabbage, there is a reddening and purpling of both upper and lower leaf surfaces; the cupped leaves turn back on themselves, presenting flattened-to-concave surfaces on the upper side. The characteristic appearance of S-deficient rapeseed is shown in Figure 3.16. Paler than normal blossoms

FIGURE 3.15   Typical S deficiency symptoms in winter
wheat.   *Courtesy of B. Wells, Univ. of Arkansas.*

FIGURE 3.16   Normal (*right*) and S-deficient (*left*) rapeseed plants.   *Courtesy of
K. D. McLachlan, CSIRO, Canberra, Australia, and The Sulphur Institute, Washington,
D.C.*

and severely impaired seed set also characterizes S deficiency symptoms in rapeseed.

S has many important functions in plant growth and metabolism. It is required for synthesis of the S-containing amino acids cystine, cysteine, and methionine, which are essential components of protein. Approximately 90% of the S in plants is found in these amino acids.

One of the main functions of S in proteins is the formation of disulfide bonds between polypeptide chains. This bridging is achieved through the reaction of two cysteine molecules, forming cystine. Linking of two cysteine units within a protein by a disulfide bond (—S—S—) will cause the protein to fold. Disulfide linkages are therefore important in stabilizing and determining the configuration of proteins. This, in turn, is directly relevant to the catalytic or structural properties of the protein.

S is needed for the synthesis of other metabolites, including coenzyme A, biotin, thiamin or vitamin B_1, and glutathione. Coenzyme A is probably the most important of these substances since it is involved in the oxidation and synthesis of fatty acids, the synthesis of amino acids, and the oxidation of certain intermediates of the tricarboxylic acid or citric acid cycle.

S is also a component of other S-containing substances, including S-adenosylmethionine, formylmethionine, lipoic acid, and sulfolipid.

Although not a constituent, S is required for the synthesis of chlorophyll. Table 3.7 shows the importance of adequate S nutrition in the occurrence of chlorophyll in red clover.

S is a vital part of the ferredoxins, a type of nonheme Fe-S protein occurring in the chloroplasts. Ferredoxin participates in oxidoreduction processes by transferring electrons and has a significant role in nitrite reduction, sulfate reduction, the assimilation of N_2 by root nodule bacteria, and free-living N-fixing soil bacteria.

S occurs in volatile compounds responsible for the characteristic taste and smell of plants in the mustard and onion families. S enhances oil formation in crops such as flax and soybeans.

The data in Table 3.8 show that, within limits, the S-containing amino acid content of plants can be altered by S fertilization. Increasing levels of S nutrition in the culture medium raised the methionine, cystine, and total S contents in both of the experimental strains of alfalfa.

TABLE 3.7   Effect of a High Level of S Nutrition on the Chlorophyll Content of Kenland Red Clover

Applied Sulfate (ppm S)	Chlorophyll Content (% dry weight)
0	0.49
5	0.54
10	0.50
20	1.02
40	1.18

SOURCE: Rendig et al., *Agron. Abstr. Annu. Meet. Am. Soc. Agron.*, p. 109 (1968).

TABLE 3.8   Effect of S on the Content of S-Containing Amino Acids in Two Strains of Alfalfa

SO_4^{2-} Ion Concentration (ppm)	Methionine (mg/g N)		Cystine (mg/g N)		Sulfur (%)	
	C_3	C_{10}	C_3	C_{10}	C_3	C_{10}
0	10.6	17.6	21.5	24.4	0.100	0.089
1	20.8	27.6	28.6	35.2	0.103	0.098
3	33.6	34.9	37.0	43.6	0.129	0.121
9	38.0	40.3	38.9	42.9	0.186	0.200
27	41.4	43.9	42.9	45.0	0.229	0.227
81	43.4	44.3	43.6	46.0	0.244	0.242

SOURCE: Tisdale et al., *Agron. J.*, **42**:221 (1950).

The results in Table 3.9 clearly show the influence of adequate S nutrition in controlling concentrations of nonprotein N in orchardgrass. Plants suffering S deficiency accumulate nonprotein N in the form of NH_2 and NO_3^-. It is also apparent that S fertilization improved the quality of this forage by narrowing the N/S ratio. An N/S ratio of between 9:1 and 12:1 is needed for effective use of recycled N by rumen microorganisms in animals. This beneficial effect of S fertilization on improving crop quality through reductions in the N/S ratio is often overlooked.

Function of Nutrients in Plants: Micronutrients

Boron

The boron (B) concentration in monocotyledons and dicotyledons generally varies between 6 and 18 ppm and 20 and 60 ppm, respectively. Levels of B in mature leaf tissue of most crops are usually adequate if over 20 ppm.

Most of the B is absorbed by plants as undissociated boric acid (H_3BO_3). Much smaller amounts of other forms, such as $B_4O_7^{2-}$, $H_2BO_3^-$, HBO_3^{2-}, and BO_3^{3-}, may be present, but they generally do not contribute significantly to plant needs.

TABLE 3.9   Effects of Elemental S Application on the Yield and Quality of Orchardgrass (113 kg/ha of N Applied After Each Cutting)

Sulfur* (kg/ha)	Yield[†] (metric tons/ha) of Cutting		Nonprotein N (%) in Cutting		Nitrate N (%) in Cutting		N/S Ratio in Cutting	
	1	3	1	3	1	3	1	3
0	3.74	1.77	1.05	1.22	0.064	0.211	21.3	21.4
23	3.72	2.55	0.64	0.85	0.037	0.184	15.3	18.7
45	3.63	2.62	0.59	0.49	0.051	0.144	14.3	14.8
90	3.40	2.89	0.51	0.44	0.037	0.137	12.2	13.4
113	3.40	2.76	0.49	0.37	0.033	0.106	10.8	10.0

*Applied in 1965 and 1967.
[†] Harvests taken in 1968.
SOURCE: Baker et al., *Sulphur Inst. J.*, **9**(1):15 (1973).

B deficiency is the most widespread micronutrient deficiency. Many fruit, vegetable, and field crops suffer from B deficiency. Examples of crops with a high B requirement include asparagus, carrots, celery, lettuce, onions, sugar beets, sunflowers, and various brassicas. B fertilizer rates for these high-requiring crops may cause injury to low-requiring ones, such as the small grains, peas, and beans.

B plays an essential role in the development and growth of new cells in the plant meristem. Since it is not readily translocated from older to actively growing tissues, the first visual deficiency symptom is cessation of terminal bud growth, followed by death of the young leaves. In B-deficient plants the youngest leaves become pale green, losing more color at the base than at the tip. The basal tissues break down, and if growth continues, the leaves have a one-sided or twisted appearance. Flowering and fruit development are also restricted by a shortage of B. Sterility and severely impaired seed set are late-season symptoms of B deficiency in both B-sensitive (rapeseed and clover) and B-insensitive (wheat) crops.

B-deficiency symptoms will often appear in the form of thickened, wilted, or curled leaves; a thickened, cracked, or water-soaked condition of petioles and stems; and a discoloration, cracking, or rotting of fruit, tubers, or roots. Internal cork of apple is caused by a deficiency of this element, and a lack of B in citrus fruits results in uneven thickness of the peel, lumpy fruit, and gummy deposits in the fruit. The breakdown of internal tissues in root crops gives rise to darkened areas referred to as *brown heart* or *black heart*. B-deficient sugar beets are shown in Figure 3.17. Some conifers exhibit striking B deficiency symptoms including distorted branches and main stems, resin bleeding, and death of major branches. Under open growing conditions the growth form will be bushy or shrubby.

Plants require B for a number of growth processes, especially the following:

· New cell development in meristematic tissue.
· Proper pollination and fruit or seed set.
· Translocation of sugars, starches, N, and P.
· Synthesis of amino acids and proteins.
· Nodule formation in legumes.
· Regulation of carbohydrate metabolism.

Regardless of our inability to explain the precise role of B in plants, it is essential in varying but usually small quantities for the growth of many important agricultural crops. Although it is required for higher plants and some algae and diatoms, B is not needed by animals, fungi, and microorganisms.

Iron

The sufficiency range of iron (Fe) in plant tissue is normally between 50 and 250 ppm. In general, when Fe contents are 50 ppm or less in the dry matter, deficiency is likely to occur. Fe is absorbed by plant roots as Fe^{2+}, as Fe^{3+}, and as organically complexed or chelated Fe, although the Fe^{2+} species is utilized in metabolic processes. The Fe^{2+} form is more mobile and available for

FIGURE 3.17   B-deficient sugar beet roots. Note the blackened heart
tissue.   *Courtesy of Professor R. L. Cook, Michigan State Univ., and reprinted from
Hunger Signs in Crops, 3rd ed., 1964, with permission of the David McKay Co., New
York.*

incorporation into biomolecular structures. Some plant tissues containing
large quantities of Fe^{3+} may exhibit Fe-deficiency symptoms.

Fe deficiency is most frequently seen in crops growing on calcareous or
alkaline soils, although some crops exhibit Fe deficiency on acid soils. Citrus
and deciduous fruits often exhibit Fe chlorosis. This is also fairly common in
blueberries (which are grown on very acid soils) and in sorghum, grown in
neutral to alkaline soils. Other crops known to display deficiencies of this
element are soybeans, bush beans, corn, strawberries, avocado, vegetable
crops, and many ornamentals. Fe deficiency of soybeans is quite common in
the western Corn Belt of the United States.

A deficiency of Fe shows up first in the young leaves of plants. It does not
appear to be translocated from older tissues to the tip meristem, and as a
result, growth ceases. The young leaves develop an interveinal chlorosis, which
progresses rapidly over the entire leaf. In severe cases the leaves turn entirely
white. Fe-deficiency symptoms in sorghum are illustrated in Figure 3.18.

The chemical properties of Fe make it an important part of oxidation-
reduction reactions in both soils and plants. Fe is a transition metal capable
of existing in more than one oxidation state, whereby it can accept or donate
electrons according to the oxidation potential of the reactants. The movement

FIGURE 3.18   Fe-deficient sorghum. Note the chlorosis and other deficiency symptoms in the Fe-deficient plot on the left.   *Courtesy of M. Marsolek and Dr. G. R. Hagstrom, Duval Sales Corp., Houston, Tex.*

of electrons between the organic molecule and Fe provides the potential for many of the enzymatic transformations in which Fe is essential. Several of these enzymes are involved in chlorophyll synthesis, and when Fe is deficient, chlorophyll production is reduced, which results in the characteristic chlorosis symptoms of Fe stress.

Fe is a structural component of porphyrin molecules: cytochromes, hemes, hematin, ferrichrome, and leghemoglobin. These substances are involved in oxidation-reduction reactions in respiration and photosynthesis. As much as 75% of the total cell Fe is associated with the chloroplasts, and up to 90% of the Fe in leaves occurs with lipoprotein of the chloroplast and mitochondria membranes.

The localization of Fe in chloroplasts reflects the presence of cytochromes for performing various photosynthetic reduction processes and of ferrodoxin as an initial electron acceptor. Ferredoxins are Fe-S proteins and are the first stable redox compound of the photosynthetic electron transport chain. Reduction of O_2 to water during respiration is the most commonly recognized function of Fe-containing compounds. Fe also is an important part of the enzyme nitrogenase, which is essential to the N_2 fixation in N-fixing microorganisms. Fe may also be capable of partial substitution for Mo as the metal cofactor necessary for the functioning of NO_3^- reductase in soybeans.

Fe toxicity causes nutritional disorders in rice grown on poorly drained, submerged soils. This condition, known as *bronzing*, is associated with Fe levels greater than 300 ppm in the leaf blade of rice at tillering.

Manganese

Manganese (Mn) is a micronutrient whose normal concentration in plants typically ranges from 20 to 500 ppm. Concentrations of Mn in upper plant parts below 15 to 20 ppm are considered deficient. Mn is absorbed by plants as Mn^{2+}, as well as in molecular combinations with certain natural and synthetic complexing agents.

Like Fe, Mn is a relatively immobile element, and deficiency symptoms usually show up first in the younger leaves. In broad-leaved plants the visual symptoms appear as an interveinal chlorosis. Leaves from Mn-deficient soybean plants are shown in Figure 3.19. Mn deficiency of several crops has been described by such terms as *gray speck of oats*, *marsh spot of peas*, and *speckled yellows of sugar beets*. Wheat plants low in Mn are often more susceptible to root rot diseases.

The involvement of Mn in photosynthesis, particularly in the evolution of O_2, is well known. It also takes part in oxidation-reduction processes and in decarboxylation and hydrolysis reactions. Mn can substitute for Mg^{2+} in many of the phosphorylating and group-transfer reactions.

FIGURE 3.19 Different stages of Mn deficiency of soybean leaves. *Courtesy of Dept. of Agronomy, Purdue Univ., and reprinted from* Hunger Signs in Crops, *3rd ed., 1964, with permission of the David McKay Co., New York.*

Although it is not specifically required, Mn is needed for maximal activity of many enzyme reactions in the citric acid cycle. In the majority of enzyme systems, Mg is as effective as Mn in promoting enzyme transformations. Mn influences auxin levels in plants, and it seems that high concentrations of this micronutrient favor the breakdown of indoleacetic acid.

Plants are injured by excessive amounts of Mn. Crinkle leaf of cotton is an Mn toxicity that is sometimes observed in highly acid red and yellow soils of the old Cotton Belt in the southern United States. Mn toxicity has also been found in tobacco, soybeans, tree fruits, and canola/rapeseed growing on extremely acid soils. Upward adjustment in soil pH by liming will readily correct this problem.

Copper

Copper (Cu) is absorbed by plants as the cupric ion, Cu^{2+}, and may be absorbed as a component of either natural or synthetic organic complexes. Its normal concentration in plant tissue ranges from 5 to 20 ppm. Deficiencies are probable when Cu levels in plants fall below 4 ppm in the dry matter.

Deficiencies of Cu have been reported in numerous plants, although they are more prevalent among crops growing in peat and muck soils. Crops most susceptible to Cu deficiency include alfalfa, wheat, barley, oats, lettuce, onions, carrots, spinach, and table beets. Other crops responding to Cu fertilization include clover, corn, and fruit trees.

Symptoms of Cu deficiency vary with the crop. In corn the youngest leaves become yellow and stunted, and as the deficiency becomes more severe, the young leaves pale and the older leaves die back. In advanced stages, dead tissue appears along the tips and edges of the leaves in a pattern similar to that of K deficiency. Cu-deficient small-grain plants lose color in the younger leaves, which eventually break, and the tips die. In many vegetable crops the leaves lack turgor. They develop a bluish-green cast, become chlorotic, and curl, and flower production fails to take place. Stem melanosis disease occurs in certain wheat varieties when Cu deficient. Also, ergot infection is associated with Cu deficiency in some wheat and barley varieties.

Cu in its reduced form readily binds and reduces O_2. In the oxidized form the metal is readily reduced, and protein complexed Cu has a high redox potential. These properties of Cu are exploited by enzymes that create complex polymers such as lignin and melanin. Cu is unique in its involvement in enzymes, and it cannot be replaced by any other metal ion.

Zinc

Zinc (Zn) is a micronutrient whose normal concentration range is 25 to 150 ppm in plants. Deficiencies of Zn are usually associated with concentrations of less than 20 ppm, and toxicities will occur when the Zn leaf concentration exceeds 400 ppm.

Plant roots absorb Zn as Zn^{2+} and as a component of synthetic and natural organic complexes. Soluble Zn salts and Zn complexes can also enter the plant system directly through leaves.

Corn and beans are particularly sensitive to Zn deficiency, as are citrus and deciduous fruit trees such as peach. Other crops classified as being very sensitive to Zn deficiency are flax, grapes, hops, onions, pecans, pine, rice, and soybeans. Some of the mildly sensitive crops are alfalfa, clovers, cotton, potatoes, sorghum, Sudan grass, sugar beets, and tomatoes.

Zn deficiency can often be identified by distinctive visual symptoms that appear most frequently in the leaves. Sometimes the deficiency symptoms will also appear in the fruit or branches or are evident in the overall development of the plant. Symptoms common to many crops include the following:

- Occurrence of light green, yellow, or white areas between the veins of leaves, particularly the older, lower leaves.
- Death of tissue in these discolored, chlorotic leaf areas.
- Shortening of the stem or stalk internodes, resulting in a bushy, rosetted appearance of the leaves.
- Small, narrow, thickened leaves. Often the leaves are malformed by continued growth of only part of the leaf tissue.
- Early loss of foliage.
- Malformation of the fruit, often with little or no yield.

Zn deficiency causes the characteristic little leaf and rosetting or clustering of leaves at the top of fruit tree branches, which have become mainly bare. In corn and sorghum, Zn deficiency is called *white bud* and in cotton it is known as *little leaf*. The deficiency is referred to as *mottle leaf* or *frenching* in citrus crops and is described as *fern leaf* in Russet Burbank potato. Zn deficiency symptoms in corn are shown in Figure 3.20.

Zn is involved in many enzymatic activities, but it is not known whether it acts as a functional, structural, or regulatory cofactor. Zn is important in the synthesis of tryptophane, a component of some proteins and a compound needed for the production of growth hormones (auxins) like indole acetic acid. Reduced growth hormone production in Zn-deficient plants causes the shortening of internodes and smaller than normal leaves.

Molybdenum

Molybdenum (Mo) is a nonmetal anion absorbed as molybdate (MoO_4^{2-}). This is a weak acid and can form complex polyanions such as phosphomolybdate. Sequestering of Mo in this form may explain why it can be taken up in relatively large amounts without any apparent toxicity.

Normally, the Mo content of plant material is less than 1 ppm, and deficient plants usually contain less than 0.2 ppm. Mo concentrations in plants are frequently low because of the extremely small amounts of MoO_4^{2-} in the soil solution. In some cases, however, Mo levels in crops may exceed the range 1,000 to 2,000 ppm.

Mo is an essential component of the enzyme NO_3^- reductase, which catalyzes the conversion of NO_3^- to NO_2^-. Most of the Mo in plants is concentrated in this enzyme, which primarily occurs in chloroplasts in leaves. The Mo requirement of plants is influenced by the form of inorganic N supplied to plants, with either NO_2^- or NH_4^+ effectively lowering its need. It is also a structural component of nitrogenase, the enzyme actively involved in N_2 fixa-

FIGURE 3.20   Typical Zn deficiency symptoms in corn.
Note the stunted nature of the Zn-deficient plant on the
right compared to the Zn-sufficient plant on the left.

tion by root-nodule bacteria of leguminous crops, by some algae and actino-myctes, and by free-living, N_2-fixing organisms such as *Azotobacter*. Mo concentrations in the nodules of legume crops of up to 10 times higher than those in leaves have been observed. Mo is also reported to have an essential role in iron absorption and translocation in plants.

Chloride

Chloride (Cl) is absorbed by plants as the Cl^- ion through both roots and aerial parts. Its normal concentration in plants is about 0.2 to 2.0%, although levels as high as 10% are not uncommon. All of these values are much greater than the physiological requirement of most plants. Concentrations of 0.5 to 2.0% in the tissues of sensitive crops can lower yield and quality. Similar reductions in yield and quality can occur when Cl^- levels approach 4% in tolerant crops like sugar beets, barley, corn, spinach, and tomatoes.

Chloride has not been found in any true metabolite in higher plants. The essential role of Cl^- seems to lie in its biochemical inertness. This inertness enables it to fill osmotic and cation neutralization roles, which may have important biochemical and/or biophysical consequences. Cl can be readily transported in plant tissues. A useful function for Cl^- is as the counterion during rapid K fluxes, thus contributing to turgor of leaves and other plant parts.

The observations that partial wilting and loss of leaf turgor are symptoms of Cl^- deficiency support the concept that Cl^- is an active osmotic agent. A high level of Cl^- nutrition will increase total leaf water potential and cell sap osmotic potential in wheat plants. Some of the favorable action of Cl^- fertilization on the growth of small grains is attributed to improved moisture relations.

Cl does appear to have a definite role in the evolution of O_2 in photosystem II in photosynthesis. Extremely high Cl^- concentrations of nearly 11% have been detected in chloroplasts.

Uptake of both NO_3^- and SO_4^{2-} can be reduced by the competitive effects of Cl^-. Lower protein concentrations in winter wheat resulting from high levels of Cl^- nutrition are attributed to the strong competitive relationship between Cl^- and NO_3^-.

Many diseases are suppressed by Cl^- fertilizers (Table 3.10). For example, Cl^--containing fertilizers greatly depress take-all root-rot infections in winter wheat. Some of the inhibitory effect of Cl^- is apparently related to restricted uptake of NO_3^- and to a less favorable rhizosphere pH for activity of the pathogen when the crop obtains much of its N requirement in the form of NH_4^+. The influence of Cl^- on water potential components within plants may also be a factor in controlling take-all, since the pathogenic fungus responsible for this disease grows best at high water potentials or under moist conditions. Increased osmotic potential of cell sap seems to be involved in greater resistance to take-all.

TABLE 3.10 Diseases Suppressed by Increased Soil Cl Levels

Location	Crop	Suppressed Disease
Oregon	Winter wheat	Take-all
Germany	Winter wheat	Take-all
North Dakota	Winter wheat	Tanspot
Oregon	Winter wheat	Stripe rust
Great Britain	Winter wheat	Stripe rust
South Dakota	Spring wheat	Leaf rust
South Dakota	Spring wheat	Tanspot
North Dakota	Barley	Common root rot
North Dakota	Barley	Spot blotch
Montana	Barley	Fusarium root rot
North Dakota	Durum	Common root rot
South Dakota	Oats	Leaf rust
New York	Corn	Stalk rot
India	Pearl millet	Downy mildew
Philippines	Coconut palm	Gray leaf spot
Oregon	Potatoes	Hollow heart
Oregon	Potatoes	Brown center
California	Celery	Fusarium yellows

SOURCE: Fixen, 2nd National Wheat research conference, (1987).

Chlorosis in younger leaves and an overall wilting of the plants are the two most common symptoms of Cl^- deficiency. Necrosis in some plant parts, leaf bronzing, and reduction in root growth may also be seen in Cl^--deficient plants. Tissue concentrations below 70 to 700 ppm are usually indicative of deficiency.

Excesses of Cl^- can be harmful, and crops vary widely in their tolerance to this condition. Tobacco, peach, avocado, and some legumes are among the most sensitive crops. Leaves of tobacco and potatoes become thickened and tend to roll when these crops accumulate excessive amounts of Cl^-. The storage quality of potato tubers are adversely affected by surplus uptake of Cl^-.

Responses to Cl^- have been observed in the field for a large number of crops, including tobacco, tomatoes, buckwheat, peas, lettuce, cabbage, carrots, sugar beets, barley, wheat, corn, potatoes, cotton, kiwi, coconut, and oil palms. The last three tropical crops are especially responsive to Cl^-. The Cl^- needs for high yields of most temperate region crops are usually satisfied by only 4 to 10 kg/ha.

Cobalt

Cobalt (Co) is essential for microorganisms fixing N_2. Co is thus needed in the nodules of both legumes and alder, as well as in N_2-fixing algae. The normal Co concentration in the dry matter of plants ranges from 0.02 to 0.5 ppm. Only 10 ppb of Co in nutrient solution was found to be adequate for N_2 fixation by alfalfa.

The essentiality of Co for the growth of symbiotic microorganisms such as rhizobia, free-living N_2-fixing bacteria, and blue-green algae is its role in the formation of vitamin B_{12}. Co forms a complex with N atoms in a porphyrin ring structure that provides a prosthetic group for association with a nucleotide in the vitamin B_{12} coenzyme. This Co complex is termed the *cobamide coenzyme*.

Other functions attributed to Co include leghemoglobin metabolism and ribonucleotide reductase in *Rhizobium*. Co is one of several metals that activate enolase and succinic kinase.

Improved growth, transpiration, and photosynthesis with applications of Co have been observed in cotton, beans, and mustard.

Vanadium

Low concentrations of vanadium (V) are beneficial for the growth of microorganisms, animals, and higher plants. Although it is considered to be essential for the green alga *Scenedesmus*, there is still no decisive evidence that V is essential for higher plants. Some workers suggest that V may partially substitute for Mo in N_2 fixation by microorganisms such as the rhizobia. It has also been speculated that it may function in biological oxidation-reduction reactions. Increases in growth attributable to V have been reported for asparagus, rice, lettuce, barley, and corn. The V requirement of plants is said to be less than 2 ppb dry weight, whereas the normal concentration in plant material averages about 1 ppm.

Sodium

Sodium (Na) is essential for halophytic plant species that accumulate salts in vacuoles to maintain turgor and growth. The succulence of such plants is sometimes increased by Na. Crops that require Na for optimum growth include celery, mangold, spinach, sugar beet, Swiss chard, table beet, and turnip. Favorable effects of Na are also reported to occur with cabbage, kale, kohlrabi, mustard, radish, and rapeseed. The increased growth produced by salt in halophytes is believed to be due to increased turgor.

Na is absorbed by plants as Na^+, and its concentration varies widely, from 0.01 to 10%, in leaf tissue. Sugar beet petioles frequently contain levels at the upper end of this range. Many plants that possess the C_4 dicarboxylic photosynthetic pathway require Na as an essential nutrient. It also has a role in inducing crassulacean acid metabolism, which is considered part of a general response to water stress. It is not definitely known how Na affects C_4 and crassulacean acid metabolism. Lack of Na will cause certain plant species to shift their CO_2-fixation pathway from C_4 to C_3. Provision of adequate Na can restore these plants to their normal C_4 fixation.

Water economy in plants seems to be related to the C_4 dicarboxylic photosynthetic pathway of plants. Many plant species that have the extremely efficient C_4 CO_2-fixing system occur naturally in arid, semiarid, and tropical conditions, where the closure of stomata to prevent wasteful water loss is essential for growth and survival. CO_2 entry must also be restricted when stomata tend to remain closed. The ratio of weights of CO_2 assimilated to water transpired by C_4 plants is often double that of C_3 plants. It is also noteworthy that C_4 plants are often found in saline habitats.

Sugar beets appear to be particularly responsive to Na. It influences water relations in this crop and increases the resistance of sugar beets to drought. In low-Na soils the beet leaves are dark green, thin, and dull in hue. The plants wilt more rapidly and may grow horizontally from the crown. There may also be an interveinal necrosis similar to that resulting from K deficiency. Some of the effects ascribed to Na may also be due to Cl^- since the usual source of Na is NaCl.

Silicon

Silicon (Si) is one of the most abundant elements in the lithosphere and is absorbed by many plants as monosilicic acid, $Si(OH)_4$. Cereals and grasses contain 0.2 to 2.0% Si, while dicotyledons may accumulate only one-tenth of this concentration. Concentrations of up to 10% occur in Si-rich plants. The involvement of Si in root functions is believed to contribute to the drought tolerance of crops such as sorghum.

Si contributes to the structure of cell walls. Grasses, sedges, nettles, and horsetails accumulate 2 to 20% of the foliage dry weight as Si. Si primarily impregnates the walls of epidermal and vascular tissues, where it appears to strengthen the tissues, reduce water loss, and retard fungal infection. Where large amounts of Si are accumulated, intracellular deposits known as *plant opals* can occur.

Although no biochemical role for Si in the development of plants has been positively identified, it has been proposed that enzyme–Si complexes form in sugar cane that act as protectors or regulators of photosynthesis and enzyme activity. Si can suppress the activity of invertase in sugar cane, resulting in greater sucrose production. A reduction in phosphatase activity is believed to provide a greater supply of essential high-energy precursors needed for optimum cane growth and sugar production.

Si additions have improved the growth of sugar cane in Florida, Hawaii, Mauritius, Puerto Rico, and Saipan. The beneficial effects of Si have been attributed to correction of soil toxicities arising from high levels of available Mn, Fe^{2+} and active Al; prevention of localized accumulations of Mn in sugar cane leaves; plant disease resistance; greater stalk strength and resistance to lodging; increased availability of P; reduced transpiration; and unknown physiological functions. Si additions to greenhouse-produced cucumbers increased yields, reduced the incidence of powdery mildew, and improved plant health and shelf life.

Freckling, a necrotic leaf spot condition, is a symptom of low Si in sugarcane receiving direct sunlight. Ultraviolet radiation seems to be the causative agent in sunlight since plants kept under plexiglass or glass do not freckle. There are suggestions that Si in the sugarcane plant filters out harmful ultraviolet radiation.

This element has also had a favorable influence on rice production in Ceylon, China, India, Japan, South Korea, and Taiwan. Si tends to maintain erectness of rice leaves, increases photosynthesis because of better light interception, and results in greater resistance to diseases and insect pests. The oxidizing power of rice roots and accompanying tolerance to high levels of Fe and Mn were found to be very dependent on Si nutrition. Supplemental Si was beneficial when the Si concentration in rice straw fell below 11%.

From previous discussions in this chapter, readers will perceive that many of the favorable effects of Si on plant growth, such as disease resistance, stalk strength, and reductions in lodging, have also been attributed to K. Since KCl is the most commonly used K fertilizer, the question arises as to the possibility that one or both components of this fertilizer salt enhance Si uptake and metabolism.

Nickel

The nickel (Ni) content of crop plants normally ranges from about 0.1 to 1.0 ppm dry weight. It is readily taken up by most species as Ni^{2+}. High levels of Ni may induce Zn or Fe deficiency because of cation competition. Application of some sewage sludge may result in elevated levels of Ni in crop plants.

Ni is the metal component of urease that catalyzes the reaction $CO\,(NH_2)_2 + H_2O \rightarrow 2\,NH_3 + CO_2$. Apparently Ni is essential for plants supplied with urea and for those in which ureides are important in N metabolism. Nodule weight and seed yield of soybeans have been stimulated by Ni.

Results clearly demonstrate the beneficial role of Ni for legumes, with their particular type of N metabolism. In order to classify Ni as an essential nutrient, further studies are needed. It might be essential for certain types of N nutrition and certain plant families.

Summary

1. All elements absorbed by plants are not necessarily essential to plant growth. The term *functional or metabolic nutrient* was introduced to include any element that functions in plant nutrition, regardless of whether its action is specific. It was suggested that this term might avoid the confusion that sometimes occurs in a definition of *essential* plant nutrients.

2. Twenty elements have been found to be essential to the growth of plants. Not all are required by all plants, but all are necessary to some plants.

3. The nutrients required by plants are C, H, O, N, P, K, Ca, Mg, S, B, Fe, Mn, Cu, Zn, Mo, and Cl. In addition, Co, V, Na, Si, and Ni also are needed by some plants. The first three, with N, P, and S, constitute their living matter or protoplasm. Elements other than C, H, and O are termed *mineral nutrients*. The elements N, P, K, Ca, Mg, and S are classed as macronutrients and the remaining mineral elements as micronutrients.

4. N is used largely in the synthesis of proteins, but structurally it is also part of the chlorophyll molecule. Many proteins are enzymes, and the role of N can be considered as both structural and metabolic.

5. P is essential in supplying phosphate, which acts as a linkage unit or binding site. The stability of phosphate enables it to participate in the numerous energy capture, transfer, and recovery reactions that are vital for plant growth.

6. K is necessary to many plant functions, including carbohydrate metabolism, enzyme activation, osmotic regulation and efficient use of water, N uptake and protein synthesis, and translocation of assimilates. It also has a role in decreasing certain plant diseases and in improving quality.

7. S plays an important part in protein synthesis and the functioning of several enzyme systems. The synthesis of chlorophyll and the activity of nitrate reductase are strongly dependent on S.

8. The remaining mineral elements are generally involved in the activation of various enzyme systems. Mg, in addition, is an essential component of the chlorophyll molecule.

9. Na and Cl are electrolytes necessary for osmotic pressure and acid–base balance. Cl activates the O_2-producing enzyme of photosynthesis.

10. Si contributes to the structure of cell walls, thereby imparting greater disease resistance, stalk strength, and resistance to lodging.

Questions

1. Can you as a commercial grower do anything to supply plants with C, H, and O? What, specifically, can you do?
2. In what ways does N function in plant growth?
3. Visually, how would you differentiate between N and K deficiencies of corn?
4. Is N a mobile element in plants? What visual proof is there?
5. P is important in many plant functions. What, however, is probably its most important overall function?

6. N, P, and K are arbitrarily classed as macroelements or major elements. In terms of their importance in plant nutrition, is this terminology justified? Why? What justification is there for such a classification?

7. Fruit growers in a certain region of the Pacific Northwest decided to change their N fertilizer and use ammonium nitrate. A year or two later, their trees turned a uniform light yellow-green. Tissue tests showed that the leaves were high in N. The trees were irrigated with water from a glacier-fed river into which no industrial wastes had been emptied. The area was far from industrial activities. P, K, and Mg levels were adequate, and soil pH was satisfactory. None of the microelements was in short supply. What element was most likely to be deficient? How would you correct this deficiency?

8. A deficiency of Ca is sometimes observed under very dry soil conditions. Can you explain this?

9. In what ways do the symptoms of Mg and K deficiencies resemble each other? In what ways are they dissimilar?

10. What function of Mg is unique?

11. Which of the essential or metabolic elements are structurally a part of protoplasm?

12. What element is specifically involved in NO_3^- reduction in plants?

13. What element is required specifically by rhizobia in the fixation of N_2?

14. S is an integral part of certain amino acids. Name the amino acids. Can ruminants synthesize these S-containing amino acids from inorganic S and N compounds?

15. What crops have responded to applications of Na?

16. What precaution must be observed in applying elements such as Cu, Zn, B, Co, Mo, and Mg to crops?

17. Name several elements, deficiencies of which are exhibited first in the apical region of the growing plant. What does this imply?

18. If you saw a field of dwarfed corn, reddish purple in color, and the corn plant tissue tested high in NO_3^-, you might suspect that these plants were deficient in what element?

19. List the essential mineral elements required in plant nutrition, and give the principal functions of each.

Selected References

EPSTEIN, E. 1972. *Mineral Nutrition of Plants: Principles and Perspectives.* John Wiley & Sons, New York.

MENGEL, K., and E. A. KIRKBY. 1987. *Principles of Plant Nutrition.* International Potash Institute, Bern, Switzerland.

RÖMHELD, V., and H. MARSCHER. 1991. Function of micronutrients in plants. In J. J. MORTREDT et al. (eds.), *Micronutrients in Agriculture.* No. 4. Soil Science Society of America, Madison, Wisc.

CHAPTER 4

Basic Soil–Plant Relationships

The interaction of numerous physical, chemical, and biological properties in soils controls the availability of plant nutrients. Understanding these processes enables us to manage selected soil properties to optimize nutrient availability and plant productivity. The purpose of this chapter is to provide a review of ion exchange reactions in soils, ion movement in soil solution, and ion uptake by plants.

Nutrient supply to plant roots is a very dynamic process (Fig. 4.1). Plants absorb nutrients (cations and anions) from the soil solution and release small quantities of ions such as H^+, OH^-, and HCO_3^- (reactions 1 and 2). Changes in ion concentrations in soil solution are "buffered" by ions adsorbed on surfaces of soil minerals (reactions 3 and 4). Ion removal from solution causes partial desorption of the same ions from these surfaces. Soils contain mineral compounds that can dissolve to resupply soil solution with many ions (reactions 5 and 6). Likewise, increases in ion concentration in soil solution caused by fertilization or other inputs can cause some minerals to precipitate.

Soil microbial processes are very dynamic (reactions 7 and 8). Soil organisms remove ions from soil solution and incorporate them into microbial tissues. When microbes or other organisms die, they release nutrients to the soil solution. Microbial activity produces and decomposes organic matter or humus in soils. These dynamic processes are very dependent on adequate energy supply from organic carbon (i.e., crop residues), inorganic ion availability, and numerous environmental conditions. Plant roots and soil organisms utilize O_2 and respire CO_2 through metabolic activity (reactions 9 and 10). As a result, CO_2 concentration in the soil air is greater than in the atmosphere. Diffusion of these gases in soils is strongly affected by soil water content and other factors that alter soil solution pH, nutrient availability, and nutrient uptake.

Numerous environmental factors and human activities can influence ion concentration in soil solution, which interacts with the mineral and biological processes in soils (reactions 11 and 12). For example, adding P fertilizer to soil initially increases the $H_2PO_4^-$ concentration in soil solution. With time, the $H_2PO_4^-$ concentration will decrease with plant uptake, $H_2PO_4^-$ adsorption on mineral surfaces, and with P mineral precipitation.

All of these processes and reactions are important to availability of plant nutrients; however, depending on the specific nutrient, some processes are more important than others. For example, microbial processes are more im-

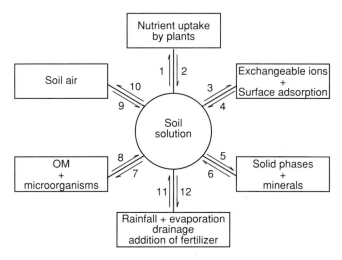

FIGURE 4.1   Relationships between the various
components of the dynamic soil system.   *Adapted from
Lindsay,* Chemical Equilibria in Soils, *Wiley Interscience,
1979.*

portant to N and S availability than mineral surface exchange reactions,
whereas the opposite is true for K, Ca, and Mg. Obviously, these processes
are complex, and only their general description and importance to plant
nutrient availability can be presented. Additional references that provide more
detail are listed at the end of the chapter.

Ion Exchange in Soils

Ion exchange in soils occurs on surfaces of clay minerals, inorganic compounds, organic matter, and roots (Fig. 4.2). The specific ion associated with
these surfaces depends on the kinds of minerals present and the solution
composition. Ion exchange is a reversible process by which a cation or anion
in the solid phase is exchanged with another cation or anion in the liquid
phase. If two solid phases are in contact, exchange of ions may also take place
between their surfaces. Cation exchange is generally considered to be more
important, since the anion exchange capacity of most agricultural soils is much
smaller than the cation exchange capacity. Ion exchange reactions in soils are
very important to plant nutrient availability. Therefore it is essential that we
understand the nature of the solid constituents and the origin of their surface
charge.

Cation Exchange

Solid materials in soils comprise about 50% of the volume, with the remaining
volume occupied by water and air. The solid portion is made up of inorganic
(mineral) material and organic matter in various stages of decay and humifica-
tion. The inorganic material consists of sand, silt, and clay. In some soils,

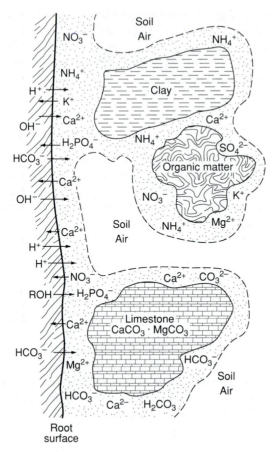

FIGURE 4.2   Diagram of the mineral
and organic exchange surfaces in soils.

coarse fragments are present in varying amounts. The clay fraction primarily
consists of layer silicate minerals made up of various combinations of silica
tetrahedra and aluminum octahedra (Fig. 4.3). The structure of a silica tetrahedra is one Si^{4+} cation bonded to four O^{2-} ions, whereas the aluminum
octahedra is one Al^{3+} cation bonded to 6 OH^- anions. The long *chains* or
layers of tetrahedra and octahedra are bonded together to form the *layer
silicates* (Fig. 4.3).

Layer silicate clay minerals in soils are of three general classes: 2:1, 2:1:1,
and 1:1. The 1:1 clays are composed of layers, each of which contains one
silica sheet and one alumina sheet. Kaolinite is the most important clay mineral
in this group (Fig. 4.4). The 2:1 clays are composed of layers, each of which
consists of two silica sheets between which is a sheet of alumina. Examples of
the 2:1 clays are smectites (montmorillonite), mica (illite), and vermiculite (Fig.
4.5). Muscovite and biotite mica are examples of 2:1 primary minerals that
are often abundant in silt and sand fractions.

Chlorites are examples of 2:1:1 layer silicates commonly found in soils. This
clay mineral consists of an interlayer hydroxide sheet in addition to the 2:1
structure referred to previously.

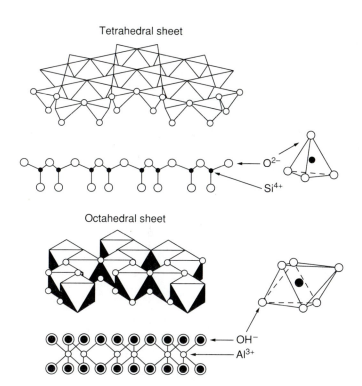

FIGURE 4.3   Chemical structure of silica tetrahedra,
Al octhedra, and the tetrahedral and octahedral
sheets.   *Adapted from G. Sposito,* The Chemistry of Soils,
Oxford University Press, 1989.

The major source of negative charge associated with layered silicates arises
from replacement of either the Si^{4+} or Al^{3+} cations with cations of lower
charge. Cation replacement in minerals is called *isomorphic substitution* and
occurs predominately in the 2:1 minerals, with very little substitution observed
in the 1:1 minerals. Isomorphic substitution occurred during the formation
of the mineral thousands of years ago, and thus is largely unaffected by
present environmental conditions.

In mica, substitution of Al^{3+} for one out of every four Si^{4+} cations in the
tetrahedral layer results in an imbalance of one negative charge. In montmorillonite, Mg^{2+} or Fe^{2+} replace some of the Al^{3+} in the octahedral layers, again
resulting in an imbalance of one negative charge for each substitution. Compare the unsubstituted 2:1 pyrophyllite mineral in Figure 4.4 with the isomorphic substitution in the 2:1 mica and montmorillonite minerals in Figure 4.5.
In the 2:1 mineral vermiculite, isomorphic substitution occurs in both the
octahedral and tetrahedral layers. The location of the isomorphic substitution
(tetrahedral, octahedral or both) imparts specific properties to the clay minerals that can affect the net quantity of negative surface charge. Table 4.1
summarizes these properties. For example, isomorphic substitution in the
tetrahedral layer locates the negative charge closer to the mineral surface
compared to octahedral substitution. The high negative surface charge combined with the unique geometry of the tetrahedral layers allows K^+ cations to

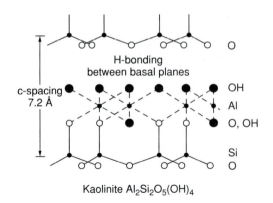

Kaolinite Al$_2$Si$_2$O$_5$(OH)$_4$

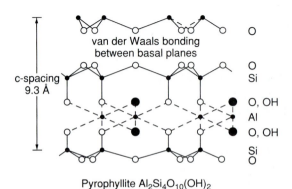

Pyrophyllite Al$_2$Si$_4$O$_{10}$(OH)$_2$

FIGURE 4.4   Structures of a 1:1 mineral, kaolinite, and a 2:1 mineral, pyrophyllite. No isomorphic substitution occurs in the tetrahedral or octahedral layers. *F.E. BEAR, Ed., Chemistry of the Soil, ASC Monograph Series No. 160, 1964.*

neutralize the negative charge between two 2:1 layers (Fig. 4.5). The resulting mica mineral exhibits a lower c-spacing and the mineral is considered "collapsed," with very little of the surface negative charge available to attract cations. Thus, mica has a lower cation exchange capacity than montmorillonite because the interlayer surfaces are not exposed (Table 4.1).

The negative charge associated with isomorphic substitution is uniformly distributed over the surface of the clay minerals and is considered permanent charge in that it is unaffected by solution pH (Fig. 4.6). Another source of negative charge on clay minerals is associated with the "broken edges" of the layer silicates. Figure 4.7 shows the exposed Si-OH and Al-OH associated with the edges of the tetrahedral and octahedral layers, respectively. The quantity of negative or positive charge on the broken edges depends on the pH of the soil solution. The broken edge charge is called a *pH-dependent* charge. Under acid conditions (low pH) the broken edge is positively charged because of the excess H$^+$ ions associated with the exposed Si-OH and Al-OH groups (Fig.

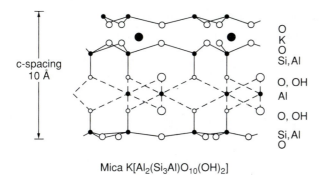

O
K
O
Si, Al
O, OH
Al
O, OH
Si, Al
O

Mica K[Al$_2$(Si$_3$Al)O$_{10}$(OH)$_2$]

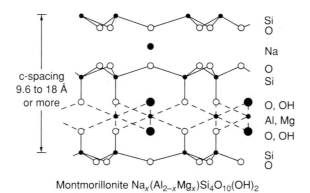

Si
O
Na
O
Si
O, OH
Al, Mg
O, OH
Si
O

Montmorillonite Na$_x$(Al$_{2-x}$Mg$_x$)Si$_4$O$_{10}$(OH)$_2$

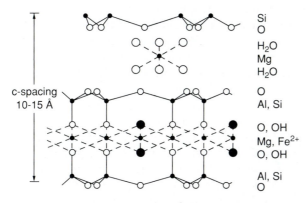

Si
O
H$_2$O
Mg
H$_2$O
O
Al, Si
O, OH
Mg, Fe^{2+}
O, OH
Al, Si
O

Vermiculite Mg$_n$(H$_2$O)$_6$(Mg, Fe^{2+})$_3$Si$_{4-n}$Al$_n$O$_{10}$(OH)$_2$

FIGURE 4.5   Structures of mica, montmorillonite, and vermiculite, all 2:1 minerals. Isomorphic substitution occurs in the tetrahedral and octahedral layers. *F.B. BEAR,Ed.*, Chemistry of the Soil, *ACS Monograph Series No. 160, 1964.*

TABLE 4.1   Common Layer Silicate Minerals in Soils

Clay Mineral	Layer Type	Layer Charge	c Spacing	CEC	pH-dependent Charge
			Å	meq/100 g	
Kaolinite	1:1	0	7.2	1–10	High
Mica (Illite)	2:1	1.0	10	20–40	Low
Vermiculite	2:1	0.8	10–15	120–150	Low
Montmorillonite	2:1	0.4	Variable	80–120	Low
Chlorite*	2:1:1	1.0	14	20–40	High
Organic matter				100–300	High

*Chlorite is a 2:1 mineral with a Mg hydroxide interlayer.

4.7). As soil solution pH increases, some of these H^+ ions are neutralized and the negative charge on the broken edge increases. Increasing the pH above 7 results in a nearly complete removal of H^+ ions on the Si-OH and Al-OH groups, which maximizes the negative charge associated with the broken edge. The increase in negative charge with increasing pH also is shown in Figure 4.6.

Only about 5 to 10% of the negative charge on 2:1 clays is pH dependent, while 50% or more of the charge developed on 1:1 clay minerals can be pH dependent.

Another source of pH-dependent charge is provided by the complex organic molecules associated with humus or soil organic matter (Fig. 4.8). Most of the negative charge originates from the dissociation of H^+ from carboxylic acid ($—COOH \leftrightarrow —COO^- + H^+$) and phenolic ($—C_6H_4OH \leftrightarrow —C_6H_4O^- + H^+$) groups. As pH increases, some of these H^+ ions are neutralized, increasing the negative charge on the surface of these large molecules.

The cation exchange capacity (CEC) of a soil represents the total quantity of negative charge available to attract positively charged ions in solution. It is one of the most important chemical properties of soils and strongly influences nutrient availability. The CEC of a soil is expressed in terms of milliequivalents

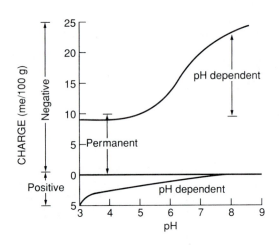

FIGURE 4.6   Permanent and pH-dependent charge associated with clay minerals. *W. Guenzi, Ed.,* Pesticides in Soil and Water, *ASA, 1974.*

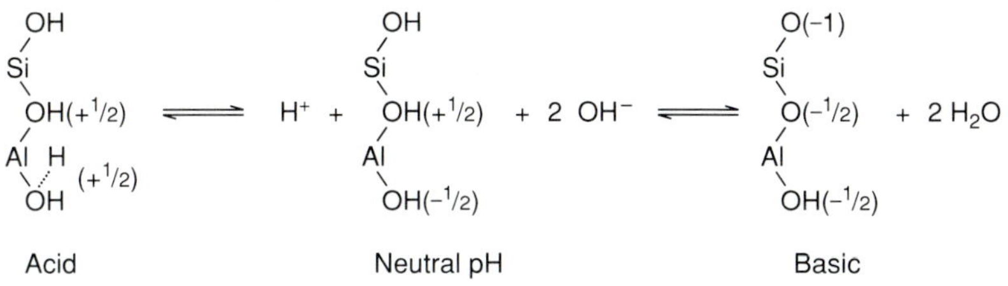

FIGURE 4.7 The pH-dependent charge associated with broken edges of kaolinite. *R. Schofield and H. Samson,* Clay Miner. Bull. *2:45, 1953.*

of negative charge per 100 g of oven-dried soil (meq/100 g).[1] The CEC also represents the total meq/100 g of cations held on the negative sites. The meq unit is used instead of mass because CEC represents total charges involved, and since the specific cations associated with the CEC will vary, it is more meaningful to simply quantify the total charges involved.

The definitions of equivalents and equivalent weight are as follows:

- **Atomic weight:** weight in grams of 6×10^{23} atoms of the substance. One *mole* of substance is 6×10^{23} atoms, molecules, ions, compounds, and so on; therefore, units of atomic weight are grams per mole (g/mole).
- **Equivalent weight:** quantity (mass) of a substance (e.g., cation, anion, compound) that will react with or displace one gram of hydrogen (H^+), which equals Avogadro's number of charges (+ or −). This is equal to the weight in grams of 6×10^{23} charges; therefore, units of equivalent weight are grams per equivalent (g/eq).

The definitions of atomic weight and equivalent weight are very similar:

$$\text{atomic wt} = \text{grams per } 6 \times 10^{23} \text{ ions or molecule}$$
$$\text{equivalent wt} = \text{grams per } 6 \times 10^{23} \text{ charges } (+ \text{ or } -)$$

The use of equivalents in soil chemistry/fertility is a convenient way to express quantities of exchangeable ions in soils. In *cation exchange* problems atomic weight and equivalent weight are related:

$$\text{equivalent wt of "A"} = \text{atomic wt of "A"/valence of "A."} \text{ For example,}$$
$$\text{equivalent wt of } K^+ = \frac{39 \text{ g/mole}}{1 \text{ eq/mole}} = 39 \text{ g/eq}$$
$$\text{equivalent wt of } Ca^{2+} = \frac{40 \text{ g/mole}}{2 \text{ eq/mole}} = 20 \text{ g/eq}$$
$$\text{equivalent wt of } Al^{3+} = \frac{27 \text{ g/mole}}{3 \text{ eq/mole}} = 9 \text{ g/eq}$$

---

[1] The SI unit system is used by the scientific community. Thus, meq/100 g becomes cmol/kg in SI units, which represent the centimoles (cmol) of charge per kilogram of soil. The conversion is 1 meq/100 g soil = 1 cmol/kg. We use meq/100 g in this text because most soil testing laboratories in the United States use meq/100 g for CEC.

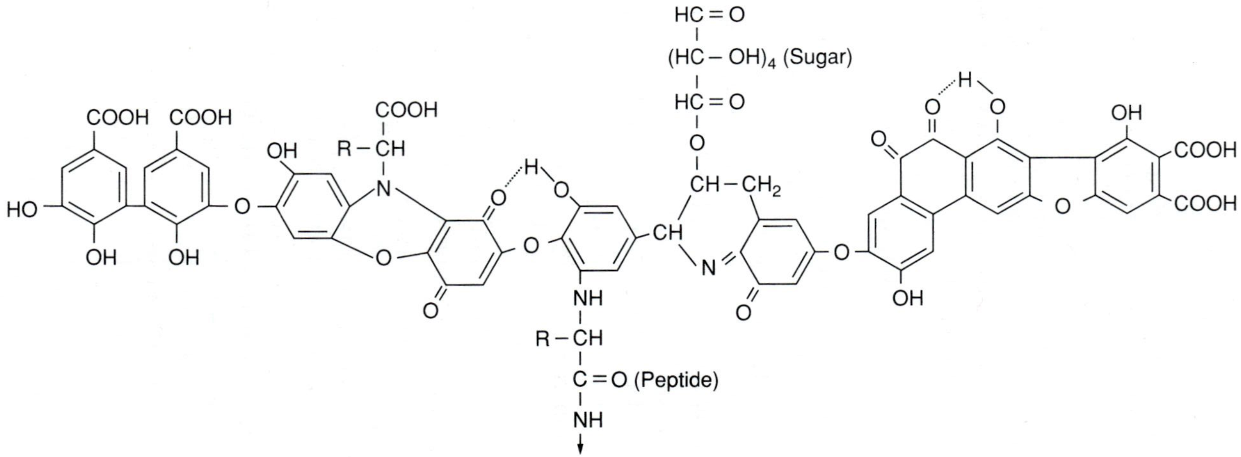

FIGURE 4.8 Suggested structure for humic acid in soil. The COOH and OH groups are the pH-dependent sites. *J. Mortvedt, P. Giordano, and W. Lindsay, Eds., Micronutrients in Agric., ASA, 1972.*

If a soil contains 1 mole of Ca^{2+} ions (6×10^{23} ions), then there are $2 \times$ (6×10^{23} charges) or 12×10^{23} charges since each Ca^{2+} ion has two charges. The definition of equivalent weight, the weight of 6×10^{23} charges, means that 1 equivalent of Ca^{2+} weighs 20 g/eq or 20 g/6×10^{23} charges. Recall that 1 mole of Ca^{2+} weighs 40 g/m or 40 g/6×10^{23} ions.

The use of equivalents to express concentrations or quantities of nutrients in soils is very convenient because of the nature of exchange reactions and cation exchange. If Ca^{2+} replaces K^+ on the exchange, then each Ca^{2+} cation can replace two K^+ cations, but one equivalent of Ca^{2+} replaces one equivalent of K^+ or one equivalent of any other cation. Thus,

1 equivalent A = 1 equivalent B, where A and B are ions, compounds, etc.

This concept is crucial to understanding and quantifying many chemical reactions in soil fertility.

The equivalent weight of a compound is determined by knowing the reaction the compound is involved in. For example:

$$CaCO_3 + 2HCl \rightarrow Ca^{2+} + 2Cl^- + H_2O + CO_2$$

What is the *equivalent weight* of $CaCO_3$ is this reaction?

Answer: one mole of $CaCO_3$ neutralizes 2 moles of HCl; therefore,

$$\text{equivalent wt} = \frac{\text{molecular weight}}{2} = \frac{100}{2} = 50 \text{ g/eq}$$

The CEC of common clay minerals and organic matter is given in Table 4.1. Soils with predominately 2:1 colloids will have higher exchange capacities than soils with mainly 1:1 mineral colloids.

The CEC is strongly affected by the nature and amount of mineral and organic colloid present in the soil. Soils with large amounts of clay and organic matter will have higher exchange capacities than sandy soils low in organic matter.

Examples of CEC values for different soil textures are as follows:

Sands (light-colored)	3–5 meq/100 g
Sands (dark colored)	10–20
Loams	10–15
Silt loams	15–25
Clay and clay loams	20–50
Organic soils	50–100

The most prevalent cations associated with the CEC of the various solid constituents are listed in Table 4.2. Except for Al^{3+}, most of the exchangeable cations are plant nutrients. In acids soils the principal cations are Al^{3+}, H^+, Ca^{2+}, Mg^{2+}, K^+, and small amounts of Na^+ (Table 4.2). In neutral and basic soils the predominant cations are Ca^{2+}, Mg^{2+}, K^+, Na^+, and very small amounts of Al^{3+}. Cations are held on the exchange sites with different adsorption strengths; therefore, the ease with which cations can be replaced or exchanged

TABLE 4.2   Cation and Anions Associated with Exchange Capacity of Soils

Element	Atomic Weight (g/mole)	Equivalent* Weight (g/eq)	Ionic Radii	
			Nonhydrated (nm)	Hydrated (nm)
Cations				
Al^{3+}	27	9	0.051	
H^+	1	1		
Ca^{2+}	40	20	0.099	0.96
Mg^{2+}	24	12	0.066	1.08
K^+	39	39	0.133	0.53
NH_4^+	18	18	0.143	0.56
Na^+	23	23	0.097	0.79
Anions				
$H_2PO_4^-$	97	97		
SO_4^{2-}	96	48		
NO_3^-	62	62		
Cl^-	35	35		
OH^-	17	17		

*g/eq or mg/meq.

with other cations also varies. For most minerals, the strength of cation adsorption, or *lyotropic series*, is:

$$Al^{3+} > Ca^{2+} > Mg^{2+} > K^+ = NH_4^+ > Na^+$$

Properties of the cations determine the strength of adsorption or ease of desorption. First, the strength of adsorption is directly proportional to the charge on the cations. The H^+ ion is unique because of its very small size and high charge density; thus, its adsorption strength is between Al^{3+} and Ca^{2+}. Second, the adsorption strength for cations with similar charge is determined by the size or radii of the hydrated cation (Table 4.2). As the size of the hydrated cation increases, the distance between the cation and the clay surface increases. Larger hydrated cations can't get as close to the exchange site as smaller cations, resulting in decreased strength of adsorption.

Exchange of one cation for another is primarily driven by the quantity of the cation added. This principle is used to determine the CEC of soils in the laboratory. However, when only small quantities of added cations are involved, the ease of desorption or exchange is influenced by the *complementary* cation involved. The *complementary cation effect* occurs when the exchange of one cation for another becomes easier as the adsorption strength of the third or complementary cation increases. For example, NH_4^+ exchange for Ca^{2+} occurs more completely when Al^{3+} is the predominant, complementary cation rather than Na^+. Because Na^+ is easily exchanged by NH_4^+, there would be less NH_4^- left to exchange for Ca^{2+}. Since NH_4^+ would not be easily exchanged for Al^{3+}, more NH_4^+ remains in solution to exchange for Ca^{2+}.

Determination of CEC

A conventional method of CEC measurement is to extract a soil sample with neutral 1 N ammonium acetate (NH_4OAc). All of the exchangeable cations are replaced by NH_4^+ ions, and the CEC becomes saturated with NH_4^+. If this NH_4^+-saturated soil is extracted with a solution of a different salt, say 1.0 N KCl, the K^+ ions will replace the NH_4^+ ions. If the soil–KCl suspension is filtered, the filtrate will contain the NH_4^+ ions that were previously adsorbed by the soil. The quantity of NH_4^+ ions in the leachate is a measure of the CEC.

$$\text{SOIL} \begin{vmatrix} H^+ \\ Ca^{2+} \\ Mg^{2+} \\ K^+ \\ Al^{3+} \end{vmatrix} + 9\,NH_4OAc^- \rightarrow \text{SOIL} \begin{vmatrix} NH_4^+ \\ NH_4^+ \\ NH_4^+ \\ NH_4^+ \\ NH_4^+ \\ NH_4^+ \\ NH_4^+ \\ NH_4^+ \\ NH_4^+ \end{vmatrix} + \begin{matrix} H^+ \\ Ca^{2+} \\ Mg^{2+} \\ K^+ \\ Al^{3+} \end{matrix} + 9\,OAc^-$$

Solution

$$\text{SOIL} \begin{vmatrix} NH_4^+ \\ NH_4^+ \\ NH_4^+ \\ NH_4^+ \\ NH_4^+ \\ NH_4^+ \\ NH_4^+ \\ NH_4^+ \\ NH_4^+ \end{vmatrix} + 9\,KCl \rightarrow \text{SOIL} \begin{vmatrix} K^+ \\ K^+ \\ K^+ \\ K^+ \\ K^+ \\ K^+ \\ K^+ \\ K^+ \\ K^+ \end{vmatrix} + 9\,NH_4^+ + 9\,Cl^-$$

Solution

(filter and measure NH_4^+)

For example, suppose that the concentration of NH_4^+ in the filtrate was 270 ppm (20 g of soil extracted with 200 ml of KCl solution). The CEC is calculated as follows:

270 ppm NH_4^+ = 270 mg NH_4^+/l
(270 mg NH_4^+/l) × (0.2 l/20 g soil) = 2.7 mg NH_4^+/g soil
(2.7 mg NH_4^+/g soil)/(18 mg NH_4^+/meq) = 0.15 meq CEC/g soil
0.15 meq CEC/g soil × 100/100 = 15 meq/100 g soil
CEC = 15 meq/100 g

The equivalent weight of NH_4^+ was given in Table 4.2.

Base Saturation

One of the important properties of a soil is its base saturation, which is defined as the percentage of total CEC occupied by basic cations (Ca^{2+}, Mg^{2+}, K^+, and Na^+). To illustrate, suppose that the following ions were measured in the 200 ml NH_4OAc extract obtained from leaching the 20 g of soil in the previous example:

$$Ca^{2+} = 100 \text{ ppm}$$
$$Mg^{2+} = 30 \text{ ppm}$$
$$K^+ = 78 \text{ ppm}$$
$$Na^+ = 23 \text{ ppm}$$

The equivalent weights of the cations are found in Table 4.2. The following calculations are used to express the cation concentrations in CEC units and determine the base saturation.

$$Ca^{2+} = 100 \text{ ppm} = 100 \text{ mg/l} \times (0.2 \text{ 1/20 g soil})/(20 \text{ mg/meq}) \times 100/100$$
$$= 5 \text{ meq } Ca^{2+}/100 \text{ g}$$
$$Mg^{2+} = 30 \text{ ppm} = 30 \text{ mg/l} \times (0.2 \text{ 1/20 g soil})/(12 \text{ mg/meq}) \times 100/100$$
$$= 2.5 \text{ meq } Mg^{2+}/100 \text{ g}$$
$$K^{+} = 78 \text{ ppm} = 78 \text{ mg/l} \times (0.2 \text{ 1/20 g soil})/(39 \text{ mg/meq}) \times 100/100$$
$$= 2 \text{ meq } K^{+}/100 \text{ g}$$
$$Na^{+} = 23 \text{ ppm} = 23 \text{ mg/l} \times (0.2 \text{ 1/20 g soil})/(23 \text{ mg/meq}) \times 100/100$$
$$= 1 \text{ meq } Na^{+}/100 \text{ g}$$
$$\text{Total} = 10.5 \text{ meq bases}/100 \text{ g}$$

$$\text{Base saturation \%} = (\text{total bases/CEC}) \times 100$$
$$= [(10.5 \text{ meq}/100 \text{ g})/(15 \text{ meq}/100 \text{ g})] \times 100$$
$$= 70 \text{ \%}$$

The percent saturation with any cation may be calculated in a similar fashion. For example, from the preceding data, % Mg saturation = (2.5 meq Mg/10.5 meq CEC) × 100 = 23.8% Mg.

As a general rule, the degree of base saturation (BS %) of normal uncultivated soils is higher for arid than for humid region soils. Although not always true, especially in humid regions, the BS % of soils formed from limestones or basic igneous rocks is greater than that of soils formed from sandstones or acid igneous rocks.

The availability of the nutrient cations such as Ca^{2+}, Mg^{2+}, and K^{+} to plants increases with increasing BS %. For example, a soil with 80% BS would provide cations to growing plants far more easily than the same soil with a BS of only 40%. The relation between BS % and cation availability is modified by the nature of the soil colloids. As a rule, soils with large amounts of organic or 1:1 colloids can supply nutrient cations to plants at a much lower BS % than soils high in 2:1 colloids.

BS % is related to soil pH (Fig. 4.9). As the percent Ca^{2+}, Mg^{2+}, and K^{+} on the exchange sites of a soil increases, the pH increases. In this example pH 5.5 equals about 50% BS and pH 7.0 equals 90% BS.

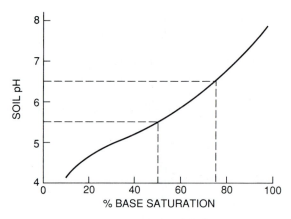

FIGURE 4.9   General relationship between soil pH and base saturation.

The shape of the curve varies slightly among different soils; however, the relationship can be helpful in evaluating lime requirements for acid soils.

For example, assume that a soil has a pH of 5.5 and a CEC of 20 meq/100 g. The grower needs to lime the soil to pH 6.5 for optimum production. Using Figure 4.9, the initial BS at pH 5.5 is about 50%. At pH 6.5 the BS is estimated to be 75%. The following calculation is used to estimate the lime ($CaCO_3$) required to raise soil pH from 5.5 to 6.5 (BS from 50% to 75%):

$$\text{Final BS} = 75\% = (0.75) \times 20 \text{ meq CEC/100 g} = 15 \text{ meq/100 g}$$
$$\text{Initial BS} = 50\% = (0.50) \times 20 \text{ meq CEC/100 g} = \underline{10 \text{ meq/100 g}}$$
$$\text{Total bases needed to raise pH} = 5 \text{ meq/100 g}$$

$$\begin{aligned}
CaCO_3 \text{ required}[2] \text{ (kg/ha)} &= 5 \text{ meq } CaCO_3/100 \text{ g} \times 50 \text{ mg } CaCO_3/\text{meq} \\
&= 250 \text{ mg } CaCO_3/100 \text{ g soil} \\
&= 0.25 \text{ g } CaCO_3/100 \text{ g}
\end{aligned}$$

$$0.25 \text{ g } CaCO_3/100 \text{ g} = X \text{ kg } CaCO_3/2 \times 10^6 \text{ kg soil}$$
$$X = 5000 \text{ kg } CaCO_3/\text{ha} - 15 \text{ cm}$$

Anion Exchange

Anions in soil solution also are subject to adsorption to positively charged sites on clay mineral surfaces and organic matter. The positive charges responsible for electrostatic adsorption and exchange of anions originate in the broken bonds, primarily in the alumina octahedral sheet, exposing OH groups on the edges of clay minerals (Fig. 4.7). Anion exchange may also occur with OH groups on the hydroxyl surface of kaolinite. Displacement of OH ions from hydrous Fe and Al oxides is considered to be an important mechanism for anion exchange, particularly in highly weathered soils of the tropics and subtropics, and it is in such soils that anion exchange is greatest.

The anion exchange capacity (AEC) increases as soil pH decreases (Fig. 4.6). Further, anion exchange is much greater in soils high in 1:1 clays and those containing hydrous oxides of Fe and Al than it is in soils with predominately 2:1 clays. Montmorillonic clay minerals usually have an AEC or AECs of less than 5 meq/100 g, while kaolinites can have an AEC as high as 43 meq/100 g at pH 4.7. The pH of most productive soils in the United States and Canada is usually too high for full development of AEC.

Anions such as Cl^- and NO_3^- may be adsorbed, although not to the extent of $H_2PO_4^-$ and SO_4^{2-}. The order of adsorption strength is $H_2PO_4^- > SO_4^{2-} > NO_3^- = Cl^-$. In most soils $H_2PO_4^-$ is the primary anion adsorbed, although some acidic soils also adsorb significant quantities of SO_4^{2-}.

The mechanisms responsible for anion retention in soils are much more complex than the simple electrostatic attractions involved in most cation exchange reactions. Anions may be retained by soil particles through specific adsorption or chemisorption reactions that are nonelectrostatic (Fig. 4.10).

---

[2] One hectare (ha) of soil to a depth of 15 cm weighs about 2×10^6 kg.
One acre (ac) of soil to a depth of 6 in. weighs about 2×10^6 lb.

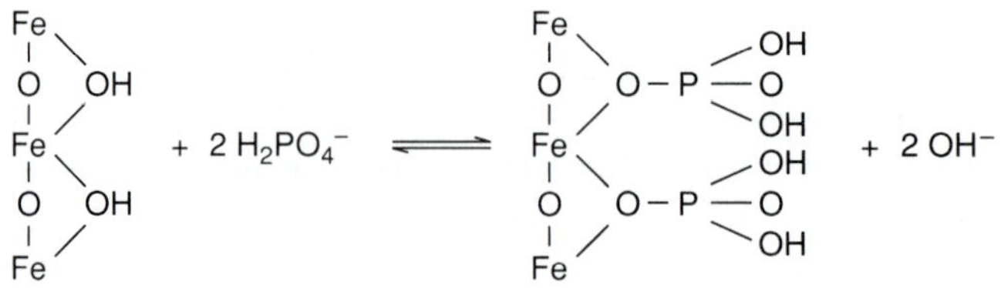

FIGURE 4.10   Chemisorption of phosphate ($H_2PO_4^-$) to iron hydroxide [$Fe(OH)_3$] minerals in soils. *Bohn et al., Soil Chemistry, John Wiley & Sons, 1979.*

Buffering Capacity

Plant nutrient availability depends on the concentration of nutrients in solution but, more important, on the capacity of the soil to maintain the concentration. The buffering capacity represents the ability of the soil to resupply an ion to the soil solution. The buffering capacity involves all the solid components in the soil system; thus, the ions must also exist in soils as solid compounds or adsorbed to cation/anion exchange sites (Fig. 4.1). For example, when H^+ ions in solution are neutralized by liming, the H^+ held on exchange will desorb from the exchange sites. The solution pH is thus buffered by exchangeable H^+ and will not increase until significant quantities of exchangeable acids have been neutralized. Similarly, as plant roots absorb or remove nutrients such as K^+, exchangeable K^+ is desorbed to resupply solution K^+. With some nutrients, such as $H_2PO_4^-$, solid P minerals dissolve to resupply or buffer the solution $H_2PO_4^-$ concentration.

Soil buffer capacity (BC) can be described by the ratio of the concentrations of absorbed (ΔQ) and solution (ΔI) ions:

$$BC = \frac{\Delta Q}{\Delta I}$$

Figure 4.11 illustrates the quantity (Q) and intensity (I) relationships between two soils. Soil A has a higher BC than Soil B, as indicated by the steeper slope ($\Delta Q/\Delta I$). Thus, increasing the concentration of adsorbed ion increases the solution concentration in Soil B much more than that in Soil A, indicating that $BC_A > BC_B$. Alternatively, decreasing the solution concentration by plant uptake decreases the quantity of ion in solution much less in Soil A than in Soil B.

The BC in soil increases with increasing CEC, organic matter, and other solid constituents in the soil. For example, the BC of montmorillonitic, high organic matter soils is greater than that of kaolinitic, low organic matter soils. Since CEC increases with increasing clay content, fine-textured soils will exhibit higher BC than coarse-textured soils. If exchangeable K^+ decreases, for example, as a result of plant uptake, the capacity of the soil to buffer further decreases in solution K^+ concentration will be reduced. The nutrient will likely become deficient, and fertilizer K^+ will be needed to increase exchangeable K^+. Addition of fertilizer P will increase the anion-exchangeable $H_2PO_4^-$, but more important, some $H_2PO_4^-$ will precipitate as solid P com-

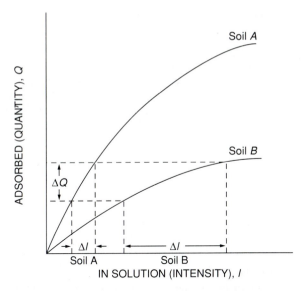

FIGURE 4.11 Relationship between quantity of adsorbed nutrient and the concentration of the nutrient in solution (intensity). BC ($\Delta Q/\Delta I$) of Soil A is greater than that of Soil B.

pounds that contribute to the BC of P in soils. BC is a very important soil property that strongly influences nutrient availability and fertilizer management.

Root Cation Exchange Capacity

Plant roots exhibit a CEC ranging from 10 to 30 meq/100 g in monocotyledenous plants such as the grasses and 40 to 100 meq/100 g in dicotyledons such as forage legumes (Table 4.3). The exchange properties of roots are attributable mainly to carboxyl groups (—COOH), similar to the exchange sites on humus (Fig. 4.8), and account for 70 to 90% of the exchange properties of roots.

Legumes and other plant species with high CEC values tend to absorb divalent cations such as Ca^{2+} preferentially over monovalent cations, whereas the reverse occurs with grasses. These cation exchange properties of roots help to explain why, in grass-legume pastures on soils containing less than adequate K^+, the grass survives but the legume disappears. The grasses are considered to be more effective absorbers of K^+ than are the legumes.

TABLE 4.3   CEC of Roots

Species	CEC meq/100 g Dry Root
Wheat	23
Corn	29
Bean	54
Tomato	62

Movement of Ions from Soils to Roots

For ions to be absorbed by plant roots, they must come in contact with the root surface. There are generally three ways in which nutrient ions in soil may reach the root surface: (1) root interception; (2) mass flow of ions in solution; and (3) diffusion of ions in the soil solution. The relative importance of these mechanisms in providing nutrients to plant roots is shown in Table 4.4. The contribution of diffusion was estimated by the difference between total nutrient needs and the amounts supplied by interception and mass flow.

Root Interception

The importance of root interception as a mechanism for ion absorption is enhanced by the growth of new roots throughout the soil mass and perhaps also by mycorrhizal infections. As the root system develops and exploits more soil, soil solution and soil surfaces retaining adsorbed ions are exposed to the root mass, and absorption of these ions occurs by a contact exchange mechanism. Ions attached to the surface of root hairs (such as H^+ ions) may exchange with ions held on the surface of clays and organic matter in soils because of the intimate contact that exists between roots and soil particles. The ions held by electrostatic forces at these sites tend to oscillate within a certain volume (Fig. 4.12). When the oscillation volumes of two ions overlap, the ions exchange places. In this way Ca^{+2} on a clay surface could then presumably be absorbed by the root and utilized by the plant.

The quantity of nutrients that can come in direct contact with the plant roots is the amount in a volume of soil equal to the volume of roots. Roots usually occupy 1% or less of the soil; however, roots growing through soil pores with higher than average nutrient content would contact a maximum of 3% of the available soil nutrients.

TABLE 4.4   Relative Significance of the Principal Ways in Which Plant Nutrient Ions Move from Soil to the Roots of Corn

Nutrient	Amount of Nutrient Required for 150 bu/a of Corn (lb/a)	Percentage Supplied by		
		Root Interception	Mass Flow	Diffusion
Nitrogen	170	1	99	0
Phosphorus	35	3	6	94
Potassium	175	2	20	78
Calcium	35	171	429	0
Magnesium	40	38	250	0
Sulfur	20	5	95	0
Copper	0.1	10	400	0
Zinc	0.3	33	33	33
Boron	0.2	10	350	0
Iron	1.9	11	53	37
Manganese	0.3	33	133	0
Molybdenum	0.01	10	200	0

SOURCE: Barber, Stanley A. *Soil bionutrient availability.* John Wiley & Sons, New York, NY (1984).

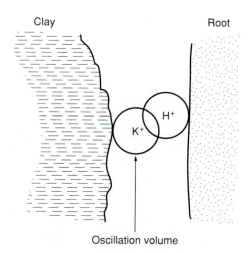

FIGURE 4.12   Conceptual model for root interception or contact exchange of nutrients between ions on soil and root exchange sites. Overlapping oscillation volumes cause exchange of H^+ on the root with K^+ on the clay mineral surface.

Root interception of nutrients can be enhanced by mycorrhiza, a symbiotic association between fungi and plant roots. This beneficial effect of mycorrhiza is greatest when plants are growing in infertile soils. The extent of mycorrhizal infection also is enhanced under conditions of slightly acid soil pH, low P, adequate N, and low soil temperatures. The hyphal threads of mycorrhizal fungi act as an extension of plants root systems, resulting in greater soil contact. (See Figure 13.9 for a diagrammatic representation of a mycorrhizal infected root.)

There are two major groups, ectomycorrhizas and endomycorrhizas. The ectomycorrhizas occur mainly in the tree species of the temperate zone but can also be found in semiarid zones.

The endomycorrhizas are more widespread. The roots of most agronomic crops have vesicular arbuscular mycorrhiza. The fungus grows into the cortex. Inside the plant cells small structures known as *arbuscules* are formed. These are considered to be the site of transfer of nutrients from fungi to host plants. The positive effect of inoculation of English oaks with ectomycorrhizas is shown in Figure 4.13.

The increased nutrient absorption is thought to be due to the larger nutrient-absorbing surface provided by the fungi. This has been calculated to be up to 10 times that of uninfected roots. Fungal hyphae extend up to 8 cm into the soil surrounding the roots, thus increasing the absorption of nutrients such as P that do not diffuse readily to the roots (Table 4.5).

Enhanced P uptake is the primary cause of improved plant growth from mycorrhiza. This improvement in growth may lead to more rapid uptake of other elements.

FIGURE 4.13 Differential growth responses of 16-week-old English oaks (*Quercus robur* L.) inoculated (left) with *Pisolithus tinctorius* (Pers.) Coker and Couch, an ectomycorrhizal former, and uninoculated (right). *H.E. Garrett, School of Forestry, Fisheries and Wildlife, University of Missouri, Columbia, Mo.*

TABLE 4.5   Effect of Inoculation of Endomycorrhiza and of Added P on the
Content of Different Elements in Corn Shoots (μg)

| Element Added | Content in Shoots (μg) | | | |
| | No P | | 25 ppm P Added | |
	No Mycorrhiza	Mycorrhiza	No Mycorrhiza	Mycorrhiza
P	750	1340	2,970	5,910
K	6000	9700	17,500	19,900
Ca	1200	1600	2,700	3,500
Mg	430	630	990	1,750
Zn	28	95	48	169
Cu	7	14	12	30
Mn	72	101	159	238
Fe	80	147	161	277

SOURCE: Lambert et al. *J. Soil Sci.* **43**:976 (1979).

Mass Flow

Movement of ions in the soil solution to the surfaces of roots by mass flow is an important factor in supplying nutrients to plants. Mass flow occurs when plant nutrient ions and other dissolved substances are transported in the flow of water to the root that results from transpirational water uptake by the plant. Some mass flow can also take place in response to evaporation and percolation of soil water.

The amounts of nutrients reaching roots by mass flow are determined by the rate of water flow or the water consumption of plants and the average nutrient concentrations in the soil water. Mass flow supplies an overabundance of Ca^{2+} and Mg^{2+} in many soils, as well as most of the mobile nutrients, such as NO_3^- and SO_4^{2-}. As soil moisture is reduced (increased soil moisture tension), water movement slows down. Thus the movement of moisture to the root surface is slowed. The movement of nutrients ions by mass flow will be reduced at low temperatures because the transpirational demands of plants will be substantially less at low temperatures than under warmer conditions. In addition, the transport of ions in the flow of water evaporated at the soil surface will diminish at low soil temperatures.

Diffusion

Diffusion occurs when an ion moves from an area of high concentration to one of low concentration. Most of the P and K moves to the root by diffusion. As plant roots absorb nutrients from the surrounding soil solution, the nutrient concentration at the root surface decreases compared to the "bulk" soil solution concentration (Fig. 4.14). Therefore, a nutrient concentration gradient is established that causes ions to diffuse toward the plant root. A high plant requirement for a nutrient results in a large concentration gradient, favoring a high rate of ion diffusion from the soil solution to the root surface.

Many soil factors influence nutrient diffusion in soils; the most important one is the magnitude of the diffusion gradient. The following equation (known as *Fick's law*) describes this relationship:

$$dC/dt = De\ A\ dC/dX$$

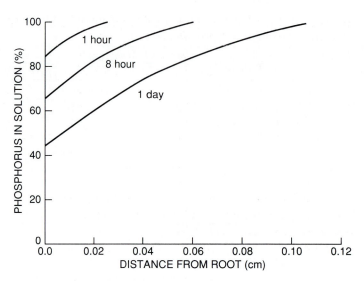

FIGURE 4.14   The influence of P uptake on the
distribution of P in the soil solution as a function of
distance from the root surface.

where dC/dt = rate of diffusion (change in concentration C with time)
     dC/dX = concentration gradient (change in concentration with
            distance)
     De = effective diffusion coefficient (defined later)
     A = cross-sectional area through which the ions diffuse

The diffusion equation shows that the rate of nutrient diffusion (dC/dt) is
directly proportional to the concentration gradient (dC/dX). As the difference
in nutrient concentration between the root surface and the bulk solution
increases, the rate of nutrient diffusion increases. Also, increasing the cross-
sectional area for diffusion increases dC/dt. The diffusion rate also is proportional to the diffusion coefficient, De, which controls how far nutrients can
diffuse to the root. For a given spacing between roots, De determines the
fraction of nutrients in the soil that can reach the roots during a specific period
of plant growth. De is described as follows:

$$De = Dw\ \theta\ (1/T)\ (1/b)$$

where Dw = diffusion coefficient in water
     θ = volumetric soil water content
     T = tortuosity factor
     b = soil BC

This relationship shows that as soil moisture content (θ) increases, De increases, which results in an increase in the diffusion rate, dC/dt. As the moisture
content of the soil is lowered, the moisture films around the soil particles

become thinner and the diffusion of ions through these films becomes more tortuous. Transport of nutrients to the root surface is probably most effective at a soil moisture content corresponding to field capacity. Therefore, raising θ reduces tortuosity or the diffusion path length, which in turn increases dC/dt.

Tortuosity (T) also is related to soil texture. Nutrients diffusing in finer-textured soils experience a more tortuous path to the root surface. As T increases with increasing clay content, $1/T$ decreases, which reduces the diffusion coefficient and, thus, dC/dt. Also, ions diffusing through soil moisture in clay soils are much more likely to be attracted to adsorption sites on the clay than in a sandy soil.

The diffusion coefficient in soil (De) is directly related to the diffusion coefficient for the same nutrient in water (Dw). Inherent in the Dw term is a temperature factor such that increasing temperature increases Dw, De, and then dC/dt. The diffusion coefficient is inversely related to the soil BC, b. Increasing the BC of the soil decreases De, which in turn decreases the rate of nutrient diffusion. Therefore, decreasing the BC by increasing the nutrient concentration in solution increases De, which then increases dC/dt. Increasing the solution concentration also increases the diffusion gradient, dC/dX, which contributes to the increased rate of diffusion.

Uptake of ions at the root surface, which is responsible for creating and maintaining diffusion gradients, is strongly influenced by temperature. Within the range of about 10 to 30°C, an increase of 10°C usually causes the rate of ion absorption to go up by a factor of 2 or more. Diffusion of nutrient ions is slow under most soil conditions and occurs over very short distances in the vicinity of the root surface. Typical average distances for diffusion to the root are 1 cm for N, 0.02 cm for P, and 0.2 cm for K. The mean distance between corn roots in the top 15 cm of soil is about 0.7 cm, indicating that some nutrients would need to diffuse half this distance, or 0.35 cm, before they would be in position for absorption by the plant root.

Roots do not absorb all nutrients at the same rate. Thus, certain ions may build up at the root surface, especially during periods of rapid absorption of water. This results in a phenomenon known as *back diffusion*, in which the concentration gradient, and hence the movement of certain ions, will be away from the root surface and back toward the soil solution. Nutrient diffusion away from the root is much less than diffusion toward the root; however, higher concentrations of some nutrients in the rhizosphere can affect the uptake of other nutrients.

The importance of diffusion and mass flow in supplying ions to the root surface depends on the ability of the solid phase of the soil to supply the liquid phase with these ions. Solution concentrations of ions will be influenced by the nature of the colloidal fraction of the soil and the degree to which these colloids are saturated with cations. For example, the ease of replacement of Ca from colloids by plant uptake varies in this order: peat > kaolinite > illite > montmorillonite. An 80% Ca-saturated, 2:1 clay provides the same percentage Ca^{2+} release as a 35% Ca-saturated kaolinite or a 25% Ca-saturated peat.

Mass flow and diffusion processes also are important in fertilizer management. Soils that may exhibit low diffusion rates because of high BC, low soil moisture, or high clay content may require the application of immobile nutrients near the roots to maximize nutrient availability and plant uptake.

Ion Absorption by Plants

Plant uptake of ions from the soil solution can be described by *passive* and *active* processes, where ions passively move to a "boundary" through which ions are actively transported to organs in plant cells that metabolize the nutrient ions. Solution composition or ion concentrations outside and inside the boundary are controlled by different processes, each essential to mineral nutrition and growth of the plant.

Passive Ion Uptake

A considerable fraction of the total volume of the root is accessible for the passive absorption of ions. The *outer* or *apparent free space*, where the diffusion and exchange of ions occur, is located in the walls of the epidermal and cortical cells of the root and in the film of moisture lining the intercellular spaces (Fig. 4.15). Walls of the cells of the cortex are the principal locale of the outer space. This extracellular space is outside the outermost membrane, the Casparian strip, which is a barrier to diffusion and exchange of ions.

Ions in soil solution enter the root tissue through diffusion and ion exchange processes. The concentration of ions in the apparent free space is normally less than the bulk solution concentration; therefore, diffusion occurs with the concentration gradient, from high to low concentration. Interior surfaces of cells in the cortex are negatively charged, attracting cations. Cation exchange readily occurs along the extracellular surfaces and explains why cation uptake

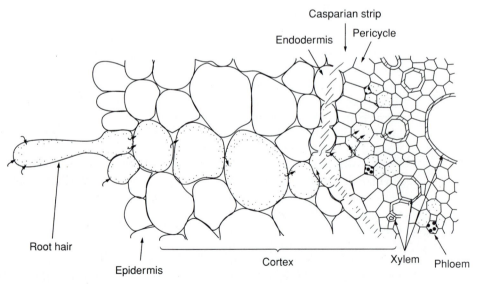

FIGURE 4.15   Cross section of a plant root. Site of passive uptake is the apparent free space, which is outside the Casparian strip in the cortex. *Epstein*, Mineral Nutrition of Plants, *Wiley-Interscience, 1972.*

TABLE 4.6   Characteristics of Ion Movement into Inner and Outer Cell Spaces

Outer	Inner
Diffusion and exchange adsorption	Ion-binding compounds or carriers
Uptake occurs quickly	Uptake occurs slowly
Ions stoichiometrically exchangeable	Ions essentially nonexchangeable
Not highly selective	Specific with regard to site and entry
Nonmetabolic	Dependent on aerobic metabolism
Ions in solution or adsorbed in outer space	Ions in vacuoles and partly in cytoplasm

SOURCE: Gauch, *Annu. Rev. Plant Physiol.*, **8**:31 (1957). Reprinted with permission of the author and the publisher, Annual Reviews, Inc., Palo Alto, Calif.

usually exceeds anion uptake. To maintain electrical neutrality, the root cells release H^+, decreasing soil solution pH near the root surface.

Diffusion and ion exchange are passive processes because uptake into the outer space is controlled by ion concentration (diffusion) and electrical (ion exchange) gradients. These processes are nonselective and do not require energy produced from metabolic reaction within the cell (Table 4.6). Passive uptake occurs outside the Casparian strip and plasmalemma, which are the boundary membranes or barriers to diffusion and ion exchange.

Extracellular spaces exist in the mesophyll cells of leaves where ions are able to diffuse and exchange. Most of the nutrient ions reach the "outer" space of leaves via the xylem from the roots. Mineral ions in rain, irrigation water, and foliar applications penetrate leaves through the stomata and cuticle to reach the interior of leaves, where they become available for absorption by mesophyll cells.

The movement of ions from roots to shoots is determined by the rates of water absorption and transpiration, suggesting that mass flow may be important in the movement of ions.

Active Ion Uptake

The membrane that provides the boundary between the apparent free space and the interior contents of the cell is the plasmalemma (Fig. 4.16). Ions that were passively absorbed occupy spaces between cells; however, the plasmalemma prevents passive transport of nutrients into the cell. Since ion concentrations are greater inside than outside the cell, transport of ions across the plasmalemma is strictly against an electrochemical gradient. Therefore, ion transport across the plasmalemma into the cytoplasm requires energy derived from cell metabolism. Other organs within the cell also are surrounded by impermeable membranes. For example, the tonoplast is the barrier membrane for the vacuole, which regulates cell water content and serves as the reservoir for inorganic ions, sugars, and amino acids.

The ion–carrier mechanism involves a metabolically produced substance that combines with free ions. This ion–carrier complex can then cross membranes and other barriers not permeable to free ions. After the transfer is accomplished, the ion–carrier complex is broken, the ion is released into the inner space of the cell, and the carrier is believed in some cases to be restored (Fig. 4.17).

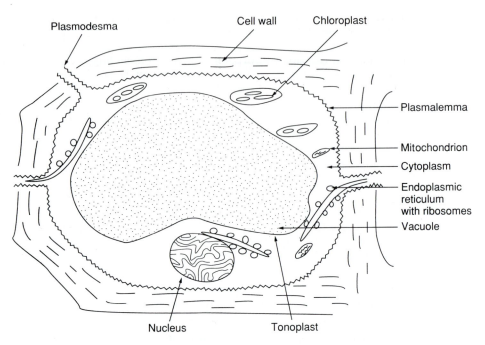

FIGURE 4.16   Diagram of a plant cell. Active ion uptake occurs at the plasmalemma. *Mengel and Kirkby*, Principles of Plant Nutrition, *IPI, 1987.*

Two different mechanisms are involved in the transport of ions into the inner space. For some ions a "mechanism 1" operates at very low concentrations, while at high concentrations above about 1 mM a "mechanism 2" with different properties comes into play. The nature of the ion–carrier compounds is not known, but it is likely that they are directly connected with proteins or are themselves proteins.

Active ion transport also is a selective process such that specific ions are transported or "carried" across the plasmalemma by specific carrier mechanisms. Although K, rubidium (Rb), and cesium (Cs) compete for the same carrier, they do not compete with elements such as Ca, strontium (Sr), and barium (Ba). The last three elements do, however, compete among themselves for another carrier. Selenium (Se) will compete with SO_4^{2-} but not with phosphate or with monovalent anions. Interestingly, $H_2PO_4^-$ and HPO_4^{2-} apparently have separate carriers and do not compete with one another for entry into the inner space.

Many aspects of absorption, transport, and utilization of mineral nutrients in plants are under genetic control. Genotypes within a species may differ in the rate of absorption and translocation of nutrients, efficiency of metabolic utilization, tolerance to high concentrations of elements, and other factors. The differential tolerance of soybean varieties to Fe stress is exhibited in Table 4.7. The Bragg cultivar was able to absorb sufficient Fe to grow satisfactorily, whereas the Forrest cultivar developed severe deficiencies of this nutrient. In addition to genetically controlled differences in mechanisms of mineral nutrition, the morphology of roots can significantly influence the uptake of nutrient ions. Some varieties are better able to exploit

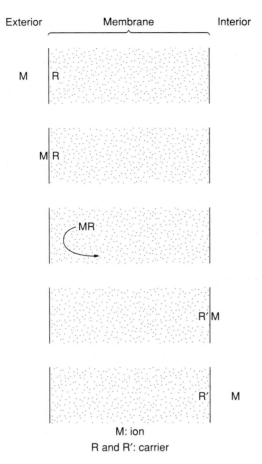

M: ion
R and R': carrier

FIGURE 4.17 Model of the ion–carrier mechanism for ion transport across the plasmalemma. *Epstein*, Mineral Nutrition of Plants, *Wiley-Interscience, 1972.*

TABLE 4.7 Differential Response of Soybean Varieties to Fe Stress

Cultivar	Yield (g Dry Weight)		Fe Concentration in Tops (ppm)	
	Quinlan	Millville	Quinlan	Millville
North				
Amsoy 71	1.17	1.50	32	38
Hodgson	1.55	1.96	43	43
South				
Forrest	1.07	1.60	20	22
Bragg	1.46	1.87	39	45

SOURCE: Brown and Jones, *Agron. J.,* **69:**401 (1977).

soil for nutrients and moisture because of larger or more finely branched root systems.

Summary

1. Ion exchange, which is defined as the reversible process by which cations and anions are exchanged between solid and liquid phases, was reviewed, with emphasis on the phenomenon as it occurs in soils.

2. The determination of cation exchange capacity (CEC) in soils was reviewed, and the factors affecting this important property were discussed. The CEC of a soil is related to the nature and amounts of the mineral and organic colloids present. It increases as soil pH rises.

3. Base saturation, which is the degree to which the exchange capacity of a soil is saturated with basic cations (i.e., Ca^{2+}, Mg^{2+}, K^+, and Na^+), was covered in relation to the nature of the charge on the exchange complex. The dependence of the CEC on the way in which this property is measured was discussed.

4. Anion exchange can take place in soils, but for all practical purposes it is confined to the phosphate and sulfate ions. Unlike cation exchange, anion exchange increases with a decrease in soil pH. The reason that other anions do not undergo adsorption in most agricultural soils was explained.

5. Contact exchange, which is the exchange of ions between the surfaces of two solids without movement through a liquid phase, was described. Its possible importance in soils was considered because plant roots themselves exhibit the property of cation exchange. The latter phenomenon was also discussed briefly.

6. Plant nutrient ions are brought into contact with the absorbing surfaces of roots by (a) root interception and contact exchange, which is enhanced by the growth of roots through the soil mass; (b) diffusion of ions in the soil solution; and (c) mass flow of soil water, which brings the nutrient ions into contact with the plant roots. Absorption of moisture by the roots is one of the principal causes of the mass flow of soil water.

7. Mass flow supplies large amounts of Ca and Mg and a large proportion of the N and S requirements of plants. Most of the P and K moves from the soil to the root surface by diffusion. Mass flow and diffusion are influenced by soil factors such as clay content, moisture level, and temperature. Nutrient concentrations in the soil solution have an important effect on the transport of ions to root surfaces.

8. Fertilizer applications can improve the mass flow and diffusion of ions to the root by increasing nutrient concentrations in the soil solution. Diffusion is slowed by increasing clay content, low soil moisture, and low temperatures. The addition of P and K helps to offset these retarding effects by increasing the diffusion gradient.

9. Some of the suggested mechanisms by which ions are absorbed by plants were considered. Ions are absorbed by both active and passive mechanisms. Active absorption is thought to take place by metabolically produced carriers that transfer the ions across otherwise impassable barriers. Passive absorption is governed largely by exchange adsorption and diffusion phenomena. Ac-

tively absorbed ions are believed to be taken into what is termed the *inner space* in roots, whereas ions that are passively absorbed move into the *outer space* of roots. It was pointed out that much is still to be learned about the mechanics of ion absorption by roots.

Questions

1. Define ion exchange.
2. From what sources does the charge on soil colloids arise?
3. Why does anion adsorption appear to be of little importance in most agricultural soils? Why are AECs of ultisols usually greater than those of mollisols?
4. Potassium acetate (KAc) solution was used to determine the cation exchange capacity of a soil. Ten grams of oven-dry soil was extracted with 200 ml of 1.0 N KAc. It was found that 0.078 g of K was retained by the soil. What is its CEC?
5. Can one determine, by the method described in Question 4, the % BC of this soil? Why?
6. In the example listed on pages 91–92, what assumption must be made if the % BC measured is to be considered valid? Under what conditions would you assume that this assumption is valid?
7. Cation exchange in soils appears to increase as the pH _____. Anion exchange in soils appears to increase as the pH _____.
8. A soil was found to have a CEC of 24 meq/100 g. If the exchange capacity were saturated with Na, to how many grams of NaCl would this be equivalent?
9. The CEC of a soil, as measured using 1.0 N neutral NH_4OAc, gives lower values than those obtained when a solution of $BaCl_2$–triethanolamine buffered at pH 8.3 is used. Why?
10. What is the origin of the effective CEC in mineral soil colloids?
11. What is contact exchange?
12. A soil has a CEC of 25 meq/100 g on an oven-dry basis. Suppose that this soil was 5% saturated with K and you wished to increase the saturation to 9%. Assuming that all of the added K would be adsorbed, how much 100% KCl would have to be added to an acre furrow slice of this soil to raise the K saturation to the desired level? Assume that an acre furrow slice of oven-dry soil weighs 2×10^6 lb.
13. What mechanisms are believed to be the cause of anion adsorption?
14. Under what soil conditions will the SO_4^{2-} ion be adsorbed to the greatest extent?
15. Based on the methods presently used to measure the CEC of roots, what type of plant root appears to have the greater CEC—grasses or legumes? Has the CEC been related to the behavior of these plants growing in the field? Explain.
16. What is your opinion of the relative importance of root interception and contact exchange, simple diffusion, and mass flow in bringing nutrient ions into contact with the absorbing surfaces of plant roots? Would the importance of these three mechanisms be altered by soil texture? By plant species? Why?
17. What soil factors influence diffusion of nutrient ions to roots? Describe and explain a practical way of improving the diffusion of nutrient ions.
18. Assume that 50 lb of S as SO_4^{2-} was lost per acre per year by leaching

and runoff. Assume that another 30 lb was lost through crop removal. How many milliequivalents of cations per 100 g of soil would have to accompany this S to maintain electrical neutrality? If Ca were the only ion involved, to how many pounds per acre of Ca would this removal amount?

19. What is active and passive absorption of elements by plant root cells? In what way are these types of absorption related to the inner and outer spaces in roots?

20. What, generally, is the mechanism that has been proposed to account for the active absorption of ions by roots?

21. What is the complementary ion effect? In what way does it influence plant uptake of ions? Is it of any practical consequence to commercial crop production? Explain. Give some specific examples to support your answer.

22. Why is cation exchange such an important factor in a study of soil fertility and commercial crop production? What is the importance of BC?

23. Assume that you are addressing a group of farmers and business managers who are well versed in crop production but who are not so conversant with the technical aspects of plant nutrition and soil fertility. Your mission is to explain to this group the nature of cation exchange and why it is important to crop production. How would you proceed?

24. During the discussion after your speech, one member of the audience asks why chlorides and nitrates will leach from soils but phosphates, which also have a negative charge, will not. What is your answer?

25. Can genotypes within a species of plant differ in various features of mineral nutrition? How will differences in root morphology influence the ability of plants to obtain moisture and nutrients?

26. Explain how mycorrhizas function.

Selected References

BOHN, H. L., B. L. MCNEAL, and G. A. O'CONNOR. 1979. *Soil Chemistry*. John Wiley & Sons, New York.

EPSTEIN, E. 1972. *Mineral Nutrition of Plants: Principles and Perspectives*. John Wiley & Sons, New York.

MENGEL, K., and E. A. KIRKBY. 1987. *Principles of Plant Nutrition*. International Potash Institute, Bern, Switzerland.

RENDIG, V. V., and H. M. TAYLOR. 1989. *Principles of Soil–Plant Relationships*. McGraw-Hill, New York.

TAN, K. H. 1982. *Principles of Soil Chemistry*. Marcel Dekker, New York.

CHAPTER 5

Soil and Fertilizer Nitrogen

The N Cycle

N is the most frequently deficient nutrient in crop production; therefore, most nonlegume cropping systems require N inputs. Understanding the behavior of N in the soil is essential for maximizing agricultural productivity and profitability while reducing the impacts of N fertilization on the environment. Many N sources are available for use in supplying N to crops. In addition to inorganic fertilizer N, organic N from animal manures and other waste products and from N_2 fixation by leguminous crops can supply sufficient N for optimum crop production.

The ultimate source of the N used by plants is N_2 gas, which constitutes 78% of the earth's atmosphere. Unfortunately, higher plants cannot metabolize N_2 directly into protein. N gas must be converted to a plant-available form by one of the following methods:

1. Fixation by microorganisms that live symbiotically on the roots of legumes and certain nonleguminous plants.
2. Fixation by free-living or nonsymbiotic soil microorganisms.
3. Fixation as oxides of N by atmospheric electrical discharges.
4. Fixation as NH_3, NO_3^-, or CN_2^{2-} by the manufacture of synthetic N fertilizers.

The unlimited supply of atmospheric N_2 is in dynamic equilibrium with the various fixed forms in the soil. As N_2 is fixed by the different processes just indicated, numerous microbial and chemical processes release N_2 to the atmosphere. The cycling of N in the soil–plant–atmosphere system involves many transformations of N between inorganic and organic forms (Fig. 5.1). The N cycle can be divided into N inputs or gains, N outputs or losses, and N cycling within the soil, where N is neither gained or lost (Table 5.1). Except for industrial and combustion fixation, all of these N transformations occur naturally; however, humans can influence many of these N processes through soil and crop management activities. The purpose of this chapter is to describe the chemical and microbial cycling of N and how humans can influence or manage these transformations to optimize plant productivity.

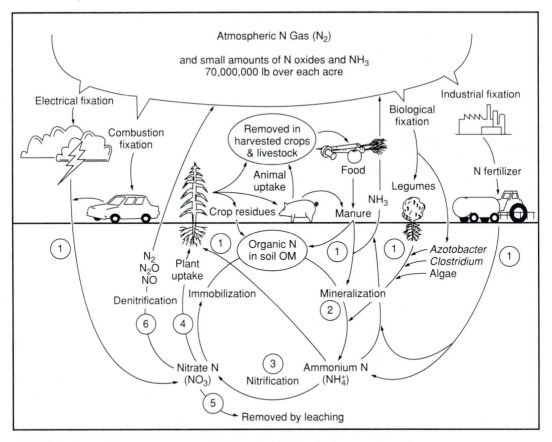

FIGURE 5.1   The N cycle. In step 1, N in plant and animal residues and N derived from the atmosphere through electrical, combustion, and industrial processes (processes by which molecular nitrogen, N_2, is combined with H_2 or O_2) is added to the soil. In step 2, N in the residues is mobilized as NH_4^+ by soil organisms as an end product of residue decomposition. Plant roots absorb a portion of the NH_4^+. In step 3, much of the NH_4^+ is converted to NO_3^- by nitrifying bacteria in a process called *nitrification*. In step 4, NO_3^- is taken up by plant roots and (along with the NH_4^+ absorbed) is used to produce the protein in crops that are eaten by humans or fed to livestock. In step 5, some NO_3^- is lost to groundwater or drainage systems as a result of downward movement through the soil in percolating water. In step 6, some NO_3^- is converted by denitrifying bacteria into N_2 and nitrogen oxides (N_2O and NO) that escape into the atmosphere, completing the cycle.   *Council for Agricultural Science and Technology.* Agriculture and Groundwater Quality, *Report No. 103, p. 23, 1985.*

N Fixation by Rhizobium and Other Symbiotic Bacteria

For centuries, legumes and animal manures supplied additional N to nonleguminous crops. Since the 1940s, their value as an N source has declined because of increased production and use of low-cost synthetic N fertilizers. Legumes

TABLE 5.1   N Inputs, Outputs, and Cycling in the
Soil–Plant–Atmosphere System*

N Inputs	N Outputs	N Cycling
Fixation:	Plant uptake	Immobilization
Biological	Denitrification	Mineralization
Industrial	Volatilization	Nitrification
Electrical	Leaching	
Combustion	Ammonium fixation[†]	
Animal manure		
Crop residue		

*Some N output and cycling components can be influenced by
management but generally are not managed.
[†] Some fixed ammonium can be released to soil; thus, N cycling
also applies.

and animal manures are still important sources of fixed N in many countries.
For example, most of the N required for crop production in New Zealand is
obtained through N_2 fixation by legumes. Many organisms have the unique
ability to fix atmospheric N_2; the most common ones are listed in Table 5.2.

Amounts of N Fixed

Estimates of total annual biological N_2 fixation range from 100 to 175×10^6
metric tons, with about 90×10^6 metric tons of N per year fixed by *Rhizobia*.
In contrast, world fertilizer N use was 77.1×10^6 metric tons in 1991. The
crops grown in the United States and Canada in 1991 received 10.1 and 1.2
million metric tons of N, respectively.

TABLE 5.2   Economically Important Microorganisms Involved in Biological
N Fixation

Organisms	General Properties	Use in Agriculture
Azotobacter	Aerobic, free fixers, live in soil, water, rhizosphere (area surrounding the roots), leaf surfaces	Proposed benefit to crops has not been confirmed; hormonal effect on root and plant growth
Azospirillum	Microaerobic, free fixers or in association with roots of grasses	Potential use in increasing yield of grasses; inoculation benefits crops
	Inside root symbiosis?	Hormonal effect on roots and plant growth
Rhizobium	Fix N in legume–*Rhizobium* symbiosis	Legume crops are benefited by inoculation with proper strains
Actinomycetes, Frankia	Fix N in symbiosis with nonlegume wood trees— alder, *Myrica, Casuarina*	Potentially important in reforestation, wood production
Blue-green algae, *Anabaena*	Contain chlorophyll, as in higher plants; aquatic and terrestrial	Enhance rice in paddy soils; *Azolla* (a water fern)– *Anabaena azolla* symbiosis; is used as green manure

SOURCE: Okon, *Phosphorus Agr.*, **82:**3 (1982).

The quantity of N fixed by properly nodulated legumes averages about 75% of the total N used for plant growth. N present in soil or additions of fertilizer must make up the difference. The amounts of N fixed by *Rhizobia* differ with the *Rhizobium* strain, the host plant, and the environmental conditions under which the two develop.

Amounts of N typically fixed by various legume crops are shown in Table 5.3. N fixation by most perennial legumes ranges from 100 to 200 lb/a/yr, although under optimum conditions, N fixation can reach two to three times these values. Short-season annual legumes will often fix between 50 and 100 lb N/a/yr.

Numerous *Rhizobium* species exist, each requiring a specific host legume plant. For example, the bacteria that live symbiotically with soybeans will not fix N_2 with alfalfa. Inoculation of the legume seed with the correct inoculum is recommended the first time a field is planted to a new legume species. For example, 40% increases in N fixation by alfalfa have been obtained by carefully matching cultivars and strains of inoculum.

The presence of nodules on the legume roots does not necessarily indicate N fixation by active *Rhizobia*. Mature effective alfalfa nodules tend to be large, elongated (2 to 4 by 4 to 8 mm), often clustered on the primary roots, and have pink to red centers. The red color is attributed to the occurrence of leghemoglobin, which is confined to those nodule cells containing rhizobia that are fixing N_2. Ineffective nodules are small (< 2 mm in diameter), are usually numerous, and are scattered over the entire root system; in some

TABLE 5.3   Amounts of N Fixed by Legumes in Temperate Climates

Legume	N Fixed [lb/a/yr]	
	Range in Reported Values	*Typical*
Alfalfa	50–450	200
Crimson clover	—	125
Ladino clover	—	180
Sweet clover	223–267	120
Red clover	76–169	115
Clovers (general)	50–300	—
Kudzu	—	110
White clover	—	100
Cowpeas	58–116	90
Lespedezas (annual)	—	85
Vetch	80–138	80
Chickpeas	–108	—
Peas	30–178	70
Soybeans	58–160	100
Trefoil	—	105
Winter peas	—	50
Peanuts	—	40
Beans	– 71	40
Fababeans	51–267	130
Fababeans (shaded)	–648	—
Lentils	–134	—

cases, they are very large (> 8 mm in diameter) and few in number. They have white or pale green centers.

Legume N Availability to Nonlegume Crops

Yields of nonlegume crops are often increased when they are grown following legumes. Some of the benefit is related to improved N availability to the nonlegume crop, although other rotation effects also are involved. Many field studies have been conducted to evaluate legume N availability to subsequent crops. In general, when corn follows soybean, the amount of N required for optimum yield is less than that required for corn after corn (Fig. 5.2a). While the difference has been attributed to increased N availability from the previous legume crop, other crop rotation effects are involved. Similar responses have been observed with corn following wheat as with corn after soybeans. These results strongly suggest that rotation benefits are not limited to legume rotations and that some of the benefits can be related to increased soil N availability.

When a perennial legume such as alfalfa is used in rotation, the response of corn to applied N varies with time (Fig. 5.2b). Little or no response of corn to N fertilization is observed in the first year; however, the amount of N required for optimum crop production increases with time as the legume N reserves are depleted. The quantity of symbiotically fixed N available in the cropping system depends on (1) the quantity of N fixed, (2) the amount and type of legume residue returned, (3) the availability of soil N to the legume, and (4) harvest management. The quantity of legume N incorporated into the soil from first-year alfalfa varies between 35 and 300 kg/ha. Generally, alfalfa can supply all or most of the N to a nonlegume crop in the first year. Several studies suggest that the N credit commonly attributed to legumes in rotation is overestimated. These contrasting results can probably be explained by soil, climate, and legume management effects. Crop utilization of N in green manure crops also is highly variable. Legume residue N availability during the first subsequent cropping year ranges between 20 and 50%.

Legume N availability to a companion crop is not well understood. Small amounts of amino acids and other organic N compounds may be excreted by the legume roots. Microbial decomposition of the sloughed-off root and nodule tissue also may contribute N to the crop growing with legumes. Under some conditions, the quantity of fixed N and/or legume N availability is not sufficient, and N fertilization is required for optimum production of both nonlegume and legume crops.

Development of improved legume varieties for increased N fixation may encourage the use of legumes in rotation. For example, "Nitro" is a nondormant alfalfa capable of 4 to 6 weeks of additional fall growth. Field results show 45% greater N yield as a result of the longer N fixation period.

Optimum utilization of legume N by a nonlegume grain crop requires that mineralization of legume N occurs over the same time as crop N uptake. Legume N mineralization by soil microbes is controlled by climate, increasing with soil temperature and moisture, as previously noted. The period of N uptake by the crop can vary, depending on the crop. Thus, for maximum utilization of legume N by the nonlegume crop, N uptake must be in synchrony

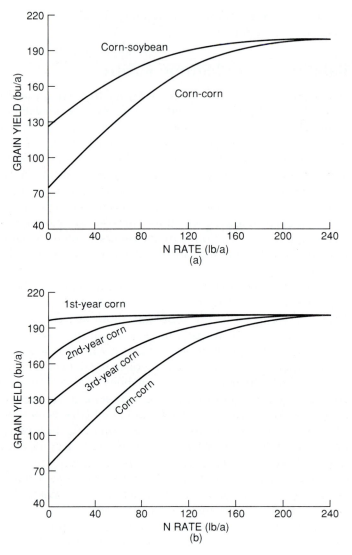

FIGURE 5.2   Generalized corn response to N
fertilization when corn follows an annual legume (a) or a
perennial legume (b) compared to continuous corn.

with N mineralization. For example, the N uptake period for winter wheat is
considerable earlier than for corn (Fig. 5.3). The hypothetical distribution of
N mineralization shows that corn N uptake is more synchronous with N
mineralization than is winter wheat. Therefore, compared to corn, winter
wheat may not utilize much legume N, and when mineralization occurs, the
inorganic N would be subject to leaching and other losses. Therefore, efficient
management of legume N requires careful crop selection.

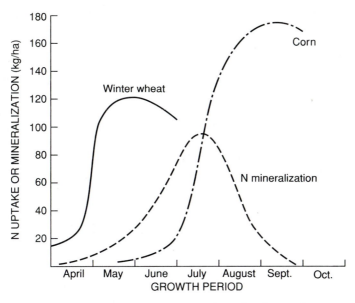

FIGURE 5.3 Synchrony of soil N mineralization and crop N uptake in corn and winter wheat.

Supplemental N for Legumes

Maximum N fixation occurs only when available soil N is at a minimum. Rhizobial activity is reduced if the plant has a readily available supply of inorganic N. However, it is sometimes advisable to include a small amount of N fertilizer at planting time to ensure that the young legume seedlings have an adequate supply until the rhizobia can become established on the roots. Early spring N application can be beneficial for legume crops where rhizobial activity is restricted by cold, wet conditions. N fixation by common bean is low and usually unreliable, and N fertilization is recommended.

Factors Affecting N Fixation

The most important factors influencing the quantity of N fixation by *Rhizobia* are soil pH, mineral nutrient status, photosynthetic activity, climate, and legume management. Any stress in the legume plant by these factors can severely reduce the legume yield and N availability to subsequent crops.

SOIL pH   Soil acidity is a major factor restricting the survival and growth of *Rhizobia* in soil and can severely affect nodulation and N fixation processes. *Rhizobia* and roots of the host legume plants can be injured by soil acidity related to Al^{3+}, Mn^{2+}, and H^+ toxicity, as well as low levels of available Ca^{2+} and $H_2PO_4^-$.

There are significant differences in the sensitivity of the various rhizobial species to soil acidity. Soil pH values below 6.0 drastically reduce the number of *Rhizobium meliloti* in the root zone of alfalfa, degree of nodulation, and yields of host alfalfa plants, whereas soil pH values between 5.0 and 7.0 have little influence on *R. trifoli* and the host red clover crop.

Application of lime to acid soils is an obvious way of improving conditions for crops such as alfalfa that are dependent on *R. meliloti*. For locations where economical sources of lime may not be available, alternative means of growing alfalfa must be used. Success in establishing alfalfa under acid soil conditions has been achieved by using special inoculation techniques. These include high levels of inoculum, where excessive amounts of soluble Mn^{2+} and Al^{3+} are not a problem, and by rolling inoculated seeds in a slurry of pulverized lime. Another approach is to select and use acid-tolerant strains of *Rhizobium*. The effectiveness of some strains of *R. meliloti* at low soil pH is illustrated in Figure 5.4.

MINERAL NUTRIENT STATUS   Except in acid soils, where Ca^{2+} and $H_2PO_4^-$ deficiencies can limit the growth of *Rhizobia*, mineral deficiencies seldom reduce N fixation. N fixation in the nodule requires more Mo than the host plant; thus, Mo deficiency is the most important micronutrient deficiency. Initiation and development of nodules can be affected by Co, B, Fe, and Cu deficiencies. Differences exist in the sensitivity of various *Rhizobia* strains to

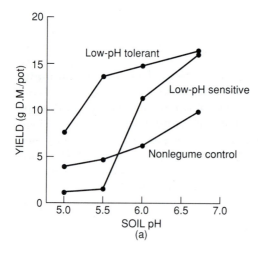

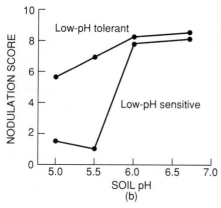

FIGURE 5.4   Forage yields (a) and nodulation scores (b) of alfalfa inoculated with low-pH-tolerant and low-pH-sensitive strains of *Rhizobium meliloti*. The results are the mean for eight low-pH-tolerant strains and four low-pH-sensitive strains. Barley was used as the nonlegume control. *Rice, Can. J. Plant Sci. 62:943, 1982.*

nutrient stress. Excess NO_3^- concentration in the soil can reduce nitrogenase activity and, thus, N fixation. Nodules lose their pink color in a high NO_3^- soil. The reduction in N fixation is related to the competition for photosynthate between NO_3^- reduction and N fixation reactions.

PHOTOSYNTHESIS AND CLIMATE   A high rate of photosynthate production is strongly related to increased N fixation by *Rhizobia*. Factors that reduce the rate of photosynthesis will reduce N fixation. These include reduced light intensity, moisture stress, and low temperature.

LEGUME MANAGEMENT   In general, any management practice that results in reduced legume stands or yield will reduce the quantity of N fixed by legumes. These include water and nutrient stress, excessive weed and insect pressure, and improper harvest management. Harvest practices vary greatly with location, but excessive cutting frequency, premature harvest, and delayed harvest, especially in the fall, can reduce legume stands and the quantity of N fixed.

Fixation by Leguminous Trees and Shrubs

N fixation by legume trees is important to the ecology of tropical and subtropical forests and to agroforestry systems in developing countries. Numerous legume tree species fix appreciable amounts of N. Well-known examples in the United States are *Mimosa, Acacia*, and *black locust*. Three woody leguminous species—*Gliricidia sepium, Leucaena leucocephala*, and *Sesbania biospinosa*—are used as green manure crops in rice-based cropping systems.

Some widely distributed nonlegume plants also fix N by a mechanism similar to legume and *Rhizobia* symbiosis. Certain members of the following plant families are known to bear root nodules and to fix N: Betulaceae, Elaegnaceae, Myricaceae, Coriariaceae, Rhamnaceae, and Casurinaceae. Alder and *Ceanothus*, two species commonly found in the Douglas fir forest region of the Pacific Northwest, can potentially contribute substantial N to the ecosystem. *Frankia*, an actinomycete, is the microorganism responsible for N fixation by these nonlegume woody plants (Table 5.2).

N Fixation by Nonsymbiotic Soil Microorganisms

N fixation in soils is also brought about by certain strains of free-living bacteria and blue-green algae (Table 5.2).

Blue-green algae are completely autotropic and require only light, water, N_2, CO_2, and the essential mineral elements. Their numbers are normally far greater in flooded than in well-drained soils. Because they need light, they probably make only minor contributions to the N supply in upland agricultural soils after closure of crop canopies. In desert or semiarid regions, blue-green algae or lichens containing them become active following occasional rains and fix considerable quantities of N during their short-lived activity. N fixation by blue-green algae is of economic significance in hot climates, particularly in tropical rice soils. The N made available to other organisms by blue-green algae is probably of considerable importance during the early stages of soil formation.

There is a noteworthy symbiotic relationship between *Anabaena azolla* (a blue-green alga) and *Azolla* (a water fern) in temperate and tropical waters. The blue-green alga located in cavities in leaves of the water fern is protected from external adverse conditions and it is capable of supplying all of the N needs of the host plant. An important feature of this association is the water fern's very large light-harvesting surface, a property that limits the N_2-fixing capacity of free-living blue-green algae. The organism *Beijerinckia*, found almost exclusively in the tropics, inhabits the leaf surfaces of many tropical plants and fixes N on these leaves rather than in the soil.

In southeast Asia, *Azolla* has been used for centuries as a green manure in wetland rice culture, as a fodder for livestock, as a compost for production of other crops, and as a weed suppressor.

In California the *Azolla–Anabaena* N-fixing association has supplied 105 kg N/ha per season, or about 75% of the N requirements of rice. When used as a green manure, it provided 50 to 60 kg N/ha and substantially increased yields over unfertilized rice.

Certain N-fixing bacteria can grow on root surfaces and to some extent within root tissues of corn, grasses, millet, rice, sorghum, wheat, and many other higher plants. *Azospirillum brasilense* is the dominant N-fixing bacterium that has been identified. Inoculation of cereal crops with *A. brasilense* has been reported to improve growth and N nutrition, although the response to inoculation has been highly variable. In most of the studies where inoculation was beneficial, the response was related to factors other than increased N fixation. Some of the possibilities are increased scavenging of plant nutrients, altered root permeability, hormonal action, and enhanced NO_3^- reduction in the roots.

Azotobacter- and *Clostridium*-inoculated seed may only provide a maximum of 5 kg N/ha; therefore, these nonsymbiotic organisms are of little value to N availability in intensive agriculture.

N Addition from the Atmosphere

N compounds in the atmosphere are returned to the earth in rainfall as NH_3, NO_3^-, NO_2^-, nitrous oxide (N_2O), and organic N. Ammonia comes largely from industrial sites where NH_3 is used or manufactured. Ammonia also escapes from the soil surface (volatilization) as a result of chemical reactions in the soil. The organic N accumulates as finely divided organic residues that are swept into the atmosphere from the earth's surface.

Soil has a pronounced capacity for adsorbing NH_3 gas from the atmosphere. In localized areas where atmospheric NH_3 concentrations are high, 50 to 70 lb NH_3/a/yr may be adsorbed by soils. Sorption is positively related to NH_3 concentration and to temperature but is independent of rainfall.

Because of the small amount of NO_2^- present in the atmosphere, NO_2^- and NO_3^- are combined and reported as NO_3^-. The presence of NO_3^- has been attributed to its formation during atmospheric electrical discharges, but recent studies suggest that only about 10 to 20% of the NO_3^- in rainfall is due to lightning. The remainder probably comes from industrial waste gases or possibly from the soil.

Atmospheric N compounds are continually being returned to the soil in rainfall. The total amount of N in rainfall ranges between 1 and 50 lb/a/yr, depending on the location. These figures are generally higher around areas of intense industrial activity and, as a rule, are greater in tropical than in polar or temperate zones.

Industrial Fixation of N

From the standpoint of commercial agriculture, the industrial fixation of N is by far the most important source of N as a plant nutrient. The production of N by industrial fixation is based on the Haber–Bosch process, in which H_2 and N_2 gases react to form NH_3:

$$3H_2 + N_2 \xrightarrow[\text{1200°C, 500 atm}]{\text{Catalyst}} 2NH_3$$

The NH_3 produced can be directly used as a fertilizer (anhydrous NH_3), although numerous other fertilizer N products are manufactured from NH_3 (see pp. 154, 162–164).

Forms of Soil N

Total N content of soils ranges from less than 0.02% in subsoils to more than 2.5% in peats. The N concentration in the top 1 ft of most cultivated soils in the United States normally varies between 0.03 and 0.4%. N in soil occurs as inorganic or organic N, with 95% or more of total N in surface soils present as organic N.

Inorganic N Compounds

The inorganic forms of soil N include ammonium (NH_4^+), nitrite (NO_2^-), nitrate (NO_3^-), nitrous oxide (N_2O), nitric oxide (NO), and elemental N (N_2), which is inert except for its utilization by *Rhizobia* and other N-fixing microorganisms.

From the standpoint of soil fertility, NH_4^+, NO_2^-, and NO_3^- are the most important and are produced from aerobic decomposition of soil organic matter or from the additions of N fertilizers. These three forms usually represent 2 to 5% of the total soil N. N_2O and NO are important forms of N lost through denitrification.

Organic N Compounds

Organic soil N occurs as proteins, amino acids, amino sugars, and other complex N compounds. The proportion of total soil N in these various fractions is as follows: bound amino acids, 20 to 40%; amino sugars such as the hexosamines, 5 to 10%; and purine and pyrimidine derivatives, 1% or less. Very little is known about the chemical nature of the 50% or so of the organic N not found in these fractions.

Proteins are commonly found in combination with clays, lignin, and other materials resistant to decomposition. Analytical techniques are now available to isolate free amino acids from soils that are not in peptide linkages or in combination with high molecular weight organic polymers, clays, or lignin. The suitability of these substrates for biological oxidation suggests that they will not accumulate in soils and that they may be an important source of NH_4^+. Relative to other forms, the quantities of free amino acids in soils are low.

Forms of N Absorbed by Plants

Plants absorb N as both NH_4^+ and NO_3^- (Fig. 5.1). NO_3^- generally occurs in higher concentrations than NH_4^+, and it is free to move to the roots by mass flow and diffusion. Some NH_4^+ is always present and will influence plant growth and metabolism in ways that are not completely understood.

Preference of plants for either NH_4^+ or NO_3^- is determined by the age and type of plant, the environment, and other factors. Cereals, corn, sugar beets, pineapple, rice, and ryegrass use either form of N. Kale, celery, bush beans, and squash grow best when provided with some NO_3^-. Some plants, such as blueberries, *Chenopodium album*, and certain rice cultivars cannot tolerate NO_3^-. Solanaceous crops, such as tobacco, tomato, and potato, prefer a high NO_3^-/NH_4^+ ratio.

Nitrate

The rate of NO_3^- uptake is usually high and is favored by low-pH conditions. When plants absorb high levels of NO_3^-, there is an increase in organic anion synthesis within the plant coupled with a corresponding increase in the accumulation of inorganic cations (Ca, Mg, K). The growth medium will become alkaline, and some HCO_3^- can be released from the roots to maintain electroneutrality in the plant and in the soil solution.

Ammonium

Ideally, NH_4^+ is the preferred N source since energy will be saved when it is used instead of NO_3^- for synthesis of protein. NO_3^- reduction is an energy-requiring process that uses two NADH molecules for each NO_3^- ion reduced in protein synthesis. Also, NH_4^+ is less subject to losses from soil by leaching and denitrification.

Plant uptake of NH_4^+ proceeds best at neutral pH values and is depressed by increasing acidity. Absorption of NH_4^+ by roots reduces Ca^{2+}, Mg^{2+}, and K^+ uptake while increasing absorption of $H_2PO_4^-$, SO_4^{2-}, and Cl^-.

Plants supplied with NH_4^+ may have increased carbohydrate and protein levels compared to NO_3^-. Rhizosphere pH decreases when plants receive NH_4^+ and are caused by H^+ exuded by the root to maintain electroneutrality or charge balance inside the plant. Differences in 2 pH units have been observed for NH_4^+ versus NO_3^- uptake in wheat. This acidification can affect both nutrient availability and biological activity in the vicinity of roots.

NH_4^+ tolerance limits are narrow, with excessive levels producing toxic reactions. High levels of NH_4^+ can retard growth, restrict uptake of K^+, and produce symptoms of K^+ deficiency. In contrast, plants tolerate large excesses of NO_3^- and accumulate it to comparatively high levels in their tissues.

NH_4^+ and NO_3^- Combinations

Plant growth is often improved when the plants are nourished with both NO_3^- and NH_4^+ compared to either NO_3^- or NH_4^+ alone. There is increasing evidence that mixtures of these forms are beneficial at certain growth stages for some genotypes of corn, sorghum, soybeans, wheat, and barley. Increased wheat, barley, and sorghum yields with NH_4^+ nutrition are associated with greater tillering. Corn yields increased from 8 to 25% with $NH_4^+ + NO_3^-$, compared to yields with NO_3^- alone, which was related to increased numbers of kernels per plant and not to heavier kernels (Table 5.4). These data also illustrate that corn genotypes differ in their physiological response to NH_4^+.

Recent research results demonstrate that NH_4^+ application during grain fill was required to maximize corn yields and that a 50:50 ratio of NH_4^+ to NO_3^- was optimum. Others have observed that 2 to 4 weeks postsilking is one of the most critical periods when NH_4^+ increases corn yields.

Form of Inorganic N and Plant Diseases

$NH_4^+ + NO_3^-$ nutrition is a major factor influencing the occurrence and severity of plant diseases. Some diseases are more severe when NH_4^+ is the

TABLE 5.4   Responses of Corn Hybrids When Supplied with Differing N Sources: Plants Grown Under Field Conditions in a Gravel-Hydroponic System

Year	Hybrid	N Source	Grain Yield	Kernel Number	Kernel Weight
			g/plant	no./plant	mg/kernel
1986	B73 × LH51	All NO_3	254	688	369
		NO_3/NH_4	275	764	361
	FS 854	All NO_3	277	818	339
		NO_3/NH_4	315	1,000	315
1987	B73 × LH51	All NO_3	154	540	285
		NO_3/NH_4	193	691	279
	B73 × LH38	All NO_3	161	603	267
		NO_3/NH_4	180	742	243
	CB59G × LH38	All NO_3	137	475	288
		NO_3/NH_4	154	545	283
	LH74 × LH51	All NO_3	181	592	306
		NO_3/NH_4	199	607	328

SOURCE: Below and Gentry, *Better Crops* **72**(2): (1988).

primary form of inorganic N in the root zone; others are more severe when NO_3^- predominates.

Two processes may be involved, starting with the direct effect of the form of N on pathogenic activity. The other is the influence of NO_3^- or NH_4^+ on the functioning of organisms capable of altering the availability of micronutrient cations. For example, a high NO_3^- supply stimulates certain bacteria, which lowers the availability of Mn to wheat. The effect of N form on rhizosphere soil pH also is partially responsible for the differences observed in the incidence and severity of diseases.

N Transformations in Soils

The quantities of NH_4^+ and NO_3^- available to plants depend largely on the amounts applied as N fertilizers and mineralized from organic soil N. The amounts released from organic N, and to some extent those existing in the soil after the addition of NH_4^+ or NO_3^-, depend on many factors affecting N mineralization, immobilization, and losses from the soil.

Soil organic matter is a term used to describe organic materials in all stages of decomposition. Broadly speaking, soil organic matter can be placed in two categories. The first is a relatively stable material, termed *humus*, that is somewhat resistant to further rapid decomposition (Fig. 4.8). The second includes those organic materials that are subject to fairly rapid decomposition, which range from fresh crop residues to relatively stable humus. The primary microbial processes involved in fresh residue and humus turnover or cycling in soils are mineralization and immobilization of soil N. These reactions, combined with other physical, chemical, and environmental factors, are important in organic matter stability in soils and in inorganic N availability to plants.

N Mineralization

N mineralization is the conversion of organic N to NH_4^+ (Fig. 5.1). Mineralization of organic N involves two reactions, aminization and ammonification, which occur through the activity of heterotrophic microorganisms. Heterotrophs require organic C compounds for their source of energy.

Mineralization increases with a rise in temperature and is enhanced by adequate, although not excessive, soil moisture and a good supply of O_2. Decomposition proceeds under waterlogged conditions, although at a slower rate, and is incomplete. Aerobic, and to a lesser extent anaerobic, respiration release the contained N in the form of NH_4^+.

AMINIZATION   Heterotrophic bacteria and fungi are responsible for one or more steps in the reactions in organic matter decomposition. Bacteria dominate in the breakdown of proteins in neutral and alkaline environments, with some involvement of fungi, while fungi predominate under acid conditions. The end products of the activities of one group furnish the substrate for the next, and so on down the line, until the material is decomposed. One of the final stages is the decomposition of proteins and the release of amines, amino acids, and urea. This step is termed *aminization* and is represented schematically by the following:

$$\text{Proteins} \xrightarrow[\substack{\text{Bacteria} \\ \text{Fungi}}]{\text{H}_2\text{O}} \quad \underset{\substack{| \\ \text{H} \\ \text{Amino acids}}}{\overset{\substack{\text{NH}_2 \\ |}}{\text{R}-\text{C}-\text{COOH}}} + \underset{\text{Amines}}{\text{R}-\text{NH}_2} + \underset{\substack{| \\ \text{NH}_2 \\ \text{Urea}}}{\overset{\substack{\text{NH}_2 \\ |}}{\text{C}=\text{O}}} + \text{CO}_2 + \text{energy}$$

AMMONIFICATION   The amines and amino acids produced by aminization of organic N are decomposed by other heterotrophs, with the release of NH_4^+. This step is termed *ammonification* and is represented as follows:

$$R-NH_2 + H_2O \rightarrow NH_3 + R-OH + \text{energy}$$
$$\searrow^{+ H_2O}$$
$$\rightarrow NH_4^+ + OH^-$$

A very diverse population of aerobic and anaerobic bacteria, fungi, and actinomycetes is capable of liberating NH_4^+. The NH_4^+ produced is subject to several fates (Fig. 5.1):

1. It may be converted to NO_2^- and NO_3^- by the process of nitrification.
2. It may be absorbed directly by higher plants.
3. It may be utilized by heterotrophic organisms in further decomposing organic C residues.
4. It may be fixed in a biologically unavailable form in the lattice of certain expanding-type clay minerals.
5. It may be slowly released back to the atmosphere as N_2.

Soil organic matter contains about 5% N, and during a single growing season, 1 to 4% of organic N is mineralized to inorganic N. The quantity of N mineralized during the growing season can be estimated. For example, if a soil contained 4% organic matter (OM) and 2% mineralization occurred, then

$$4\% \text{ OM} \times (2 \times 10^6 \text{ lb)soil/a} - 6 \text{ in.}) \times (5\% \text{ N}) \times (2\% \text{ N mineralized}) = 80 \text{ lb N/a}$$

Thus, each year, 80 lb N/a as NH_4^+ are mineralized, which enter the soil solution and can be utilized by plants or other soil N processes (Fig. 5.1).

N Immobilization

N immobilization is the conversion of inorganic N (NH_4^+ or NO_3^-) to organic N and is basically the reverse of N mineralization (Fig. 5.1). If decomposing OM contains low N relative to C, the microorganisms will immobilize NH_4^+ or NO_3^- in the soil. The microbes need N in a C:N ratio of about 8:1; therefore, inorganic N in the soil is utilized by the rapidly growing population. N immobilization during crop residue decomposition can reduce NH_4^+ or NO_3^- concentrations in the soil to very low levels. Soil microorganisms compete very effectively with plants for NH_4^+ or NO_3^- during immobilization, and plants can readily become N deficient. Fortunately, in most cropping systems, sufficient fertilizer N is applied to compensate for immobilization and crop requirements. After decomposition of the low N residue, microbial activity subsides and the immobilized N, which occurs as proteins in the microbes, can be mineralized back to NH_4^+ (Fig. 5.1).

If added organic material contains high N relative to C, N immobilization will not proceed because the residue contains sufficient N to meet the microbial demand during decomposition. Inorganic N in solution will actually increase from mineralization of some of the organic N in the residue material.

C/N Ratio Effects on N Mineralization and Immobilization

N in some form, as well as other nutrients, are needed by heterotrophic soil microorganisms that decompose OM. If the decomposing OM has a small amount of N in relation to the C present (e.g. wheat straw, mature corn stalks), the microorganisms will utilize NH_4^+ or NO_3^- present in the soil to further the decomposition. This N is needed to permit rapid growth of the microbial population that accompanies the addition to the soil of a large supply of carbonaceous material.

If, on the other hand, the material added contains much N in proportion to the C present (alfalfa or clover), there will normally be no decrease of mineral N in the soil. There may even be a fairly rapid increase in N caused by its release from the decomposing organic material. The ratio of %C to %N (C/N ratio) defines the relative quantities of these two elements in crop residues and other fresh organic materials, soil OM, and soil microorganisms (Table 5.5). The N content of humus or stable soil OM ranges from 5.0 to 5.5%, while C ranges from 50 to 58%, giving a C/N ratio ranging between 9 and 12.

Whether N is mineralized or immobilized depends on the C/N ratio of the OM being decomposed by soil microorganisms. For example, a typical soil mineralized 0.294 mg N, as measured by plant uptake (Table 5.6). When residues of variable C/N ratio are added to the soil, N mineralization or immobilization would be indicated if plant uptake was greater or less than 0.294 mg N, respectively. In this study, a C/N ratio of approximately 20:1 was the dividing line between immobilization and mineralization.

TABLE 5.5   C/N Ratios in a Selection of Organic Materials

Organic Substances	C/N Ratio	Organic Substances	C/N Ratio
Soil microorganisms	8:1	Bitumens and asphalts	94:1
Soil organic matter	10:1	Coal liquids and shale oils	124:1
Sweet clover (young)	12:1	Oak	200:1
Barnyard manure (rotted)	20:1	Pine	286:1
Clover residues	23:1	Crude oil	388:1
Green rye	36:1	Sawdust (generally)	400:1
Corn/sorghum stover	60:1	Spruce	1,000:1
Grain straw	80:1	Fir	1,257:1
Timothy	80:1		

SOURCE: Beaton, "Land Reclamation Short Course," Univ. of British Columbia, pp. Ba–B24 (1974); McGill et al. and Paul and Ladd, Eds., *Soil Biochemistry*, Vol. 5, p. 238. New York: Marcel Dekker, 1980.

TABLE 5.6   N Mineralized from Various Vegetable
Residues as Measured by Plant Uptake Following
Incubation Under Laboratory Conditions

Plant Residue*	C/N Ratio	N Uptake (mg)
Check soil	1.8	0.294
Tomato stems	45.3	0.051
Corn roots	48.1	0.007
Corn stalks	33.4	0.038
Corn leaves	31.9	0.020
Tomato roots	27.2	0.029
Collard roots	19.6	0.311
Bean stems	17.3	0.823
Tomato leaves	15.6	0.835
Bean stems	12.1	1.209
Collard stems	11.2	2.254
Collard leaves	9.7	1.781

*Residues above the dashed line have a C/N ratio > 20:1.
Residues below the dashed line have a C/N ratio < 20:1.
SOURCE: Iritani and Arnold, *Soil Sci.*, **89**:74 (1960).

The influence of the C/N ratio of residues on N mineralization and immobilization in soils can be illustrated in a generalized diagram (Fig. 5.5). During the initial stages of the decomposition of fresh organic material there is a rapid increase in the number of heterotrophic organisms, accompanied by the evolution of large amounts of CO_2. If the C/N ratio of the fresh material is greater than 30:1, N immobilization occurs, as shown in the shaded area under the top curve. As decay proceeds, the residue C/N ratio narrows and the energy supply diminishes. Some of the microbial population dies because of the decreased food supply, and ultimately a new equilibrium is reached, accompanied by mineralization of N (indicated by the crosshatched area under the top curve). The result is that the final soil level of inorganic N may be higher than the original level.

Generally, when organic substances with C/N ratios greater than 30:1 are added to soil, there is immobilization of soil N during the initial decomposition process. For ratios between 20 and 30, there may be neither immobilization nor release of mineral N. If the organic materials have a C/N ratio of less than 20, there is usually a release of mineral N early in the decomposition process. There may also be an increase in OM or humus, depending on the quantity and type of fresh organic material added. The time required for this decomposition cycle to run its course depends on the quantity of OM added, the supply of inorganic N, the resistance of the material to microbial attack (a function of the amount of lignins, waxes, and fats present), temperature, and soil moisture level.

The N content of the residue being added to soil also can be used to predict whether N is immobilized or mineralized. Concentrations of between 1.5 and 1.7% N are usually sufficient to minimize immobilization of soil N under aerobic conditions. Under anaerobic conditions in submerged soils, the N requirement for decomposition of crop residues may be only about 0.5%.

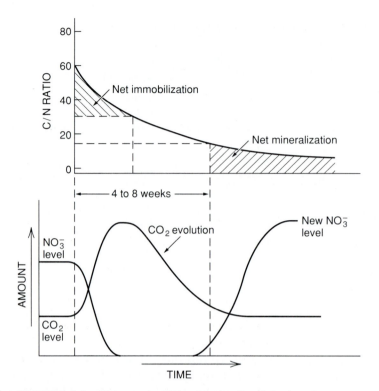

FIGURE 5.5   Changes in NO_3^- levels of soil during the decomposition of low-N crop residues.   *Courtesy of B. R. Sabey, Univ. of Illinois.*

When high C/N ratio residues are added to soil, N in the residue and inorganic soil N are used by the microorganisms during residue decomposition. The quantity of inorganic soil N immobilized by the microbes can be estimated. For example, incorporating 2,000 lb/a residue containing 45% C and 0.75% N (C/N ratio of 60:1) into the soil represents 900 lb C.

$$2,000 \text{ lb residue} \times 45\% \text{ C} = 900 \text{ lb C in the residue}$$

Increasing microbial activity will utilize 35% of the residue C, while the remaining 65% is respired as CO_2 (Fig. 5.5). Thus, the microbes will use 315 lb C in the residue.

$$900 \text{ lb C} \times 35\% = 315 \text{ lb C used by microbes}$$

The increasing microbial population will require N governed by the microbe C/N ratio of 8:1 (Table 5.5).

$$\frac{315 \text{ lb C}}{X \text{ lb N}} = \frac{8}{1}$$

$$X = 39 \text{ lb N need by microbes}$$

The microbes will readily use the 15 lb N in the residue during decomposition.

$$\frac{900 \text{ lb C}}{X \text{ lb N}} = \frac{60}{1}$$

$$X = 15 \text{ lb N in residue}$$

The residue N content also can be calculated by: 2,000 lb residue × 0.75% N = 15 lb N. Thus, the quantity of N immobilized is

$$39 \text{ lb N needed} - 15 \text{ lb N in residue} = 24 \text{ lb N immobilized}$$

Therefore, at least 24 lb N/a will be needed to compensate for immobilization of inorganic N. Routine fertilizer N recommendations will usually account for N immobilization requirements.

N Mineralization and Immobilization Effects on Soil OM

In virgin (uncultivated) soil the humus content is determined by soil texture, topography, and climatic conditions. Generally, OM content is higher in cooler than in warmer climates, and, with similar annual temperature and vegetation, increases with an increase in effective precipitation. These differences are related to reduced potential for OM oxidation with cooler temperatures and increased biomass production with increased rainfall. Humus content is greater in fine-textured than in coarse-textured soils and is related to increased biomass production in finer-textured soils because of improved soil water storage and reduced humus oxidation potential. OM contents are higher under grassland vegetation than under forest cover. These relations are generally true for well-drained soil conditions. Under conditions of poor drainage or waterlogging, aerobic decomposition is impeded and organic residues build up to high levels, regardless of temperature or soil texture.

The C/N ratio of the undisturbed *topsoil* in equilibrium with its environment is about 10 or 12. Generally, it narrows in the subsoil because of the lower amounts of C. An uncultivated soil has a relatively stable soil microbial population, a relatively constant amount of plant residue returned to the soil, and usually a low rate of N mineralization. If the soil is disturbed with tillage, there is an immediate and rapid increase in N mineralization. Continued cultivation without the return of adequate crop residues will ultimately lead to a decline in the humus content of soils.

Many studies have documented the decline in soil OM content with cultivation. Three Kansas soils were sampled at 10-year intervals, and at each site the total N content of the surface soil decreased from 20 to 50% in 40 years (Fig. 5.6). Other studies have suggested that under continuous cultivation, soil OM declines approximately 50% in 40 to 70 years, depending on the environment and the quantity of residue returned to the soil.

The rate of OM oxidation and decline decreases with time. The steady-state OM level depends on the soil and crop management practices utilized. If these practices are changed, a new OM level is attained that may be lower or higher than the previous level and depends entirely on management. These general relationships are illustrated in Figure 5.7.

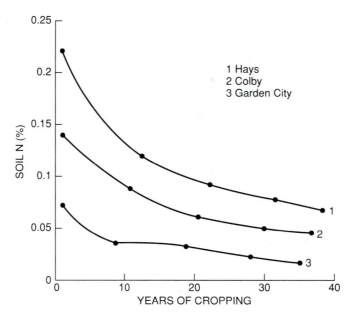

FIGURE 5.6   Decline in total soil N with years of cropping at three locations in Kansas. Each site was in wheat-fallow-wheat, with all residues incorporated with tillage. Total soil N represents soil organic matter, since 95% of total N is organic N. *Haas and Evans*, USDA Tech. Bill. No. 1164, *1957.*

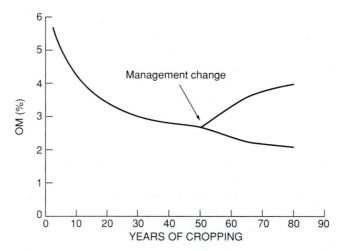

FIGURE 5.7   General decline in soil OM with cultivation. A steady-state equilibrium OM level is attained after 40 to 50 years, which depends on the cropping system, environment, and many soil factors. If the cropping system or residue management is changed, the OM content may increase or decrease until a new equilibrium is attained.

The change in OM content will dramatically reduce the quantity of N mineralization and, thus, soil N availability to crops. The differences in N mineralization can be readily calculated. For example, suppose that a virgin soil has a 5% OM content, and as the soil is cultivated (conventional tillage) the rate of OM loss is 4% per year. Therefore, the quantity of N mineralized in the first year is

$$5\% \text{ OM} \times (2 \times 10^6 \text{ lb/a-6 in.}) \times 4\% \text{ OM loss/yr} = 4{,}000 \text{ lb/a OM loss/yr}$$

Since OM contains about 5% N, the amount of N mineralized is

$$4{,}000 \text{ lb OM/a} \times 5\% \text{ N} = 200 \text{ lb N/a/yr}$$

After 50 years of cultivation, the organic matter has declined to 2.5% (Fig. 5.6). Assume that 2% of the OM oxidizes per year; thus, the quantity of N mineralized is

$$2.5\% \text{ OM} \times (2 \times 10^6 \text{ lb/a-6 in.}) \times 2\% \text{ OM loss/yr} \times 5\% \text{ N} = 50 \text{ lb N/a/yr}$$

These estimates of the change in N mineralized illustrate that cultivation of virgin soils mineralized sufficient N to optimize yields of most crops, especially at the low yield levels experienced 50 years ago. The excess N not utilized by the crop was subject to several losses, which include leaching and denitrification. However, at present yield levels, mineralization of 50 lb N/a is insufficient and fertilizer N is needed to optimize yields. For example, in the Great Plains, fertilizer N has been used for only the last 15 to 20 years in dryland cropping systems, although these soils have been cultivated for about 70 years. In this region OM levels have declined from about 3% to less than 1.5%.

Currently, about 30% of the cropland in North America (1990 estimate) is under some form of conservation tillage in order to reduce wind and water erosion, to increase precipitation efficiency, and to reduce fuel, labor, and equipment costs. Compared to conventional tillage, there are fewer operations and less thorough incorporation of crop residues under conservation tillage. With zero tillage, all crop residues remain on the soil surface rather than being worked into the soil. The insulating and shading effect of crop residues lying on the surface will reduce N and S mineralization because of lower soil temperatures. The physical nature of loose, coarse accumulations of crop debris is also far from ideal for rapid turnover of OM and release of N and S. This slow cycling or turnover of N and S under no tillage favors maintenance of soil OM.

The relationships between tillage and residue on accumulation of OM have been well documented (Fig. 5.8). In this study three crop rotations—continuous soybean, continuous sorghum, and sorghum-soybean—were managed for 12 years under conventional and no-tillage systems (0 and 100% surface residue cover, respectively). The total residue returned to the soil increased with increasing frequency of sorghum in the rotation (continuous sorghum > sorghum-soybean > continuous soybean). Under conventional tillage, soil OM increased only slightly compared to no tillage, where all the residue was left on the soil surface. Under no tillage, soil organic matter increased 45% as the

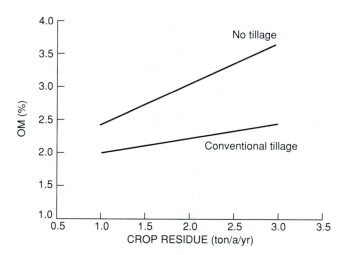

FIGURE 5.8   Influence of tillage and rotation on soil
OM content. Crop residue returned per year for the
three rotations is about 1, 2, and 3 tons/a/yr for
continuous soybean, soybean-sorghum, and continuous
sorghum, respectively. No tillage and conventional tillage
represent 100% and 0% surface residue cover,
respectively.   *Havlin et al.* SSSAJ, *54:448–452, 1990.*

level of residue increased from 1 to 3 ton/a/yr. These data illustrate that the
quantity of residue returned is important to maintaining or increasing organic
matter; however, reducing oxidation of the organic residues with no tillage
had a greater effect on increasing OM. Although soybean residue has a lower
C/N ratio (higher N content) than sorghum residue (Table 5.5), the quantity
of residue added is more important to increasing OM.

The long-term benefits of returning straw residues on organic C and total
N in soil are shown in Table 5.7. Not only was there a greater buildup of OM
and N from the heavier rates of straw, there were also substantial increases in
mineralizable N, P, and K. Adequate use of nutrients, coupled with the return
of crop residues, not only can maintain the level of soil OM but may actually
increase it (Fig. 5.9).

Generally, slightly higher N rates are required for crops grown under no
tillage than conventional tillage systems because of reduced N availability in
no-till fields due to increased N immobilization, leaching, and/or denitrification. A number of studies have demonstrated more efficient utilization of
fertilizer with no-till production compared to conventionally grown corn. Crop
yields under no tillage are often equal to or better than those of conventionally
tilled crops, particularly with optimum management of fertilizer N and other
plant nutrients. In situations where crop yields are initially reduced with no
tillage, yields may improve with time because of reduced erosion; improved
weed control; increased levels and/or quality of soil OM; improved availability
of N, P, and other nutrients; higher CEC; greater soil water-holding capacity;
and better soil structure.

Maintaining adequate OM levels influences many other soil properties that

TABLE 5.7   Cumulative Effects of Straw Residue at Different Rates During Fallow on Soil Properties in an 8-Year Wheat-Fallow Rotation

Soil Property and Depth	Residue Rate (kg/ha)			
	0	680	1360	2725
Organic C (%)				
0–7.6 cm	1.79	1.99	2.11	2.20
7.6–15.2 cm	1.33	1.40	1.50	1.71
15.2–30.5 cm	1.11	1.12	1.25	1.32
Total soil N (%)				
0–7.6 cm	0.089	0.097	0.096	0.102
7.6–15.2 cm	0.072	0.074	0.083	0.087
15.2–30.5 cm	0.063	0.068	0.069	0.068
N mineralization (ppm) (8 weeks)	18.2	20.9	22.7	24.4
NaHCO$_3$ soluble P (ppm)	7.8	8.4	8.8	9.6
Exchangeable K (meq/100 g)	0.71	0.82	0.91	1.01

SOURCE: Black and Siddoway, *J. Soil Water Conserv.*, **34:**220 (1979).

affect soil productivity. OM is important for maintaining good soil structure, especially in fine-textured soils. It increases CEC, which reduces potential leaching losses of elements such as K^+, Ca^{2+}, and Mg^{2+}. As already demonstrated, OM mineralization provides a continuous, although limited, supply of plant-available N, in addition to P and S. Improved water-holding capacity has also been observed with increasing OM content. The ultimate objective of any farming enterprise is *sustained* maximum economical production. The judicious use of lime, fertilizers, and sound management practices will lead to this objective and, incidentally, will help maintain and even increase soil OM levels. Sustaining the productive capacity of soil for future generations ultimately depends on maintaining optimum soil OM levels.

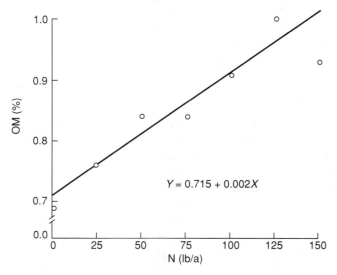

FIGURE 5.9   Effect of fertilizer N rates applied to cotton on soil OM content over an 11-year period in Arkansas.   Crops and Soils *37(9):34, 1985.*

Nitrification

Some of the NH_4^+ released during mineralization of organic N is converted to NO_3^-, a process called *nitrification* (Fig. 5.1). Nitrification is a two-step process in which NH_4^+ is converted first to NO_2^- and then to NO_3^-. Biological oxidation of NH_4^+ to NO_2^- is represented by

$$2NH_4^+ + 3O_2 \xrightarrow[\textit{Nitrosomonas}]{} 2NO_2^- + 2H_2O + 4H^+$$

Nitrosomonas are obligate autotrophic bacteria that obtain their energy from the oxidation of N and their C from CO_2. Other autotrophic bacteria (*nitrosolobus, nitrospira,* and *nitrosovibrio*), and to some extent heterotrophic bacteria, also can oxidize NH_4^+ and other reduced N compounds (i.e., amines) to NO_2^-.

In the second reaction, NO_2^- is further oxidized to NO_3^- by

$$2NO_2^- + O_2 \xrightarrow[\textit{Nitrobacter}]{} 2NO_3^-$$

Nitrite oxidation occurs with autotrophic bacteria called *nitrobacter*, although some heterotrophs also are involved.

The source of NH_4^+ can be from mineralization of organic N or from N fertilizers containing or forming NH_4^+. The reaction rates associated with nitrification in most well-drained soils are NO_2^- to $NO_3^- > NH_4^+$ to NO_2^-. As a result, NO_2^- generally does not accumulate in soils, which is fortunate, since NO_2^- is toxic to plant roots. Both reactions require molecular O_2; thus, nitrification readily takes place in well-aerated soils. The reactions also show that nitrification of 1 mole of NH_4^+ produces 2 moles of H^+. Increasing soil acidity with nitrification is a natural process, although soil acidification is accelerated with continued application of NH_4^+-containing or -forming fertilizers. Since the NO_3^- anion is readily produced, and is very mobile and subject to leaching losses, understanding the factors affecting nitrification in soils will provide insight into those management practices that minimize NO_3^- loss by leaching.

FACTORS AFFECTING NITRIFICATION   Because microbial activity is involved, the rapidity and extent of nitrification will be greatly influenced by soil environmental conditions. Generally the environmental factors favoring the growth of most upland agricultural plants are those that also favor the activity of the nitrifying bacteria. Factors affecting nitrification in soils are (1) supply of NH_4^+, (2) population of nitrifying organisms, (3) soil pH, (4) soil aeration, (5) soil moisture, and (6) temperature.

Supply of NH_4^+. The supply of NH_4^+ is the first requirement for nitrification. If conditions do not favor mineralization of NH_4^+ from organic matter (or if NH_4^+-containing/forming fertilizers are not added to the soils), nitrification does not occur. Temperature and moisture levels that enhance nitrification are also favorable to ammonification.

If large amounts of small grain straw, corn stalks, or similar materials with a wide C/N ratio are plowed into soils with only limited quantities of inorganic N, this will result in N immobilization by microorganisms in the decomposition of the residues. If crops are planted immediately after plowing, they may become N deficient. Deficiencies can be prevented by adding sufficient fertilizer N to supply the needs of the microorganisms and the growing crop.

Population of Nitrifying Organisms. Soils differ in their ability to nitrify NH_4^+ even under similar conditions of temperature, moisture, and level of added NH_4^+. One factor that may be responsible is the variation in the numbers of nitrifying organisms present in the different soils.

The presence of different-sized populations of nitrifiers results in differences in the lag time between the addition of the NH_4^+ and the buildup of NO_3^- in the soil. Because of the tendency of microbial populations to multiply rapidly in the presence of an adequate supply of C, the total amount of nitrification is not affected by the number of organisms initially present, provided that temperature and moisture conditions are favorable for sustained nitrification.

Others have suggested that differences in the nitrification patterns of soils may be attributed in part to volatile losses of N resulting from the accumulation of NO_2^- and its subsequent decomposition.

Soil pH. Nitrification takes place over a wide range in pH (4.5 to 10), although the optimum pH is 8.5. The nitrifying bacteria need an adequate supply of Ca^{2+} and $H_2PO_4^-$ and a proper balance of micronutrients. The exact requirement for these mineral elements has not been determined. The influence of both soil pH and available Ca^{2+} on the activity of the nitrifying organisms suggests the importance of liming in the farming enterprise.

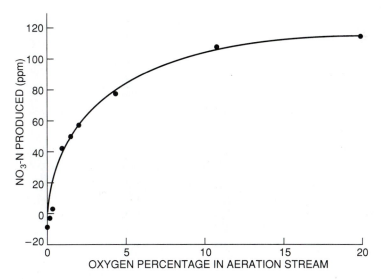

FIGURE 5.10   Production of NO_3^- in a Carrington loam incubated with added $(NH_4)_2 SO_4$ and aerated with air–N mixtures with varying O_2 percentages.   *Black,* Soil–Plant Relationships, *1957. Reprinted with permission of John Wiley & Sons, Inc., New York.*

Soil Aeration. The aerobic nitrobacteria will not produce NO_3^- in the absence of O_2 (Fig. 5.10). Maximum nitrification in this experiment occurred at 20% O_2, which is about the same O_2 concentration in the aboveground atmosphere.

This example illustrates the importance of maintaining conditions that permit rapid diffusion of gases into and out of the soil. Soils that are coarse-textured or possess good structure facilitate this rapid exchange of gases and ensure an adequate supply of O_2 for the nitrobacteria. Return of crop residues and other organic amendments will help maintain or improve soil aeration.

Soil Moisture. Nitrobacterial activity is sensitive to soil moisture. The rates of nitrification are generally highest at soil water contents equal to ⅓ bar matric suction. Water occupies about 80 to 90% of the total pore space at this matric suction. N mineralization and nitrification are reduced in wet soils with moisture contents exceeding ⅓ bar or field capacity. Between 15 bars and air dryness, mineralization and nitrification continue to decline gradually.

In a soil incubated at the wilting point of 15 bars, more than half of the NH_4^+ is nitrified in 28 days (Fig. 5.11). At 7 bars, 100% of the NH_4^+ is converted to NO_3^- at the end of 21 days. Apparently, the *Nitrobacter* are able to function well even in dry soils. Obviously, soil moisture and soil aeration are closely related in their effects on nitrification.

Temperature. Most biological reactions are influenced by temperature. The temperature coefficient, Q_{10}, of N mineralization is 2 over the range 5° to 35°C. Thus a twofold change in the mineralization rate is associated with a

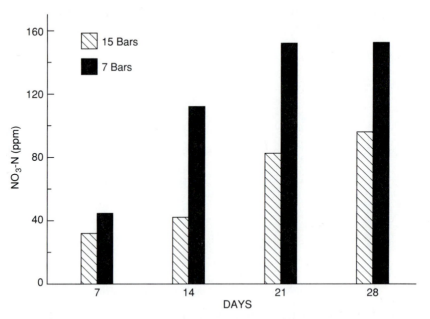

FIGURE 5.11   Effect of moisture levels near the wilting point on the nitrification of 150 ppm of N applied as $(NH_4), SO_4$ to a Millville loam and incubated at 25°C. *Justice et al., SSSA Proc., **26**:246, 1962.*

shift of 10°C within this temperature range (Fig. 5.12). Below 5°C and above 40°C the rate of N mineralization declines, with the optimum commonly lying between 30° and 35°C, although significant amounts of NO_3^- form in just 2 months when the temperature ranges between 0° and 2°C. Optimum soil temperature for nitrification of NH_4^+ to NO_3^- is between 25° and 35°C, although nitrification can occur over a wide temperature range.

Constant temperatures do not persist under most field conditions. Temperature fluctuations will determine the extent of nitrification during the winter months. Thus, if NH_4^+ fertilizer is applied in the winter in an area in which the mean temperature during the cold months is 37°F (2.8°C), there may be fluctuations in soil temperature that will allow appreciable nitrification. The occurrence of high temperatures preceding low temperatures results in greater nitrification than if the reverse situation occurs (Fig. 5.13).

A strong interactive effect of soil moisture and temperature on N mineralization exists; thus, these factors should not be considered independently. In areas with low soil temperatures and/or limited precipitation during the winter, off-season application of ammoniacal fertilizers can save the grower both time and money. It is important that winter soil temperatures be low enough to retard formation of NO_3^-, thereby reducing the risk of leaching and denitrification losses of fertilizer N. Fall applications of NH_4^+-containing or -forming fertilizers are expected to be most efficient when daily minimum air temperatures are below 40°F (4.4°C) or when soil temperatures are 50°F (10°C) or lower.

Even if temperatures are occasionally high enough to permit nitrification of fall-applied ammoniacal fertilizers, this itself is not detrimental if leaching does not occur. In many areas of the eastern and North Central states, moisture movement through the soil profile during the winter months is insuffi-

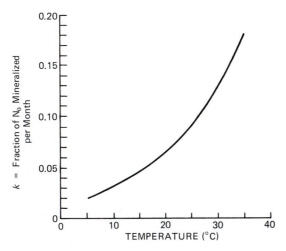

FIGURE 5.12   Fraction of N mineralized per month, k, in relation to temperature (k was estimated graphically for observed average monthly air temperatures). *Stanford et al. Agron. J., 69:303, 1977.*

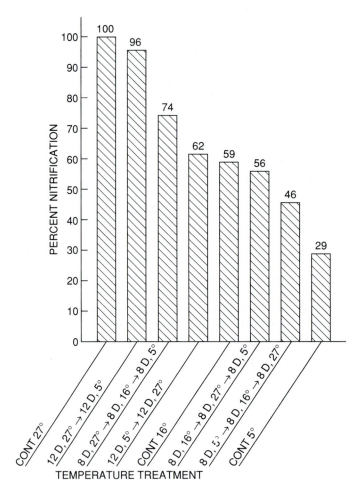

FIGURE 5.13 Nitrification as affected by
time–temperature relationships. *Chandra,* Can. J. Soil
Sci., *42:314, 1962.*

cient to remove any NO_3^- that may accumulate because of temperature
fluctuations. For example, NH_4^+ may be applied in late summer or early fall
in the Great Plains to meet the total needs of winter wheat. The same is true
farther north for spring cereal crops in most of the Prairie Provinces in
Canada. Improved positioning and distribution of N will often result from its
overwinter movement in dry regions. In other areas of the United States,
water movement through the soil profile is excessive, and NO_3^- losses will
occur. Whether or not ammoniacal fertilizers can be applied in the fall without
significant NO_3^- loss depends on local soil and weather conditions.

It is possible to apply NH_4^+ fertilizers in the fall in cool and/or dry climates
to soils of fine texture without appreciable loss by leaching, provided that
temperatures remain below 37° to 40°F (2.8 to 4.4°C). The presence of N in
the cationic form, however, does not ensure its loss against leaching. It is
necessary that the soil have a sufficiently high exchange capacity to retain the

added NH_4^+ and prevent its loss in percolating water. Sandy soils with low exchange capacities permit appreciable movement of NH_4^+ into the subsoil.

NO_3^- MOBILITY   The NO_3^- anion is very soluble in water and is not influenced by soil colloids. Consequently, it is highly mobile and subject to major leaching losses when both soil NO_3^- content and water movement are high. NO_3^- leaching is generally a major N loss mechanism from field soils in humid climates.

   NO_3^- leaching from field soils must be carefully controlled because of the serious impact that it can have on the environment. High NO_3^- (and $H_2PO_4^-$) levels in surface runoff and water percolating through the soil can pollute drinking water sources and stimulate unwanted plant and algae growth in lakes and reservoirs.

   Some of the factors that influence the magnitude of NO_3^- leaching losses are (1) rate, time, source, and method of N fertilization, (2) use of nitrification inhibitors, (3) intensity of cropping and crop uptake of N, (4) soil characteristics that affect percolation, and (5) quantity, pattern, and time of precipitation and/or supplemental irrigation.

   Examples of the movement of NO_3^- from Corn Belt soils into water draining from tile lines located several feet below the soil surface are given in Figure 5.14. NO_3^- losses by leaching occurred at all of these locations for 3 years or longer. It is important to match crop needs for N with total supplies of soil and fertilizer N so that the quantities of leachable NO_3^- are minimized.

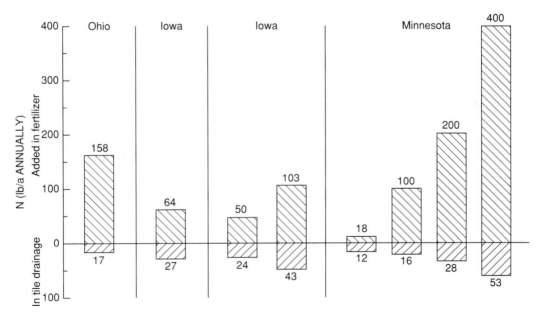

FIGURE 5.14   N added in fertilizer and lost as NO_3^- in tile drainage water in experiments in Ohio, Iowa, and Minnesota. *Council for Agricultural Science and Technology,* Agriculture and Groundwater Quality, *Report No. 103, p. 23, Ames, Iowa, 1985.*

NO_3^- previously leached to lower soil depths can be returned to upper soil horizons by capillary movement of soil water during periods of dry weather. In some cases NO_3^- will even accumulate on the soil surface.

Ammonium Fixation

Certain clay minerals, particularly vermiculite and illite, are capable of fixing NH_4^+ by a replacement of NH_4^+ for interlayer cations in the expanded lattices of clay minerals (Fig. 4.4). The fixed NH_4^+ can be replaced by cations that expand the lattice (Ca^{2+}, Mg^{2+}, Na^+, H^+) but not by those that contract it (K^+). Fixation of freshly applied NH_4^+ can occur in clay-, silt-, and sand-size particles if they contain substantial amounts of vermiculite. Coarse clay (0.2 to 2 μm) and fine silt (2 to 5 μm) are the most important fractions in fixing added NH_4^+.

The moisture content and temperature of the soil will affect the fixation of NH_4^+ (Table 5.8). The data indicate that, at least in the soil types included in this study, appreciable quantities of native fixed NH_4^+ were present and that freezing and drying increased the fixation. Alternate cycles of wetting–drying and freezing–thawing are believed to contribute to the stability of recently fixed NH_4^+.

The presence of K^+ will often restrict NH_4^+ fixation since K^+ can also fill fixation sites. Consequently, it has been suggested that K fertilization prior to NH_4^+ application is a practical way of reducing NH_4^+ fixation where it is a problem in the field.

Clay fixation of fertilizer NH_4^+ can occur relatively quickly in some soils when it is released slowly. The availability of fixed NH_4^+ ranges from negligible to relatively high. Clay fixation of NH_4^+ will provide some degree of protection against rapid nitrification and subsequent leaching, which can be important in management of fertilizer N.

There is evidence that fixed NH_4^+ is in equilibrium with exchangeable NH_4^+. It has been demonstrated that recently fixed NH_4^+ can at least partially

TABLE 5.8   Average Amounts of Native Fixed NH_4^+ and Added NH_4^+ Fixed Under Moist, Frozen, and Oven-Dry Conditions in Several Wisconsin Soils

Horizon Groupings	*Average Native Fixed NH_4^+ (meq/100 g)*	*Average Fixation of Applied NH_4^+ (meq/100 g) Under Three Conditions*		
		Moist	*Frozen*	*Oven-Dried*
Gray-brown podzolic soils				
Ap + A$_1$	0.54	0.08	0.14	0.68
A$_2$ + A$_3$	0.41	0.06	0.06	0.35
B$_1$ + B$_2$	0.60	0.15	0.25	0.82
Brunizem soils				
Ap + A$_1$	0.64	0.07	0.10	0.56
A$_3$	0.65	0.07	0.11	0.72
B$_1$ + B$_2$	0.60	0.15	0.16	0.67

SOURCE: Walsh et al., *Soil Sci.*, **89**:183 (1960). Reprinted with the permission of The Williams & Wilkins Co., Baltimore.

replace native fixed NH_4^+. Furthermore, it has been found that the nitrification inhibitor N-Serve reduces the plant availablility of recently fixed NH_4^+, which lends support to the theory that some fixed NH_4^+ is normally exchangeable and converted to readily utilizable NO_3^-.

Although the agricultural significance of NH_4^+ fixation is not generally considered to be great, it can be of importance in certain soils. A group of 10 soils from Oregon and 7 from Washington were found to fix anhydrous NH_3 in a nonexchangeable or difficult-to-exchange form. Of the NH_3 retained by these 17 soils, 1 to 8% in the surface and 2 to 31% in the subsurface horizons was fixed by the mineral fraction.

In certain soils of eastern Canada, relatively large portions of fertilizer NH_4^+ are clay fixed, often ranging from 14 to 60% in surface soil and as high as 70% in subsurface soil. Native fixed NH_4^+ is significant in many of these soils, and it can amount to about 10 to 31% of the total fixation capacity.

Gaseous Losses of N

The major losses of N from the soil are due to crop removal and leaching; however, under certain conditions, inorganic N ions can be converted to gases and lost to the atmosphere (Fig. 5.1, Table 5.1). The primary pathways of gaseous N losses are by denitrification and NH_3 volatilization (Table 5.9). The following discussion will focus on the mechanisms considered to be of greatest importance.

DENITRIFICATION   When soils become waterlogged, O_2 is excluded and anaerobic decomposition takes place. Some anaerobic organisms have the ability to obtain their O_2 from NO_2^- and NO_3^-, with the accompanying release of N_2 and N_2O. The most probable biochemical pathway leading to these losses is indicated in the following equation:

$$2HNO_3 \xrightarrow[-2H_2O]{+4H} 2HNO_2 \xrightarrow[-2H_2O]{+2H} 2NO \xrightarrow[-H_2O]{+2H} N_2O\uparrow \xrightarrow[-H_2O]{+2H} N_2\uparrow$$

Examples of the loss of NO_2^- and NO_3^- and the formation of N_2 and N_2O by denitrification in an acid and alkaline soil are shown in Figure 5.15. Only a few particular kinds of facultative aerobic bacteria are responsible for denitrification, and the active species belong to the genera *Pseudomonas*, *Bacillus*, and *Paracoccus*. Several autotrophs also involved in denitrification include *Thiobacillus denitrificans* and *T. thioparus*.

There are large populations of these denitrifying organisms in arable soils, and they are most numerous in the vicinity of plant roots. Carbonaceous exudates from actively functioning roots are believed to support the growth of denitrifying bacteria in the rhizosphere. The potential for denitrification is immense in most field soils, but conditions must arise that cause these organisms to shift from aerobic respiration to a denitrifying type of metabolism involving the use of NO_3^- as an electron acceptor in the absence of O_2.

Amounts of gaseous N lost by denitrification are variable because of the fluctuations in environmental conditions from season to season and year to year. The proportions of the two major products of denitrification, N_2 and

TABLE 5.9 Gaseous Losses of N from Soils

Form of N Lost	Source of N	General Reaction
N and NO_2 gases	A. Denitrification	$NO_3^- \rightarrow NO_2^- \rightarrow NO \rightarrow N_2O\uparrow \rightarrow N_2\uparrow$
	B. Nitrification	$NH_4^- \rightarrow NH_2OH \rightarrow$ (e.g., $H_2N_2O_2$) $\rightarrow NO_2^- \rightarrow NO_3^-$
		$\downarrow$
		N_2O
	C. Chemical reactions of nitrites with:	
	Ammonium	$NH_4^+ + NO_2^- \rightarrow N_2\uparrow + 2H_2O$
	α-Amino acids	$HNO_2 + NH_2R \rightarrow N_2\uparrow + ROH + H_2O$
		(Van Slyke reaction)
	D. Lignin	$HNO_2 + lignin \rightarrow N_2\uparrow + N_2O\uparrow + CH_3ONO$
	E. Phenol	
	Decomposition of nitrous acid with transition metal cation	$3HNO_2 \rightarrow 2NO + HNO_3 + H_2O$
		$Mn^{2+} + HNO_2 + H^+ \rightarrow Mn^{3+} + NO + H_2O$
		$Fe^{2+} + HNO_2 + H^+ \rightarrow Fe^{3+} + NO + H_2O$
NH_3	A. Fertilizers	
	anhydrous NH_3	$NH_3(liquid) \rightarrow NH_3\uparrow$ (gas)
	urea	$(NH_2)_2CO + H_2O \rightarrow 2NH_3\uparrow + CO_2$
	NH_4^+ salts	$(pH > 7)\ NH_4^+ + OH^- \rightarrow NH_3\uparrow + H_2O$
	B. Decomposition of residues and manures	Release and volatilization of NH_3

For the Phenol reaction:

$$OH\ \underset{pH > 5}{\overset{HNO_2}{\longrightarrow}}\ \underset{N=O}{\overset{OH}{\bigcirc}} \longrightarrow \underset{N-OH}{\overset{O}{\bigcirc}} \underset{}{\overset{HNO_2}{\longrightarrow}} N_2\uparrow + N_2O\uparrow + \text{organic residue}$$

SOURCE: Modified from Kurtz, ASA Spec. Publ. **38**, p. 5 (1980).

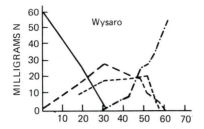
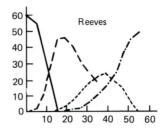

FIGURE 5.15   Sequence and magnitude of N products formed and utilized during anaerobic denitrification of Reeves loam (pH 7.8) and Wysaro clay (pH 6.1) at 30°C.   *Cooper and Smith,* Soil Sci. Soc. Am. J., ***27:659, 1963.***

N_2O, also vary, and to date no simple relationship between them has been identified. However, it has been reported that N_2 predominates, sometimes accounting for about 90% of the total N lost. The occurrence of N_2O becomes greater as soil O_2 supplies improve.

Factors Affecting Denitrification.   The magnitude and rate of denitrification are strongly influenced by several soil and environmental factors, the most important of which are the amount and nature of OM present, moisture content, aeration, soil pH, soil temperature, and level and form of inorganic N (i.e., NO_3^- vs. NH_4^+).

DECOMPOSABLE OM.   The amount of readily decomposable soil OM strongly influences denitrification in soil (Fig. 5.16). Extractable glucose C has been used as an index of the quantity of C sources associated with loss of NO_3^- during anaerobic incubation. The following equations illustrate the

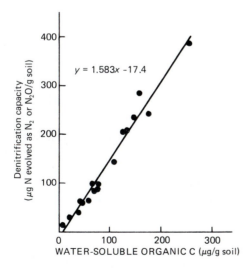

FIGURE 5.16   Relationship between denitrification capacity and water-soluble organic C.   *Burford and Bremner,* Soil Biol. Biochem., ***7:389, 1975.***

amount of available C required for microbial reduction of NO_3^- to N_2O or N_2:

$$4(CH_2O) + 4NO_3^- + 4H^+ = 4CO_2 + 2N_2O + 6H_2O$$
$$5(CH_2O) + 4NO_3^- + 4H^+ = 5CO_2 + 2N_2 + 7H_2O$$

According to these equations, 1 ppm of available C is required for the production of 1.17 ppm of N as N_2O or of 0.99 ppm of N as N_2.

Most of the basic information related to denitrification in soils has been obtained from laboratory investigations with air-dried samples stored for varying lengths of time prior to use. Drying and air storage of soils greatly increase their ability to denitrify NO_3^- under anaerobic conditions. These pretreatments substantially increase the amount of soil OM readily utilized by denitrifying microorganisms.

Under field conditions, freshly added crop residues can stimulate denitrification.

SOIL WATER CONTENT. Of the various environmental conditions, soil water content is one of the most important in determining denitrification losses. Waterlogging of soil results in rapid denitrification by impeding the diffusion of O_2 to sites of microbiological activity. The effect of increasing the degree of waterlogging on denitrification is clearly shown in Figure 5.17. English scientists demonstrated that for each 25 mm of rain that fell in a 4-week period

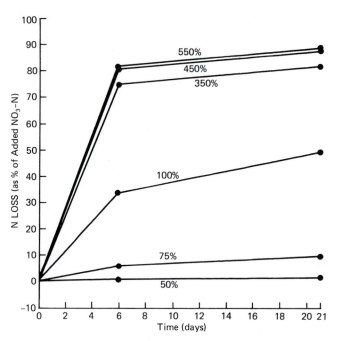

FIGURE 5.17   Effect of moisture, expressed as percent of water-holding capacity, on denitrification in soil receiving glucose. *Bremner and Shaw,* J. Agr. Sci. (Camb.), *51:40, 1958.*

following spring application of N fertilizer, about 8% of the applied N was lost. Rapid conversion of NO_3^- to N_2O and N_2 is induced when rain saturates a warm, biologically active soil. Potential denitrification losses of up to 16 kg N/ha on the first day following saturation have been measured.

Saturation of soil with water during snowmelt in the spring also is suspected of causing major denitrification losses of N. The duration of snow cover on fields and the time when the melting takes place are two factors that seem to affect denitrification associated with spring thawing.

In flooded rice soils, NO_3^--containing fertilizers are ineffective because of N lost by denitrification. Some NO_3^- is always present in such soils, however, since a portion of the NH_4^+ in the aerobic zone of the plant-soil-water system is converted to NO_3^-. When this NO_3^- diffuses into anaerobic parts of the soil, it is rapidly and completely denitrified.

AERATION. Aeration or O_2 availability affects denitrification in two apparently contrasting ways. Formation of NO_3^- and NO_2^- is dependent on an ample supply of O_2. Their denitrification, however, proceeds only when the O_2 supply is too low to meet microbiological requirements. The denitrification process can operate in seemingly well-aerated soil, presumably in anaerobic microsites where the biological O_2 demand exceeds the supply. Large losses of N by denitrification are possible with the simultaneous occurrence of a low rate of O_2 diffusion into the soil and a high respiratory demand within it.

Decreased partial pressure of O_2 will increase denitrification losses. These losses do not become appreciable, however, until the O_2 level is drastically reduced to concentrations of 10% or less. Figure 5.18 illustrates the dramatic effect of declining soil O_2 levels on evolution of N_2 gas from a soil treated with

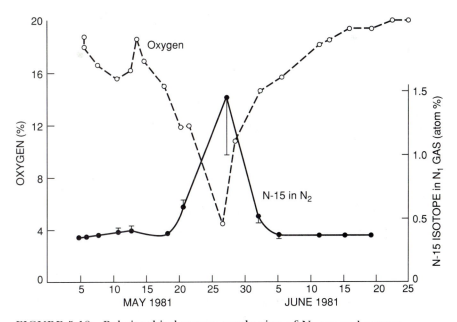

FIGURE 5.18   Relationship between production of N_2 gas and oxygen concentration in soil at a depth of 15 cm, May–June 1981. _Modified from Colburn et al., J. Soil Sci. **35(4)**:542–543, 1984._

100 kg/ha of N as $Ca(NO_3)_2$ on May 5. Release of N_2 peaked when soil O_2 dropped to 5% after 60 mm of rain. Total N loss averaged about 20% of the N applied.

SOIL pH. Soil acidity can have a marked influence on dentrification since many of the bacteria responsible for dentrification are sensitive to low pH values. As a result, many acid soils contain small populations of denitrifiers. Denitrification is negligible in soils of pH below 5.0 but very rapid in high-pH soils (Fig. 5.19).

Acidity also regulates the sequence and relative abundance of the various N gases formed during denitrification. At pH values below 6.0 to 6.5, N_2O predominates, and it frequently represents more than half of the N gases released in acid environments. Formation of NO is usually confined to low-pH conditions, usually less than about pH 5.5. NO_2 may be the first gas detectable in a neutral or slightly acid reaction, but it is reduced microbiologically, so that N_2 tends to be the principal product above pH 6. The occurrence of N_2O under acid conditions is believed to be due to its resistance to further reduction to N_2.

TEMPERATURE. Denitrification is very sensitive to soil temperature, and its rate increases rapidly in the 2°C to 25°C range. Denitrification will proceed at slightly higher rates when the temperature is increased in the range 25° to 60°C. It is inhibited by temperatures above 60°C. The rapid increase in denitrification at elevated soil temperatures suggests that thermophilic microorganisms play a major role in denitrification.

It would seem that the serious denitrification losses coinciding with spring thawing can be related to the greatly accelerated rate of denitrification when soils are quickly warmed from about 2 to 5°C to 12°C or higher.

NO_3^- LEVELS. A supply of NO_3^- and/or NO_2^- in soil is a prerequisite for denitrification. High NO_3^- concentrations increase the rate of denitrification and exert a strong influence on the ratio of N_2O to N_2 in the gases released

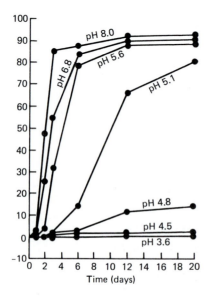

FIGURE 5.19 Effect of soil pH on denitrification. Five-gram samples of soil previously adjusted to different pH values by addition of calcium hydroxide were incubated at 25°C with 11 ml of water containing 5 mg of NO_3-N (as KNO_3) and 15 mg of C (as glucose). *Bremner and Shaw*, J. Agr. Sci. (Camb.), *51:40, 1958.*

from soil by denitrification. An example of how NO_3^-, NO_2^-, and mixtures of these two forms of N affected the accumulation of N_2O under anaerobic conditions is given in Table 5.10. Although NO_2^- inhibited reduction of N_2O to N_2, the data showed that NO_3^- had a much greater depressive action.

PRESENCE OF PLANTS. Although research results showing that plants can either encourage or impede denitrification losses are inconclusive, it is generally agreed that under field conditions denitrification rates are increased by plants because of their release of readily available C in root exudates and sloughed-off root tissues. Denitrification in most fertilized soil-cropping systems is believed to be controlled by the supply of organic C.

Plants may also increase denitrification by (1) consuming O_2 through root activity and (2) stimulating high microbial populations in the rhizosphere. On the other hand, they can restrict denitrification by (1) uptake of NO_3^-, (2) utilization of NH_4^+, which can usually be readily converted to NO_3^-, (3) reducing soil water content, with resultant improvement in the O_2 supply, and (4) directly increasing O_2 levels in the rhizospheres of certain plants that transport O_2 (e.g., paddy rice).

Agricultural and Environmental Significance of Denitrification. Fertilizer N enters a labile "pool" of soil N that must be subject to continuous denitrification losses to the atmosphere. Since the earth's atmosphere is largely N_2, while its oceans are virtually NO_3^--free, denitrification is probably the process responsible for returning N to the atmosphere, thus offsetting gains from biological N_2 fixation.

There appear to be two categories of N loss by denitrification: (1) rapid and extensive flushes associated with heavy rains, irrigation, and snow melt, and

TABLE 5.10 Effects of Different Amounts of NO_3^- and (or) NO_2^- on Amounts of N_2 Produced on Exposure of Soils to N_2O Under Anaerobic Conditions*

Soil	pH	NO_3-N Added	NO_2-N Added	N_2-N Produced in 4 Hours	Inhibition of N_2O Reduction
		———	μg/g soil	———	%
Clarion	7.2	0	0	43	—
		20	0	17	60
		10	10	20	53
		5	15	21	51
		0	20	27	37
Tama	6.6	0	0	33	—
		20	0	5	85
		10	10	8	76
		5	15	12	64

*Samples of air-dried soil (30 g) were placed in 1.2-l flasks, treated with 8 ml of H_2O, and preincubated (30°C) under He for 15 hours. They were then treated with 10 ml of H_2O or with 10 ml of H_2O containing KNO_3, and (or) $NaNO_3$ (20 μg N/g soil) and incubated (30°C) for 4 hours under He containing N_2O (1.000 μg N_2O-N/g soil).

SOURCE: Gaskell et al., SSSA J. **45**:1124–1127 (1981).

(2) continuous small losses over extended periods in anaerobic microsites. Such losses may account for 0 to > 70% of applied fertilizer N with percentages of 10 to 30% being more typical. Both the rate and the extent of denitrification losses of N under field conditions are still approximate despite much research.

Serious reductions in the effectiveness of fall-applied N for winter wheat have been observed in years when heavy winter snows persist into late spring. In spite of the fall-applied N, severe N deficiencies show up following these unusual snow conditions. The deficient areas may occupy as much as 75% of a field, with yields ranging from 5 to 10 bu/a, while adjacent normal areas may yield 35 to 50 bu/a. Leaching of N is not responsible for these shortages; the loss appears to be due to denitrification. Similar results have been obtained in field studies conducted on soils of northern Alberta, where losses of between 25 and 50% in efficiency of N fertilizers are attributed mainly to denitrification.

Contrasted with these reports of rather substantial N losses by denitrification are the findings by other scientists that total volatile loss of $N_2O + N_2$ from moderately well-drained, irrigated clay loam soil was about 2.5% of the N fertilizer applied to corn. Approximately 70% of the gaseous loss occurred as N_2O. In another study with barley, these scientists observed that only about 1% of the fertilizer N was emitted, in approximately equal proportions of N_2O and N_2. Gaseous losses of N_2O and N_2 from N fertilizer applied to corn in New York amounted to less than 3% of the amount supplied. Estimates of annual losses by denitrification in Britain range from 0.9 to 9 kg/ha N for arable land, from 11 to 29 kg/ha N for grazed grassland, and from 4 to 8 kg/ha N for cut grassland.

There is concern that increased use of N fertilizers may substantially increase emissions of N_2O from soils and thereby lead to partial destruction of the stratospheric ozone layer protecting the biosphere from biologically harmful ultraviolet radiation from the sun. Although there is evidence that denitrification of fertilizer-derived NO_3^- is responsible for emission of N_2O, contributions from NO_3^- produced by the natural transformations of soil OM and fresh crop residues have been largely ignored or discounted.

Denitrification can be useful for removal of excessive amounts of NO_3^- from irrigation water and from various wastewaters. For direct treatment of water, it may be necessary to inoculate with denitrifying organisms and provide sufficient readily mineralizable C in forms such as methanol. Where treatment systems involve disposal of contaminated wastewaters on soil, measures must also be taken to ensure that levels of mineralizable C are adequate in the soil areas being treated.

Chemical Reactions Involving NO_2^-. In addition to microbial denitrification, there are certain conditions in which losses of soil and fertilizer N can occur through chemical reactions involving NO_2^- (Table 5.9). Although NO_2^- does not usually accumulate in soil, detectable amounts occur in calcareous soils and in localized soil zones influenced by additions of NH_4^+-containing or -forming fertilizers. The conditions favoring NO_2^- buildup are discussed later.

Reaction of NO_2^- with Soil OM. Losses of N from NO_2^- by chemodenitrification increase with increasing OM content. Phenolic sites in soil OM may be responsible for the reduction of NO_2^- to N_2 and N_2O, with nitrosophenols formed as intermediates. All the N_2O or N_2 evolved is thought to come entirely from the accumulated NO_2^-, although total N losses by this mechanism are minor.

Part of the NO_2^- reacting with soil OM constituents such as lignin becomes organically bound or fixed, and it is resistant to mineralization.

Factors Favoring NO_2^- Accumulation. NO_2^- does not usually accumulate in soil, but when it does, it can adversely affect plants and microorganisms. Toxic levels of NO_2^- are generally caused by reduced *Nitrobacter* activity related to high pH and NH_4^+ levels. At 7.5 to 8.0 pH the potential for converting NH_4^+ to NO_2^- exceeds that for converting NO_2^- to NO_3^-, but at neutral pH the reverse is true. Although buildup of NO_2^- in soil is favored by high pH, its breakdown into N_2O and N_2 is restricted by high soil pH (Fig. 5.20).

NO_2^- formed in the fall may undergo chemical denitrification, even if soils freeze (Fig. 5.21). The rise in the chemodenitrification rate in frozen soil could be the result of forcing dissolved salts, including NO_2^-, into a narrow unfrozen water layer near the surface of soil colloids. This effectively increases NO_2^- concentration, which in turn enhances chemical denitrification.

Influence of Fertilizers on NO_2^- Accumulation. High rates of band applied urea, anhydrous NH_3, aqua NH_3, and $(NH_4)_2HPO_4$ fertilizers cause temporary elevation of NH_4^+ and pH, which encourages NO_2^- accumulation in the band, regardless of initial soil pH.

As might be expected, band spacing also will influence the rate of NH_4^+ disappearance and the appearance of NO_2^- and NO_3^-. Extremely large particles of urea or special application techniques that place regular-size urea particles in nests or clusters are similarly expected to intensify transitory rises in pH and NO_2^- concentration.

Diffusion and/or dilution of the NH_4^+ in fertilizer bands will restore conditions suitable for conversion of NO_2^- to NO_3^-. NO_2^- can diffuse beyond the microsite region high in pH and NH_4^+ to reach a soil environment, where the normal functioning of *Nitrobacter* will quickly convert it to NO_3^-. Small quantities of N_2O can be generated by autotrophic nitrification of NH_4^+ from fertilizers. Research results indicate that about 0.15% of the N applied is lost as N_2O. However, anhydrous NH_3 produces substantially more N_2O than other NH_4^+ sources.

In summary, the possible mechanisms for N loss involving NO_2^- include the following:

1. Decomposition of NH_4NO_2.
2. Self-decomposition of HNO_2 at pH values below 5.0, with resultant formation of NO plus NO_2^-.
3. Dissimilation of NO_2^- by reducing organic compounds.
4. Fixation of NO_2^- by soil OM and partial conversion of some NO_2^- to N_2 and N_2O.
5. Catalytic reaction of NO_2^- with metals such as Cu, Fe, and Mn.

The relative importance of these loss mechanisms will probably vary among soils and according to the N management systems practiced. From a practical

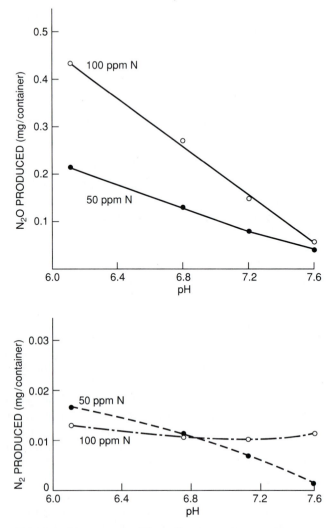

FIGURE 5.20   (Top) Effect of pH on N_2O production as
the $NaNO_2$ concentration is increased from 50 to 400
ppm N (from 1 to 8 mg per container); $t = 3$ days.
(Bottom) Effect of pH on N_2 production as the $NaNO_2$
concentration is increased from 50 to 400 ppm; $Nt = 3$
days.   *Christianson et al.*, Can. J. Soil Sci., ***59:147, 1979.***

standpoint, chemodenitrification losses under field conditions are probably
small.

VOLATILIZATION OF NH_3   Volatilization of NH_3 is a mechanism of N loss
that occurs naturally in all soils (Fig. 5.1). However, compared to NH_3 volatilization from N fertilizers, NH_3 loss from N mineralized from organic N is
relatively small. Thus, NH_3 volatilization will be discussed relative to surface
application of N fertilizers. Numerous soil, environment, and N fertilizer
management factors influence the quantity of NH_3 volatilized from fertilizers.

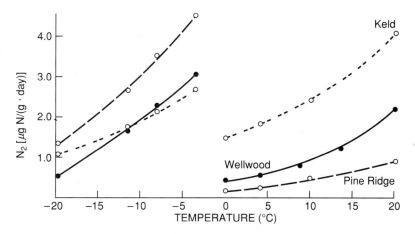

FIGURE 5.21 Chemical denitrification intensity of three soils under frozen and unfrozen conditions; 100 ppm NO_2-N. *Christianson and Cho, 23rd Annu. Manitoba Soil Sci. Meet., p. 109, Univ. of Manitoba, 1979.*

Understanding how these factors interact requires an understanding of the chemical reactions of N fertilizers with soil.

Volatilization of NH_3 ultimately depends on the quantity of NH_3 and NH_4^+ in the soil solution, which is highly dependent on pH (Fig. 5.22). The relationship is described as follows:

$$NH_4^+ \rightarrow NH_3 + H^+ \qquad (pK_a\ 9.3) \qquad (1)$$

Appreciable quantities of NH_3 appear only when soil solution pH exceeds 7.5. For example, at pH 8 and 9.3, NH_3 represents 10 and 50% of the total NH_3

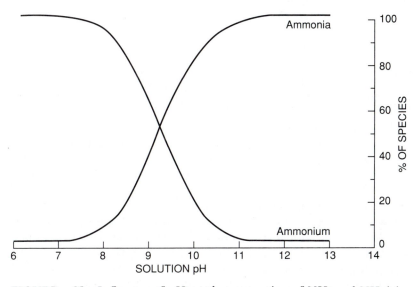

FIGURE 5.22 Influence of pH on the proportion of NH_3 and NH_4^+ in solution.

+ NH_4^+ in solution, respectively. Therefore, NH_3 loss is favored by naturally high soil pH or by reactions that temporarily raise the pH. When NH_4^+-containing fertilizers are added to acid or neutral soils, little or no NH_3 volatilization occurs because soil solution pH is not increased. Recall that soil pH will decrease slightly when the NH_4^+ is nitrified to NO_3^-. When NH_4^+-forming fertilizers (i.e., urea) are added to acid or neutral soils, solution pH around the urea granule increases during hydrolysis, as shown by the following equation:

$$CO(NH_2)_2 + H^+ + 2H_2O \rightarrow 2NH_4^+ + HCO_3 \qquad (2)$$

Solution pH increases above 7 because H^+ is consumed in the reaction; thus the $NH_4^+ - NH_3$ equilibrium shifts to the right (eq. 1) to favor NH_3 volatilization loss. Therefore, in neutral and acid soils, NH_4^+-containing fertilizers are less subject to NH_3 loss than urea and urea-containing fertilizers. Urea hydrolysis will be discussed in more detail later (pp. 165–168).

In calcareous soils, solution pH is buffered at about 7.5; thus NH_4^+-containing fertilizers may be subject to NH_3 volatilization losses. For example, when $(NH_4)_2SO_4$ is applied to a calcareous soil, it reacts according to the following equations:

$$(NH_4)_2SO_4 + 2CaCO_3 + 2H_2O \rightarrow \qquad (3)$$
$$2NH_4^+ + 2HCO_3^- + Ca^{2+} + 2OH^- + CaSO_4$$

$$NH_4^+ + HCO_3^- \rightarrow NH_3 + CO_2 + H_2O \qquad (4)$$

The solution pH is increased because of the OH^- produced. The Ca^{2+} and OH^- may further combine with $(NH_4)_2SO_4$ as follows:

$$(NH_4)_2SO_4 + Ca^{2+} + 2OH^- \rightarrow 2NH_3 + H_2O + CaSO_4 \qquad (5)$$

When all three equations (eq. 3, 4, and 5) are combined, the overall reaction can be represented as follows:

$$(NH_4)_2SO_4 + CaCO_3 \rightarrow 2NH_3 + CO_2 + H_2O + CaSO_4 \qquad (6)$$

Since the $CaSO_4$ produced is only slightly soluble, the reaction proceeds to the right and NH_3 volatilization is favored. Similar reactions occur with other NH_4^+-containing fertilizers that produce insoluble Ca precipitates [e.g., $(NH_4)_2HPO_4$]. In comparison, volatilization losses are reduced with NH_4^+-containing fertilizers that produce soluble Ca reaction products (i.e., NH_4NO_3, NH_4Cl).

Figure 5.23 demonstrates the impact that anions of various NH_4^+ salts can have on NH_3 volatilization. It is noteworthy that a rise in soil pH accompanied the formation of insoluble precipitates.

Generally NH_3 volatilization losses in calcareous soils are greater with urea fertilizers than with NH_4^+ salts, except those forming insoluble Ca precipitates. NH_3 losses also increase with increasing fertilizer rate and with liquid compared to dry N sources.

Volatilization of NH_3 is much greater with broadcast applications compared

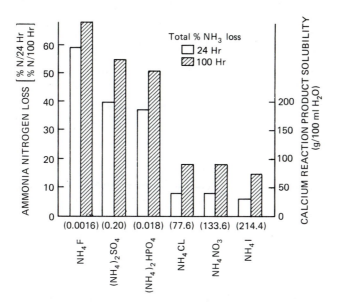

FIGURE 5.23   Total NH$_3$ loss at the end of 24 and 100
hours, as influenced by the anions of several ammonium
salts. The solubility of the Ca reaction product is shown
in parentheses above the chemical formulas. NH$_4$-N was
applied on the soil surface at the rate of 550 kg/ha of
N. *Fenn and Kissel*, Soil Sci. Soc. Am. J., *37:855, 1973.*

to subsurface or surface band methods (Table 5.11). These data show increased crop response to fertilizer N when urea-ammonium nitrate (UAN) is
band applied compared to surface broadcast. Immediate incorporation of
broadcast N will greatly reduce the NH$_3$ volatilization potential.

The BC of the soil greatly influences the quantity of NH$_3$ volatilization loss
(Fig. 5.24). Soil pH and subsequent NH$_3$ loss will be much less in a soil with
high buffering compared to one with low buffering. BC works in two ways:
(1) to resist the increase in pH with fertilizer addition and (2) to remove part
of the NH$_4^+$ and NH$_3$ from solution. Soil BC will increase with increasing
CEC and OM content.

TABLE 5.11   Mean (1979–1980) of No-till Corn Grain
Yield as Affected by N Rate and Method of Application
for UAN Solution

N Rate	Broadcast Spray	Surface Band	Incorporate Band
kg/ha	------------------- Yield, mg/ha -------------------		
90	5.61	7.40	7.87
180	6.77	8.34	8.84
270	7.18	8.69	9.66
Mean	6.52	8.14	8.46

SOURCE: Touchton and Hargrove, *Agron. J.*, **74**:825 (1982).

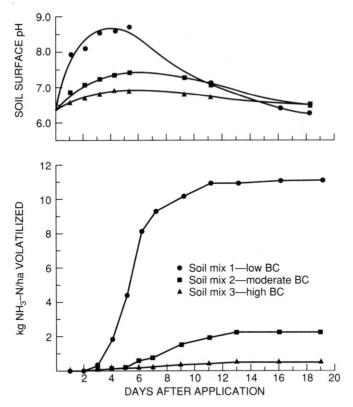

FIGURE 5.24   Soil BC effects on soil pH and NH_3 volatilization after N fertilizer application.   *Ferguson et al.,* SSSAJ, **48**:578, 1984

NH_3 losses also are influenced by environmental conditions during the reaction period of urea and NH_4^+ salts with soil. In general, volatilization increases with increasing temperature up to about 45° C, which is related to higher reaction rates and urease activity. If the surface soil is dry, the microbial and chemical reactions involved in NH_3 volatilization do not readily take place. Maximum NH_3 loss occurs when the soil surface is at or near field capacity moisture content and when slow drying conditions exist for several days. Water evaporation from the soil surface encourages NH_3 volatilization.

The presence of surface crop residues can greatly increase the potential for NH_3 volatilization. Crop residues increase NH_3 losses by maintaining wet, humid conditions at the soil surface and by reducing the quantity of urea diffusing into the soil. Crop residues also have a high urease activity. Partial incorporation of the residue can significantly reduce NH_3 losses from surface-applied urea fertilizer.

Although substantial losses of NH_3 have been measured in laboratory studies, their validity should be closely examined. It should be recognized that experimental systems will impose artificial conditions of air movement, temperature, and relative humidities quite different from those occurring naturally.

For example, NH_3 volatilization losses as high as 70% of the N fertilizer applied have been reported from laboratory studies. Field studies conducted under a wide range of conditions show that volatilization losses with $(NH_4)_2SO_4$ broadcast on a calcareous soil can be about 50% of the fertilizer N applied, while NH_3 volatilization losses can be as high as 25% with urea. In an acid soil, NH_3 losses will be greater for urea than for $(NH_4)_2SO_4$. The quantity of NH_3 loss depends on the interaction of many soil, environment, and N fertilizer factors. NH_3 volatilization losses will be greatest in coarse-textured, calcareous soils with a surface residue cover.

NH$_3$ EXCHANGE BY PLANTS   Field crops exposed to air containing normal atmospheric concentrations of NH_3 may obtain as much as 10% of their N requirement by direct absorption of NH_3. Researchers have demonstrated that corn seedlings are a natural sink for atmospheric NH_3, absorbing up to 43% of the NH_3 from air containing 1 ppm NH_3. NH_3 produced near the ground surface of grass-clover pasture can be completely absorbed by the plant cover.

The opposite reaction, one of NH_3 volatilization from plant foliage, also has been observed from a number of crops, including alfalfa pasture, corn, Rhodes grass, and winter wheat. NH_3 release was related to the stage of plant growth, with losses occurring during ripening and senescence. Others have suggested that as much as one-third of the N in a wheat crop is volatilized as NH_3 after anthesis. Losses have also been reported for rice and soybeans.

Research results indicate that both absorption and loss of NH_3 can occur in field crops. The quantity depends on the wetness of the soil surface and the extent of evaporation, which both influence the amount of NH_3 released into the air coming in contact with plant canopies.

Fertilizer N Materials

Both organic and inorganic N sources are available to supply the N required for optimum crop productivity. From a management standpoint, it is important to understand that the processes and reactions of N in the soil (nitrification, volatilization, denitrification, leaching, etc.) occur regardless of the N source used. Therefore, management practices that minimize N loss mechanisms and that increase the quantity of applied N recovered by the crop will increase production efficiency and reduce potential impacts of N use on the environment.

Organic N Forms

Before 1850 virtually all of the fertilizer N consumed in the United States was in the form of natural organic materials, primarily animal manure and legume N. Presently these materials account for only about 0.1% or less of the total N use in the United States. However, depending on the rate of manure applied, considerable quantities of N and other nutrients are added with manure. A complete discussion of fertilization with manure is found in Chapter 15. The average N concentration in natural organics is typically between 1 and 13%.

Natural organic materials at one time were thought to release their N slowly, thereby supplying the crop with N while avoiding excessive uptake and reducing potential losses by leaching and denitrification. This was shown not to be the case, however, as most of the N becomes available within the first 2 to 4 weeks after application.

Under conditions optimum for mineralization and nitrification, only about half of the total N is converted to a plant-available form at the end of 2 to 3 months. In addition, of the N mineralized during the 2 to 3 months, 80% is converted to NO_3^- at the end of the first 3 weeks. It is obvious that under warm, moist conditions, slow release of N from these materials is not affected, and the amount becoming available to the crop is but a fraction of the total amount the crop contains.

Synthetic Fertilizer N Sources

Synthetic or chemical fertilizers are the most important sources of N. Over the last 20 years, world N consumption has increased from 22 to 79 million metric tons. This trend is likely to continue into the twenty-first century.

The changes in N sources used in the United States over the last several decades are shown in Figure 5.25. Anhydrous NH_3 is the basic building block for almost all chemically derived N fertilizer materials. Most of the NH_3 in the world is produced synthetically by reacting N_2 and H_2 gases (Haber–Bosch process; see p. 119). From NH_3, many different fertilizer N compounds are manufactured. A few materials do not originate from synthetic NH_3, but they constitute only a small percentage of N fertilizers. For convenience the various N compounds are grouped into three categories: ammoniacal, nitrate, and slowly available. The composition of some common chemical sources of N is shown in Table 5.12.

AMMONIACAL SOURCES   In the United States, anhydrous NH_3, urea, and aqua NH_3 represented about 52% of total N use in 1991.

Anhydrous NH_3. Anhydrous NH_3 contains approximately 82% N, the highest amount of any N fertilizer (Table 5.12). In some respects it resembles water in its behavior, since they both have solid, liquid, and gaseous states. The great affinity of anhydrous NH_3 for water is apparent from its solubility (Table 5.13). This strong attraction of NH_3 to water is featured in its behavior in the soil. As a result of this property, NH_3 is rapidly absorbed by water in human tissue. Because NH_3 is very irritating to the eyes, lungs, and skin, safety precautions must always be taken with anhydrous NH_3 use. Safety goggles, rubber gloves, and an NH_3 gas mask are required safety equipment. A large container of water attached to the NH_3 tank also is required for washing skin and eyes exposed to NH_3. Current regulations also require that anyone applying NH_3 be licensed or certified.

Under normal atmospheric conditions, anhydrous NH_3 in an open vessel will be constantly boiling and escaping into the atmosphere. To prevent escape, it is stored under pressure. Storage in low-pressure tanks at atmospheric pressure is possible under refrigeration ($-28°F$), as is often done at large modern bulk storage facilities. When liquid NH_3 is released from a pressure

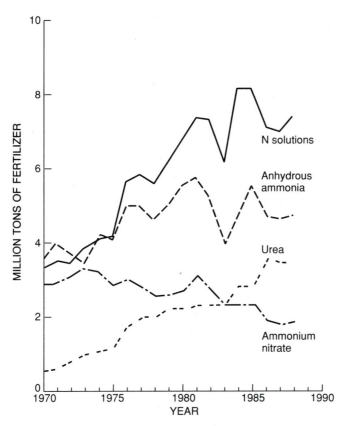

FIGURE 5.25   Changes in the most common N sources
used in the United States over the last two decades. TVA-
NFDC, *1988*.

vessel, it expands rapidly, vaporizes, and produces a white cloud of water
vapor. This cloud is formed by the condensation of water in the air surrounding the liquid NH_3 as it vaporizes.

All equipment involved in the handling of NH_3 must be designed to withstand pressure. Because NH_3 vaporizes quickly, it must be injected 3 to 8 in.
below the soil surface. Power requirements and time spent in application are
lowered by shallower application depths.

Equipment utilized for direct application of NH_3 consists of a nurse tank,
an applicator, a transfer system to fill the applicator from the nurse tank, and
a tractor to pull the applicator. Applicators may vary in size from 6-ft tool
bars with 60-gal tanks to others 55 ft or more wide equipped with tanks of
3,000-gal capacity. To reduce the number of field operations, tillage implements such as cultivators, disks, harrows, and plows are often fitted for simultaneous tillage and application of anhydrous NH_3.

Because anhydrous NH_3 is a gas at atmospheric pressure, some may be
lost to the aboveground atmosphere during and after application. Factors
associated with this loss are the physical condition of the soil during application, soil texture and moisture content, and depth and spacing of placement.
If the soil is hard or full of clods during application, the slit behind the

TABLE 5.12   Typical Composition of Some Common Chemical Sources of Fertilizer N

	Percent:						
Source	N	P_2O_5	K_2O	CaO	MgO	S	Cl
Ammonium sulfate	21.0	—	—	—	—	24.0	—
Anhydrous ammonia	82.0	—	—	—	—	—	—
Ammonium chloride	25.0–26.0	—	—	—	—	—	66
Ammonium nitrate	33.0–34.0	—	—	—	—	—	—
Ammonium nitrate-sulfate	30.0	—	—	—	—	5.0–6.0	—
Ammonium nitrate with lime (ANL)	20.5	—	—	10.0	7.0	0.6	—
Ammoniated ordinary superphosphate	4.0	16.0	—	23.0	0.5	10.0	0.3
Monoammonium phosphate	11.0	48.0–55.0	—	2.0	0.5	1.0–3.0	—
Diammonium phosphate	18.0–21.0	46.0–54.0	—	—	—	—	—
Ammonium phosphate-sulfate	13.0–16.0	20.0–39.0	—	—	—	3.0–14.0	—
Ammonium polyphosphate solution	10.0–11.0	34.0–37.0	—	—	—	—	—
Ammonium thiosulfate solution	12.0	—	—	—	—	26.0	—
Calcium nitrate	15.0	—	—	34.0	—	—	—
Potassium nitrate	13.0	—	44.0	0.5	0.5	0.2	1.2
Sodium nitrate	16.0	—	—	—	—	—	0.6
Urea	45.0–46.0	—	—	—	—	—	—
Urea-sulfate	30.0–40.0	—	—	—	—	6.0–11.0	—
Urea-ammonium nitrate (solution)	28.0–32.0	—	—	—	—	—	—
Urea-ammonium phosphate	21.0–38.0	13.0–42.0	—	—	—	—	—
Urea phosphate	17.0	43.0–44.0	—	—	—	—	—

applicator blade will not close or fill, and some NH_3 will escape to the atmosphere.

Anhydrous NH_3 convertors are often used to reduce the need for deep injection and preapplication tillage. The convertors serve as depressurization chambers for compressed anhydrous NH_3 stored in the applicator or nurse tank. Anhydrous NH_3 freezes as it expands in the convertors, separating the liquid NH_3 from the vapor and greatly reducing the pressure. The temperature of the liquid NH_3 is about $-32°C$ ($-26°F$). About 85% of the anhydrous NH_3 turns to liquid; the remainder stays in vapor form. The liquid flows by gravity through regular application equipment into the soil. Vapor collected at the top of the convertor is injected into the soil in the usual manner.

NH$_3$ RETENTION ZONES. Immediately after injection of NH_3 into soil, a localized zone high in both NH_3 and NH_4^+ is created. The horizontal, roughly circular to oval shaped zone is about 1½ to 5 in. (3 to 13 cm) in diameter, depending on the method and rate of application, spacing, CEC, soil texture, and soil moisture content. Vertical movement is normally about 2 in. (5 cm), with most of it directed toward the soil surface.

A number of temporary yet dramatic changes occur in NH_3 retention zones

TABLE 5.13   Properties of Anhydrous NH_3

Color	Colorless
Odor	Pungent, sharp
Chemical formula	NH_3
Molecular weight	17.03
Weight per gallon of liquid at 60°F	5.15 lb
Specific gravity of the gas (air = 1)	0.588
Specific gravity of the liquid (water = 1)	0.617
Boiling point	− 28°F
Vapor pressure at 0°, 68°, and 100°F	16, 110, and 198 psig, respectively
One gallon of liquid at 60°F expands to	113 standard ft^3 of vapor
One pound of liquid at 60°F expands to	22 standard ft^3 of vapor
One cubic foot of liquid at 60°F expands to	850 standard ft^3 of vapor
Solubility in water at 60.8°F	0.578 lb/lb of water

	ppm
Slight detectable odor	1
Detectable odor but no adverse effects on unprotected workers for exposure periods of up to 8 hours	25
Noticeable irritation of the eyes and nasal passages within a few minutes	100
Irritation to eyes and throat; no direct adverse effects, but exposure should be avoided	400–700
May be fatal after short exposure	2000
Convulsive coughing, respiratory spasms, strangulation, and asphyxiation	5000 +

SOURCE: Sharp, in *Agricultural Anhydrous Ammonia: Technology and Use.* Madison, Wisc.: American Society of Agronomy and Soil Science Society of America, 1966; and *Agricultural Anhydrous Ammonia Operator's Manual.* Washington, D.C.: The Fertilizer Institute, 1973.

that markedly influence the chemical, biological, and physical conditions of the soil. Some of the conditions that develop include the following:

1. Increased concentrations of NH_3 and NH_4^+, reaching levels of 1,000 to 3,000 ppm.
2. pH increases to 9 or above.
3. NO_2^- increases to 100 ppm or more.
4. Osmotic suction of soil solution exceeding 10 bar.
5. Lower populations of soil microorganisms.
6. Solubilization of OM.

Free NH_3 is extremely toxic to microorganisms, higher plants, and animals. It can readily penetrate cell membranes, while these tissue barriers are relatively impermeable to NH_4^+. There is a very close relationship between pH and concentration of free or non-ionized NH_3 and NH_4^+. Between pH 6.0 and 9.0, there is a 500-fold increase in NH_3 concentration (Fig. 5.22).

Figure 5.26 summarizes schematically the effects of pH, osmotic suction, and/or NH_4^+ concentration on the formation of NO_2^- and NO_3^-. The influence of high osmotic suction or NH_4^+ in the soil solution is primarily on

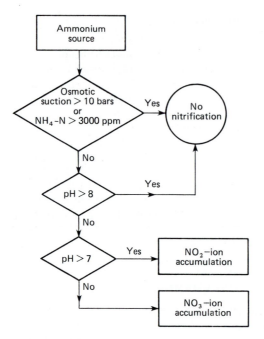

FIGURE 5.26   Schematic diagram
showing the effects of osmotic suction
and pH on nitrification.   *Wetselaar et al.,*
Plant Soil, 36:168, 1972.

Nitrosomonas bacteria. Activity is retarded by pH values above 8.0, especially in the presence of high amounts of NH_3. NO_2^- will accumulate at pH values between 7 and 8, while below pH 7, NO_3^- becomes abundant.

NH$_3$ would be lost to the atmosphere if it does not react rapidly with various organic and inorganic soil components. Possible NH_3 retention mechanisms are as follows:

1. Chemical
   a. $NH_3 + H^+ \rightarrow NH_4^+$.
   b. $NH_3 + H_2O \rightarrow NH_4^+ + OH^-$.
   c. Reaction of NH_3 with OH^- groups and tightly bound water of clay minerals.
   d. Reaction with water of hydration around the exchangeable cations on the exchange complex.
   e. Precipitation of Ca^{2+} and Mg^{2+} as carbonates in the presence of CO_2 and freeing of the exchange sites for reaction with NH_4^+.
   f. Reaction with OM.
2. Physical
   a. NH_4^+ fixation by expanding clay minerals.
   b. Adsorption by clay minerals and organic components through H bonding.

The relative importance of these mechanisms will vary from soil to soil, and they will also be influenced by environmental conditions.

The capacity of soils to retain NH_3 increases with soil moisture content, with maximum NH_3 retention occurring at or near field capacity. As soils become either drier or wetter than field capacity, they lose their ability to hold NH_3 (Figs. 5.27 and 5.28). The size of the initial NH_3 retention zone will decrease with increasing soil moisture. Diffusion of NH_3 from the injection zone is impeded by high soil moisture. The strong affinity of NH_3 for water may also be a factor.

The NH_3-holding capacity of soils increases with the clay content. NH_3 movement is greater in sandy soils than in clay soils since NH_3 can diffuse more freely in the larger pores found in coarse-textured soils. Soil textural differences in NH_3 retention are often obscured by other properties, such as the type and amount of soil minerals, OM, and moisture content.

As might be expected, NH_3 retention increases with increasing depth of injection and varies considerably, depending on soil properties and conditions. Studies have shown that an injection depth of 5 cm was effective for a silt loam soil, but placement at 10 cm was necessary in a fine, sandy loam soil. In dry soil, NH_3 loss declines with increasing placement depth (Fig. 5.27).

At a given rate, the NH_3 applied per unit volume of soil decreases with decreasing injection spacing. With the greater retention achieved with narrow spacings, there is less chance of for NH_3 loss, particularly in sandy soils with limited capacity for holding NH_3.

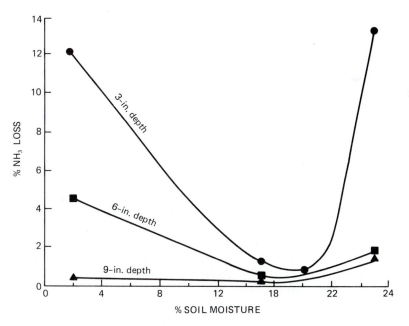

FIGURE 5.27   Losses of NH_3 from a Putnam silt loam soil as influenced by depth of application and soil moisture. Anhydrous NH_3 was applied at the rate of 100 lb/a of N in 40-in. spacings. *Stanley and Smith*, SSSAJ, **20**:*557, 1956.*

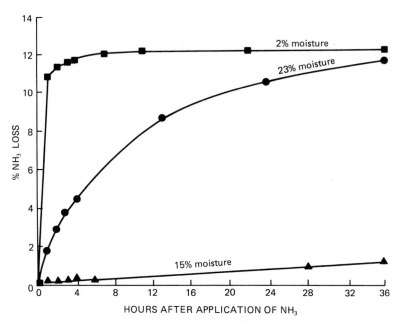

FIGURE 5.28   Rates of NH$_3$ loss from Putnam silt loam
at different soil moistures when applied 3 in. below the
surface. Anydrous NH$_3$ was applied at the rate of 100
lb/a of N in 40-in. spacings.   *Stanley and Smith*, SSSAJ,
20:557, 1956.

The OM component of soils contributes significantly to NH$_3$ retention. At
least 50% of the NH$_3$-holding capacity of soils is attributed to OM.

The nature and extent of changes in soil properties with NH$_3$ applications
can have an important bearing on crop responses to N fertilizers. The high
concentration of NH$_3$ and NH$_4{}^+$, which produces high soil pH and high
osmotic potential, results in a partial and temporary sterilization of soil within
the retention zone (Table 5.14). Bacterial activity is probably affected most by
free ammonia, while fungi are depressed by high pH. Partially sterilized
conditions at the center of the retention zone are known to persist for as long
as several weeks. A rapid recovery in the activity of bacteria and actinomycetes
generally occurs. As a consequence of reduced microbial activity, nitrification
of NH$_4{}^+$ to NO$_2{}^-$ and NO$_3{}^-$ will be reduced until conditions return to normal.

High concentrations of NH$_3$, NH$_4{}^+$, and NO$_2{}^-$ can severely damage germinating seedlings. An example of how increasing rates of anhydrous NH$_3$
reduced a stand of corn is shown in Figure 5.29. Concentrations in excess of
1,000 ppm of NH$_3$ near the seed were associated with substantial reductions
in numbers of corn plants. Deeper injection offset the harmful effects of high
rates of NH$_3$, more than extending the time for the fertilizer effects to dissi-
pate. Closer spacing of the NH$_3$ injection also would reduce the injurious
effect of large amounts of NH$_3$.

The OH$^-$ produced by the reaction of anhydrous NH$_3$ in soil will dissolve

TABLE 5.14   Numbers of Fungi, Bacteria, and
Actinomycetes in Arredondo Loamy Fine Sand in the
NH_3 Injector Row Compared with Untreated Areas

Day After Treatment	Bacteria ($\times 10^6/g$)		Actinomycetes ($\times 10^6/g$)		Fungi ($\times 10^3/g$)	
	Check	NH_3	Check	NH_3	Check	NH_3
0	2.3	0.3	1.5	0.4	20.1	5.1
3	1.3	6.3	0.9	1.0	20.2	10.4
10	3.1	9.2	0.9	2.0	15.0	9.3
24	1.3	4.2	0.5	1.3	22.7	9.2
31	4.5	3.4	0.4	1.0	20.0	13.3
38	0.9	0.9	0.3	0.7	24.0	4.0

SOURCE: Eno and Blue, *Soil Sci. Soc. Am. J.*, **18:**178 (1954).

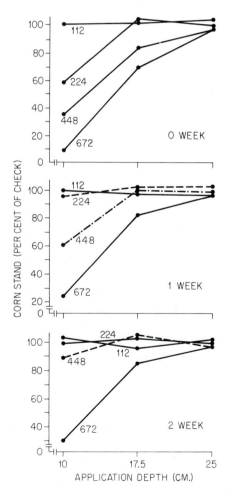

FIGURE 5.29   Effect of time, depth, and rate of NH_3 application on a stand 27 days after planting (numbers by lines are kg/ha of N).   *Colliver and Welch,* Agron. J., *62:341, 1970.*

and hydrolyze certain fractions of soil OM. Most of these effects on OM are considered to be only temporary and disappear with time. Solubilization of OM may temporarily increase the availability of nutrients associated with the organic fraction of soils. There have been reports that anhydrous NH_3 applications resulted in substantial but temporary increases in extractable P.

Contrasting beneficial and harmful effects on soil structure have been reported following the use of anhydrous NH_3. Several long-term studies have shown no difference among N sources in effects on the physical properties of soil. Impairment of soil structure is not expected to be serious or lasting except in situations involving low-OM soils, where any alteration or loss of organic matter would likely be harmful.

Aqua NH_3 (20 to 25% N). The simplest N solution is aqua NH_3, which is made by forcing compressed NH_3 gas into a closed container of water. It has a pressure of less than 10 lb/in.^{2} and usually is composed of 25 to 29% NH_3 by weight.

Transportation and delivery costs limit aqua NH_3 production to small, local fluid fertilizer plants. Aqua NH_3 is used for direct soil applications, or it can be used in the production of other liquid fertilizers.

The NH_3 will volatilize quickly at temperatures above 50°F (10°C); thus, aqua NH_3 is usually injected in soil to depths of 2 to 4 in. At temperatures over 50°F, surface applications of aqua NH_3 should be immediately incorporated into the soil. Because of its high water content, aqua NH_3 can be injected into soils regardless of their moisture content.

Nonpressure N Solutions. Of the liquid N fertilizers used for direct application in the United States, nonpressure N solutions are next to anhydrous NH_3 in popularity. Consumption of these solutions has been increasing at a faster rate than that of anhydrous NH_3. In 1991 usage of N solutions was over 7.6 million tons, equivalent to approximately 20% of the total N consumed.

N solutions are usually produced from urea, NH_4NO_3, and water, and are referred to as *UAN solutions.* The composition and properties of three of the principal nonpressure solutions are shown in Table 5.15.

Each UAN solution has a specific *salting out* temperature, which is the temperature below which dissolved salts begin to precipitate out of solution. The salting-out temperature determines the extent to which outside winter storage may be practiced and the time of year at which these solutions may be field

TABLE 5.15   Physical and Chemical Characteristics of Urea-Ammonium Nitrate Nonpressure N Solutions

	Grade (% N)		
	28	30	32
Composition by weight (%)			
Ammonium nitrate	40.1	42.2	43.3
Urea	30.0	32.7	35.4
Water	29.9	25.1	20.3
Specific gravity at 15.6°C (60°F)	1.283	1.303	1.32
Salt-out temperature, °C (°F)	−18 (+1)	−10 (+14)	−2 (+28)

SOURCE: International Fertilizer Development Center and United Nations Industrial Development Organization, *Fertilizer Manual.* Muscle Shoals, Ala.: IFDC, 1979.

applied. Salting-out temperatures vary directly with the concentration of plant nutrients in solution (Table 5.15).

Some of the main reasons for the rapid growth in use of N solutions are as follows:

1. N solutions are easier to handle and apply than other N fertilizers.
2. They can be applied more uniformly and accurately than can solid N sources.
3. Many pesticides are compatible with N solutions and both can be applied simultaneously, thus eliminating one pass across the field.
4. N solutions can be applied through various types of irrigation systems and are well suited for use in center pivots.
5. They can be safely transported in pipelines, barges, and railcars, which are less expensive and hazardous than those required for anhydrous NH_3.
6. Low-cost storage facilities can be used to store them more economically than those required for most other N products.
7. The nonpressure N solutions are excellent sources of N for use in formulation of fluid N, P, K, and S fertilizers.
8. Their cost of production is lower than that of most solid N sources.
9. They are safer to handle than anhydrous NH_3.

The one feature of nonpressure solutions that particularly stands out is their ease of handling and application. Most of the N solutions are used for direct application and are broadcast or band applied.

Application equipment for N solutions varies in size from small field sprayers pulled with a farm tractor to large, self-propelled, high-flotation applicators with tank capacities ranging from 1,200 to 2,400 gal (U.S.) and spray booms 45 to 85 ft wide. These large flotation applicators minimize soil compaction and are particularly useful under wet soil conditions.

N solutions are often added directly to grasses and small grains. When grasslands are not dormant, spray applications of UAN can cause scorching of foliage. A temporary leaf burn, usually lasting for less than a week, will sometimes occur when broadleaf herbicides and N combinations are sprayed on small grains.

Ammonium Nitrate (NH_4NO_3). Fertilizer-grade NH_4NO_3 contains between 33 and 34% N and is a more popular fertilizer in Europe than in North America. In 1991, about 1.7 million metric tons of NH_4NO_3 were used in the United States.

The NO_3^- component of NH_4NO_3 is readily available to crops, and as a consequence, this N fertilizer is widely used in the United States in cropping situations, where topdressing N to growing crops is practiced.

NH_4NO_3 has some disadvantages, which include the following:

1. It is quite hygroscopic, and care must be taken to prevent caking and physical deterioration in storage and handling.
2. There is some risk of fire or even explosions unless suitable precautions are taken. When in intimate contact with oxidizable forms of C such as fuel oil, it forms an explosive mixture that is widely used as a blasting agent.
3. It is less effective for flooded rice than urea or NH_4^+ fertilizers.
4. It is more prone to leaching and denitrification than NH_4^+ products.

Ammonium Nitrate-Sulfate. The Tennessee Valley Authority has produced two grades with analyses of 30-0-0-5(S) and 27-0-0-11(S). Of the two grades, the former, which contains about 21% $(NH_4)_2SO_4$ and 79% NH_4NO_3 was the more popular. Both grades were granular products made by neutralizing nitric and sulfuric acids with NH_3. Ammonium nitrate-sulfate is less hygroscopic than either constituent individually.

A 30-0-0-5(S) is manufactured commercially in the western United States for both bulk blending and direct application. It has been used very successfully for direct application to forage, grass seed crops, and small grains.

Ammonium Sulfate [$(NH_4)_2SO_4$]. This is one of the oldest N fertilizers, and it accounts for approximately 2.8 million tons of N worldwide. In addition to the $(NH_4)_2SO_4$ made by recovery of coke-oven gas from the steel industry, it is also a by-product from metallurgical and chemical operations.

The main advantages of $(NH_4)_2SO_4$ are low hygroscopicity and chemical stability. It is a good source of both N and S. The strongly acid-forming reaction of $(NH_4)_2SO_4$ in soil can be advantageous in high-pH soils and for acid-requiring crops. Its use can be undesirable in acidic soils already in need of liming.

The main disadvantage of $(NH_4)_2SO_4$ is its relatively low N content (21% N) and it is generally too expensive to use as a N source. It can, however, be an economical source of when transportation costs are low, when it is a relatively inexpensive by-product, and when it is used with crops requiring S.

Ammonium Phosphates. Mono-$(NH_4H_2PO_4)$ and diammonium phosphate [$(NH_4)_2HPO_4$] and ammonium phosphate-sulfate are generally considered to be more important as sources of P than of N. Therefore their properties and reactions in the soil are covered in Chapter 6.

Ammonium Chloride (NH_4Cl). Fertilizer-grade NH_4Cl usually contains 25% N. About two-thirds of the world capacity for manufacture of this material is located in Japan, with the remaining one-third situated in India. Most of it is produced by the dual-salt process, in which NH_4Cl and $(Na)_2CO_3$ are formed simultaneously. Another production method is the direct neutralization of NH_3 with HCl.

Some of its advantages include a higher N concentration than $(NH_4)_2SO_4$. Ammonium chloride is generally superior to $(NH_4)_2SO_4$ for rice, it is also suitable for a variety of other crops, including barley, wheat, corn, sorghum, fiber crops, and sugarcane. Ammonium chloride is an excellent source of both N and Cl^- for coconut, oil palm, and kiwi fruit which are Cl^--responsive crops.

Ammonium chloride is as acid-forming as $(NH_4)_2SO_4$ per unit of N, and this effect will be undesirable in acid soil, especially if liming costs are excessive. Other shortcomings are its low N analysis in comparison to urea or NH_4NO_3, and its high Cl^- content will limit its use to tolerant crops.

Ammonium Bicarbonate (NH_4HCO_3). At present this low-analysis (17%) material is the major source produced in China, accounting for more than 50% of the country's annual production of 15 million metric tons of N. It is hygroscopic and is chemically unstable resulting in often less than 30% recovery of applied N by the crop.

Urea [CO(NH₂)₂]. Although urea holds a historic position in the annals of organic chemistry, having been isolated from urine in 1773, only within the past 40 years has it received attention as a fertilizer material. Urea became the first organic compound to be synthesized from inorganic substances when in 1828 the German chemist Wöhler showed that it could be formed by heating ammonium cyanate, a compound made from NH_3 and CO_2. Commercial production of urea began in Germany in 1922 and in the United States in 1932.

Doubts about the agronomic suitability of urea impeded its adoption. Many agriculturists had reservations about using urea because of potential problems related to (1) harmful effects of biuret, an impurity normally found at low concentrations, on germination and early growth of seedlings; (2) phytotoxicity of urea to seed and seedlings due to the high concentration of NH_3 released during hydrolysis and/or the accumulation of NO_2^- during nitrification; and (3) NH_3 loss from urea exposed on the soil surface. Practical experience with urea during the past 30 years has shown that it is as good as any other fertilizer if used properly.

The satisfactory properties of the physically improved granular urea plus favorable economics of manufacturing, handling, storage, and transportation have made it a very competitive source of fertilizer N. Worldwide urea use was three times that of NH_4NO_3 in 1990. Urea is the principal form of dry fertilizer N in the United States, accounting for 14% of total N use in 1991. Its anticipated rise in popularity throughout the world is reflected in the projections of production capacities for principal dry N fertilizers, as shown in Figure 5.25.

In addition to the marked improvements in size, strength, and density of granular urea, this fertilizer material has a number of other noteworthy characteristics. These include (1) less tendency to stick and cake than NH_4NO_3, (2) lack of sensitivity to fire and explosion, and (3) less corrosiveness to handling and application equipment.

Substantial savings in handling, storage, transportation, and application costs are possible because of urea's high N content. A given volume of urea, even after allowance is made for differences in bulk density, contains considerably more N than either NH_4NO_3 or $(NH_4)_2SO_4$.

BIURET LEVELS. The concentration of biuret (NH_2-CO-NH-CO-NH_2) is of special concern because of its phytotoxicity. Biuret levels of 2% can be tolerated in most fertilizer programs. Because citrus and other crops, including pineapple, are sensitive to biuret in urea applied as a foliar spray, a product containing less than 0.25% biuret is recommended. Solutions made from urea containing 1.5% biuret are acceptable for foliar dressings on corn and soybeans. Placement of urea high in biuret near or in the seed row should be avoided.

BEHAVIOR OF UREA IN SOILS. When applied to soil, urea is hydrolyzed by the enzyme urease to NH_4^+. Depending on soil pH, the NH_4^+ may form NH_3, which can be volatilized at the soil surface, as represented in the following equations:

$$CO(NH_2)_2 + H^+ + 2H_2O \xrightarrow{\text{Urease}} 2NH_4^+ + HCO_3^-$$

$$NH_4^+ \longrightarrow NH_3 + H^+$$

The NH_3 released will react in soil the same way as NH_3 from anhydrous and aqua NH_3. There is also a marked rise in soil pH in close proximity to the urea particles. These changes in soil properties influence N transformations and crop development just as they do in the vicinity of an anhydrous NH_3 retention zone. In the presence of adequate water or other H^+ donors, NH_4^+ is retained in the soil.

Urea hydrolysis proceeds rapidly when soil conditions are also favorable for crop growth. In warm, moist soils, most of the urea will be transformed to NH_4^+ in several days.

Urease, which catalyzes the hydrolysis of urea, is abundant in soils. Large numbers of bacteria, fungi, and actinomycetes in soils possess urease. A small group of bacteria, known as *urea bacteria*, have exceptional ability to decompose urea. Activity of urease increases with the size of the soil microbial population and with OM content. The presence of fresh plant residues often results in abundant supplies of urease.

Urease activity is highest in the rhizosphere, where microbial activity is high and where it can accumulate from plant roots. Activity of rhizosphere urease varies, depending on the plant species and the season of the year. Although temperatures up to 37°C favor urease activity, hydrolysis of urea occurs at temperatures down to 2°C and lower. This evidence of urease functioning at low temperatures, combined with urea's ability to melt ice at temperatures down to 11°F (-12°C), suggests that a portion of fall- or early winter-applied urea may be converted to NH_3 or NH_4^+ prior to the spring.

The effects of soil moisture on urease activity are generally small in comparison to the influence of temperature and pH. Hydrolysis rates are probably highest at soil moisture contents optimum for plants. Soil moisture contents between 24 and 100% have little effect on the hydrolysis rate of urea.

Free NH_3 inhibits the enzymatic action of urease. Since significant concentrations of free NH_3 can occur at pH values above 7, some temporary inhibition of urease by free NH_3 occurs after the addition of urea because soil pH in the immediate vicinity of the urea source may reach values of up to 9.0. High rates of urea fertilization and its confinement to bands and other methods of localized placement could thus create conditions restrictive to the enzymatic action of urease.

MANAGEMENT OF UREA FERTILIZERS. Careful management of urea and urea-based fertilizers will reduce the potential for NH_3 volatilization losses and increase the effectiveness of urea fertilizers.

Surface applications of urea are most efficient when they are washed into the soil or applied to soils with low potential for volatilization. Conditions for best performance of surface-applied urea are cold or dry soils at the time of application and/or the occurrence of significant precipitation, probably more than 0.25 cm (0.1 in.), within the first 3 to 6 days following fertilization. Movement of soil moisture containing dissolved NH_3 and diffusion of moisture vapor to the soil surface during the drying process probably contribute to NH_3 volatilization at or near the soil surface.

Incorporation of broadcast urea into soil will minimize NH_3 losses by increasing the volume of soil to retain NH_3. Also, NH_3 not converted in the soil must diffuse over much greater distances before reaching the atmosphere. If soil and other environmental conditions appear favorable for NH_3 volatilization, deep incorporation is preferred over shallow surface tillage.

Band placement of urea will probably result in soil changes comparable to those produced by applications of anhydrous NH_3. Diffusion of urea from banded applications can be 2.5 cm (1 in.) within 2 days after its addition, while appreciable amounts of NH_4^+ can be observed at distances of 3.8 cm (1.5 in.) from the band. After dilution or dispersion of the band by moisture movement, hydrolysis begins within 3 to 4 days or less under favorable temperature conditions. These findings support suggestions made earlier that although there is little or no difference in effectiveness between surface applications of urea in solid or solution forms, volatilization of NH_3 after application of urea may decrease with increasing urea particle size. The effectiveness of fall-applied solid urea is greatly increased by large particles placed at 5-cm depths or, alternatively, nests of regular-size urea granules spaced on a grid 30 cm by 30 cm.

For example, on the average, fall-applied urea in nests or as large particles was superior to standard urea incorporated in the fall; however, urea incorporated in the spring increased yields compared to those of fall-applied urea (Table 5.16).

Similarly, point placement of large urea granules will substantially increase the efficiency of urea for rice production. The usual practice on small rice farms in developing countries is to broadcast small urea granules into the floodwater, which frequently results in only 20 to 30% utilization by the plant. Incorporating the initial dressing of urea into puddled soil can increase efficiency 35 to 44%. Hand placement of large urea particles can double the efficiency to 75 to 85%.

Placement of urea with the seed at planting is *not* recommended because of the toxic effects of free NH_3 on germinating seedlings (see reactions, p. 165). The harmful effects of urea placed in the seed row can be eliminated or greatly reduced by banding at least 2.5 cm (1 in.) directly below and/or to the side of the seed row of most crops.

The effect on germination of urea placed near small grain seeds is influenced by available soil moisture. With adequate soil moisture in medium-textured loam soils at seeding time, urea at 30 lb N/a can be used without reducing germination and crop emergence. However, in low-moisture, coarse-textured (sandy loams) soils, urea at 10 to 20 lb N/a often reduces both

TABLE 5.16   Effect on Barley Yields of Fall Applications of Urea in Large Particles or Nests

Urea Applied at Rates of 56 and 84 kg/ha N	*Average Yield Increase of Barley Grain at Six Locations During 1975–1978 (kg/ha)*
Incorporated in soil in fall	1120
Large particles or nests in fall	1790
Incorporated in soil in spring	1980

SOURCE: Nyborg et al., "Placement of urea in big pellets or nests," *in Effective Use of Nutrient Resources in Crop Production,* Proc. 1979 Alberta Soil Sci. Workshop, pp. 99–112 (1979).

germination and crop yields. Seedbed moisture is less critical in fine-textured (clay and clay loam) soils, and urea can usually be drilled in at rates of up to 30 lb N/a.

Inclusion of acid-forming fertilizers with urea may help reduce soil pH and NH_3 produced during urea hydrolysis. The advantage of acid-forming fertilizers may be reduced in calcareous soils because the $CaCO_3$ may neutralize the H^+ before urea hydrolysis occurs. Volatilization losses of N from surface applications can be reduced if solid urea is applied with solid Ca salts [i.e., $CaCl_2$, $Ca(NO_3)_2$], or with P fertilizers [i.e., $(NH_4)_2H_2PO_4$, $Ca(H_2PO_4)_2$].

To summarize, the effectiveness of urea depends on the interaction of many factors, which cause some variability in the crop response to urea. However, if managed properly, urea will be about as effective as the other N sources.

UREA-BASED FERTILIZERS. Urea phosphate [$CO(NH_2)_2H_3PO_4$] is a crystalline product formed by the reaction of urea with orthophosphoric acid. The common grade is 17-44-0, and it is primarily used to produce other grades of lower analysis. Urea phosphates with lower purity standards may be adequate for production of suspension fertilizers and for fertigation. Urea also has been combined with $(NH_4)_2HPO_4$ into a solid 28-28-0.

Granular urea sulfate with grades ranging from 40-0-0-4 to 30-0-0-13 have been produced. The N/S ratio in this product may vary from 3:1 to 7:1, thus providing enough scope to correct N and S deficiencies in most soils. Although numerous urea-based fertilizers have been produced in pilot plants, they are not commonly used in North America.

NO_3^- SOURCES In addition to NH_4NO_3, several other NO_3^--containing fertilizers, including sodium nitrate ($NaNO_3$), potassium nitrate (KNO_3), and calcium nitrate $Ca(NO_3)_2$, should be mentioned because of their importance in certain regions. These NO_3^- sources are quite soluble and thus very mobile in the soil solution. They are quickly available to crops and are susceptible to leaching under conditions of high rainfall. They also may be immobilized by soil microorganisms in the decomposition of organic residues. Like all other NO_3^- sources, they are subject to denitrification.

In general, the NO_3^- salts of Na^+, K^+, and Ca^{2+} are not acid forming as NH_4^+ fertilizers. Because the NO_3^- is often absorbed by crops more rapidly than the accompanying cation, HCO_3^- and organic anions are exuded from roots, resulting in a slightly higher soil solution pH. Prolonged use of $NaNO_3$, for example, will maintain or even raise the original soil pH.

At one time sodium nitrate [$NaNO_3$ (16% N)] was the major source of nitric acid and N fertilizer in many countries. Most of it originated in a large ore body on the Chilean coastal range, and NO_3^- production continues to be a major industry in Chile. Substantial amounts of synthetic $NaNO_3$ were once made in Europe and the United States. Its manufacture and use have declined since 1950, and now only small amounts are produced from by-product sources.

Potassium nitrate (KNO_3, 13% N) contains two essential nutrients and is manufactured by reaction of concentrated HNO_3 with KCl. Now commercially available in Chile, Israel, the United States, and certain other countries, KNO_3 finds its greatest use in fertilizers for intensively grown crops such as tomatoes, potatoes, tobacco, leafy vegetables, citrus fruits, peaches, and other crops. The properties of KNO_3 that make it attractive for these crops include moderate salt index, rapid NO_3^- uptake, favorable N/K_2O ratio, negligible Cl^- content, and alkaline reaction in soil. Its low hygroscopicity allows considerable flexibility in its use for direct application and in mixtures.

Calcium nitrate [$Ca(NO_3)_2$, 15% N] originates mainly in Europe, where it is produced by treating $CaCO_3$ with HNO_3. It is extremely hygroscopic, which detracts from its utility as a fertilizer. Except in very dry climates, $Ca(NO_3)_2$ is prone to liquefication, and storage in moisture-proof bags is usually mandatory. As with other NO_3^- fertilizer salts, its sensitization by impregnation with carbonaceous substances should be avoided.

Because of its fast-acting NO_3^- component, $Ca(NO_3)_2$ is a useful fertilizer for winter-season vegetable production. It is sometimes used in foliar sprays for celery, tomatoes, and apples. On sodium-affected soils, $Ca(NO_3)_2$ can be used as a Ca^{2+} source to displace Na^+ on the CEC.

SLOWLY AVAILABLE N COMPOUNDS In addition to N uptake by crops, commercial fertilizers are subject to many different fates in soil; thus, crop recovery of applied N seldom exceeds 60 to 70%. Development of N fertilizers with greater efficiency is desirable because their production is energy intensive and because of environmental concerns over excessive movement of N into surface waters and groundwaters, as well as the effects of gaseous N losses on the upper atmosphere.

It would be desirable to have sources capable of releasing N over an extended period, thus avoiding the need for repeated applications of conventional water-soluble products. These materials also would reduce hazards of injury to germinating crops when used at high rates with or near the seed.

The ideal product would be one that releases N in accordance with crop needs throughout the growing season. Most of the materials that have been developed for controlled N availability can be grouped as follows:

1. Substances of low water solubility that must undergo chemical and/or microbial decomposition to release plant-available N.
2. Nitrification and urease inhibitors.

Substances of Low Water Solubility Requiring Decomposition. This group is composed of chemical compounds that are only slightly soluble in water or in the soil solution. The rate of N liberation from them is related to their water solubility and to the rate of microbiological action and chemical hydrolysis. The rate of microbial and chemical decomposition is related to the rate of solution, which is dependent on solubility, particle size, and other factors.

The best-known products in this category are urea-formaldehydes or ureaforms. They are white, odorless solids containing about 38% N that are made by reacting urea with formaldehyde in the presence of a catalyst.

A typical ureaform may contain 30% of its N in forms that are soluble in cold water (25°C). N in the cold-water fraction nitrifies almost as quickly as urea. Solubility in hot boiling water is a measure of the quality of the remaining 70% of its N. At least 40% of the N insoluble in cold water should be soluble in hot water for an acceptable agronomic response; typical values are 50 to

70%. The activity index used to evaluate the suitability of urea-formaldehyde compounds is defined as follows:

$$AI = \frac{\% \ CWIN - \% \ HWIN}{\% \ CWIN} \times 100$$

where AI is the activity index, CWIN is the %N insoluble in cold water (25°C), and HWIN is the %N insoluble in hot water (98 to 100°C).

The suitability of these compounds as fertilizers is dependent on the following:

1. The *quantity* of cold-water insoluble N, which is the source of the slowly available N.
2. The *quality* of the cold-water insoluble N determined by its activity index, which reflects the rate at which the cold-water insoluble N will become available.

Consumption of ureaform in the United States is approximately 50,000 tons annually. Most of it is used in nonfarm markets for turfgrass, landscaping, ornamental use, horticulture, greenhouse crops, and as an aid in overcoming planting shock of transplanted coniferous seedlings. Several other urea-based compounds include crotonylidene diurea (CDU, 30% N), isobutylidene diurea (IBDU, 30% N), and urea-Z (UZ, 35% N). These compounds are not used to any great extent.

Sulfur-coated urea (SCU) is a controlled-release N fertilizer consisting of an S shell around each urea particle. The N concentration is between 36 and 38%, and all of it is supplied as urea. The release rate of SCU can be adjusted by changing the quantity of S used for coating. The S coating must be oxidized by soil microorganisms before the urea is exposed and subsequently hydrolyzed.

SCU has the greatest potential for use in situations where multiple applications of soluble N sources are needed during the growing season, particularly on sandy soils under high rainfall or irrigation. It is advantageous for use on sugarcane, pineapple, grass forages, turf, ornamentals, fruits such as cranberries and strawberries, and rice under intermittent or delayed flooding. SCU might also find general use under conditions where decomposition losses are significant.

Another advantage of SCU is its S content. Although S in the coating may not be sufficiently available to correct deficiencies during the first year after application, it can be an important source of plant available S in succeeding years.

Nitrification and Urease Inhibitors. Certain substances are toxic to the nitrifying bacteria and will, when added to the soil, temporarily inhibit nitrification. Many chemicals have been tested in recent years for their ability to inhibit nitrification in soils and thereby manage additions of fertilizer N more effectively. A nitrification inhibitor should ideally (1) be nontoxic to plants, other soil organisms, fish, and mammals; (2) block the conversion of NH_4^+ to NO_3^- by specifically inhibiting *Nitrosomonas* growth or activity; (3) not interfere with the transformation of NO_2 by *Nitrobacter*; (4) be able to move with the fertilizer so that it will be distributed uniformly throughout the soil; (5) be able to maintain inhibitory action for periods ranging from several weeks to months; and (6) be relatively inexpensive.

The two best-known and more generally effective compounds are N-Serve, or 2-chloro-6(trichloromethyl)pyridine, frequently referred to as *nitrapyrin* and *AM*, which is a substituted pyrimidine (2-amino-4-chloro-6-methylpyrimidine). About 5 million acres of U.S. cropland are being treated annually with nitrapyrin. Dwell or Etridiazol [5-ethoxy-3-(trichloromethyl)-1,2,4-thiadiazole] also has shown promise as a nitrification inhibitor when added to NH_3, urea, and UAN solutions.

Dicyandiamide,

$$NH_2 - \overset{\overset{\displaystyle NH}{\|}}{C} - NH - C = N \text{ or DCD,}$$

has been tested both as a nitrification inhibitor and as a slow-release N source. In Japan it has been added to mixed fertilizers and a product containing ureaform plus 10% by weight of DCD has been produced. Dicyandiamide is used in both cases to provide nitrification control and to increase the content of water-soluble N.

A fertilizer composed of urea and DCD in a 4:1 ratio is commercially available in West Germany. DCD is readily soluble and stable in anhydrous NH_3, and nitrification is effectively inhibited for up to 3 months by the addition of 15 kg/ha of DCD. Urea containing 1.4% DCD (by weight) and UAN solution with 0.8% DCD are currently available in North America for increasing effectiveness of fertilizer N.

Nitrification inhibitors will prevent N losses only when conditions suitable for unwanted transformations to NO_3^- coincide with the effective period of the inhibitor. If soil and environmental conditions are favorable for NO_3^- losses, treatment with an inhibitor will often increase fertilizer N efficiency. Generally, coarse-textured, low-OM soils are responsive to nitrification inhibitors added to N fertilizers. Although the circumstances favoring loss of NO_3-N are generally known, it is very difficult to predict accurately when and how much N will be lost. Also, protective action is unlikely when the situations for NO_3^- loss develop after the effects of the inhibitor have dissipated.

A large number of urease inhibitors have been evaluated for their ability to control urea hydrolysis in soils. All of the known inhibitors fall into three categories. The first group of substances inhibits urease activity by blocking essential sulfhydryl groups at active sites on the enzyme.

Metal ions including Ag^+, Hg^{2+}, and Cu^{2+} belong in this first category, and inhibition is inversely proportional to the solubility product of the metal–sulfide complex. Benzoquinones, quinones, and dihydric phenols, when present in quinone form, also react with the sulfhydryl groups of urease. Heterocyclic S compounds influence urease similarly by combining with sulfhydryl groups. The quinones, which are effective inhibitors of soil urease, are mild irritants to human beings. Offsetting the promising inhibitory action of the heterocyclic S compounds is their complete ineffectiveness under reducing conditions.

The second class of inhibitors are structural analogues of urea and include thiourea, methylurea, and other substituted ureas. They inhibit urease by competing for the same active site on the enzyme and therefore are ineffective at high urea concentrations.

Compounds that react with Ni in the urease molecule comprise the third group of urease inhibitors, although not all compounds that react with Ni will inhibit urease. The hydroxamic acids are specific, noncompetitive inhibitors of urease, and they are the most thoroughly studied of the known inhibitors. Caprylohydroxamic acid is the most potent member of this class.

Phenylphosphorodiamidate (PPD) also has been evaluated as a proposed urea inhibitor. Several cyclotriphosphazatriene derivatives have been patented as highly effective inhibitors of urease in soils.

Recently, ammonium thiosulfate (ATS) has been evaluated as a urease inhibitor. Laboratory results show that ATS mixed with UAN (10% by volume) will inhibit urease for about a month, depending on soil properties and environmental conditions. Results from field studies have been inconsistent in demonstrating increased N fertilizer efficiency with ATS.

Compounds that are effective urease inhibitors should be (1) effective at low concentrations, (2) relatively nontoxic to higher forms of life, (3) inexpensive, (4) compatible with urea, and (5) as mobile in soil as urea. Prospects for improving the effectiveness of urea through urease inhibition do not appear as promising as the more direct alternative of concentrating urea in the soil by using large particles or by some form of localized placement.

Summary

1. Atmospheric N is fixed in soils by various free-living and symbiotic bacteria. The amounts fixed by these organisms are generally inadequate for the sustained high yields of crops in commercial farming.

2. The various forms of soil N and their turnover in soils were discussed. Important to the immobilization and release of N are such factors as the supply of C, P, and S, soil aeration, and temperature.

3. Considerable attention was given to the reactions of inorganic N compounds in soils, especially ammonification and nitrification. The retention of the various forms of inorganic N in soils, as well as gaseous losses of this element, were covered.

4. The four general classes of N fertilizer were discussed: ammoniacal, NO_3^-, slowly available forms, and miscellaneous materials. Most of the ammoniacal forms are acid forming, and their continued use will lower soil pH values. The NO_3^- form is subject to loss by leaching. In coarse-textured soils under high rainfall, such losses can be serious.

5. Crop responses to the various forms of N were discussed. It is generally concluded that when the N alone is considered, the results from one form are as good as those from another. However, method of application, accompanying elements in the carrier, and placement in the soil may cause differences in crop responses to the various carriers. The cost per unit of N applied to the land is an important item in determining the selection of the N fertilizer.

Questions

1. What are the ways, exclusive of synthetic N fixation, by which atmospheric N is made usable to higher plants?

2. What are the various microorganisms responsible for N fixation?

3. What soil property can exercise considerable influence on the survival and growth of *Rhizobia* in soil? Describe at least two practical ways of improving the effectiveness of growth and performance of *Rhizobia*.

4. Is it possible to distinguish between effective and noneffective nodules on the roots of legume plants? Describe the location and appearance of effective nodules.

5. Can N be fixed in association with crop plants such as corn, wheat, sorghum, rice, and so on? If so, how is it accomplished? What is the contribution of the crop plant, and does this contribution influence yields?

6. Define ammonification and nitrification. What are the factors affecting these reactions in soils?

7. Do crops utilize both NH_4^+ and NO_3^-? Which is the preferred form of N? Does the stage of growth influence crop uptake of either NH_4^+ or NO_3^-?

8. Does uptake of either NH_4^+ or NO_3^- influence the chemical composition of crop plants? Discuss the major differences in both organic and inorganic constituents.

9. Identify an important soil property that can be altered by uptake of NH_4^+ and NO_3^-. Describe at least two beneficial side effects resulting from NH_4^+ uptake.

10. Is energy consumed during the reduction of NO_3^- prior to protein formation in plants? Does NH_4^+ behave similarly?

11. Are high concentrations of NH_3 and NH_4^+ detrimental to crop growth? If so, briefly describe the harmful effects.

12. If leaching losses of N are to be minimized after the fall application of ammoniacal N, soil temperatures during these winter months should not rise above what point?

13. As a general rule, is the fall application of nitrate fertilizers a sound practice? Why?

14. How, specifically, is nitrification defined? It is a two-step reaction. What are the two steps, and what organisms are responsible for each?

15. Why is nitrification important? Would you consider this phenomenon a mixed blessing? Why? Be precise.

16. What is ammonia fixation? What are the soil conditions under which it occurs? Discuss the role that K and NH_3 play in the fixation or release of each other. How important do you consider this factor to be in the overall N fertilization picture?

17. Describe the environmental and soil conditions under which you would expect to get significantly lower leaching losses of NH_4^+ N in contrast to NO_3^- N. Under what soil and environmental conditions would you expect not to get these differences?

18. You disked in a large amount of barley straw just about a week before planting fall wheat. At planting time you applied fertilizer, which supplied 20 lb of N, 20 lb of P, and 40 lb of K. The wheat germinates and shortly thereafter turns yellow. Tests show no NO_3^--N in the tissue. What is

wrong with the wheat, and why? The farmer on whose field this is observed asks you, as the county farm adviser, what to do. What is your answer?

19. In what forms may N as a gas be lost from soil? Discuss the conditions under which each form is lost, and indicate the reactions that are thought to take place.

20. How would you prevent or minimize the various gaseous losses of N?

21. Describe a set of conditions that are now believed to result in serious denitrification losses in northern soils. Are there practical ways of largely eliminating such losses? If so, identify them.

22. Classify the various forms of N fertilizers.

23. What is the single most important *original* source of fertilizer N today?

24. What is the commercial N fertilizer with the highest percentage of N?

25. What solid nitrogenous material has the highest percentage of N?

26. What developments have resulted in the great increase in the popularity of urea?

27. List the major changes in soil properties that occur in the injection zones of anhydrous and aqua NH_3. Also, identify the major changes in soil conditions within or near the retention zone of dry ammoniacal N sources.

28. What conditions favor NO_2^- accumulation? Describe the harmful effects, if any, of NO_2^- on crops.

29. Can there be serious losses of N through NO_2^- as an intermediate substance? If so, list the pathways of N loss.

30. Identify at least three ways that NO_2^- will react with the OM fraction in soils.

31. Why is it sometimes unwise to apply urea to the surface of the soil?

32. What are the practical ways of improving the agronomic effectiveness of urea?

33. What is perhaps the most important factor governing the selection of the source of fertilizer N?

34. Is urea (46% N) at a price of $225 per ton more or less expensive per pound of N than anhydrous NH_3 (82% N) at $325 per ton?

35. What precaution should be observed in applying ammoniacal N fertilizers to calcareous soils?

36. Intensive cultivation of land leads to a rapid decomposition of the OM and a more rapid rate of nitrification. Why?

37. What is the difference between N fixation and nitrification?

38. Why, specifically, do ammonia forms of N have an acidifying effect on the soil?

39. In general, what types of N solutions can be dribbled onto the soil surface without losses of N to the atmosphere?

40. Describe the ideal source of fertilizer N. What are the major products and approaches used for controlling N fertilizer availability?

41. List some of the more important nitrification inhibitors.

42. Describe the conditions in which nitrification inhibitors have the greatest potential for increasing the efficiency of N fertilizer management.

43. What is urease, and why is it important? List some of the chemicals that can control urease activity.

44. Can N fertilizers affect soil pH? What is the effect of the most widely used N products on soil reaction?

Selected References

BARBER, S. A. 1984. *Soil Nutrient Bioavailability—A Mechanistic Approach.* John Wiley & Sons, New York.

BOSWELL, F. C., J. J. MEISINGER, and N. L. CASE. 1985. Production, marketing, and use of nitrogen fertilizers, pp. 229–292. *In* O. P. ENGELSTAD (Ed.), *Fertilizer Technology and Use.* Soil Science Society of America, Madison, Wisc.

FOLLETT, R. F. (Ed.). 1989. Nitrogen management and groundwater protection. *In Developments in Agricultural and Managed-Forest Ecology #21.* Elsevier, New York.

FOLLETT, R. H., L. S. MURPHY, and R. L. DONAHUE. 1981. *Fertilizers and Soil Amendments.* Prentice-Hall, Inc., Englewood Cliffs, N.J.

HAUCK, R. D. 1984. *Nitrogen in Crop Production.* American Society of Agronomy, Madison, Wisc.

HAUCK, R. D. 1985. Slow-release and bioinhibitor-amended nitrogen fertilizers, pp. 293–322. *In* O. P. ENGELSTAD (Ed.), *Fertilizer Technology and Use.* Soil Science Society of America, Madison, Wisc.

POWER, J. F., and R. I. PAPENDICK. 1985. Organic sources of nitrogen, pp. 503–520. *In* O. P. ENGELSTAD (Ed.), *Fertilizer Technology and Use.* Soil Science Society of America, Madison, Wisc.

STEVENSON, F. J. (Ed.). 1982. *Nitrogen in Agricultural Soils.* American Society of Agronomy, Madison, Wisc.

STEVENSON, F. J. 1986. *Cycles of Soil.* John Wiley & Sons, New York.

CHAPTER 6

Soil and Fertilizer Phosphorus

Phosphorus (P) does not occur as abundantly in soils as N and K. Total concentration in surface soils varies between about 0.02 and 0.10%. The average total P content of virgin U.S. soils is lower in the humid Southeast than in the prairie and western states (Fig. 6.1). Unfortunately, the quantity of total P in soils has little or no relationship to the availability of P to plants. Although prairie soils are often high in total P, many of them are characteristically low in plant-available P. This condition is often aggravated by low soil moisture and low soil temperatures early in the growing season. Therefore, understanding the relationships and interactions of the various forms of P in soils and the numerous factors that influence P availability is essential to efficient P management.

The P Cycle

Figure 6.2 illustrates the interrelationships between the various forms of P in soils. The decrease in soil solution P concentration with absorption by plant roots is buffered by both inorganic and organic P fractions in soils. Primary and secondary P minerals dissolve to resupply $H_2PO_4^-/HPO_4^{2-}$ in solution. Inorganic P adsorbed on mineral and clay surfaces as $H_2PO_4^-$ or HPO_4^{2-} (labile inorganic P) also can desorb to buffer decreases in solution P. Numerous soil microorganisms digest plant residues containing P and produce many organic P compounds in soil. These organic P compounds can be mineralized through microbial activity to supply solution P.

Water-soluble fertilizer P applied to soil readily dissolves and increases the concentration of soil solution P. Again, the inorganic and organic P fractions can buffer the increase in solution P. In addition to P uptake by roots, solution P can be adsorbed on mineral surfaces and precipitated as secondary P minerals. Soil microbes immobilize solution P as microbial P, eventually producing readily mineralizable P compounds (labile organic P) and organic P compounds more resistant to microbial degradation.

Soil solution P is often called the *intensity factor*, while the inorganic and organic labile P fractions are collectively called the *quantity factor*. Maintenance

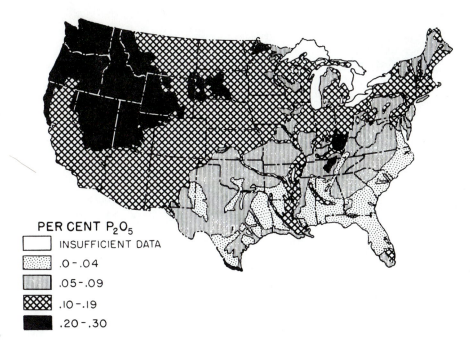

FIGURE 6.1   Phosphate content in the surface foot of soils in the United States.
Pierre and Norman, Eds., Agronomy, *Vol. 4, p. 401, 1953.*

of solution P concentration or intensity for adequate P nutrition in the plant depends on the ability of labile P (quantity) to replace soil solution P taken up by the plant. The ratio of quantity to intensity factors is called the *capacity factor,* which expresses the relative ability of the soil to buffer changes in soil solution P. Generally, the larger the capacity factor, the greater the ability to buffer solution P.

The P cycle can be simplified to the following relationship:

$$\text{Soil solution} \longleftrightarrow \text{labile P} \longleftrightarrow \text{nonlabile P}$$

where labile and nonlabile P represents both inorganic and organic fractions (Fig. 6.2). Labile P is the readily available portion of the quantity factor that exhibits a high dissociation rate and rapidly replenishes solution P. Depletion of labile P causes some nonlabile P to become labile, but at a slow rate. Thus, the quantity factor comprises both labile and nonlabile P fractions.

The interrelationships among the various P fractions are complex; however, understanding the dynamics of P transformations in soils will provide the basis for sound management of soil and fertilizer P to ensure adequate P availability to plants.

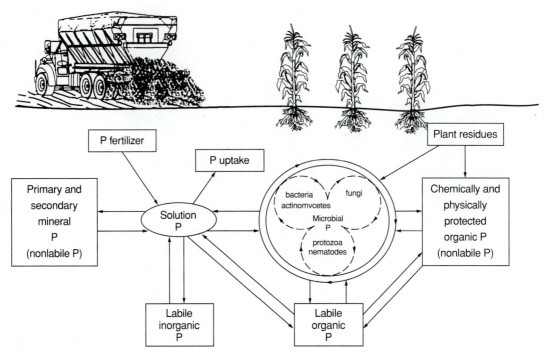

FIGURE 6.2   Schematic representation of the P cycle in soil.   *Adapted from Chauhan et al.,* Can. J. Soil Sci., **61**:373, 1981.

Soil Solution P

P is absorbed by plants largely as orthophosphate ions ($H_2PO_4^-$ and HPO_4^{2-}), which are present in the soil solution. The amount of each form present depends on soil solution pH (Fig. 6.3). At pH 7.2 there are approximately equal amounts of $H_2PO_4^-$ and HPO_4^{2-}. Below this pH, $H_2PO_4^-$ is the major form in soil solution, whereas HPO_4^{2-} is the predominant form above pH 7.2. Plant uptake of HPO_4^{2-} is much slower than with $H_2PO_4^-$. Some low molecular weight, soluble organic P compounds exist in soil solution and may be absorbed, but generally they are of minor importance.

The actively absorbing surface of plant roots is the young tissue near the root tips. Relatively high concentrations of P accumulate in root tips, followed by a zone of lesser accumulation where cells are elongating and then by a second region of higher concentration where the root hairs are developed. Rapid replenishment of soil solution P is important where roots are actively absorbing P.

The average soil solution P concentration is about 0.05 ppm and varies widely among soils. The solution P concentration required by most plants varies from 0.003 to 0.3 ppm and depends on the crop species and level of production (Table 6.1). Maximum corn grain yields may be obtained with 0.01 ppm P if the yield potential is low, but 0.05 ppm P is needed under high yield potential (Fig. 6.4). The average P requirement for wheat is slightly greater than for corn and sorghum. Soybean has a much higher requirement than corn. Optimum solution P concentrations are probably not constant for

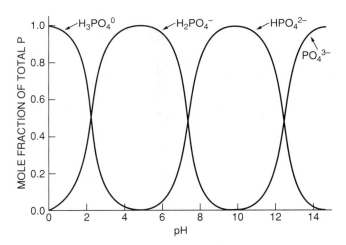

FIGURE 6.3 Influence of pH on the distribution of orthophosphate species in solution.

TABLE 6.1 Estimated Concentration of P in Soil Solution Associated with 75% and 95% of Maximum Yield of Selected Crops

| Crop | Approximate P in Soil Solution for Yield Indicated (ppm) | |
	75% of Max.	95% of Max.
Cassava	0.003	0.005
Peanuts	0.003	0.01
Corn	0.008	0.025
Wheat*	0.009	0.028
Cabbage	0.012	0.04
Potatoes	0.02	0.18
Soybeans	0.025	0.20
Tomatoes	0.05	0.20
Head lettuce	0.10	0.30

*Unpublished data of K. S. Memon, University of Hawaii.
SOURCE: Fox, *Better Crops Plant Food,* **66**:24 (Winter 1981–1982).

a specific crop. Stage of growth and plant stress caused by disease and adverse climatic conditions can greatly influence the amount of P required in soil solution.

As roots absorb P from soil solution, diffusion and mass flow transport additional P to the root surface (see Chapter 4). Mass flow in low-P soils will provide only a small portion of the requirement. For example, assume a transpiration ratio[1] of 400 and 0.2% P concentration in the crop. If the average solution concentration is 0.05 ppm P, then the quantity of P moving to the plant by mass flow is estimated by

$$\frac{400 \text{ g H}_2\text{O}}{\text{g plant}} \times \frac{100 \text{ g plant}}{0.2 \text{ g P}} \times \frac{0.05 \text{ g P}}{10^{-6} \text{ g H}_2\text{O}} \times 100 = 1\%$$

[1] Transpiration ratio = weight of water transpired per unit weight of plant.

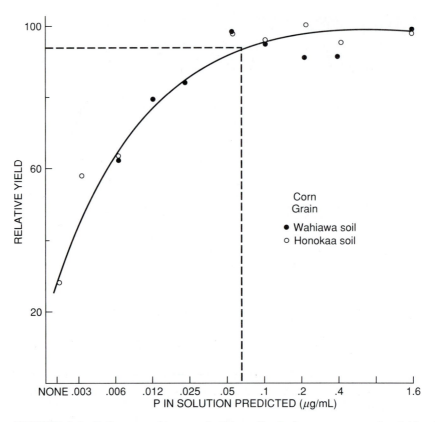

FIGURE 6.4   Influence of inorganic P in soil solution on corn grain yield.   *Fox,*
Chemistry in the Soil Environment, p. 232, ASA, 1981.

In fertilized soil with a solution concentration of 1 ppm P, mass flow contributes 20% of the total requirement.

The very high P concentrations that exist temporarily in and near fertilizer
bands are expected to encourage further P uptake by mass flow, as well as P
diffusion. For example, P concentrations between 2 and 14 ppm have been
found to occur in soil-fertilizer reaction zones.

Since mass flow contributes 20% or less to P transport to the root surface,
P diffusion is the primary mechanism of P transport (see Chapter 4). A number of soil factors can profoundly influence P diffusion. The principal factors
are (1) percentage of the soil volume that is occupied by soil water, (2) tortuosity of the diffusion path, (3) phosphate-buffering capacity of the soil, and (4)
temperature.

Organic Soil P

Organic P represents about 50% of the total P in soils and typically varies
between 15 and 80% in most soils (Table 6.2). Like OM, soil organic P decreases with depth, and the distribution with depth also varies among soils
(Fig. 6.5). The P content of soil OM ranges from about 1 to 3%. Therefore,

TABLE 6.2   Range in Organic P Levels in Various Soils

Location	Organic P	
	μg/g	% of Total P
Australia	40– 900	—
Canada	80– 710	9–54
Denmark	354	61
England	200– 920	22–74
New Zealand	120–1360	30–77
Nigeria	160–1160	—
Scotland	200– 920	22–74
Tanzania	5–1200	27–90
United States	4– 85	3–52

SOURCE: Stevenson, *Cycles of Soil*, p. 260, John Wiley & Sons, 1986.

if a soil contains 4% OM in the surface 6 in., the organic P content (assume 1% of OM) is:

$$2 \times 10^6 \text{ lb soil/a} - 6 \text{ in.} \times 0.01 \times 0.04 = 800 \text{ lb organic P/a} - 6 \text{ in.}$$

The quantity of organic P in soils generally increases with increasing organic C and/or N; however, the C/P and N/P ratios are more variable among soils that the C/N ratio. Soils also have been characterized by their C/N/P/S ratio, which also is variable among soils (Table 6.3). On the average, the C/N/P/S ratio in soil is 140:10:1.3:1.3.

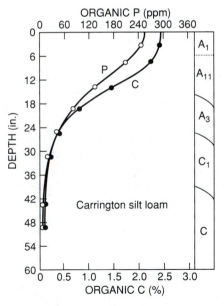

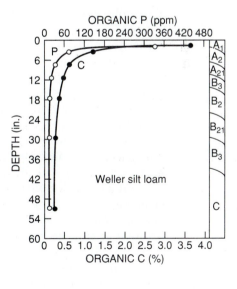

FIGURE 6.5   Distribution of organic P and C with depth in two Mollisol soils in Iowa.   *From Stevenson,* Cycles of Soil, *p. 261, John Wiley & Sons, 1986.*

TABLE 6.3   Organic C, N, P, and S Ratios in
Selected Soils

Location	Number of Soils	C/N/P/S
Iowa	6	110:10:1.4:1.3
Brazil	6	194:10:1.2:1.6
Scotland*		
Calcareous	10	113:10:1.3:1.3
Noncalcareous	40	147:10:2.5:1.4
New Zealand†	22	140:10:2.1:2.1
India	9	144:10:1.9:1.8

*Values for S given as total S.
†Values for subsurface layers (35–53 cm) were 105:10:3.5:1.1.
SOURCE: Stevenson, *Cycles of Soil*, p. 262, John Wiley & Sons, 1986.

Many of the organic P compounds in soils have not been characterized. Most organic P compounds are esters of orthophosphoric acid ($H_2PO_4^-$) and have been identified primarily as inositol phosphates, phospholipids, and nucleic acids. The approximate proportion of these compounds in total organic P is as follows:

Inositol phosphates	10–50%
Phospholipids	1–5%
Nucleic acids	0.2–2.5%

Thus, on the average, only about 50% of organic P compounds in soils are known.

Inositol phosphates represent a series of phosphate esters ranging from monophosphates up to hexaphosphate. Phytic acid (myoinositol hexaphosphate) has six orthophosphate ($H_2PO_4^-$) groups attached to each C atom in the benzene ring (Fig. 6.6). Successive replacement of $H_2PO_4^-$ with OH^- represents the other five phosphate esters. For example, the pentaphosphate ester has five $H_2PO_4^-$ groups and one OH^-. Inositol hexaphosphate is the most common phosphate ester and comprises as much as 50% of total organic P in soils (Table 6.2). Most of the inositol phosphates in soils are products of microbial activity and the degradation of plant residues.

Inositol hexaphosphate forms strong complexes with proteins. It also forms insoluble salts with Fe^{3+} and Al^{3+} under acid conditions and with Ca^{2+} in alkaline soils. In these various complexes, inositol hexaphosphate is more resistant to enzyme attack, which may explain its predominance in soils compared to other phosphate esters.

Clay minerals such as montmorillonite and Fe and Al oxides will strongly adsorb inositol hexaphosphate. The degree of adsorption decreases with declining numbers of phosphate groups. Adsorption of inositol penta- and hexaphosphates can occur at the same active sites involved in sorption of inorganic orthophosphate ions.

Nucleic acids occur in all living cells and are produced during the decomposition of residues by soil microorganisms. Two distinct forms of nucleic acids,

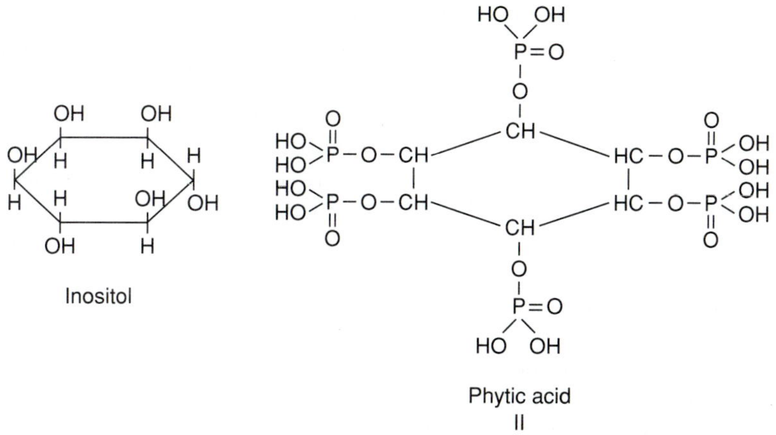

FIGURE 6.6 Chemical structure of inositol and inositol phosphate (phytic acid).

ribonucleic acid (RNA) and deoxyribonucleic acid (DNA), are released into soil in greater quantities than inositol phosphates, and they are broken down more quickly. Therefore, nucleic acids represent only a small portion of total organic P in soils, approximately 2.5% or less. Organic P compounds in soils called *phospholipids* are insoluble in water but are readily utilized and synthesized by soil microorganisms. Some of the most common phospholipids are derivatives of glycerol. The rate of release of phospholipids from organic sources in soils is rapid. Thus, the phospholipid content in soils is also low, about 5% or less of total organic P.

The remaining organic P compounds in soils are believed to originate from microorganisms, especially from bacterial cell walls, which are known to contain a number of very stable esters.

ORGANIC P TURNOVER IN SOILS In general, P mineralization and immobilization are similar to those of N in that both reactions or processes occur simultaneously in soils and can be depicted as follows:

$$\text{Organic P} \underset{\text{Immobilization}}{\overset{\text{Mineralization}}{\rightleftarrows}} \text{inorganic P}(H_2PO_4^-/HPO_4^{2-})$$

The initial source of soil organic P is plant and animal residues, which are degraded by microorganisms to produce other organic compounds and release inorganic P (Fig. 6.2). Some organic P is resistant to microbial degradation and is most likely associated with humic acids. The inositol phosphates, nucleic acids, and phospholipids also can be mineralized in soils by a reaction catalyzed by the enzyme phosphatase.

$$R{-}O{-}\overset{\overset{\displaystyle O}{\|}}{\underset{\underset{\displaystyle O^-}{|}}{P}}{-}O^- + H_2O \xrightarrow[\text{Enzyme}]{\text{Phosphatase}} H{-}O{-}\overset{\overset{\displaystyle O}{\|}}{\underset{\underset{\displaystyle O^-}{|}}{P}}{-}O^- + ROH$$

Phosphatase enzymes play a major role in the mineralization of organic phosphates in soil. There exists in soil a wide range of microorganisms that through their phosphatase activities are capable of dephosphorylating (mineralizing) all known organic phosphates of plant origin. Phosphatase activity of a soil is due to the combined functioning of the soil microflora and any free enzymes present. Phosphatase activity in soils increases with increasing organic C content but also is affected by pH, moisture, temperature, and other factors. Evidence of organic P mineralization can be provided by measuring changes in soil organic P during the growing season (Fig. 6.7). Organic P content decreases with crop growth and development and increases again after harvest.

Many factors influence the total quantity of P mineralized in soil. In most soils, total organic P is highly correlated with soil organic C; thus, P mineralization increases with increasing total organic P (Fig. 6.8). In contrast, the quantity of inorganic P immobilized is inversely related to soil organic P (Fig. 6.9). These data show that as the ratio of soil organic C/P increases (i.e., decreasing organic P), P immobilization increases.

The C/P ratio of the decomposing residues regulates the predominance of P mineralization over immobilization, just as the C/N ratio regulates N mineralization/immobilization. The following guidelines have been suggested:

C/P Ratio	Mineralization/Immobilization
> 200	Net mineralization of organic P
200–300	No gain/loss of inorganic P
< 300	Net immobilization of inorganic P

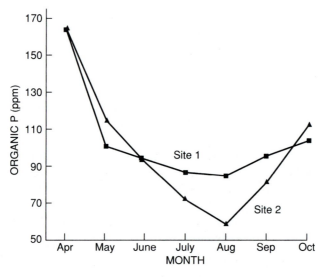

FIGURE 6.7 Changes in soil organic P over a cropping season for two locations. *Dormaar*. Can. J. Soil. Sci., *52:107, 1972.*

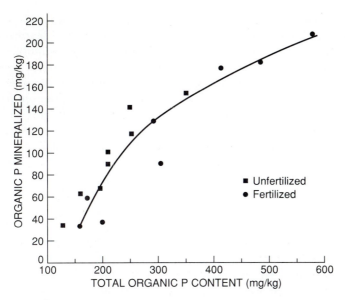

FIGURE 6.8   Mineralization of organic P in soil as
influenced by total organic P.   *Sharpley*, SSSAJ, *49:907,*
1985.

Expressed as %P in the degrading residue, net P immobilization occurs
when %P < 0.2% and net mineralization occurs with > 0.3% P. When residues
are added to soil, net P immobilization occurs during the early stages of
decomposition followed by net P mineralization as the C/P ratio of the residue
decreases. Thus, P mineralization/immobilization processes are similar to
those described for N (Fig. 5.5).

The N/P ratio also is related to mineralization and immobilization of P, such
that the decreased supply of one results in the increased mineralization of the
other. Thus, if N were limiting, inorganic P might accumulate in the soil and
the formation of soil OM would be inhibited. The addition of fertilizer N

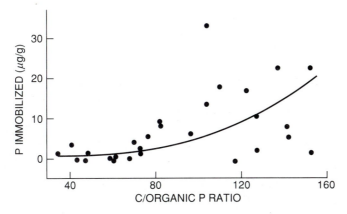

FIGURE 6.9   Relationship between inorganic P
immobilization and C/P ratio in the soil.   *Enwezor*, Soil
Sci., *103:62, 1967.*

under such conditions could result in the immobilization not only of some of the accumulated inorganic P but also of some of the added fertilizer N.

Other factors affecting the quantity of P mineralization/immobilization are temperature, moisture, aeration, pH, cultivation intensity, and P fertilization. The environmental effects are similar to those described for N mineralization/ immobilization, since both are microbial processes (see Chapter 5). Several studies illustrated that organic P mineralization increased with soil pH, but organic C and N mineralization did not, as measured by increasing C/P and C/N ratios with increasing soil pH. The effect of soil pH on P mineralized as a function of increasing organic C is illustrated in Figure 6.10. The pH influence is related to (1) OH^- competing with $H_2PO_4^-$ or HPO_4^{2-} for bonding sites, (2) greater microbial activity at neutral pH levels, and (3) increased precipitation of Ca-P minerals at pH levels above 7.

Inorganic fertilizer P can be immobilized to organic P by microorganisms. The quantity of fertilizer P immobilized varies widely, with values of 25 to 100% of applied fertilizer P reported. On some soils, continued fertilizer P applications can increase the organic P content and subsequently increase P mineralization. Several studies reported 3 to 10 lb/a/yr increases in organic P with continued heavy P fertilization. In general, organic P will accumulate with P fertilization when C and N are available in quantities relative to the C/N/P ratio of the soil OM. Inorganic P will likely accumulate if C and N are limiting.

Further evidence of organic P mineralization is clearly demonstrated by the decrease in organic P with continued cultivation. When virgin soils are brought under cultivation, the OM content decreases, as described in Chapter 5. With this decrease in OM there is an initial increase in extractable inorganic P, but

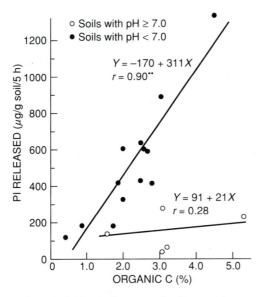

FIGURE 6.10   Influence of soil organic C and pH on P mineralization.
Tabatabai and Dick, Soil Biol. Biochem., *11:655, 1979.*

within a few years this also decreases. In the northern Great Plains, organic C and P decreased an average of 38 and 21% after 60 to 70 years of cultivation, respectively, with no decrease in inorganic P (Table 6.4). Studies in the Midwest showed that after 25 years of cultivation, organic P mineralization reduced organic P 24% in the surface soil, which was less than the loss in organic C and N. In the southern Plains states, losses of organic P are greater because of increased soil temperature. Generally in temperate regions, the decline in organic P with cultivation is less than that of organic C and N because of fewer loss mechanisms for P, resulting in comparatively greater conservation of organic P. Under higher temperature and moisture regimes, equal losses of organic C, N, and P have been observed.

Measuring organic P cycling in soils is more difficult than for N because inorganic P produced through mineralization can be removed from solution by (1) P adsorption to clay and other mineral surfaces and (2) P precipitation as secondary Al^{3+}, Fe^{3+}, or Ca-P minerals. However, the quantity of P mineralized during a growing season has been estimated in numerous studies and, as expected, varies widely among soils (Table 6.5). These data illustrate the large quantities of organic P mineralized in tropical, high-temperature environments. In most conventional cropping systems in the Midwest, organic P mineralization probably contributes about 4 to 10 lb/a/yr of plant-available P. In a northeastern U.S. study, similar quantities of P mineralization, approximately 4.5 lb/a/yr, were observed in both native and cultivated soils; however, only 31% of the organic P mineralized was taken up by the crop in the cultivated system compared to 83% in the previously uncropped soil.

Much has yet to be learned about the immobilization and mineralization of

TABLE 6.4   Loss of Organic P with Continued Cultivation in Three Soils of the Canadian Prairie

Soil Association	Native Prairie	60–70 Years of Cultivation	C or P Loss, %
	--------------------------------- Content ---------------------------------		
Blaine Lake			
Organic C, mg/g	47.9 ± 10.2	32.8 ± 5.2	32
Total P, μg/g*	823 ± 92	724 ± 53	12
Organic P, μg/g	645 ± 125	528 ± 54	18
Inorganic P, μg/g	178 ± 47	196 ± 8	NS‡
Sutherland			
Organic C, mg/g†	37.7 ± 6.5	23.7 ± 1.8	37
Total P, μg/g	756 ± 28	661 ± 31	12
Organic P, μg/g	492 ± 5.2	407 ± 30	17
Inorganic P, μg/g	256 ± 44	254 ± 19	NS
Bradwell			
Organic C, mg/g	32.2 ± 8	17.4 ± 1.6	46
Total P, μg/g	746 ± 101	527 ± 15	29
Organic P, μg/g	446 ± 46	315 ± 21	29
Inorganic P, μg/g	300 ± 84	212 ± 46	29

*μg/g = ppm.
†mg/g = ppm × 10^3.
‡NS = not significant.
SOURCE: Tiessen et al., *Agron. J,* **74**:831 (1982).

TABLE 6.5   Quantities of Organic P Mineralized in a Growing Season for Several Soils

Location	Land Use	Soil	Study Period	Organic P Mineralized	Percent Organic P Mineralized/Year
			yr	kg/ha/yr	% yr
	Slightly weathered, temperate soils				
Australia	Grass	—	4	6	4
Australia	Wheat	—	55	0.3	0.3
Canada	Wheat	Silt loam	90	7	0.4
		Sandy loam	65	5	0.3
England	Grassland	Silt and sandy loam	1	7–40	1.3–4.4
	Arable	Silt and sandy loam	1	2–11	0.5–1.7
	Woodland	Silt loam	1	22	2.8
England	Cereal crop			0.5–8.5	
England	Deciduous forest	Brown earth	1	9	1.2
	Grass	Brown earth	1	14	1.0
Iowa	Row crops	Clay loam	80	9	0.7
Maine	Potatoes	Silt loam	50	6	0.9
Minnesota	Alfalfa	Silty clay loam	60	12	1.2
Mississippi	Cotton	Silt loam	60	5	1.0
	Soybean	Silty clay loam	40	8	1.0
New Mexico	Row crops	Loam	30	2	0.4
Texas	Sorghum	Clay	60	7	1.0
	Weathered, tropical soils				
Honduras	Corn	Clay	2	6–27	5.9–11.9
		Clay	2	10–22	6.9–8.8
Nigeria	Bush	Sandy loam	1	123	24
	Cocoa	Sandy loam	1	91	28
Ghana					
—Cleared Shaded		Ochrosol fine	3	141	6
—Tropical half shaded		Sandy loam	3	336	17
—Rainforest exposed			3	396	17

SOURCE: Stewart and Sharpley, *SSSA Spec. Publ. No. 19*, p. 111, 1987.

P in soils and its relation to C, N, and S cycling. These reactions could have an important bearing on fertilizer practices, but the extent of their influence is not known. It is probably safe to assume the following:

1. If adequate amounts of N, P, and S are added to soils to which crop residues are returned, some of the added nutrients may be immobilized in fairly stable organic C compounds.
2. Continued cropping of soils without the addition of supplemental N, P, and S will result in the mineralization of these elements and their subsequent depletion in soils.
3. If N, P, or S is present in insufficient amounts, the synthesis of soil OM may be reduced. All of these reactions presuppose the presence of adequate C and conditions conducive to the synthesis and breakdown of soil organic matter.

In spite of the fact that in some mineral soils one-half to two-thirds of the total P is organic, this important fraction is usually ignored in measurements of available soil P. Most soil extractants only measure a proportion of the labile inorganic P pool.

Inorganic Soil P

As organic P is mineralized to inorganic P or as fertilizer P is added to soil, the inorganic P in solution not absorbed by plant roots or immobilized by microorganisms can be adsorbed to mineral surfaces (labile P) or precipitated as secondary P compounds (Fig. 6.2). Terms frequently used to describe surface adsorption and precipitation reactions collectively are *P fixation* or *retention*. The nature and extent of inorganic P fixation or retention reactions, discussed in the following sections, depend on many factors, the most important of which is soil pH. The simplified relationship between pH and P retention is illustrated in Figure 6.11. In acid soils, inorganic P precipitates as Fe/Al-P secondary minerals and/or is adsorbed to surfaces of Fe/Al oxide and clay minerals. In neutral and calcareous soils, inorganic P precipitates as Ca-P secondary minerals and/or is adsorbed to surfaces of clay minerals and $CaCO_3$.

There is considerable evidence supporting a wide range of adsorption and precipitation mechanisms in soils, with no consensus as to the relative magnitude of their contributions. The most probable conclusion is that P retention is a continuous sequence of precipitation and adsorption. With low-solution P concentrations, adsorption probably dominates, while precipitation reactions proceed when the concentration of P and associated cations in the soil solution exceeds that of the solubility product (K_{sp}) of the mineral. Where water-soluble fertilizers are applied to soils, the soil solution concentrations of P and accompanying cations are initially very high. Initially, P precipitation reactions occur because the solution P and accompanying cation concentrations exceed

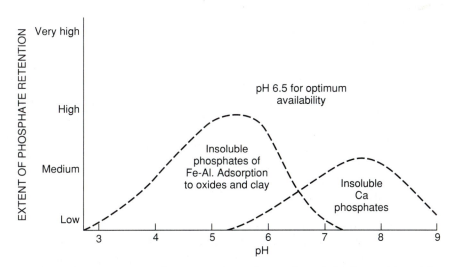

FIGURE 6.11   Soil pH effect on P adsorption and precipitation.   *Adapted from Stevenson,* Cycles of Soil, *p. 250, John Wiley & Sons, 1986.*

a specific mineral solubility. As the solution P level declines, P adsorption to reactive surface sites occurs. It is probable that both reactions proceed, to some extent, immediately following fertilizer P addition. Regardless of the relative contributions of adsorption and precipitation reactions, understanding these P fixation processes is important for optimum P nutrition and efficient management of fertilizer P.

PRIMARY AND SECONDARY MINERAL P SOLUBILITY   As discussed before, the form of inorganic P in soil solution ($H_2PO_4^-$ or HPO_4^{2-}) depends on solution pH. Since plants absorb both P forms, the concentration of P in solution and the ability of the various P compounds to resupply solution P in soils are more important to P availability to plants. The P cycle (Fig. 6.2) illustrates that solution P levels are buffered by the release of adsorbed P from mineral surfaces (labile P), mineralization of organic P, and dissolution of solid P minerals. Ultimately, the P concentration in solution is controlled by the solubility of inorganic P minerals in soil. The most common P minerals found in acid soils are Al- and Fe-P minerals, while Ca-P minerals predominate in neutral and calcareous soils (Table 6.6).

Mineral solubility represents the concentration of ions contained in the mineral that is maintained in solution. Each P mineral will support specific ion concentrations that depend on the solubility product of the mineral. For example, $FePO_4 \cdot 2H_2O$ will dissolve according to the following equation:

$$FePO_4 \cdot 2H_2O + H_2O \leftrightarrows H_2PO_4^- + H^+ + Fe(OH)_3 \qquad (1)$$

As solution $H_2PO_4^-$ decreases with P uptake by the crop, strengite in the soil can dissolve to resupply or maintain the $H_2PO_4^-$ concentration in solution. This reaction also shows that as H^+ increases (decreasing pH), the $H_2PO_4^-$ concentration will decrease. Therefore, the specific P minerals present in soil and the concentration of solution P supported by these minerals are highly dependent on solution pH.

The relationship between the solubility of the various P minerals and soil solution pH is shown in Figure 6.12. In this diagram, the y-axis represents the concentration of $H_2PO_4^-$ or HPO_4^{2-} in soil solution. From Figure 6.3, $H_2PO_4^-$ is the predominant ion below pH 7.2 and HPO_4^{2-} is the predominant ion

TABLE 6.6   Common P Minerals Found in Acid, Neutral, and Calcareous Soils

Acid soils*	
Variscite	$AlPO_4 \cdot 2H_2O$
Strengite	$FePO_4 \cdot 2H_2O$
Neutral and calcareous soils	
Dicalcium phosphate dihydrate (DCPD)	$CaHPO_4 \cdot 2H_2O$
Dicalcium phosphate (DCP)	$CaHPO_4$
Octacalcium phosphate (OCP)	$Ca_4H(PO_4)_3 \cdot 2.5H_2O$
ß-tricalcium phosphate (ßTCP)	$Ca_3(PO_4)_2$
Hydroxyapatite (HA)	$Ca_5(PO_4)_3OH$
Fluorapatite (FA)	$Ca_5(PO_4)_3F$

*Minerals are listed in order of decreasing solubility.

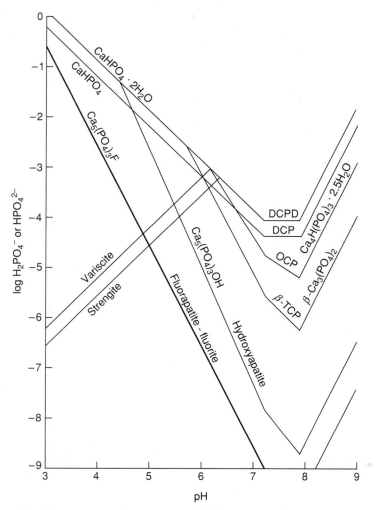

FIGURE 6.12   Solubility of Ca, Al, and Fe phosphate
minerals in soils.   *Lindsay,* Chemical Equilibria in Soils,
Wiley Interscience, p. 181, 1979

above pH 7.2. The *x*-axis expresses the soil solution pH. At pH 4.5,
$AlPO_4 \cdot 2H_2O$ and/or $FePO_4 \cdot 2H_2O$ control the concentration of $H_2PO_4^-$ in
solution. Increasing pH increases the $H_2PO_4^-$ concentration in solution because the minerals dissolve according to Eq. 1, which also is depicted in the
diagram as a positive slope. Increasing P availability is often observed when
acid soils are limed. Also, hydroxyapatite or fluorapatite can be used as a
fertilizer in very low pH soils, as shown by the their high solubility at low pH
(Fig. 6.12). In contrast, they cannot be used to supply plant-available P in
neutral or calcareous soils because of their low solubility.

   As pH is increased, the variscite and strengite solubility lines intersect several
lines representing the solubility of Ca-P minerals. For example, at pH 4.8,
both strengite and fluorapatite can exist in soil, supporting $10^{-4.5}$ *M* $H_2PO_4^-$
in solution. Between pH 6.0 and 6.5, the Al-P and Fe-P minerals can coexist

with β-tricalcium phosphate (β-TCP), octacalcium phosphate (OCP), dical-cium phosphate (DCP), and dicalcium phosphate dihydrate (DCPD) at about $10^{-3.2} M$ $H_2PO_4^-$ in solution, which is about the highest solution P concentra-tion that can exist in most unfertilized soils.

The solubility of Ca-P minerals is affected much differently than that of the Al-P and Fe-P minerals, as shown by the negative slopes of the Ca-P lines in Figure 6.12. As pH increases, the $H_2PO_4^-$ concentration decreases as the Ca-P precipitates, as described by the following equation for DCPD:

$$CaHPO_4 \cdot 2H_2O + H^+ \leftrightarrows Ca^{+2} + H_2PO_4^- + 2H_2O \qquad (2)$$

For example, assume that a soil contains β-TCP and the pH is 7.0. If pH decreases, the concentration of $H_2PO_4^-$ increases until about pH 6.0 as β-TCP dissolves. If pH decreases below 6.0, strengite and/or variscite will precip-itate and the $H_2PO_4^-$ concentration in solution will decrease.

The slopes of the Ca-P lines in Figure 6.12 change above pH 7.2 because HPO_4^{2-} becomes the dominant species in solution compared to $H_2PO_4^-$. Thus, the mineral solubility lines represent only HPO_4^{2-} above pH 7.2. Above pH 7.8, the slope of the Ca-P solubility lines becomes positive, which means that as pH increases above 7.8, HPO_4^{2-} concentration increases. The change in solubility is due to the competing reaction of $CaCO_3$ solubility given by

$$
\begin{aligned}
CaHPO_4 \cdot 2H_2O + H^+ &\rightarrow Ca^{2+} + H_2PO_4^- + 2H_2O \\
\underline{Ca^{2+} + CO_2 + H_2O} &\underline{\leftrightarrows CaCO_3 + 2H^+} \qquad (3) \\
CaHPO_4 \cdot 2H_2O + CO_2 &\leftrightarrows H_2PO_4^- + H^+ + H_2O + CaCO_3 \qquad (4)
\end{aligned}
$$

The precipitation of $CaCO_3$ in soil occurs at pH 7.8 and above. As solution Ca^{2+} decreases with $CaCO_3$ precipitation in soils (Eq. 3), the DCPD will dissolve (Eq. 2) to resupply solution Ca^{2+}. Therefore, when DCPD dissolves, $H_2PO_4^-$ also increases, as shown in Eq. 4, which is the sum of Eqs. 2 and 3. All the Ca-P minerals listed in Table 6.6 behave similarly in soils that contain $CaCO_3$.

Although not readily apparent from the P solubility diagram (Fig. 6.12), P minerals that support the lowest concentration of P in solution (i.e., the lowest P solubility) are the most stable in soils. The apatite minerals, TCP, and OCP, are more stable than DCPD in slightly acid and neutral soils, for example. Therefore, the P mineral solubility relationships shown in Figure 6.12 can be used to understand the fate of fertilizer P applied to soils.

An important fertilizer P source is monocalcium phosphate [MCP, $Ca(H_2PO_4)_2$], which is very soluble in soil. When MCP applied to soil dissolves, the concentration of $H_2PO_4^-$ in soil solution is much higher than the P concen-trations supported by the minerals shown in Figure 6.12. Because the soil P minerals have lower solubility, the $H_2PO_4^-$ from the fertilizer will likely precipitate as these minerals. For example, in an acid soil, fertilizer $H_2PO_4^-$ reacts with Fe^{3+} and Al^{3+} in solution to form $FePO_4$ and $AlPO_4$ compounds, respectively. As a result, the $H_2PO_4^-$ concentration in solution decreases once the precipitation reactions begin. In neutral and calcareous soils, fertilizer $H_2PO_4^-$ initially precipitates as DCDP and DCP within the first few weeks after application. After 3 to 5 months, OCP begins to precipitate, with β-TCP forming after 8 to 10 months. After long periods of time, apatite minerals eventually form.

Thus, after MCP is applied to soil, a series of reactions occur that decrease the elevated concentration of $H_2PO_4^-$ in solution as the insoluble minerals precipitate. These reactions in soils cannot be altered and help explain why plant recovery of fertilizer P is generally lower than recovery of soluble nutrients like NO_3^- and SO_4^{2-}.

P ADSORPTION REACTIONS   Labile inorganic P represents $H_2PO_4^-$ and/or HPO_4^{2-} adsorbed to mineral surfaces (Fig. 6.2). The terms *adsorption* and *chemisorption* have been used to describe P reactions with mineral surfaces, where chemisorption generally represents a greater degree of bonding to the mineral surface. The term *sorption* has been used to describe adsorption and chemisorption collectively. *Adsorption* is the preferred term.

In acid soils, Al and Fe oxide and hydroxide minerals are primarily involved in adsorption of inorganic P. Since the soil solution is acidic the surface of these minerals has a net positive charge, although both ($+$) and ($-$) sites exist. The predominance of ($+$) charges readily attracts $H_2PO_4^-$ and other anions. P ions adsorb to the Fe/Al oxide surface by exchanging with OH^- and/or OH_2^+ groups on the mineral surface (Fig. 6.13). When the orthophosphate ion is bonded through one Al-O-P bond, the $H_2PO_4^-$ is considered *labile* and can be readily desorbed from the mineral surface to soil solution. When two Al-O bonds with $H_2PO_4^-$ occur, a stable six-member ring is formed (Fig. 6.13). Consequently, desorption is more difficult and the $H_2PO_4^-$ is considered chemisorbed or *nonlabile*. Another proposed mechanism for P adsorption on Fe or Al oxide surfaces is shown in Figure 6.14. Regardless of the reaction, electrical neutrality or charge balance must be maintained.

In acid soils, P adsorption also readily occurs on the broken edges of kaolinite clay minerals (see Fig. 4.7). Again, exposed OH^- groups on these edges can exchange for $H_2PO_4^-$ similarly to exchange with OH^- on the surface of Fe/Al oxides. Cations held to the surface of silicate clay minerals also influence P adsorption by developing a small ($+$) charge near the mineral surface saturated with cations. This small ($+$) charge will attract and hold small quantities of anions like $H_2PO_4^-$. As discussed earlier, precipitation of Al-P minerals

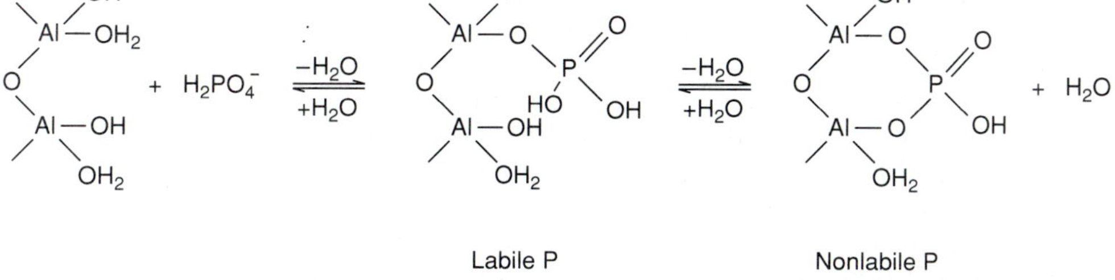

Labile P   Nonlabile P

FIGURE 6.13   Mechanism of P adsorption to Al or Fe oxide surface. Phosphate bonding through one Al-O bond results in labile P; however, bonding through two Al-O bonds produces a stable structure that results in very little desorption of P.

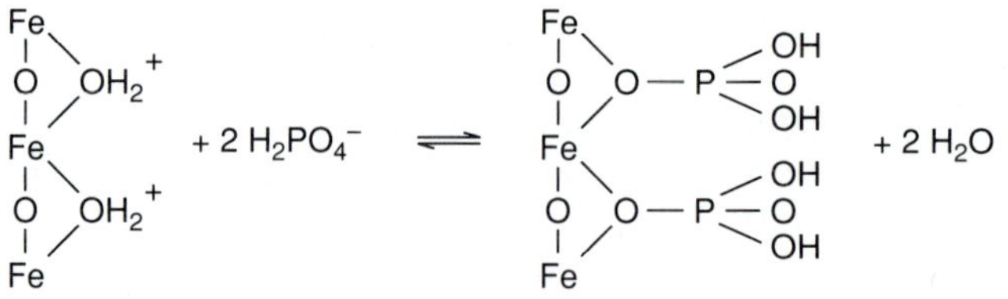

$$\begin{array}{c}
\text{Fe} \\
| \\
\text{O} \quad \text{OH}_2{}^+ \\
| \\
\text{Fe} \\
| \\
\text{O} \quad \text{OH}_2{}^+ \\
| \\
\text{Fe}
\end{array}
\quad + 2\,H_2PO_4{}^- \quad \rightleftharpoons \quad
\begin{array}{c}
\text{Fe} \\
| \\
\text{O} \quad \text{O} - P \overset{OH}{\underset{OH}{\big<}} O \\
| \\
\text{Fe} \\
| \\
\text{O} \quad \text{O} - P \overset{OH}{\underset{OH}{\big<}} O \\
| \\
\text{Fe}
\end{array}
\quad + 2\,H_2O$$

FIGURE 6.14   Another possible mechanism of P adsorption to Fe and/or Al oxide surfaces. *Bohn et al.*, Soil Chemistry, *p. 177, John Wiley & Sons, 1979.*

in acid soils and Ca-P minerals in neutral and calcareous soils will occur at high P concentrations.

In calcareous soils, small quantities of P can be adsorbed through replacement of $CO_3{}^-$ on the surface of $CaCO_3$. At low P concentrations, surface adsorption predominates; however, at high P concentrations, Ca-P minerals will precipitate on the $CaCO_3$ surface. Other minerals, mostly $Al(OH)_3$ and $Fe(OH)_3$, contribute to adsorption of solution P in calcareous soils.

Retention of P is frequently a problem in acid soils high in finely divided Fe and Al oxides. Representative acid soils of the United States fix two times more P per unit surface area of soil than neutral or calcareous soils. The P adsorbed is held with five times more bonding energy in acid soils than in calcareous soils.

Adsorption Equations. Several equations, or adsorption isotherms, have been developed to describe P adsorption in soils. The Freundlich and Langmuir equations, or modifications of them, have been used most frequently. These equations are helpful for understanding the relationship between the quantity of P adsorbed per unit soil weight and the concentration of P in solution. Therefore, all of the equations have the general form

$$q = f(c)$$

where q is the quantity of P adsorbed and is a function (f) of the solution P concentration (c).

The *Freundlich equation* was one of the first used in *P* adsorption studies and is represented by

$$q = ac^b$$

where a and b are coefficients that vary among soils, and q and c are the same as defined for the previous equation. This equation does not predict or include a maximum adsorption capacity and, therefore, works best with low solution P concentrations (Fig. 6.15). According to the Freundlich equation, energy of adsorption decreases as the amount of adsorption increases.

Since P adsorption data often exhibit a maximum P adsorption capacity at some solution P concentration, another equation is needed to describe situations where the adsorption sites are saturated with P. The *Langmuir equation* includes a term for the maximum P adsorption and is described by

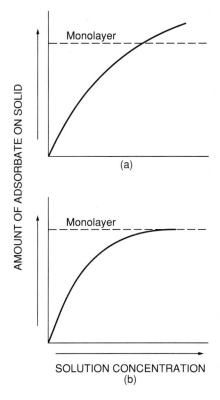

AMOUNT OF ADSORBATE ON SOLID

Monolayer

(a)

Monolayer

SOLUTION CONCENTRATION
(b)

FIGURE 6.15   Graphical
representation of adsorption isotherms
of the Freundlich (a) and Langmuir (b)
equations used to describe adsorption
reactions in soils.

$$q = \frac{abc}{1 + ac}$$

where q, c, and a are defined as before and b is the P adsorption maximum (Fig. 6.15). The P adsorption maximum in the Langmuir equation implies that a monolayer of P ions is adsorbed on the surface of the mineral, which occurs at relatively higher solution P concentrations than described by the Freudlich equation. The equation also shows that further increases in solution P concentration do not increase P adsorption. Although this does not occur with all soils, the Langmuir equation frequently has been used to quantify P adsorption maxima.

Use of adsorption isotherms in describing P adsorption is generally limited to a relatively narrow range of solution P concentration. Under a wide range of solution P concentration, more complex equations have been used or the Freundlich and Langmuir equations have been modified to describe the adsorption data.

These adsorption equations provide virtually no information about the mechanisms responsible for adsorption. They are incapable of showing whether Fe and Al oxides, silicate clays, or $CaCO_3$ dominate the adsorption reactions. In addition, the equations do not indicate whether P adsorption involves replacement of either hydroxyl, silica, or carbonate.

There is interest in knowing if nonlabile P can become available to plants.

Desorption isotherms that characterize the release of adsorbed P have shown that the process is extremely slow. It is not usually completed within hours or days. The rate of desorption decreases with time. In general, desorption appears to become very slow after about 2 days.

As might be expected, the extent of desorption is dependent on the nature of the adsorption complex at the surface of the Fe and Al oxides. Formation of six-membered ring structures, illustrated in Figure 6.13, will prevent the desorption of P.

FACTORS INFLUENCING P RETENTION IN SOILS   Many soil physical and chemical properties influence the P solubility and adsorption reactions (P fixation) in soils. Consequently, these soil properties also affect solution P concentration, P availability to plants, and recovery of P fertilizer by crops. Understanding these influences will enable us to manage soil and fertilizer P efficiently for optimum production.

Nature and Amount of Soil Minerals.   Adsorption and desorption reactions are affected by the type of mineral surfaces in contact with P in the soil solution. Al and Fe oxides have the capacity to adsorb large amounts of P in solution. Although present in most all soils, they are most abundant in weathered, acid soils. Al and Fe oxides can occur as discrete particles in soils or as coatings or films on other soil particles. They also exist as amorphous Al hydroxy compounds between the layers of expandable Al silicates. In soils with significant Fe and Al oxide contents, the less crystalline the oxides are or the more amorphous they are, the larger their P fixation capacity because of their greater surface areas. However, crystalline hydrous metal oxides are usually capable of retaining more P than amorphous forms or layer silicates.

P is adsorbed to a greater extent by 1:1 (e.g., kaolinite) than by 2:1 clays (e.g., montmorillonite). The greater amount of P fixed by 1:1 clays is probably due to the higher amounts of Fe and Al oxides associated with kaolinitic clays that predominate in highly weathered soils. Kaolinite, with a 1:1 Si/Al lattice, has a larger number of exposed OH groups in the Al layer that can exchange with P. In addition, kaolinite develops pH-dependent charges on its edges that can adsorb P (see Fig. 4.7).

Figure 6.16 shows the influence of clay mineralogy on P adsorption. First, compare the three soils that have more than 70% clay content. Compared to the Oxisol and Andept soils, very little P adsorption occurred in the Mollisol, composed mainly of montmorillonite, with only small amounts of kaolinite and Fe/Al oxides. The Oxisol soils contained Fe and Al oxides and exhibited considerably more P adsorption capacity compared to the Mollisol. Greatest P adsorption occurred with the Andept soils, composed principally of Fe and Al oxides and other minerals.

Soils containing large quantities of clay will fix more P than soils with low clay content (Fig. 6.16). In other words, the more surface area exposed with a given type of clay, the greater the tendency to adsorb P. For example, compare the three Ultisol soils containing 6, 10, and 38% clay. A similar relationship is evident in the Oxisol soils (36, 45, and 70% clay) and the Andept soils (11 and 70% clay).

In calcareous soils, P adsorption to $CaCO_3$ surfaces occurs; however, much of the adsorption is attributed to Fe oxide impurities. The amount and reactiv-

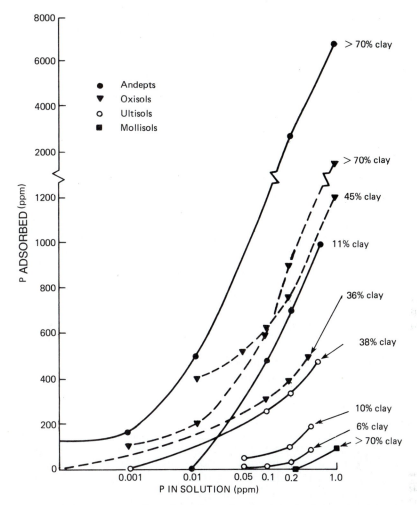

FIGURE 6.16   Examples of P adsorption isotherms determined by the method of Fox and Kamprath.   *Sanchez and Uehara, in F. E. Khasawneh, E. C. Sample, and E. J. Kamprath, Eds.,* The Role of Phosphorus in Agriculture, *p. 480. Madison, Wisc.: American Society of Agronomy, 1980.*

ity of $CaCO_3$ will influence P fixation. Impure $CaCO_3$ with large surface area exhibits greater P adsorption and more rapid precipitation of Ca-P minerals. Calcareous soils with highly reactive $CaCO_3$ and a high Ca-saturated clay content will exhibit low solution P levels, since P can readily precipitate or adsorb.

Conversely, in order to maintain a given level of solution P in soils with a high retention capacity, it is necessary to add larger quantities of P fertilizers. Figure 6.17 illustrates the effect of texture and amount of added P on solution P concentration. In any one soil, the P concentration in solution increased with increasing P additions. Larger additions of P were required to reach a given level of solution P in fine-textured compared to coarse-textured soils. Consequently, high-clay, calcareous soils will often require more fertilizer P to optimize yields compared to loam soils.

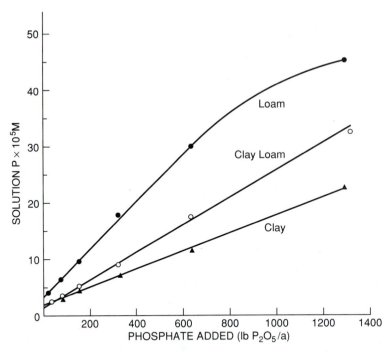

FIGURE 6.17   P solubility (the mean activity of DCP in
solution) as a function of the amounts of CSP added to
three calcareous soils of different texture.   *Cole et al.,
SSSA Proc.,* **23**:*119, 1959.*

Soil pH. Soil pH has a profound influence on the quantity of P adsorption and precipitation in soils. Adsorption of P by Fe and Al oxides declines
with increasing pH. Gibbsite [γ-Al(OH)$_3$] adsorbs the greatest amount of P at
pH 4 to 5. P adsorption by goethite (α-FeOOH) decreases steadily between
pH 3 and 12 (Fig. 6.18).

P availability in most soils is at a maximum in the pH range 5.5 to 6.5. At
low pH values, the retention results largely from the reaction with Fe and Al
and precipitation as AlPO$_4$ and FePO$_4$ oxides. As pH increases, the activity of
Fe and Al decreases, which results in lower P adsorption/precipitation and
higher P concentration in solution. Above pH 7.0, Ca^{2+} can precipitate with
P as Ca-P minerals (Fig. 6.12) and P availability again decreases. The pH
range of minimum P adsorption (pH 6.0 to 6.5) shown in Figure 6.11 also
corresponds with the pH range of maximum P solubility (Fig. 6.12).

As will be discussed in Chapter 10, liming acid soils generally increases
the solubility of P. Overliming can depress P solubility, however, due to the
formation of more insoluble Ca-P minerals similar to those occurring in basic
soils naturally high in Ca^{2+}.

Cation Effects. Divalent cations enhance P adsorption relative to monovalent cations. For example, clays saturated with Ca^{2+} retain greater amounts
of P than those saturated with Na$^+$ or other monovalent ions. Current explanations for this effect of Ca^{2+} involve making positively charged edge sites of
clay minerals more accessible to P anions. This action of Ca^{2+} is possible at

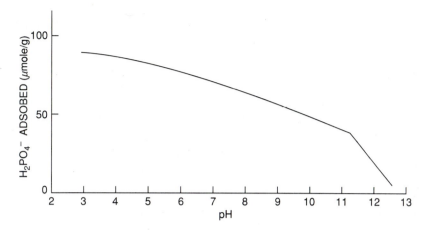

FIGURE 6.18   The adsorption of P by Fe oxide (goethite) as influenced by soil pH.   *Adapted from Hingston et al., Trans. 9th Int. Cong. Soil Sci., 1:1459–1461, 1968.*

pH values slightly less than 6.5, but in soils more basic than this, Ca-P minerals would directly precipitate from solution.

Concentration of exchangeable Al^{3+} is also an important factor in P adsorption in soils since 1 meq of exchangeable Al^{3+} per 100 g of soil, when completely hydrolyzed, may precipitate up to 100 ppm P in solution. The following illustrates one of the possible ways that hydrolyzed Al^{3+} can adsorb soluble P.

Cation Exchange:

$$\text{clay surface} \begin{bmatrix} - \\ - \\ - \\ - \end{bmatrix} \begin{matrix} Al^{3+} \\ \\ Al^{3+} \end{matrix} + 3Ca^{2+} \rightleftarrows \text{clay surface} \begin{bmatrix} - \\ - \\ - \\ - \end{bmatrix} \begin{matrix} Ca^{2+} \\ Ca^{2+} \\ Ca^{2+} \end{matrix} + 2Al^{3-}$$

Hydrolysis:

$$Al^{3+} + 2H_2O \rightleftarrows Al(OH)_2^+ + 2H^+$$

Precipitation and/or Adsorption:

$$Al(OH)_2^+ + H_2PO_4^- \rightleftarrows Al(OH)_2H_2PO_4 \ (K_{sp} = 10^{-29})$$

Strong correlations between P adsorption and exchangeable Al^{3+} have been reported when there is appreciable hydrolysis of Al^{3+} (Fig. 6-19).

Anion Effects. Both inorganic and organic anions can compete with P for adsorption sites, resulting in decreased adsorption of added P. Weakly held inorganic anions such as NO_3^- and Cl^- are of little consequence, whereas specifically adsorbed anions and acids such as OH^-, $H_3SiO_4^-$, SO_4^{2-}, and MoO_4^{2-} can be competitive. The strength of bonding of the anion with the mineral surface determines the competitive ability of that anion. For example,

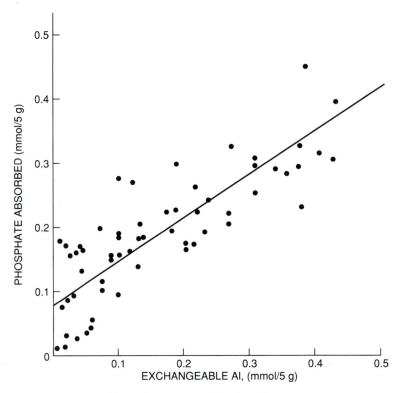

FIGURE 6.19   Effect of exchangeable Al on the amount
of P adsorbed by suspended clay.   *Coleman et al., Soil Sci.,*
90:*1, 1960. Reprinted with permission of The Williams &*
Wilkins Company, Baltimore.

SO_4^{2-} is unable to desorb much $H_2PO_4^-$, since $H_2PO_4^-$ is capable of forming
a stronger bond than is SO_4^{2-}.

Organic anions from sources such as organic waste materials and wastewater
treatment can affect the P adsorption-desorption reactions in soils. The impact
of organic anions on reduction of adsorbed P is related to their molecular
structure and pH. Organic anions form stable complexes with Fe and Al,
which reduces adsorbed P. Anions of tricarboxylic acids are more effective in
reducing P adsorption than dicarboxylic or monocarboxylic acids. Oxalate,
citrate, and polygalacturonate can be adsorbed at soil surfaces similar to that
of $H_2PO_4^-$. Some of the effects of organic anions on P adsorption are partially
responsible for the beneficial action of organic matter on availability.

Extent of P Saturation.   In general, P adsorption is greater in soils, with
little P adsorbed to mineral surfaces. As fertilizer P is added to soil and the
quantity of P adsorption subsequently increases, the potential for additional
P adsorption decreases. When all adsorption sites are saturated with $H_2PO_4^-$,
further adsorption will not occur and recovery of applied fertilizer P should
increase.

Organic Matter. Organic P compounds can move in soils to a greater depth than can inorganic P in solution. Several studies have demonstrated appreciable downward movement of P following the field application of manure. Continued application of manure can result in elevated P levels at 2- to 4-ft depths. In contrast, application of the same quantity of P as inorganic fertilizer P results in much less downward movement of P.

The effect on P availability of other compounds arising from the decomposition of organic residues has received considerable attention. Numerous workers have reported that organic compounds in soils increased P availability by (1) the formation of organophosphate complexes that are more easily assimilated by plants, (2) anion replacement of $H_2PO_4^-$ on adsorption sites, and (3) the coating of Fe and Al particles by humus to form a protective cover and thus reduce P adsorption.

Organic anions produced from the decomposition of organic matter also may form stable complexes with Fe and Al, thus preventing their reaction with $H_2PO_4^-$. These complex ions also may release P previously fixed by Fe and Al by the same mechanism. The anions that are most effective in replacing $H_2PO_4^-$ are citrate, oxalate, tartrate, malate, and malonate, which occur as organic matter degradation products.

In some instances, P adsorption has been found to correlate with soil organic C, although this is probably a minor contribution to P adsorption in soils. The Al and Fe adsorbed by the organic colloids are most likely active in P adsorption. Generally, incorporation of green manures results in better utilization of P by subsequent crops. Part of this favorable effect is likely related to the decomposition of organic residues accompanied by the evolution of CO_2. When dissolved in water, CO_2 forms H_2CO_3, which is capable of decomposing certain primary soil minerals. In neutral and calcareous soils, increasing CO_2 increases the solubility of Ca-P minerals. Thus, on the basis of the available evidence, it is clear that the addition of organic materials to mineral soils may increase the availability of soil P. Recall that soil OM content also is correlated with the quantity of organic P mineralized to inorganic P.

Leaching of inorganic P can occur in soil, but only under certain conditions. Soils that consist largely of quartz sand and soils that are primarily muck and peat are subject to leaching losses of added P fertilizer. The data in Table 6.7 show very little adsorption of added P by either the surface or subsurface horizon of this soil. This is due to the absence of Al and Fe compounds that are largely responsible for the P adsorption in acid soils. When $AlCl_3$ was added to the two soils, the leaching of added P was almost completely stopped.

Reaction Time and Temperature. The rate of most chemical and biological reactions increases with increasing temperature. Mineralization of P from soil OM or crop residues is dependent on soil biological activity, and increases in temperature stimulate biological activity up to the optima for the predominant biological systems. The dissolution of fertilizer P granules and resultant reactions with soil components to produce less soluble reaction products are hastened by higher temperatures. For example, the solution P concentration decreases with increasing soil temperature following the addition of several fertilizer P sources (Fig. 6.20). Results of most studies show that P adsorption generally increases with higher temperatures. Other studies show that P

TABLE 6.7   Leaching of Fertilizer P from an Acid Organic (Muck) Soil

P added (mg/column)	pH	Exchangeable Al (meq/100 g)	P Leached (mg/column)	Fertilizer P Absorbed (mg/column)
		Surface soil		
0	4.6	0.02	0.48	—
2.5	4.6	0.02	2.95	0.03
10.0	4.6	0.02	8.67	1.81
		Subsurface soil		
0	3.3	—	0.54	—
10.0	3.3	—	10.22	0.32

SOURCE: Fox and Kamprath, *SSSA Proc.*, **35**:154 (1971).

extracted from soils to which fertilizer had been added decreased when the soil incubation temperatures were above 59°F (15°C). However, on soils with no added P, the amounts of P extracted were not affected by soil incubation temperatures. P adsorption in soils of warm regions in the world is generally greater than in soils of temperate regions. These warmer climates also give rise to soils with higher contents of the Fe and Al oxides.

P adsorption in soils follows two rather distinct patterns: an initial rapid reaction followed by a much slower reaction. The adsorption reactions involving exchange of P for anions on the surface of Fe and Al oxides are extremely rapid. The much slower reactions are believed to involve (1) formation of

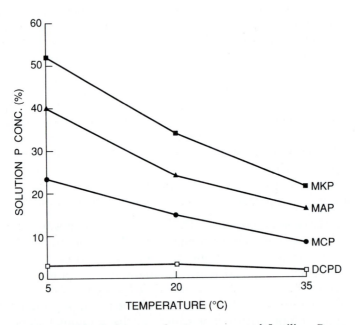

FIGURE 6.20   Influence of temperature and fertilizer P source on solution P concentration. MKP = monopotassium phosphate; MAP = monoammonium phosphate; MCP = monocalcium phosphate; DCPD = dicalcium phosphate dihydrate. *Beaton et al., SSSAJ, 29:194, 1965.*

covalent Fe-P or Al-P bonds on Fe and Al oxide surfaces (Figs. 6.13 and 6.14), and (2) precipitation of a P compound for which the solubility product has been exceeded. These slow reactions involve a transition from more loosely bound to more tightly bound adsorbed P, which is less accessible to plants.

The initial compounds precipitated during the reaction of fertilizer P salts in soils are relatively unstable (metastable), and will usually change with time into more stable and less soluble compounds. Table 6.8 shows the percentage of DCPD converted to OCP as a function of time and temperature. Conversion of 70% or more of the DCPD occurred after 10 months at 10°C and after 4 months at 20° and 30°C.

Flooding. In most soils there is an increase in available P after flooding, largely due to a conversion of Fe^{3+} phosphate to soluble Fe^{2+} phosphate and hydrolysis of Al phosphate. Other mechanisms resulting in increased P availability following submergence include dissolution of occluded P, hydrolysis of Fe phosphate, increased mineralization of organic P in acid soils, increased solubility of Ca phosphate in calcareous soils, and greater diffusion of P. These changes in P availability explain why the response to applied P by irrigated rice is usually less than the response of an upland crop grown on the same soil.

Fertilizer P Management Considerations. An important practical consequence of P adsorption and precipitation reactions is the time after application during which the plant is best able to utilize the added fertilizer P. On some soils with a high fixing capacity, this period may be short, whereas with other soils it may last for months or even years. This time period will determine whether the fertilizer P should be applied at one time in the rotation or in smaller, more frequent applications. Adsorption of fertilizer P is greater in fine-textured soils because the amount of reactive mineral surface is greater than in coarse-textured soils.

Also important is the placement of P in the soil. If fertilizer P is broadcast and incorporated, the P is exposed to a greater amount of surface; hence, more fixation takes place than if the same amount of fertilizer had been band applied. Band placement reduces the contact between the soil and fertilizer, with a subsequent reduction in P adsorption (see Chapter 12). Although this is not the only factor to consider in P fertilizer placement, it is very important for crops grown on low-P soils with a high P adsorption capacity. Thus, band placement generally increases the plant utilization of the water-soluble P fertilizer such as the superphosphates and ammonium phosphates. There are,

TABLE 6.8   Percentage of DCPD Hydrolyzed to OCP
as a Function of Time and Temperature

Temperature (°C)	*Percent OCP Present at:*			
	1 Month	*2 Months*	*4 Months*	*10 Months*
10	< 5	20	20	70
20	< 5	40	75	100
30	< 5	30	80	100

SOURCE: Sheppard and Racz, *Western Canada Phosphate Symp.*, p. 170 (1980).

however, water-insoluble P fertilizers, for which plant utilization is greatest when mixed with the soil rather than when applied in bands.

Soil Tests for P

The soil tests commonly used for P are based on chemical principles that relate to inorganic P minerals. When solution P decreases with plant uptake, P minerals can dissolve or adsorbed P can be released to resupply soil solution P (Fig. 6.12). The chemical extractants for P simulate this process, as they reduce solution Al or Ca through precipitation as Al-P or Ca-P minerals. As solution Al or Ca decreases during extraction, native Al-P or Ca-P minerals dissolve to resupply solution Al or Ca. Solution P then concurrently increases, which provides a measure of the ability of the soil to supply or buffer plant available P.

The Bray-1 extractant was developed for use in acid soils and contains 0.025 M HCl + 0.03 M NH$_4$F. From Figure 6.12 we see that AlPO$_4$ is the primary P mineral controlling solution P concentration. The F complexes Al^{3+} in solution, and as the Al^{3+} concentration in solution decreases, AlPO$_4$ dissolves to buffer or resupply solution Al and release P into solution. The subsequent increase in solution P is measured, which represents an estimate of the capacity of the soil to supply plant-available P. The HCl in the extractant also dissolves Ca-P minerals present in slightly acid and neutral soils.

The Olsen (Bicarb-P) extractant was developed for use in neutral and calcareous soils and contains 0.5 M NaHCO$_3$ buffered at pH 8.5. In these soils, Ca-P minerals control the solution P concentration (Fig. 6.12). The HCO$_3^-$ ion causes CaCO$_3$ to precipitate during extraction, which reduces the Ca^{2+} concentration in solution. Consequently, Ca-P minerals dissolve to buffer solution Ca^{2+} and release P into solution. As with the Bray-1 soil test, the increase in solution P provides a measure of the ability of the soil to supply plant available P.

Although the Bray-1 and Olsen soil tests were developed for acid and calcareous soils, respectively, both have been used to quantify plant-available P in both soil types. For example, the Bray-1 soil test extracts P in quantities that are not equal to, but highly correlated with, Olsen-extractable P in calcareous soils. Provided that a given soil test has been carefully calibrated with the crop response to fertilizer P, either soil test can be used.

Other P soil tests have been developed and are being used in various regions. The Mehlich test, containing NH$_4$F + CH$_3$COOH + NH$_4$NO$_3$/HNO$_3$ or NH$_4$Cl/HCl, has been used extensively in the United States. This test extracts P in the same manner as the Bray-1 test.

Although soil test calibrations will differ among regions and crops, general sufficiency levels for the common P soil tests are shown in Table 6.9. These categories show that the Bray-1 and Mehlich P tests extract similar quantities of P, while the Olsen P test extracts about half as much P.

Soil tests for P reflect the relative responsiveness of crops to P fertilization on soils with varying extractable P levels. As soil test P increases, the response to P fertilization decreases, as measured by percent yield (Fig. 6.21). These data also show that the critical Bray-1 P level is about 25 ppm, which is the soil test P level above which no response to P fertilization is expected. Similarly,

TABLE 6.9 Calibrations for the Bray-1, Mehlich III, and Olsen Soil Tests

P Sufficiency Level	Bray-1	Mehlich III	Olsen	Fertilizer P Recommendation	
	--------- ppm ---------			lb P_2O_5/a	kg P/ha
Very low	<−05	<−07	<−03	50	25
Low	06−12	08−14	04−07	30	15
Medium	13−25	15−28	08−11	15	8
High	>−25	>−28	>−12	0	0

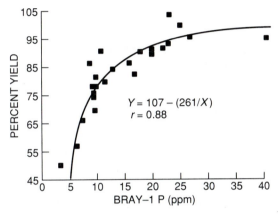

FIGURE 6.21 Relationship between soil test P level and percent yield for winter wheat. Percent yield is the ratio of unfertilized grain yield to fertilized grain yield.

the soil test predicts or estimates the fertilizer P rate required for optimum crop yield. As the Bray-1 P level increases, the optimum P rate decreases (Fig. 6.22). At or above the critical soil test P level, no fertilizer P would be recommended. These data demonstrate that soil testing for P estimates the P sufficiency level for a given soil and the recommended P rate for optimum production. Soil testing for P should be included in every soil/crop management system to insure optimum production.

P Fertilizers

The development of the modern P fertilizer industry began with the demonstration by Liebig in 1840 that the fertilizing value of bones could be increased by treatment with H_2SO_4. In 1842, John B. Lawes patented a process by which phosphate rock was acidulated with H_2SO_4, and commercial production of this material began in England in 1843. The sale of single superphosphate manufactured in the United States did not take place until 1852. Today, the P fertilizer industry uses essentially the same principle for the manufacture of soluble phosphates employed by John Lawes in 1842.

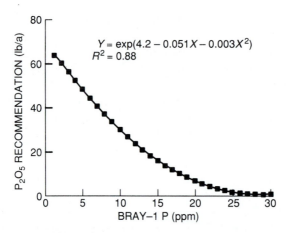

$$Y = \exp(4.2 - 0.051X - 0.003X^2)$$
$$R^2 = 0.88$$

FIGURE 6.22   Relationship between soil test P level and
P recommendation for winter wheat.

P Content of Fertilizers

Historically, the P content of fertilizers has been expressed as the oxide, P_2O_5, instead of as elemental P. Although attempts have been made to change from $\%P_2O_5$ to $\%P$, most industries still express P concentration in fertilizers as $\%P_2O_5$. Similarly, the concentration of K in fertilizers is usually expressed as $\%K_2O$ instead of $\%K$. As a matter of interest, N was formerly guaranteed as $\%NH_3$ rather than as $\%N$, as is now done. Fertilizer P concentration expressed as $\%P$ is much easier to work with, calculate, and discuss than oxide concentration.

The conversion of $\%P$ to $\%P_2O_5$, and vice versa, is simple and is given by

$$\%P = \%P_2O_5 \times 0.43$$
$$\%P_2O_5 = \%P \times 2.29$$

The conversion factors are derived from the ratio of molecular weights of P and P_2O_5:

$$\frac{2 \times M \ wt \ P}{M \ wt \ P_2O_5} = \frac{2 \times 31}{142} = 0.43$$

For example, the $\%P$ concentration in $CaHPO_4$ (DCP) is

$$\frac{M \ wt \ of \ P}{M \ wt \ CaHPO_4} = \frac{31}{136} \times 100 = 23 \ \%P$$

Thus, the $\%P_2O_5$ is calculated by

$$\%P_2O_5 = 23\% \times 2.29 = 53\%$$

In this text, $\%P$ will be used to express P concentration in fertilizers, although $\%P_2O_5$ also may be given in parentheses.

P Fertilizer Terminology

The solubility of the P in the different P fertilizers is variable. When P fertilizers are added to soil, only part may be readily soluble in water; thus, the water solubility of fertilizer P is not always the best estimate of P availability to plants. The most accurate measure of plant available P is the extent to which it is absorbed by plants under conditions favorable to growth. Such determinations are not easily made when P availability has to be determined on large numbers of samples. Chemical methods have been developed that rapidly estimate the water-soluble, available, and total P content of P fertilizers.

The terms used to describe the P content in fertilizers are *water-soluble, citrate-soluble, citrate-insoluble, available,* and *total P* (as P_2O_5). A small sample is first extracted with water for a prescribed period of time. The slurry is then filtered, and the amount of P contained in the filtrate is determined. Expressed as a percentage by weight of the sample, it represents the fraction of the sample that is *water-soluble.*

The remaining water-insoluble material is extracted with 1 *N* ammonium citrate. The P content of the filtrate is termed the *citrate-soluble* P. The sum of the water-soluble and citrate-soluble P represents an estimate of the fraction available to plants and is termed *available P.* The amount of P in the residue remaining from the water and citrate extractions is *citrate-insoluble* P. The sum of available and citrate-insoluble P represents the total amount present.

P Fertilizer Sources

The original source of P in the early manufacture of P fertilizers was bones, but the supply was soon exhausted. Today, rock phosphate (RP) is the only important raw material for P fertilizers. The general formula for pure RP is $Ca_{10}(PO_4)_6(X)_2$, where X is either F^-, OH^-, or Cl^-. These minerals are called *apatites*; the most common RP mined is fluorapatite (the F^- form of RP). RP currently mined contains numerous impurities; the most common ones are CO_3, Na, and Mg. The most significant commercial phosphate deposits in the world contain sedimentary apatites. Carbonate-fluorapatite (francolite) is the primary apatite mineral in the majority of phosphate rocks. The high reactivity of some phosphate rocks is due to the occurrence of francolite. The major deposits are found in the United States, Morocco, Russia, South Africa, and China. The United States produces about 40% of the world's RP, although nearly 50% of the world reserves are in Morocco. The common P fertilizers are produced from either acid-treated or heat-treated RP to break the apatite bond and to increase the water-soluble P content. The common commercially available P fertilizers are listed in Table 6.10.

ROCK PHOSPHATE   After several processing and purification steps, RP contains between 11.5 and 17.5% total P (27 to 41% P_2O_5). None of the P is water-soluble, although the citrate solubility varies from 5 to 17% of the total P. RP can be used directly as a P fertilizer under certain conditions; however, in most situations, P fertilizers processed from RP are more cost effective. Dissolution of RP in soil is mainly influenced by its chemical reactivity and soil properties. Soil pH, exchangeable Ca^{2+} and Al^{3+}, and OM can affect RP dissolution.

TABLE 6.10 Common Commercially Available P Fertilizers

Material	Frequently Used Abbreviations	Analysis (%)				Form of P	Percentage Total P Available	Formula of Main P Compound
		N	P_2O_5	K_2O	S			
Rock phosphate	RP	—	25–40	—	—	Orthophosphate	14–65	$[Ca_3(PO_4)_2]_3 \cdot CaF_x \cdot (CaCO_3)_x \cdot (Ca(OH)_2)_x$
Single superphosphate	SSP	—	16–22	—	11–12	Orthophosphate	97–100	$Ca(H_2PO_4)_2$
Wet process phosphoric acid	–	—	48–53	—	—	Orthophosphate	100	H_3PO_4
Triple superphosphate	TSP or CSP	—	44–53	—	1–1.5	Orthophosphate	97–100	$Ca(H_2PO_4)_2$
Ammonium phosphates								
Monoammonium phosphate	MAP	11–13	48–62	—	0–2	Orthophosphate	100	$NH_4H_2PO_4$
Diammonium phosphate	DAP	18–21	46–53	—	0–2	Orthophosphate	100	$(NH_4)_2HPO_4$
Ammonium polyphosphate	APP	10–15	35–62	—	—	Mixture of ortho and polyphosphates	100	$(NH_4)_3HP_2O_7 + NH_4H_2PO_4$ + others
Urea-ammonium phosphate	UAP or UAPP	21–34	16–42	—	—	Mixture of ortho and polyphosphates	100	$NH_4H_2PO_4 \cdot (NH_4)_3HP_2O_7$
Nitric phosphates	NP	14–29	14–28	0–20	—	Orthophosphate	80–100	$CaHPO_4 \cdot NH_4H_2PO_4$
Ammoniated normal superphosphate	–	2–5	14–21	—	9–11	Orthophosphate	97–100	$NH_4H_2PO_4 \cdot CaHPO_4$
Ammoniated triple superphosphate	–	4–6	44–53	—	0–1	Orthophosphate	96–100	$CaHPO_4 \cdot NH_4H_2PO_4$
Potassium phosphates								
Monopotassium phosphate	–	—	51	35	—	Orthophosphate	100	KH_2PO_4
Dipotassium phosphate	–	—	41	54	—	Orthophosphate	100	K_2HPO_4
Potassium polyphosphate	–	—	51	40	—	Polyphosphate and orthophosphate	100	$K_3HP_2O_7 \cdot KH_2PO_4$ · others

Follett et al., Fertilizers and Soil Amendments, p. 131, Prentice-Hall, 1981.

RPs with solubilities in 2% formic acid, 2% citric acid, and neutral ammonium citrate of more than 65, 40, and 18%, respectively, are considered highly reactive and almost as effective as water-soluble P under favorable conditions. There are indications that 2% formic acid may be the most reliable of these solubility tests.

RP has only limited value to plants, unless finely ground. It must also be thoroughly mixed into the soil and applied at three to five times the P normally provided in conventional water-soluble fertilizers.

Finely ground apatitic RPs are effective only on acid soils, with pH 6 or less. On low-P, acid soils, RP application may be profitable, but the treated phosphates are generally more economical. Generally, RP produces greater crop yields than does single super-phosphate, but *only* when the rock is supplied in quantities that furnish two to three times more P. RP has been reported to give better residual effects than superphosphate, but whenever this was so, it was found that the rates of applied RP were considerably in excess of those of superphosphate.

Environmental conditions such as warm climates, moist soils, and long growing seasons will increase the effectiveness of RP. RP is extensively used for plantation crops such as rubber, oil palm, and cacao grown on very acid soils (pH < 5) in Southeast Asia. Ground RP is sometimes used for restoration of low-P soils on abandoned farms and on newly broken lands. For these purposes, a heavy initial application is recommended, such as 1 to 3 tons/a, which may be repeated at 5- to 10-year intervals. Addition of 1,000 lb/a of reactive RP is a key step in the technology used to rehabilitate tropical savanna land in Southeast Asia.

Factors limiting the use of RP include uncertain agronomic value; inconvenience of handling and applying the fine, dusty material; and relatively low P content compared with triple superphosphate or ammonium phosphate. In situations where the reactivity of RP is inadequate for immediate crop response and where the P-fixation capacity of the soil quickly renders soluble P fertilizer unavailable to plants, partially acidulated RP can increase the water-soluble P content and improve the short-term crop response to RP. Partially acidulated RP is produced by treating RP with 10 to 20% of the quantity of H_3PO_4 used for the manufacture of triple superphosphate or by reacting it with 40 to 50% of the amount of H_2SO_4 normally used in the production of single superphosphate.

PHOSPHORIC ACID   Phosphoric acid (H_3PO_4), used largely in the fertilizer industry, is manufactured by treating RP with H_2SO_4, and is called *green* or *wet process acid*. A higher concentration of H_2SO_4 is used in the wet process acid reaction than is used to produce single superphosphate. Both reactions produce gypsum ($CaSO_4 \cdot 2H_2O$), which can be used for other industrial purposes. Gypsum also is used as a soil amendment for sodic soils and as an S source. Phosphoric acid also can be made by heating RP in an electric arc furnace to produce elemental P, which is then reacted with O_2 to produce P_2O_5 and subsequently with H_2O to form H_3PO_4.

Phosphoric acid made by burning is termed *white* or *furnace acid* and is used almost entirely by the nonfertilizer segment of the chemical industry. White acid has a much higher degree of purity than green acid; however, the high energy cost involved in manufacturing makes it expensive and limits its use in the fertilizer industry.

Agricultural-grade green acid, containing 17 to 24% P (39 to 55% P_2O_5), is used to acidulate RP to make TSP and is neutralized with NH_3 in the manufacture of NH_4^+ phosphates and liquid fertilizers. It can also be applied by injection in the soil or to irrigation water, particularly in alkaline and calcareous areas, but this method requires special handling and equipment. Almost all wet acid is used to manufacture other P fertilizers.

CALCIUM ORTHOPHOSPHATES   The Ca orthophosphate fertilizers—single superphosphate, 7 to 9.5% P (16 to 22% P_2O_5); triple or concentrated superphosphate, 17 to 23% P (44 to 52% P_2O_5); and enriched superphosphates, 11 to 13% P (25 to 30% P_2O_5)—were once the most important P sources.

The superphosphates are *neutral fertilizers* in that they have no appreciable effect on soil pH, as do H_3PO_4 and the NH_4^+-containing fertilizers. The ammoniated superphosphates have a slightly acid reaction, depending, of course, on the extent to which they have been ammoniated. Single superphosphate (SSP) is manufactured by reacting H_2SO_4 with RP:

$$[Ca_3(PO_4)_2]_3 \cdot CaF_2 \; + \; 7\;H_2SO_4 \; \rightarrow \; 3\;Ca(H_2PO_4)_2 \; + \; 7\;CaSO_4 \; + \; 2\;HF$$

| Rock phosphate | Sulfuric acid | Monocalcium phosphate | Gypsum | Hydrofluoric acid |

SSP contains 7 to 9.5% P (16 to 22% P_2O_5), of which about 90% is water-soluble and essentially all is plant *available*. In addition, SSP contains about 12% S and is an excellent source of this nutrient.

The P component of superphosphate reacts with the soil, as does any water-soluble orthophosphate, in keeping with the solubility and adsorption reactions already discussed. It is an excellent source of P, but its low P analysis is a main disadvantage in its use. As a result, SSP is not stocked at many fertilizer outlets in the United States and Canada. Because of the low energy requirements for production, use of H_2SO_4 by-products of other industries, and increasing demand for S and micronutrients, SSP use may increase in the future.

Triple or concentrated superphosphate (TSP or CSP), containing 17 to 23% P (44 to 52% P_2O_5), is manufactured by treating RP with H_3PO_4:

$$[Ca_3(PO_4)_2]_3 \cdot CaF_2 \; + \; 12\;H_3PO_4 \; + \; 9\;H_2O \; \rightarrow \; 9\;Ca(H_2PO_4)_2 \; + \; CaF_2$$

| Rock phosphate | Phosphoric acid | Water | Monocalcium phosphate | Calcium fluoride |

TSP was manufactured to increase the P content above that of SSP, although TSP contains very little S (0 to 1%).

TSP is an excellent source of fertilizer P and was the most common source of fertilizer P used in the United States until the early 1960s, when ammonium phosphates became more popular (Fig. 6.23). Its high P content makes it particularly attractive when transportation, storage, and handling charges make up a large fraction of the total fertilizer cost. TSP is manufactured in granular form and is used in mixing and blending with other materials and

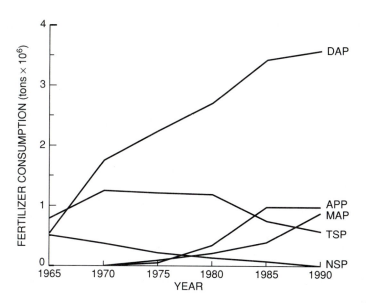

FIGURE 6.23   Use of common P fertilizers in the United States. DAP = diammonium phosphate; APP = ammonium polyphosphate; MAP = monoammonium phosphate; TSP = triple superphosphate; NSP = normal superphosphate. TVA-NFDC, *1990*.

in direct soil application. SSP and TSP can be ammoniated to produce monoammonium phosphate or MAP ($NH_4H_2PO_4$). The reactions shown in Fig. 6.24 illustrate the ammoniation of superphosphates but also reveal that excessive ammoniation produces very insoluble P (see Fig. 6.12). The total P content of the end product is decreased in proportion to the weight of the NH_3 added. Ammoniation of superphosphates offers the advantage of inexpensive N but decreases the amount of water-soluble P in the product.

AMMONIUM PHOSPHATES   Ammonium phosphates are produced by reacting wet process H_3PO_4 with NH_3 (Fig. 6.25). Monoammonium phosphate (MAP) contains 11 to 13% N and 21 to 24% P (48 to 55% P_2O_5); however, the most common grade is 11-22-0 (11-52-0). Diammonium phosphate (DAP) contains 18 to 21% N and 20 to 23% P (46 to 53% P_2O_5); the most common grade is 18-20-0 (18-46-0). Although MAP use has increased significantly in the last decade, DAP is more widely used than any other P fertilizer (Fig. 6.23). The increased interest in and use of NH_4 phosphates result from considerable evidence supporting increased $H_2PO_4^-$ uptake when NH_4^+ is placed with P fertilizer.

Both MAP and DAP are granular fertilizers and completely water-soluble. Ammonium phosphates have the advantage of a high plant-food content, which minimizes shipping, handling, and storage costs. They also can be used for formulating solid fertilizers by bulk blending or in manufacturing suspension fertilizers. MAP and DAP also are used for direct application as starter fertilizers.

Care must be taken with row or seed placement of DAP since free NH_3 can be produced, causing seedling injury and inhibiting root growth. This is

Ammoniation of normal superphosphate

$$Ca(H_2PO_4)_2 + NH_3 \longrightarrow CaHPO_4 + NH_4H_2PO_4$$

Monocalcium phosphate   Ammonia   Dicalcium phosphate   Monoammonium phosphate

$$NH_4H_2PO_4 + CaSO_4 + NH_3 \longrightarrow CaHPO_4 + (NH_4)_2SO_4$$

Monoammonium phosphate   Calcium sulfate   Ammonia   Dicalcium phosphate   Ammonium sulfate

$$2\,CaHPO_4 + CaSO_4 + 2\,NH_3 \longrightarrow Ca_3(PO_4)_2 + (NH_4)_2SO_4$$

Dicalcium phosphate   Calcium sulfate   Ammonia   Tricalcium phosphate   Ammonium sulfate

Ammoniation of triple superphosphate

$$Ca(H_2PO_4)_2 + NH_3 \longrightarrow CaHPO_4 + NH_4H_2PO_4$$

Monocalcium phosphate   Ammonia   Dicalcium phosphate   Monoammonium phosphate

$$3\,CaHPO_4 + NH_3 \longrightarrow NH_4H_2PO_4 + Ca_3(PO_4)_2$$

Dicalcium phosphate   Ammonia   Monoammonium phosphate   Tricalcium phosphate

---

Ammoniation usually expressed in terms of kg NH_3 per 20 kg of P_2O_5. For normal superphosphate, the normal range is 4–6 kg NH_3 per 20 kg P_2O_5; for triple superphosphate, 3–4 kg NH_3 per 20 kg P_2O_5.

FIGURE 6.24   Ammoniation of SSP and TSP to produce MAP. The reactions are carefully controlled to prevent excessive ammoniation and formation of water-insoluble tricalcium phosphate and DCP. *Follett et al.*, Fertilizers and Soil Amendments, *p. 122. Prentice-Hall, 1980.*

$$NH_3 + H_3PO_4 \longrightarrow NH_4H_2PO_4$$

Ammonia   Orthophosphoric acid   Monoammonium phosphate

$$2\,NH_3 + H_3PO_4 \longrightarrow (NH_4)_2HPO_4$$

Ammonia   Orthophosphoric acid   Diammonium phosphate

$$3\,NH_3 + H_4P_2O_7 \longrightarrow (NH_4)_3HP_2O_7$$

Ammonia   Pyrophosphoric acid   Triammonium pyrophosphate

FIGURE 6.25   Reactions of ammonia with ortho- and pyrophosphate to produce monoammonium phosphate (MAP), diammonium phosphate (DAP), and ammonium polyphosphate (APP). *Follett et al.*, Fertilizers and Soil Amendments, *p. 127, Prentice-Hall, 1981.*

especially true in calcareous or high-pH soils. Adequate separation of seed from DAP will usually be all that is required to eliminate seedling damage. In most cases, the N rate should not exceed 15 to 20 lb N/a as DAP applied with the seed. Seedling injury with MAP is seldom observed except in sensitive crops such as canola/rapeseed and flax.

The initial soil reaction pH of DAP is about 8.0, which favors NH_3 production (see Fig. 5.22), whereas the reaction pH with MAP is 3.5. Except for the differences in reaction pH and seedling injury when applied with the seed, few agronomic differences exist between MAP and DAP. Reports of improved crop response to MAP compared to DAP on high-pH or calcareous soils are generally not substantiated. Low-reaction pH with MAP has been claimed to suggest increased micronutrient availability in calcareous soils, but this has not been demonstrated.

AMMONIUM POLYPHOSPHATE   Ammonium polyphosphate (APP) is manufactured by reacting pyrophosphoric acid, $H_4P_2O_7$, with NH_3 (Fig. 6.26). Pyrophosphoric acid is produced from dehydration of wet process acid. *Polyphosphate* is a term used to describe two or more orthophosphate ions ($H_2PO_4^-$) combined together, with the loss of one H_2O molecule per two $H_2PO_4^-$ ions (Fig. 6.26). APP is a liquid containing 10 to 15% N and 15 to 16% P (34 to 37% P_2O_5), with about 75 and 25% of the P present as polyphosphate and orthophosphate, respectively. The most common APP grade is 10-15-0 (10-34-0).

Granulation during the manufacture of APP will result in a solid product analyzing 11-24-0 (11-55-0). Upon the addition of 99.5% urea solution, a granular urea-APP with a grade of 28-28-0 (12% P) can be made. The liquid APP is more popular and can be directly applied if mixed with other liquid fertilizers. Commonly, UAN and APP will be combined and subsurface band applied. Granular APP also can be applied directly or blended with other granular fertilizers.

One unique property of APP is the chelation or sequestering reaction with

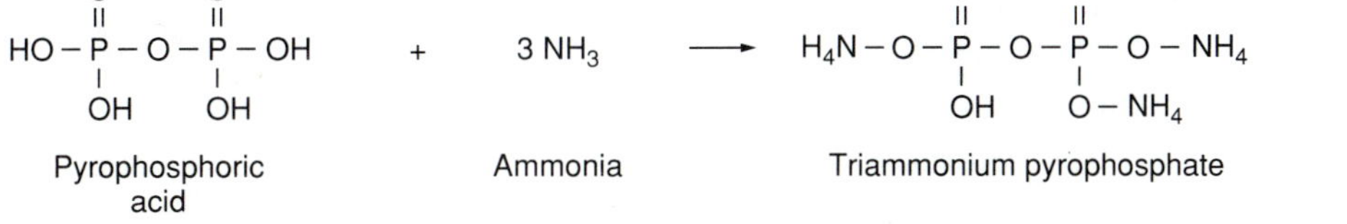

FIGURE 6.26   Reaction of two orthophosphate molecules to produce pyrophosphate. The reaction can continue to form longer chain products called *polyphosphate*. Ammoniation of pyro- and polyphosphates produces APP. *Follett et al.*, Fertilizers and Soil Amendments, *p. 130, Prentice-Hall, 1981.*

metal cations, which maintains higher concentrations of micronutrients in APP than is possible with orthophosphate solutions. For example, APP can maintain 2% Zn in solution compared with only 0.05% Zn with orthophosphate. The sequestering of Zn in polyphosphate solutions is shown in Figure 6.27.

A relatively new granular fertilizer, urea-ammonium phosphate (UAP), has been developed and is produced by reacting urea with APP. The fertilizer grade is 28-12-0 (28-28-0), containing 20 to 40% polyphosphate, and 100% of the P is water-soluble. UAP can be easily blended with other granular fertilizers. Like DAP, seedling damage will occur when UAP is applied with the seed.

NITRIC PHOSPHATES   Nitric phosphates are manufactured by reacting HNO_3 with RP (Fig. 6.28). The reaction product, $Ca(NO_3)_2$, is very hygroscopic and must be converted to $CaHPO_4$ or $CaSO_4$ by reaction with H_3PO_4 or H_2SO_4, respectively. The $Ca(NO_3)_2$ also can be removed by filtration and used as an N fertilizer.

Nitric phosphates are granular materials containing 14 to 28% N and 6 to 10% P (14 to 28% P_2O_5). The most common grade is a 20-9-0 (20-20-0). The major disadvantage of this fertilizer is that only 50% of the P content is water-soluble. Nitric phosphates are not produced widely in the United States but are used extensively in Europe.

Results of numerous agronomic tests have shown that these materials are generally satisfactory sources of fertilizer P. When used on crops responding to water-soluble P, nitric phosphates may be inferior to those materials containing a high degree of water-soluble P.

Nitric phosphates, in general, will give the best results on acid soils and under crops with a relatively long growing season, such as turf and sod crops. It must be reemphasized, however, that when the degree of water-soluble P in these materials is kept high (60% or greater), they are usually just as effective as sources of P for most crops as the super- and ammonium phosphates.

Nitric phosphates are potentially important fertilizers in the United States

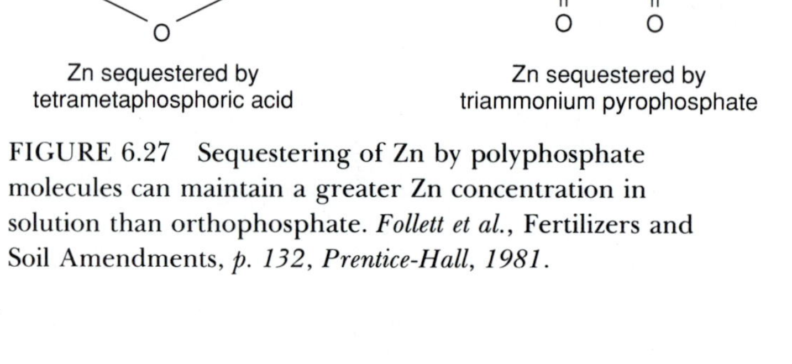

Zn sequestered by
tetrametaphosphoric acid

Zn sequestered by
triammonium pyrophosphate

FIGURE 6.27   Sequestering of Zn by polyphosphate molecules can maintain a greater Zn concentration in solution than orthophosphate. *Follett et al.*, Fertilizers and Soil Amendments, *p. 132, Prentice-Hall, 1981.*

INITIAL STEP

$Ca_{10}F_2(PO_4)_6$ + $20\ HNO_3$ + $4\ H_3PO_4$ $\xrightarrow{H_2O}$ $10\ H_3PO_4$ + $10\ Ca(NO_3)_2$ + $2\ HF$
Rock phosphate   Nitric   Phosphoric   Phosphoric   Calcium   Hydrofluoric
                    acid   acid   acid   nitrate   acid

SECOND STEP, REMOVAL OF CALCIUM NITRATE

$10H_3PO_4$ + $10\ Ca(NO_3)_2$ + $2\ HF$ + $21\ NH_3$ $\longrightarrow$ $9\ CaHPO_4$ + $NH_4H_2PO_4$ + $20\ NH_4NO_3$ + CaF_2
Phosphoric   Calcium   Hydrofluoric   Ammonia   Dicalcium   Monoammonium   Ammonium   Calcium
acid   nitrate   acid   phosphate   phosphate   nitrate   fluoride

FIGURE 6.28   Chemical reactions involved in manufacturing nitric phosphate fertilizers. *Follett et al., Fertilizers and Soil Amendments, p. 136, Prentice-Hall, 1981.*

because of the increasing supply of NH_3, and hence of HNO_3. The economics of producing these materials and the ability to increase their N content, in addition to the degree of water-soluble P, are the two factors that will probably determine the future of nitric phosphate fertilizers in the United States.

POTASSIUM PHOSPHATE   Potassium phosphate is represented by two salts, KH_2PO_4 and K_2HPO_4, which have the grades 0-52-35 (22% P, 29% K) and 0-41-54 (18% P, 45% K), respectively. They are completely water-soluble and find their greatest market in soluble fertilizers sold in small packets for home and garden use. Their high content of P and K makes them attractive possibilities for commercial application on a farm scale. Developments in the economics of producing these salts will determine whether they can be manufactured on a large scale for use as commercial fertilizers.

In addition to the high plant nutrient content of the K phosphates, they have other desirable characteristics. As they contain no Cl^-, K phosphates are ideally suited for "solanum"-type crops such as potatoes, tomatoes, and many leafy vegetables that are sensitive to high levels of Cl^-. Their low salt index reduces the risk of injury to germinating seeds and to young seedlings when they are placed in or close to the seed row.

Microbial P Fertilization

The use of microorganisms to increase plant-available P has been documented. Since the late 1950s, bacteria collectively called *phosphobacterins* have been soil applied to increase the P uptake and yield of crops in Russia and Eastern Europe. In controlled experiments, an average of 10% yield increases were reported. The mode of action was initially thought to be enhanced microbial degradation of organic P. However, more recent studies suggest that this mechanism may not explain the increased P availability. Studies in the United States showed no response to phosphobacterin applied to numerous crops over a wide geographic area. In addition, the use of OM-degrading organisms was discouraged in the United States, because of the obvious beneficial role of organic matter on soil productivity.

In the 1980s, several fungi, in particular *Penicillium bilaji*, were shown to increase P uptake, especially in high-pH, calcareous soils. Increased solubilization of native soil mineral P and added RP have been observed. Phosphate-solubilizing organisms apparently release organic acids that may dissolve P minerals. The future use of P-solubilizing organisms is uncertain, but it will probably be promoted in countries that do not produce or have limited access to P fertilizers and where RP is the only available P source. Research is continuing in the United States, Canada, and other countries.

Behavior of P Fertilizers in Soils

The chemical characteristics of the soil and the P fertilizer source determine the soil-fertilizer reactions, which influence fertilizer P availability to plants. Many of the same factors that affect native P availability, discussed earlier, also influence fertilizer P reaction product chemistry and availability. As seen in Figure 6.2, P fertilizer added to soil initially increases solution P but subsequently influences mineral, adsorbed (labile), and organic P fractions.

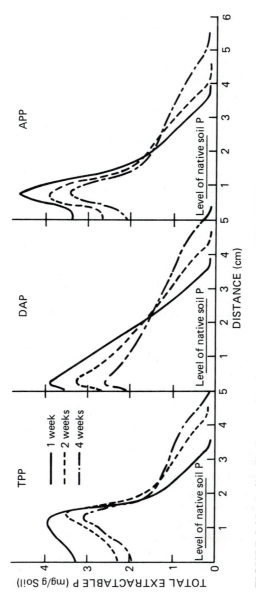

FIGURE 6.29 P distribution profiles in columns treated with TPP, DAP, or APP. *Khasawneh et al.*, Soil Sci. Soc. Am. J., *38:446, 1974.*

The commonly used granular P fertilizers are 90 to 100% water-soluble and, thus, dissolve rapidly when placed in moist soil. Water sufficient to initiate dissolution moves to the granule by either capillarity or vapor transport. A nearly saturated solution of the P fertilizer material forms in and around fertilizer granules or droplets.

While water is drawn into this fertilizer zone by vapor transport, the fertilizer solution moves into the surrounding soil. Movement of water inward and fertilizer solution outward continues to maintain a nearly saturated solution as long as the original salt remains. Initial movement of P away from fertilizer application site seldom exceeds 3 to 5 cm (Fig. 6.29).

Diffusion of fertilizer P reaction products increases with increasing soil moisture content. For example, studies at the Tennessee Valley Authority showed P diffusion of 18, 25, and 34 mm when the moisture content was 6.7, 9.6, and 19%, respectively. Extensive reaction zones combined with thorough distribution of the reaction products are factors that should enhance absorption of P by plant roots encountering reaction zones. These conditions may, in fact, offset the low water solubility of many of the P compounds that are precipitated in reaction zones.

As the saturated P solutions move into the first increments of soil, the chemical environment is dominated by the solution properties rather than by the soil properties. Solutions formed from water-soluble P fertilizers have pH values between 1 and 8 and contain from 2.9 to 6.8 mol/l of P (Table 6.11). The concentration of the accompanying cations ranges from 1.3 to 10.2 mol/l.

When the concentrated P solution leaves the granule, droplet, or band site and moves into the surrounding soil, the soil components are altered by the

TABLE 6.11   Phosphate Compounds Commonly Found in Fertilizers and Compositions of Their Saturated Solutions

Compound	Formula	Composition of Saturated Solution				
		Solution Symbol	pH	P (mol/liter)	Accompanying Cation (mol/liter)	
Highly water-soluble						
Monocalcium phosphate	$Ca(H_2PO_4)_2 \cdot H_2O$	TPS	1.5	4.5	Ca	1.3
Monoammonium phosphate	$NH_4H_2PO_4$	MAP	3.5	2.9	NH_4	2.9
Triammonium pyrophosphate	$(NH_4)_3HP_2O_7 \cdot H_2O$	TPP	6.0	6.8	NH_4	10.2
Diammonium phosphate	$(NH_4)_2HPO_4$	DAP	8.0	3.8	NH_4	7.6
Sparingly soluble						
Dicalcium phosphate	$CaHPO_4$ $CaHPO_4 \cdot 2H_2O$	DCP	6.5	~0.002	Ca	0.001
Hydroxyapatite	$Ca_{10}(PO_4)_6(OH)_2$	HAP	6.5	~10^{-5}	Ca	0.001

SOURCE: Sample et al., in F. E. Khasawneh et al., Eds., *The Role of Phosphorus in Agriculture*, p. 275. Madison, Wis.: American Society of Agronomy, 1980.

solution; at the same time, the solution's composition is changed by its contact with soil. Some soil minerals may actually be dissolved by the concentrated P solution, resulting in the release of large quantities of reactive cations such as Fe^{3+}, Al^{3+}, Mn^{2+}, K^+, Ca^{2+}, and Mg^{2+}. Cations from exchange sites may also be displaced by these concentrated solutions. P in the concentrated solutions reacts with these cations to form specific compounds, referred to as *soil-fertilizer reaction products.*

An example of fertilizer P dissolution and reaction product formation is shown in Figure 6.30. Monocalcium phosphate [MCP, $Ca(H_2PO_4)_2$] is added to soil, and water diffuses toward the granule. As the MCP dissolves, H_3PO_4 is formed, resulting in a solution pH of 1.5 near the granule (Table 6.11). Other soil minerals in contact with the H_3PO_4 may be dissolved, increasing the cation concentration near the granule. Subsequently, the solution pH will increase as the H_3PO_4 is neutralized. Within a few days or weeks, DCP and/ or DCPD will precipitate as the initial fertilizer reaction product. Depending on the native P minerals initially present in the soil, OCP, TCP, HA, or $Fe/AlPO_4$ may eventually precipitate (see Fig. 6.12).

Precipitation reactions are favored by the very high P concentrations existing in close proximity to granules, droplets, and bands of fertilizer. Adsorption reactions are expected to be most important at the periphery of the soil-fertilizer reaction zone, where P concentrations are much lower. Although both precipitation and adsorption occur at the application site, precipitation reactions usually account for most of the P being retained in that vicinity. The precipitation of DCP at the application site of MCP is readily apparent in Figure 6.31. From 20 to 34% of the applied P will remain as this reaction product at the granule site.

Although the initial reaction products are metastable and are usually transformed with time into more stable but less water-soluble compounds, they will have a favorable influence on the P nutrition of crops. Some of the initial reaction products will provide P concentrations in solution 1,000 times those in untreated soil. The rate of change of the initial reaction products is influenced by soil properties and environmental factors. For example, after initial DCP formation (a few weeks), formation of OCP may take 3 to 5 months.

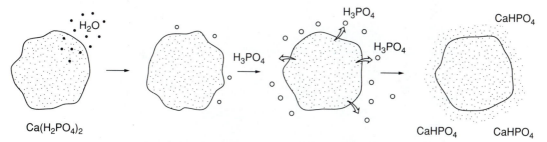

FIGURE 6.30   Reaction of an MCP granule in soil. Water vapor moves toward the granule, which begins to dissolve. Phosphoric acid forms around the granule, resulting in a solution pH of 1.5. The acidic solution causes other soil minerals to dissolve, increasing the cation (and anion) concentration near the granule. With time the granule dissolves completely and the solution pH increases, with subsequent precipitation of a DCP reaction product.

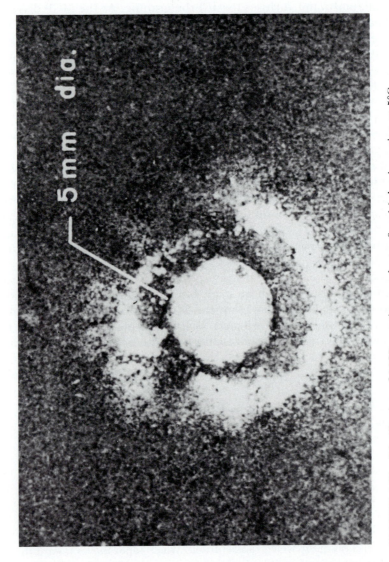

5 mm dia.

FIGURE 6.31 Distribution of MCP reaction products after 14 days' reaction at 5°C in the Bradwell very fine sandy loam. *Hinman et al.*, Can. J. Soil Sci., **42**:229, *1962.*

Further transformation to TCP or HA may take 1 year or longer. The residual value of fertilizer P is dependent on the nature and reactivity of long-term reaction products.

In acid soils, reaction products formed from MCP include DCP and eventually $AlPO_4$ and $FePO_4$ precipitates (Fig. 6.12). If the soil is very acid with low Ca^{2+}, then $AlPO_4$ may precipitate first. Similarly, in calcareous soils, DCP is the dominant initial reaction product. Also, in the presence of extremely large amounts of calcium carbonate, OCP may form.

Inclusion of other fertilizer salts, such as $(NH_4)_2SO_4$, NH_4NO_3, NH_4Cl, KNO_3, K_2SO_4, and KCl with MCP, will significantly reduce the amount of P precipitated in the vicinity of the application site. Because MAP has a reaction pH of 3.5 compared to pH 8.0 for DAP, P should be more soluble near the dissolving granule. The acid pH with MAP may temporarily reduce the rate of P reaction product precipitation in calcareous soils. Although differences in pH among the various P fertilizers cause differences in reaction product chemistry, the overall effect is temporary because the volume of soil influenced by the P granule or droplet is small. Differences in availability of P sources to crops are small compared to differences in other P management factors like P rate and placement.

Fluid ammonium polyphosphate (10-15-0) applied to soil will react similarly to the granular P fertilizers. The reaction pH is 6.2, and both precipitation and adsorption of the polyphosphate and the orthophosphate present initially, plus that formed by hydrolysis of the polyphosphate, occur with soils similar to those described earlier for orthophosphate.

Hydrolysis or the reaction of H_2O with polyphosphate results in a stepwise breakdown, producing orthophosphates and various shortened polyphosphate fragments. The shortened polyphosphate fragments then undergo further hydrolysis. Reactions of polyphosphates in soil and the nature of the substances formed are dependent on the rate of their reversion back to orthophosphates. Slow hydrolysis rates permit condensed phosphates to sequester or form soluble complexes with soil cations and thus avoid or reduce P retention reactions.

Transformation of polyphosphates back to $H_2PO_4^-$ occurs by two principal pathways, either chemically or biologically. Chemical hydrolysis of condensed phosphates proceeds very slowly in sterile, neutral solutions at room temperature. Clay minerals and hydrous oxides, particularly iron oxide, are reported to make minor contributions to the chemical hydrolysis of polyphosphates. In soils, where both mechanisms can function, hydrolysis is usually rapid.

Several factors control hydrolysis rates in soils, with enzymatic activity provided by plant roots and microorganisms probably being the most important. Phosphatases associated with plant roots and rhizosphere organisms are responsible for the biological hydrolysis of pyro- and polyphosphates.

Temperature, moisture, soil C, pH, and various conditions that encourage microbial and root development favor phosphatase activity and hydrolysis of polyphosphates. Temperature is probably the most important environmental factor influencing the rate of polyphosphate hydrolysis. The extent of hydrolysis of pyro- and polyphosphate was increased substantially by elevating the temperature from 5 to 35°C (Fig. 6.32).

Retention of pyrophosphate by soil constituents will substantially lower

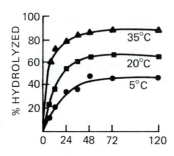

FIGURE 6.32   Effects of temperature on hydrolysis of water-soluble pyrophosphate (200 ppm). *Chang and Racz*, Can. J. Soil Sci., *57:271, 1977.*

hydrolysis rates by decreasing its accessibility for conversion by phosphatase. There are indications that pyrophosphates are adsorbed more strongly on clays and soils than is $H_2PO_4^-$.

Diffusion of P from the application site is accompanied by two reactions: (1) hydrolysis of pyro- and polyphosphates to orthophosphate and (2) precipitation of phosphate anions. Hydrolysis of the water-soluble fraction of polyphosphate is rapid, and diffusion of the pyro- and polyphosphates occurs largely as $H_2PO_4^-$. The precipitation reactions of pyro- and polyphosphates also are rapid and localized within the application zone or band.

Polyphosphates are as effective as $H_2PO_4^-$ as sources of P for crops. Plants can absorb and utilize the polyphosphates directly.

Because polyphosphates have the ability to form metal ion complexes, it has been suggested that they may be effective in mobilizing Zn in soils in which deficiencies have been induced by high pH or high P levels. After the addition of a high rate of pyrophosphate, only slight increases in Zn in the soil solution have been observed. This short-lived effect probably results from either sequestering of Zn by the pyrophosphate or solubilization of soil organic matter. Any complexing of Zn by polyphosphate can only be transitory because hydrolysis is usually very rapid.

INTERACTION OF N WITH P   Since N accounts for at least one-half of the total number of ions absorbed, it is reasonable that P uptake is influenced by the presence of fertilizer N. N promotes P uptake by plants by (1) increasing top and root growth, (2) altering plant metabolism, and (3) increasing the solubility and availability of P. Increased root mass is largely responsible for increased crop uptake of P. Ammoniacal fertilizers have a greater stimulating effect on absorption than NO_3^-.

Greater effectiveness of fertilizer P has been reported from many areas of the United States and Canada when fertilizer application systems place P in close association with NH_4^+-N sources. For example, agronomic advantages, often resulting in 5 to 6 bu/a yield increases of winter wheat, can be gained by simultaneously injecting anhydrous NH_3 and APP solution into the soil. This and other application methods are discussed at greater length in Chapter 12.

EFFECT OF GRANULE OR DROPLET SIZE   The relative effectiveness of P fertilizers is influenced by the size of the granule or droplet and the water solubility of the P fertilizer. Generally, the plant response to P is improved with water-insoluble or slightly soluble phosphates on both acid and calcareous soils when they are applied in powdered form or in very fine granules (less than 35 mesh) and mixed thoroughly with the soil of the root zone.

Since water-soluble P is rapidly converted to less soluble P reaction products, decreasing the contact between soil and fertilizer will generally improve the plant response to P fertilizer. Increasing the granule size and/or band application of the fertilizer will decrease soil–fertilizer contact and maintain a higher solution P concentration for a longer time compared to broadcast P and/or fine particle size.

Most fertilizer P sold in the United States contains 90% or more water-soluble P, and the particle size for MAP and DAP ranges between 6 and 20 mesh (3.4 to 0.84 mm). Thus, compared to broadcast P, band-applied P generally increases plant-available P, especially on low-P soils. With fluid P fertilizers (i.e., APP), droplet size varies with the type of applicator, but in most cases, band-applied fluid P increases P availability compared to broadcast applications.

SOIL MOISTURE   Moisture content of the soil influences the effectiveness and availability of applied P in various forms. When the soil water content is at field capacity, 50 to 80% of the water-soluble P can be expected to move out of the fertilizer granule within a 24-hour period. Even in soils with only 2 to 4% moisture, 20 to 50% of the water-soluble P will move out of the granule within the same time.

RATE OF APPLICATION   At low rates of P water solubility may be much more important than at high rates. When optimum application rates cannot be used, it is important, that materials of high water solubility be used for full benefit from the limited amounts of fertilizer applied. The effect of the degree of water-soluble P on corn yields in Iowa is shown in Figure 6.33.

Even though fertilizer P eventually forms less soluble P compounds, the P concentration in solution increases with P application. With time the P concentration decreases as less soluble P compounds precipitate. The duration of elevated solution P levels depends on the rate of P fertilizer applied, the method of P placement, the quantity of P removed by the crop, and the soil properties that influence P availability.

RESIDUAL P   When the amount of fertilizer P added to soil exceeds removal by cropping, the fertilizer P residues gradually increase, with a corresponding rise in P concentration in the soil solution. In both acidic and basic soils, substantial benefits from residual P can persist for as long as 5 to 10 years or more. The duration of the response will, of course, be influenced by the amount of residual P.

The effect of large initial additions of P on available soil P is demonstrated in (Figure 6.34). At all three rates there was a substantial increase in available P in these deficient soils. Following a rapid decline in available P in the first year, there was a gradual decrease of 2 to 7 ppm soil P per year, depending on the P rate. At the end of the experiment, soil test P was about two, four, and eight times greater than that of the unfertilized soil. Phosphate fertilizer is usually recommended when $NaHCO_3$ soil test levels are below 15 ppm and

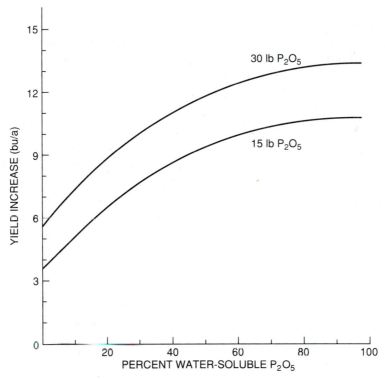

FIGURE 6.33   Effect of rate and water solubility of
applied phosphate fertilizer on the yield increase of
corn.   *Webb et al.,* SSSA Proc., ***22:533, 1958.***

large, economical yield increases are expected from applied P when soil tests
are below 10 ppm.

These data demonstrate that relatively high P rates are needed to substantially increase and maintain residual available P over a long time period. The
data shown in Figure 6.35 illustrate the change in plant-available P with P
rates based on crop need. First, plant removal of P in the unfertilized soil
caused initial soil test P to decrease substantially over the 6-year study period.
Annual application of 50 kg/ha P maintained the soil test P at 2 to 3 ppm
above the initial soil test level, whereas the intermediate P rate (25 kg/ha P)
resulted in soil test levels midway between the 0 and 50 kg/ha P annual rates.
Triennial application of 75 kg/ha P increased available P in the first year;
however, soil test P subsequently decreased below the initial soil test level until
the next triennial application. Similarly, 75 kg/ha P applied only in the first
year maintained soil test P at or above the initial level during the first 3 years,
followed by decreasing soil test P in subsequent years. These data illustrate
the importance of soil testing for accurately determining when additional
fertilizer P is needed for optimum production.

P placement also influences the quantity of residual fertilizer P. The data
shown in Figure 6.36 illustrate the influence of both P rate and placement
effects on soil test P 23 months after application. On this low-P soil, soil test

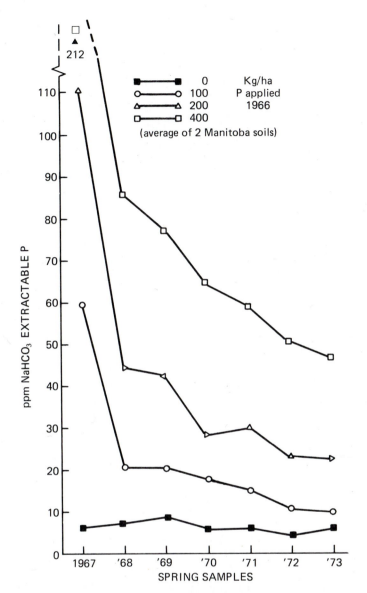

FIGURE 6.34   Effect of single applications of P in 1966
on the $NaHCO_3$ extractable P levels in the soils while
being cropped alternately with wheat and flax from 1967
to 1973.   *Spratt*, Better Crops Plant Food, *62:24, 1978.*

P for broadcast (BC) P applied at 15 and 45 lb P_2O_5/a was no different from
that of the unfertilized soil, indicating that the fertilizer P applied and not
taken up by the crop had been converted to P compounds with a solubility
similar to that of the native P minerals. However, increasing band-applied
(KN) P from 15 to 75 lb/a P_2O_5 dramatically increased soil test P in the band,
indicating that the solubility of the P reaction products is greater than that of
the native P minerals and that they persist for several years after application.

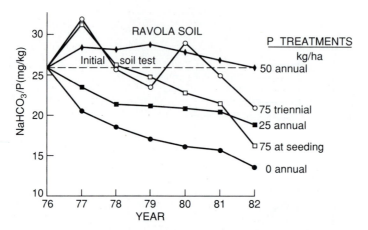

FIGURE 6.35   Influence of broadcast fertilizer P on buildup or decline in soil test P over 6 years.   *Havlin et al., SSSAJ, 48:332, 1984.*

There is some question about the need for additional P even when residual P levels are high. Low rates of P in starter fertilizers placed with or near the seed row are potentially beneficial on high-P soils when the crop is stressed by cold, wet conditions, and diseases such as root rots. Although residual P contributes significantly to crop yields, additional banding of P may be required to maximize crop production.

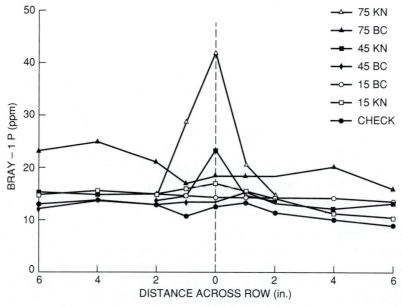

FIGURE 6.36   Influence of band-applied fertilizer P on soil test P in the band 23 months after application.   *Havlin et al., Proc. FFF Symposium, p. 213, 1990.*

Summary

1. P occurs in soils in both inorganic and organic forms. The concentration of the inorganic forms ($H_2PO_4^-$, HPO_4^{2-}) in the soil solution is the most important single factor governing the availability of this element to plants. Uptake of $H_2PO_4^-$ is more rapid than that of HPO_4^{2-}, with the former being most abundant at pH values below 7.2.

2. The concentration of phosphate ion in the soil solution is influenced by the rate and extent to which this element is immobilized by biological factors and by reaction with the mineral fraction of soils. Soils high in soluble Fe and Al react with ortho- and polyphosphates to form a variety of insoluble compounds, including variscite and strengite, which are largely unavailable to plants. Soluble phosphates also undergo reactions in soils high in clays (especially those of the 1:1 type and accompanying hydrated oxides of Fe and Al), which convert them to forms of limited availability to plants.

3. Availability of soluble phosphates is reduced by the high Ca activity prevailing in most basic soils. In such soils, phosphate is precipitated as relatively insoluble DCP and other, more basic calcium phosphates such as OCP and hydroxyapatite.

4. The availability of added water-soluble fertilizer phosphates can be considerably extended by placing them in a band in the soil. Similar results can be obtained by granulating the phosphate materials.

5. The terminology distinctive to phosphate fertilizers was discussed, and the terms *water-soluble, citrate-soluble, available*, and *total phosphorus* were defined. The adequacy of the current methods for assaying the availability of phosphate fertilizers was covered.

6. Acid-treated phosphates are those in which the RP is treated with a strong acid such as sulfuric, phosphoric, or nitric acid. The P in unammoniated, acid-treated phosphates is largely water-soluble.

7. The phosphobacterins, so widely publicized by Russian scientists, were discussed. Although these bacterial cultures may be required to increase native soil phosphates under conditions of Soviet agriculture, they do not appear to be helpful in areas in which large amounts of inorganic phosphate fertilizers are used.

8. The accumulation and benefits of residual P were reviewed. Formation of OCP is an important factor in the value of residual P in basic soils.

9. The behavior and properties of the various phosphate fertilizers in the soil were discussed. The reaction of soluble phosphate fertilizers with various soil components gives rise to what is termed *fertilizer-soil reaction products*, and it is the solubility of these compounds that largely governs plant availability of added phosphate fertilizers.

10. The suitability of the various materials as sources of fertilizer P was covered. As a general rule, those materials that have a high percentage of the contained phosphate in the water-soluble form are more generally acceptable than those with none or only a small amount. There are, however, certain crops and certain soil conditions with and on which the less water-soluble forms perform as well as those that are water-soluble. As with N, the cost per unit of contained P should loom large as one of the determining factors in

the selection of a phosphate fertilizer, but this decision should be tempered by the response of the crop to water-soluble forms.

Questions

1. What is the original source of soil P?
2. Is the soil P in organic combinations available to plants?
3. What are the factors affecting the retention of P in soils?
4. How is P availability influenced by soil pH?
5. What are soil P reaction products?
6. What are probably the two most important factors that influence the uptake of P by plants?
7. What is phosphate retention or fixation? Why is it important agriculturally? Is fixed P totally lost to plants?
8. What are the various mechanisms of phosphate retention in acid mineral soils?
9. What soil properties influence the retention or fixation of added fertilizer P?
10. What can be done to reduce the amount of fixation of fertilizer P?
11. What is the original source of most fertilizer P?
12. A fertilizer contains 46% P_2O_5. To what percent of P does this correspond?
13. Derive the conversion factor

$$\%P = \frac{\%P_2O_5}{2.29}$$

14. What is meant by ammoniating superphosphates? What is the effect on the water- and citrate-soluble contents of ammoniating SSP and TSP? Why the difference?
15. What acids are commonly used to acidify RP? Why, specifically, does acid treatment of RP render the P more plant-available?
16. Describe the soil conditions under which you might expect an appreciable downward movement of P through the soil profile.
17. What is the significance of a large R_2O_3/P_2O_5 ratio? A narrow ratio? Is this important to the grower? Why?
18. Under what soil conditions would the band placement of P result in its greatest utilization by the plant? If there were no such thing as P fixation, what method of fertilizer placement would probably result in the greatest utilization of this element by plants? Why?
19. Phosphates held in organic combination are generally considered to be of little value to plants during cold weather. Why?
20. A soil was reported to contain 20 ppm of available P. To how many pounds per acre of concentrated superphosphate (8.5% P) does this correspond?
21. Chemically speaking, on what is the stability of RP based?
22. Under what types of soil and cropping conditions might the use of RP give satisfactory results? Explain.
23. With what type of crop would the use of a high-water-soluble phosphate be particularly recommended?
24. What types of phosphate fertilizer are not recommended for use on alkaline and calcareous soils?
25. On what basis should phosphate fertilizers be purchased—total or available P?
26. What advantages are offered by the high-analysis phosphates such as DAP, MAP, and TSP? What disadvantages?
27. What is residual P? Why is it important agriculturally?

28. Why are DCP and OCP important reaction products?

29. What is an adsorption isotherm? Write the equations for two of the most widely used adsorption isotherms.

30. What information is not provided by adsorption isotherms? What factor is mainly responsible for the shape of isotherms?

31. What is known about desorption rates?

32. Give brief descriptions of the main pathways of transporting soil P to plant roots. Can phosphate fertilization alter the importance of any of these pathways?

33. Compare typical soil solution concentrations of P in unfertilized soil with the required soil solution levels for top yields of corn, wheat, and head lettuce.

34. Define P intensity and quantity factors. What is labile soil P?

35. What are polyphosphates? Are they stable in biologically active soils?

36. Briefly describe the sequence of events that takes place during the dissolution of water-soluble phosphate fertilizers.

37. What is meant by biocycling of P?

38. Describe how the presence of N improves plant utilization of phosphate fertilizers. Which of the two forms, NH_4^+ or NO_3^-, is more beneficial?

39. What occurs during the hydrolysis of polyphosphates, and what agents are responsible for this reaction?

40. What are typical distances for the initial movement of P from fertilizer application sites? Will P in the reaction zones eventually become more uniformly distributed in the soil?

Selected References

BARBER, S. A. 1984. *Soil Nutrient Bioavailability: A Mechanistic Approach.* John Wiley & Sons, New York.

FOLLETT, R. H., L. S. MURPHY, and R. L. DONAHUE. 1981. *Fertilizers and Soil Amendments.* Prentice-Hall, Englewood Cliffs, N.J.

KHASAWNEH, F. E. (Ed.). 1980. *The Role of Phosphorus in Agriculture.* American Society of Agronomy, Crop Science Society of America, Soil Science Society of America. Madison, Wisc.

LINDSAY, W. L. 1979. *Chemical Equilibria in Soils.* John Wiley & Sons, New York.

STEVENSON, F. J. 1986. *Cycles of Soil: Carbon, Nitrogen, Phosphorus, Sulfur, Micronutrients.* John Wiley & Sons, New York.

YOUNG, R. D., D. G. WESTFALL, and G. W. COLLIVER. 1985. Production, marketing, and use of phosphorus fertilizers, pp. 324–376. In O. P. Englestad (Ed.), *Fertilizer Technology and Use.* Soil Science Society of America, Madison, Wisc.

CHAPTER 7

Soil and Fertilizer Potassium

Potassium (K) is absorbed by plants in larger amounts than any other nutrient except N. Although the total K content of soil is usually many times greater than the amount taken up by a crop during a growing season, in most cases only a small fraction of it is available to plants. Potassium—soil mineral relationships are consequently of major significance.

K Content of Soils

Potassium is present in relatively large quantities in most soils, averaging about 1.9%. The total K content of soils may range from only a few hundred 1b/a-6 in. in coarse-textured soils formed from sandstone or quartzite to 50,000 lb/a or more in fine-textured soils formed from rocks high in the K-bearing minerals.

Soils of the southeastern and southern coastal plain areas of the United States are formed from marine sediments that have been highly leached and generally have a low content of plant nutrients. Soils of the Midsouth are formed from igneous, sedimentary, and metamorphic rocks. Because of their age and the climate in which they were formed, these soils are low in K even though the parent rocks are frequently high in K-bearing minerals. In contrast, the soils of the middle western and far western states generally have a high K content because these soils are formed from geologically young parent materials and under conditions of lower rainfall. The low K content in the coastal soils of the Pacific Northwest is accounted for by the high rainfall in that area.

In tropical soils, the total K content may be quite low because of the origin of the soils, high rainfall, and continued high temperatures. Unlike N and P, which are immediately deficient in most tropical soils due to leaching and/or fixation, the need for K frequently arises only after a few years of cropping a virgin soil. From 70 to 90% of the total K is contained in the forest vegetation, and it takes only few crops to remove the K in forest residues.

Exclusive of that added in fertilizers, the K in soils originates from the slow weathering of K-bearing minerals. The minerals considered to be original sources of K are the K feldspars orthoclase and microcline ($KAlSi_3O_8$), muscovite [$KAl_3Si_3O_{10}(OH)_2$], biotite [$K(Mg,Fe)_3AlSi_3O_{10}(OH)_2$], and phlogopite

[$KMg_2Al_2Si_3O_{10}(OH)_2$]. The ease with which these K minerals weather depend on their properties and the environment. As far as plant response is concerned, the availability of K in these minerals, although slight, is of the order biotite > muscovite > potassium feldspars.

K is also found in secondary or clay minerals in the soil: (1) illites or hydrous micas, (2) vermiculites, (3) chlorites, and (4) interstratified minerals in which two or more of the preceding types occur in a more or less random arrangement in the same particle (see Chapter 4).

Forms of Soil K

Soil K exists in four forms, each differing in its availability to crops. These forms, in increasing order of availability, along with estimates of the approximate amounts in each, are as follows: mineral, 5000 to 25,000 ppm; nonexchangeable (fixed or difficultly available), 50 to 750 ppm; exchangeable, 40 to 600 ppm; and solution, 1 to 10 ppm. The relationships and transformations among the various forms of K in soils are depicted in Figure 7.1. The relative importance of the four groupings depends on the mineralogical composition of the soil.

K cycling or transformations among the K forms in soils are dynamic, and equilibrium is generally not attained. Exchangeable and solution K equilibrate rapidly, whereas difficultly available or fixed K equilibrates very slowly with the exchangeable and solution forms. Transfer of K from the mineral or structural fraction to any of the other three forms is extremely slow in most soils, and this K is considered essentially unavailable to crops during a single growing season.

K is held tightly in feldspars and micas, which are very resistant to weathering. Fixed or nonexchangeable K is present mainly within clay minerals such as illite, vermiculite, and chlorite. The small particle size of clays facilitates K release.

Because of the continuous removal of K by crop uptake and leaching, a static equilibrium probably never occurs. There is a continuous but slow transfer of K in the primary minerals to the exchangeable and slowly available forms. Under some soil conditions, including applications of large amounts of fertilizer K, some reversion to the slowly available form will occur. The unavailable form accounts for 90 to 98% of the total soil K, the slowly available form, 1 to 10%, and the readily available form, 0.1 to 2%.

Soil Solution K

Plants take up the K^+ ion from the soil solution. The concentration of K needed in the soil solution will vary considerably, depending on the type of crop and the amount of growth. The optimum K level in the soil solution is between 10 and 60 ppm, depending on the nature of the crop, soil structure, general fertility level, and moisture supply.

Levels of solution K in humid region soils commonly range between 1 and 80 ppm, with 4 ppm being representative. The K concentration in soil saturation extracts usually varies from 3 to 156 ppm, and the higher figures

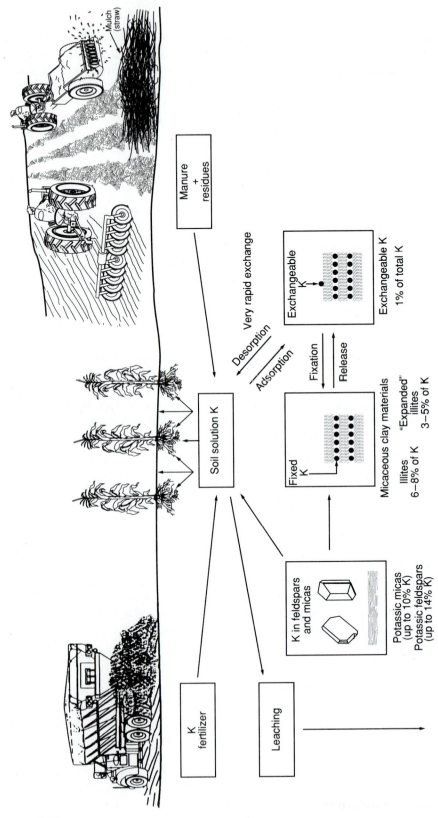

FIGURE 7.1 K equilibria and cycling in soils.

232

are found in arid or saline soils. Under field conditions, the K concentration of the soil solution varies considerably due to the concentration and dilution processes brought about by evaporation and precipitation, respectively.

The effectiveness of soil solution K^+ for crop uptake is influenced by the presence of other cations, particularly Ca^{2+} and Mg^{2+}. It may also be desirable to consider Al^{3+} in very acidic soils and Na^+ in salt-affected soils. The activity ratio (AR_e^K)

$$\frac{\text{activity of } K^+}{\sqrt{\text{activity of } Ca^{2+} \text{ and } Mg^{2+}}} \quad \text{or} \quad \frac{(a_K)}{\sqrt{(a_{Ca + Mg})}}$$

in a solution in equilibrium with a soil provides a satisfactory estimate of the availability of K. This ratio is a measure of the "intensity" of labile K in the soil and represents the K that is immediately available to crop roots.

A single AR_e^K measurement will characterize only momentary availability. Soils with similar AR_e^K values may have quite different capacities for maintaining AR_e^K while K^+ is being depleted by plant uptake or by leaching. Thus, to describe the K status of soils, it is necessary to specify not only the current potential of K in the labile pool but also the way in which the intensity depends on the quantity of labile K present. The quantity–intensity relationships (Q/I) are discussed in the section "Exchangeable K," since this fraction has the principal role in replenishing solution K.

K ABSORPTION BY PLANTS   Diffusion and mass flow of K to plant roots account for the majority of K absorbed. The amount of K that can diffuse is directly related to the intensity of K in the soil solution.

Mass flow depends on the amount of water used by plants and the concentration in solution. The relative contribution of mass flow to K absorption by the plant can be estimated by assuming that the average K concentration in the crop is 2.5% and that the transpiration ratio is 400 g H_2O/g plant. On this basis, water moving to the root would need to contain in excess of 60 ppm K. Since most soils, particularly those in humid regions, contain only about one-tenth this amount, mass flow contributes only 10% of the K requirement. It is apparent, however, that mass flow could supply considerably more K to crops grown in soils naturally high in water-soluble K or where fertilizer K has elevated K in the soil solution. Involvement of mass flow in the transport of K from around a fertilizer source is identified in Table 7.1.

Diffusion of K occurs in response to a concentration gradient, resulting in K transport from a zone of high concentration to one of lower concentration. It is a slow process compared to mass flow (Table 7.1). K diffusion takes place

TABLE 7.1   Mechanisms and Speed of K Transport in Soils

Situation	Mechanism	Speed (cm/day)
In profile	Mainly mass flow	Up to 10
Around fertilizer source	Mass flow and diffusion	≈ 0.1
Around root	Mainly diffusion	0.01–0.1
Out of clay interlayers	Diffusion	10^{-7}

SOURCE: Tinker, in G. S. Sekhon, Ed., *Potassium in Soils and Crops.* New Delhi: Potash Research Institute of India (1978).

in the moisture films surrounding soil particles and is influenced by soil and environmental conditions, including moisture content, tortuosity of the diffusion path, and temperature, which influence the diffusion rates of ions such as K^+ (see Chapter 4).

K diffusion to roots is limited to very short distances in soil, usually only 1 to 4 mm from the root surface during a growing season. The nature of K diffusion to roots can be seen from autoradiographs (Figure 7.2). These were made using ^{86}Rb, an element very similar to K but having a more stable radioactive isotope. The behavior of Rb in soil closely resembles that of K. Since K absorption occurs within only a few millimeters of the root, K farther away, although possibly plant available, is not positionally available.

Diffusion in many soils accounts for 88 to 96% of K absorption by roots. Other factors that influence K diffusion, and thus K availability to crops, include the K concentration gradient, the rate of diffusion, and the surface area of roots.

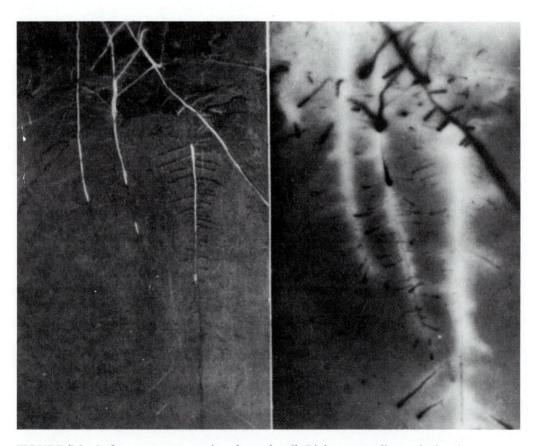

FIGURE 7.2   Left, corn roots growing through soil. Right, autoradiograph showing the effect of corn roots on ^{86}Rb distribution in the soil. Lighter areas are where ^{86}Rb concentration is reduced by root uptake of ^{86}Rb.   *Barber, Potassium availability at the soil–root interface and factors influencing K uptake, in R. D. Munson, Ed.,* Potassium in Agriculture. *Madison, Wisc. ASA, CSSA, SSSA, 1985.*

LUXURY CONSUMPTION   The term *luxury consumption* is often misused. It means that plants will continue to absorb a nutrient in amounts in excess of that required for optimum growth. This results in an accumulation of the nutrient in the plant without a corresponding increase in growth and suggests inefficient and uneconomical use of that particular nutrient. However, with higher crop yields, a greater concentration of K, and of other nutrients as well, is required. As an example, 1.0 to 1.2% K in alfalfa was formerly thought to be adequate. Now a K level of 2 to 3% is considered necessary to maintain consistently high yields and good stands, and to enable alfalfa plants to survive stress periods with a minimum decrease in growth.

Exchangeable K

Like other exchangeable cations, K^+ is held around negatively charged soil colloids by electrostatic attraction. Cations held in this manner are easily exchanged when the soil is brought into contact with neutral salt solutions. In many laboratories, neutral ammonium acetate (in NH_4OAc) solution is the standard extractant for exchangeable K in soils. Usually, less than 1% of the total K in soils occurs in this form.

The distribution of K between negatively charged sites on soil colloids and the soil solution is a function of the kinds and amounts of complementary cations, the anion concentration, and the properties of the soil cation exchange materials. Ca is commonly the major cation in the soil solution and on the exchange complex.

Some of the principles of K^+ exchange from soil colloids are summarized by the two equations that follow. Consider first the reaction

$$\text{clay}\begin{bmatrix} Al \\ K \\ K \\ K \\ K \end{bmatrix} + CaSO_4 \rightarrow \begin{bmatrix} Al \\ Ca \\ K \\ K \end{bmatrix}\text{clay} + K_2SO_4$$

If a soil colloid is saturated with K^+ and a neutral salt such as $CaSO_4$ is added, some of the adsorbed K^+ will be replaced by Ca^{2+}. The amount of replacement will depend on the nature and amount of the added salt, as well as the quantity of K^+ adsorbed on the clays. On some soils used for production of perennial crops, $CaSO_4$ is applied to encourage K^+ displacement and movement into the subsoil, where it becomes available to roots deeper in the profile. Suppose that a soil condition is represented by the following equation:

$$\text{clay}\begin{bmatrix} Ca \\ Al \\ Al \end{bmatrix} + KCl \rightarrow \begin{bmatrix} \frac{1}{2}Ca \\ K \\ Al \\ Al \end{bmatrix}\text{clay} + \frac{1}{2}CaCl_2$$

This soil clay contains adsorbed Ca^{2+} and Al^{3+} to which KCl has been added. Because the Ca^{2+} is more easily replaced than the Al^{3+}, the added K^+ will replace some of the Ca^{2+} and will itself be adsorbed onto the surface of the clay. This reaction illustrates an important point: the greater the degree of Ca^{2+} saturation, the greater the adsorption of K^+ from the soil solution. This is consistent with the previous example, in which the Ca^{2+} from $CaSO_4$ replaced K^+ from the colloid. Ca, when added as a neutral salt, replaces Al^{3+} only with great difficulty, and if a soil clay contains, K^+, Na^+, and NH_4^+ in addition to Al^{3+}, these ions, rather than the Al^{3+}, will be replaced. In such cases, there will be a net transfer of K^+ to the soil solution. The difference between cations in ease of displacement is due to the *lyotropic* series for exchangeable cations; that is, the strength of adsorption is $Al^{3+} > Ca^{2+} > Mg^{2+} > K^+ = NH_4^+ > Na^+$ (see Chapter 4).

Sandy soils with a high base saturation lose less of their exchangeable K^+ by leaching than soils with a low base saturation. Liming is the common means by which the base saturation of soils is increased, and it must follow that liming decreases the loss of exchangeable K^+. Part of this may be due to an increase in pH-dependent CEC.

Exchangeable K^+ on soil colloids is held at three types of exchange sites or binding positions (Fig. 7.3). The planar position (*p*) on the outside surfaces of some clay minerals such as mica is rather unspecific for K. By contrast, the edge position (*e*) and the inner position (*i*) in particular have a rather high specificity for K.

Under field conditions, soil solution K concentrations are buffered more readily by K^+ held to *p* positions; however, K held on all three positions contributes to solution K. A high proportion of adsorption by clay minerals tends to saturate the specific binding sites, resulting in higher concentrations in the equilibrated soil solution.

Because of the major role of exchangeable K in replenishing soil solution K removed by cropping or lost by leaching, there is much interest in defining the relationship between exchangeable K (*Q* for quantity) and the activity of soil solution K (*I* for intensity). The *Q/I* ratio, is used to quantify the K status of soils.

Labile soil K (held in *p* positions) may be more reliably estimated by *Q/I* than by the measurement of exchangeable K with $1\,N\,NH_4OAc$. Higher values

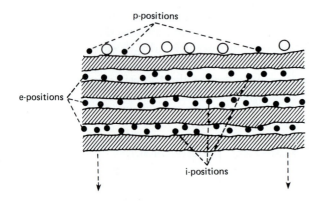

FIGURE 7.3   Binding sites for K on 2:1 clay minerals such as illite, vermiculite, and chlorite. *Mengel and Haeder*, Potash Rev., **11:1, 1973.**

of labile K indicate a greater K release into soil solution, resulting in a larger pool of labile K. Fertilizer K also will increase labile K.

Q/I measures the ability of soil to maintain the intensity of soil solution K^+ and is proportional to the CEC. A high value signifies good K-supplying power or BC, while a low figure suggests a need for K fertilization. Liming can increase *Q/I*, presumably as a result of the increase in pH-dependent CEC.

When *Q/I* values are low, small changes in exchangeable K produce large differences in soil solution K^+. Potential BC is extremely small in sandy soils in which the CEC is due mainly to OM. In such soils, intense leaching or rapid plant growth can seriously deplete available K in just a few days.

The difference in soil solution K^+ between kaolinitic and illitic clays is indicated in Figure 7.4. The steeper slope of the kaolinitic soil indicates lower BC. The effect of a lower clay content and hence lower BC, on increasing the K^+ concentration in the soil solution around the root is shown in Figure 7.5.

In general, the relation between exchangeable and solution K^+ is a good measure of the availability of the more labile K in soils to plants. The ability of a soil to maintain the activity ratio against depletion by plant roots and leaching is governed partly by the labile K pool, partly by the rate of release of fixed K, and partly by the diffusion and transport of K ions in the soil solution.

Nonexchangeable and Mineral K

The remaining soil K is generally referred to as *nonexchangeable* and *mineral* or *reserve* K, which is largely nonexchangeable and considered to be slowly available. Although nonexchangeable K reserves are not always immediately available, they can contribute significantly to maintenance of the labile K pool in soil. In some soils, nonexchangeable K becomes available as the exchange-

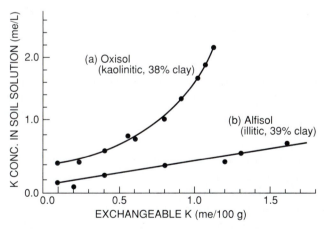

FIGURE 7.4   Relationship between exchangeable K and K concentration in the soil solution of two soils with the same clay content but different clay mineralogy. The steeper slope of the kaolinitic soil indicates less buffer capacity.   *Nemeth, unpublished.*

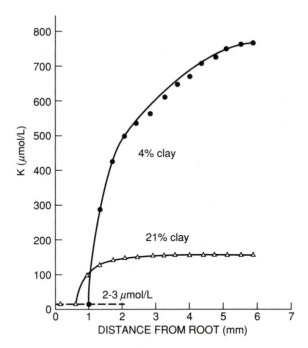

FIGURE 7.5   K concentration in the soil
solution around a maize root in two soils after 3
days. The initial exchangeable K content was
0.17 meq/100 g in the loamy soil (21% clay) and
0.37 meq/100 g in the sandy soil (4% clay).
Because of the lower BC of the sandy soil, the
difference of concentrations in the soil solution
were higher than that of exchangeable K.
Claassen and Jungk, *1982.*

able and solution K^+ are removed by cropping or lost by leaching. In other
soils, release from nonexchangeable K is too slow to meet crop requirements.

Nonexchangeable K in illitic clays, especially in vermiculite and 2:1 in-
tergrade clay minerals, is often determined by extraction with a strong acid
such as 1 N boiling nitric acid (HNO_3). With certain soils, nonexchangeable K
release helps to explain crop responses or the absence of responses to fertilizer K.

The rate of K supply or release to solution and exchangeable K is largely
governed by the weathering of K-bearing micas and feldspars. Feldspars have
a three-dimensional crystal structure, with K located at the interlayers
throughout the mineral lattice (see Chapter 4). K can be released from feldspars only by destruction of the mineral. In micas, interlayer K can be released
by exchange with other cations, without fundamental alteration of the mineral.

K feldspars are the largest natural reserve of K in many soils. In moderately
weathered soils, there are usually considerable quantities of K feldspars. They
often occur in much smaller amounts or may even be absent in strongly
weathered soils such as those in humid tropical areas. K-bearing minerals may
be present in the clay fraction, as well as in the silt and sand fractions.

Figure 7.6 shows the percent distribution of feldspar and mica K in clay, silt, and sand fractions of two Saskatchewan soils. Most of the K feldspar was found in the silt and sand fractions. The observed partition of both K-containing minerals is due to the effects of soil-forming processes and topography.

The micas are 2:1 layer-structured silicates composed of a sheet of Al octahedra between two sheets of Si tetrahedra (see Chapter 4). In muscovite (dioctahedral) only two out of the three octahedral positions are occupied by trivalent ions (Al^{3+}). Biotite and phlogopite, the other major types of mica in soils, are classified as trioctahedral and have all three octahedral positions filled by Mg^{2+} and Fe^{2+}. Potassium ions reside mainly between the silicate layers. Bonding of interlayer K is stronger in dioctahedral than in trioctahedral micas; therefore, K release generally occurs more readily with biotite than with muscovite. The rate of K release from biotite is 13 to 16, 75 to 105, and 118 to 190 times faster than that from phlogopite, muscovite, and microcline feldspar, respectively.

The gradual release of K from positions in the mica lattice results in the formation of illite (hydrous mica) and eventually vermiculite, with an accompanying gain of water or OH_3^+ and swelling of the lattice (Fig. 7.7). There is also an increase in the specific surface charge and CEC of the clay minerals formed during the weathering and transformation of mica.

K release from mica is both a cation exchange and a diffusion process, requiring time for the exchanging cation to reach the site and for the exchanged ion (K^+) to diffuse from it. A low K concentration or activity in the soil solution favors the liberation of interlayer K. Thus depletion of K by the

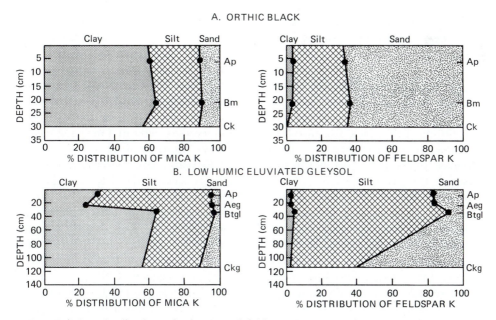

FIGURE 7.6 Distribution of mica K and feldspar K among clay, silt, and sand fractions of the Orthic Black (A) and Low Humic Eluviated Gleysol (B) profiles in the Oxbow catena. *Somasiri et al., Soil Sci. Soc. Am. J., **35**:500, 1971.*

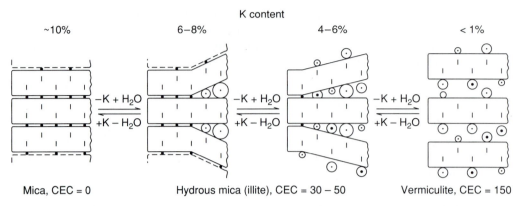

FIGURE 7.7   Schematic weathering of micas and their transformation into clay minerals: a matter of K release and fixation.   *McLean, in G. S. Sekhon, Ed., Potassium in Soil and Crops, pp. 1–13. New Delhi: Potash Research Institute of India, 1979.*

plant or leaching may induce release of K from nonexchangeable interlayer positions. It is possible for K^+ to be progressively released and diffused from all interlayer locations, or it may come only from alternate interlayers, leading to the formation of interstratified mica-vermiculite.

K FIXATION   K fixation does not occur to the same extent in all soils. It reaches its maximum, however, in soils high in 2:1 clays and with large amounts of illite (Fig. 7.7). Fixation of K is the result of reentrapment of K^+ ions between the layers of the 2:1 clays. The 1:1-type minerals such as kaolinite do not fix K.

K ions are sufficiently small to enter the silica sheets, where they are held very firmly by electrostatic forces. The NH_4^+ ion has nearly the same ionic radius as the K^+ ion and is subject to similar fixation. Cations such as Ca^{2+} and Na^+ have larger ionic radii than K^+ and do not move into the interlayer positions of the clays. Because NH_4^+ can be fixed by clays in a manner similar to that of K^+, its presence will alter both the fixation of added K and the release of fixed K. Just as the presence of K^+ can block the release of fixed NH_4^+, the presence of NH_4^+ can block the release of fixed K. This is illustrated in Figure 7.8, which shows a reduction in the release of nonexchangeable K with increasing amounts of added NH_4^+. The NH_4^+ ions evidently are held in the interlayer positions, further trapping the K^+ ions already present.

Soils may differ greatly in the rate at which nonexchangeable K is released. Considerably more K is released from Marshall than from Clarion soil; however, this release is blocked in both soils by the presence of NH_4^+.

K fixation is generally more important in fine-textured soils which have a high fixation capacity for both K^+ and NH_4^+. Although it is not generally considered to be a serious factor in limiting the crop response to either applied NH_4^+ or K^+, increasing the concentration of K^+ in soils with a high fixation capacity will obviously encourage greater fixation.

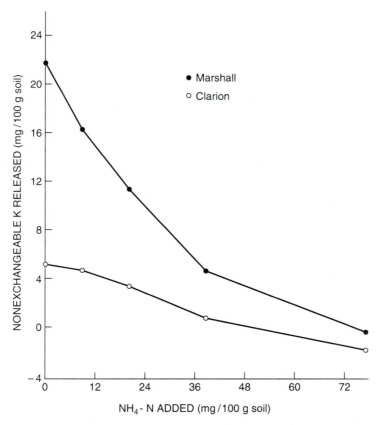

FIGURE 7.8   Nonexchangeable K released by Marshall and Clarion surface soils during a 10-day cropping period as influenced by the amount of added NH_4^+.   *Welch and Scott, SSSA Proc., 25:102, 1961.*

K fixation capacity can be reduced by the presence of Al^{3+} and their polymers that form under acid conditions. These Al^{3+} cations will occupy the K selective binding sites. The occurrence of hydroxyl Al-Fe interlayer groups under acidic conditions will prevent the collapse of Si layers in highly expanded clays. This separation prevents K ions from being trapped and subsequently fixed by collapsing Si layers.

Air-drying some soils high in exchangeable K can result in fixation and a decline in exchangeable K. In contrast, drying of field-moist soils low in exchangeable K, particularly subsoils, will frequently increase exchangeable K. In some cases, the exchangeable K in subsoils will be increased severalfold by drying. The release of K upon drying is thought to be caused by cracking of the clay edges and exposure of interlayer K, which can then be released to exchange sites. Most of the K fixation observed during drying of moist, high exchangeable K soils is associated with vermiculite or other minerals containing expanded mica (beidellite, illite).

The effects of wetting and drying on the availability of K under field conditions are not known. They are important, however, in soil testing. Soil test procedures call for the air-drying of samples before analysis. This drying treatment can substantially modify soil test K values—and subsequent recommendations for K fertilization.

The freezing and thawing of moist soils may also be important in K release and fixation. With alternate freezing and thawing, certain soils will release K, while in other soils, particularly those high in exchangeable K, no K release is observed. This phenomenon was observed in soils that contained appreciable quantities of illite. It seems probable that freezing and thawing play a significant role in the K supply of certain soils, depending on their clay mineralogy and degree of weathering.

Retention of K in less available or fixed forms is of considerable significance to the practical aspects of farming. As with P, the conversion of K to slowly available or fixed forms reduces its immediate value as a plant nutrient. However, it must not be assumed that K fixation is completely unfavorable. K fixation results in conservation of K, which can become available over a long period of time and is thus not entirely lost to plants, although crop plants do vary in their ability to utilize slowly available K.

Soil Factors Affecting K Availability

Kinds of Clay Minerals

The greater the proportion of clay minerals high in K, the greater the potential K availability in a soil. For example, soils containing vermiculite or montmorillonite will have more K than soils containing predominantly kaolinitic clays, which are more highly weathered and very low in K. However, intensively cropped vermiculitic and montmorillonitic soils may also be low in K and require K fertilization for optimum crop production. Occasionally, sandy soils low in K, may show little yield response to applied K which is attributed to K release from fixed and mineral forms rapidly enough to maintain adequate exchangeable and solution K levels.

Cation Exchange Capacity (CEC)

Finer-textured soils usually have a higher CEC and can hold more exchangeable K; however, a higher level of exchangeable K does not always mean that a higher level of K will be maintained in the soil solution. In fact, soil solution K^+ in the finer-textured soils (loams and silt loams) may be considerably lower than that in a coarse-textured (sandy) soil at any given level of exchangeable K (Fig. 7.5).

Amount of Exchangeable K

Determination of exchangeable K is the universal measure for predicting K availability and fertilizer requirements. Many studies show the relationship between soil test K and response to applied K. What is implied is that fertilizer applications of K can be adjusted downward with increasing levels of available soil K (Fig. 7.9). It should be noted, however, that profitable responses of small grains and other crops to applied K occur in several northern Plains

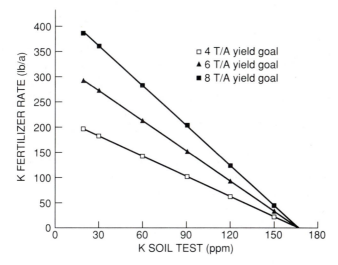

FIGURE 7.9 K fertilizer recommendations for alfalfa in the northern Plains. Fertilizer K rate decreases with increasing soil test K and decreasing alfalfa yield goal.

states and Prairie Provinces on soil testing high in exchangeable K. In Illinois, profitable responses by corn to K on high K soils have been obtained. A combination of higher crop yields with a higher K demand, plus occasional periods of low soil moisture or temperature, could help explain such responses.

Wet conditions in the spring may reduce K availability. Reducing conditions in the soil bring about reduction of Fe in the swelling clay minerals. This reduced Fe causes the clay mineral to fix K, thereby making it unavailable to plants during the early growing season. Drying during the growing season will bring about a gradual reoxidation of the structural Fe and subsequent release of the fixed K.

Capacity to Fix K

In general, the amount of K needed to increase exchangeable K 1 ppm may vary from 1 to 45 lb K/a or more, depending on the soil. The wide difference is related, in part, to the variation in K fixation potential among soils. Fortunately, some of the K that is fixed may be subsequently released to crops, but the release may be too slow for high levels of crop production.

Subsoil K and Rooting Depth

Exchangeable K in subsoils can vary with the soil type; however, little progress has been made in relating subsoil K to crop responses to K. In northern climates, low soil temperature may inhibit K release and diffusion, thus increasing the probability of a crop response to K fertilization in these soils. Temperature effects on K supply are discussed in more detail later.

Soil Moisture

With low soil moisture, water films around soil particles are thinner and discontinuous, resulting in a more tortuous path for K^+ diffusion to roots. Increasing K levels or moisture content in the soil will accelerate K diffusion.

Soil moisture can have substantial effects on K transport in soil (Fig. 7.10). Increasing soil moisture from 10 to 28% increases total transport by up to 175%. Figure 7.11 shows that low soil moisture content impedes diffusion and reduces corn growth. Corn yields decline as the moisture content of the soil drops from 38% to 22%, but these yield depressions are partially offset by high levels of exchangeable K. The yield reduction is due to insufficient K diffusion from soil to the corn roots because an increase in exchangeable K increases yield.

Soil Temperature

The effect of temperature on K uptake is due to changes in both availability of soil K and to changes in root activity and rate of plant physiological processes. It is generally agreed that reduced temperature slows down plant processes, plant growth, and rate of K uptake. For example, K influx into

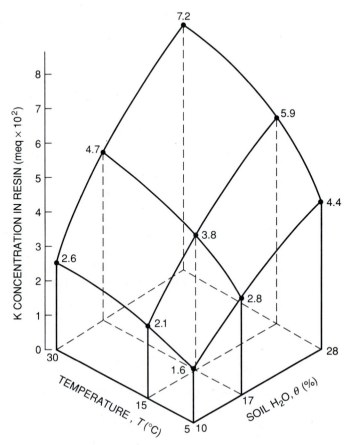

FIGURE 7.10 Diffusion of K to an ion exchange resin sink in Bozeman silt loam during 96 hours, as influenced by temperature and soil moisture. *Skogley*, Proc. 32nd Annu. Northwest Fert. Conf., *Billings, Mont., 1981.*

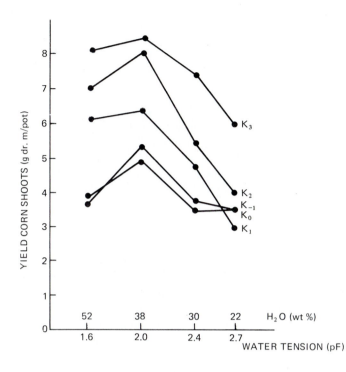

FIGURE 7.11   Effect of soil moisture and K status of the soil on growth of corn. Exchangeable K increasing from $K_0 \rightarrow K_3$.   *Mengel and Haeder*, Potash Rev., *11:1, 1973.*

corn roots at 15°C (59°F) was only about one-half of that at 29°C (84°F) (Fig. 7.12).

In the same study, the root length increase over a 6-day period was eight times greater at 29°C than at 15°C. K concentration in the shoot was 8.1% at 29°C and 3.7% at 15°C.

K uptake of added K can be strongly influenced by soil temperature (Fig. 7.13). The sloping part of each curve shows that K uptake per unit of K applied was 2.6 times higher at 29°C than at 15°C. The soil ceased to limit uptake at approximately the same K level at both temperatures. The supple-

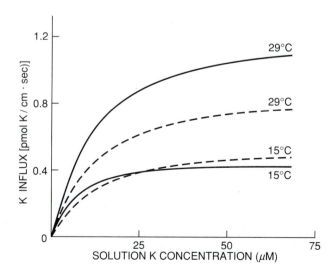

FIGURE 7.12   Rate of K influx into young corn roots is increased by higher temperature and K concentration in solution. Solid lines are for 11-day-old and dashed lines are for 16-day-old corn roots. *Ching and Barber*, Agron. J., *71:1040, 1979.*

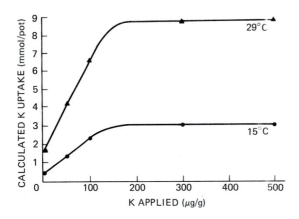

FIGURE 7.13   Total K uptake by corn plants growing at 15 and 29°C root temperatures as a function of K level in Raub soil. Predicted K uptake was highly correlated with observed K uptake ($r^2 = 0.98$).   *Ching and Barber, Agron. J., **71**:1040, 1979.*

mental K needed to increase K uptake at low temperatures overcomes some of the adverse effect that low temperature has on rate of diffusion.

As indicated previously, the capacity of the soil to supply K to roots is reduced by the effect of low temperature on diffusion. The K diffusion study shown in Figure 7.10 is further evidence of the strong effect that temperature has on this phenomenon. Increasing the temperature from 5°C to 30°C increased K accumulation in the resin sink by about 65% at each level of soil moisture.

Also influencing the capability of soil to satisfy plant requirements for K is the concentration of K in soil solution. The values in Table 7.2 show less K in soil solution when soils were equilibrated at 15°C compared to 29°C. This effect of temperature was greatest at low K levels. Also of special interest is the situation that where no K was added, the effective diffusion coefficient at 15°C was only 0.4% of what it was at 29°C. The divergences in effective diffusion coefficients at the two temperatures tended to diminish at the higher soil K levels (see Chapter 4).

Providing high levels of K is a practical way of overcoming some of the problems of low temperature. Temperature effects are probably a major reason for crop responses to row-applied fertilizer for early-planted crops such as corn. The beneficial effect of K fertilization for early seedings of barley

TABLE 7.2   Effect of the Soil K Level and Temperature on the Initial Concentration of K in Soil Solution (C_{li}), Buffer Power of K on the Solid Phase for K in Solution (b), and Effective Diffusion Coefficient ($\bar{D}e$) in the Raub Silt Loam

K Added ($\mu g/g$)	$C_{li}(\mu mol/ml)$ at:		b at:		$\bar{D}e \times 10^7$ (cm^2/sec) at:	
	15°C	29°C	15°C	29°C	15°C	29°C
0	0.046	0.089	32	23	0.15	0.39
50	0.174	0.256	12	9.5	0.50	0.94
100	0.355	0.516	8.7	3.3	0.69	2.7
300	1.97	2.66	2.2	1.8	2.7	5.0
500	6.26	8.10	1.2	1.2	5.0	7.5
700	11.90	13.90	1.2	1.2	5.0	7.5

SOURCE: Ching and Barber, *Agron. J.*, **71**:1040 (1979).

grown on soils high in available K is mainly attributed to improvement in the K supply under cool soil conditions (Table 7.3).

Soil Aeration

Respiration and the normal functioning of roots are strongly dependent on an adequate O_2 supply. Under high moisture levels or in compact soils, root growth is restricted, O_2 supply is lowered, and absorption of K and other nutrients is slowed. The inhibitory action of poor aeration on nutrient uptake is most pronounced with K (Table 7.4).

Cropping systems that are detrimental to soil tilth and that cause reduced soil porosity and increased compaction have been found to impair K uptake. Applications of K fertilizer to increase the concentration of K in soil have helped overcome the depressive influence of poor aeration on K absorption by plants. For example, marked decreases in K content in sugar beet petioles were associated with high soil moisture and reduced soil air space. K applied in a band beside the row was effective in minimizing yield reductions due to compaction (Table 7.5).

Soil pH

In very acid soils, toxic amounts of exchangeable Al^{3+} and Mn^{2+} create an unfavorable root environment for uptake of K^+ and other nutrients. When acid soils are limed, exchangeable Al^{3+} and hydroxyaluminum cations such as $Al(OH)_2^+$ are converted to insoluble $Al(OH)_3$. This change removes the Al^{3+} from cation exchange competition with K^+, and it frees blocked binding sites so that K^+ can compete with Ca^{2+} for them. As a consequence, much greater amounts of K^+ can be held by clay colloids and removed from the soil solution. Leaching losses of K will also likely be reduced.

Raising soil pH from 5.5 to 7.0 will favor the collapse of silicate layers of expanded clays and trap K^+ already present in the interlayers. Hydroxyaluminum cations keep the clay layers wedged apart but lose this ability when they are changed to $Al(OH)_3$. K trapped in this manner is unaccessible to plants.

The use of lime on acid soils low in exchangeable K can induce a K deficiency

TABLE 7.3 Effect of K Fertilization on Early Seedings of Barley Grown on Montana Soils High in Available K

Seeding Date	K_2O* (lb/a)	Yield (bu/a)
April 6	0	48
	20	55
May 6	0	36
	20	42
June 3	0	30
	20	33

*N at 60 lb/a/yr and P_2O_5 at 25 lb/a/yr.

SOURCE: Dubbs, *Better Crops Plant Food*, **65:**27 (1981).

TABLE 7.4   Uptake of Nutrients by
Corn Grown in Nonaerated and
Aerated Cultures of a Silt Loam Soil
Containing 50% Water

Component Measured	Relative Uptake: Nonaerated/Aerated
K	0.3
N	0.7
Mg	0.8
Ca	0.9
P	1.3
Dry matter	0.6

SOURCE: Lawton, *Soil Sci. Soc. Am. J.*,
10:263 (1946).

through ion competition effects. Liming soils already at pH 6.0 to 7.5 will usually decrease exchangeable K and reduce K uptake by plants. Removal by liming of the restrictive effect of Al^{3+} on root growth and vigor is expected to more than offset the competitive effect of the added Ca^{2+} on K^+ uptake by plants.

When acid soils are limed, there is usually a substantial increase in pH-dependent CEC. Raising soil pH from 5 to 6 will increase the effective CEC by as much as 50%. The competitive effects of Ca^{2+} and Mg^{2+} will increase, however, if K levels are not increased accordingly.

Applications of high rates of KCl to acid soils can result in large increases in the concentration of potentially toxic elements such as Al^{3+} and Mn^{2+} in the soil solution. The toxicity of Al^{3+} and Mn^{2+} can be intensified and the benefits of increasing the K supply nullified. The extremely high rates of KCl used to induce these changes will, under field conditions, most likely occur in a fertilizer band or in the soil immediately adjacent to a granule or droplet of fertilizer K.

TABLE 7.5   Partial Reduction of Yield Losses Due to
Compaction as a Result of Row-Applied K—Soil Test K,
204 lb/a

Row K_2O	Compaction		Loss
	Low	High	
lb/a	-------- bu/a --------		
0	151	129	22
45	169	164	5
Response	18	35	

SOURCE: Wolkowski et al., Proceedings of 1987 Fertilizer,
Aglime and Pest Management Conference. University of
Wisconsin. **26:**142–150 (1987).

Ca and Mg

Both Ca^{2+} and Mg^{2+} compete with K^+ for entry into plants; thus, soils high in one or both of these cations may require high levels of K for satisfactory nutrition of crops. According to the activity ratio defined earlier, K uptake would be reduced as Ca^{2+} and Mg^{2+} are increased; conversely, uptake of these two cations would be reduced as the available supply of K is increased. Thus, the availability of K is somewhat more dependent on its concentration relative to that of Ca^{2+} and Mg^{2+} than on the total quantity of K present. Because the activity ratio does not always agree with K^+, Ca^{2+}, and Mg^{2+} uptake measurements, consideration must also be given to the absolute amount of K.

Although the depressive action of Ca^{2+} in calcareous soils on K uptake has been widely recognized (Table 7.6), many soil testing laboratories fail to make appropriate adjustments in K fertilizer recommendations. Consistent economic responses in wheat and barley yield have been reported from the addition of K to calcareous soils that test adequate in exchangeable K.

Relative Amounts of Other Nutrients

If nutrients other than K are limiting, the plant need for K is not as great because of reduced growth. For example, in a low-P soil, N fertilization had little effect on K uptake by wheat even though 289 lb of K_2O/a had been applied (Fig. 7.14). K uptake increased dramatically with increasing N rate when 131 lb P_2O_5/a was applied. In situations where other plant nutrients are low, applications of K may reduce yields. For example, corn yields decreased with added K and little or no N; however, with adequate N, yields increased when K was applied (Table 7.7).

Plants receiving high NH_4^+ and inadequate K may develop toxicity symptoms remedied by K applications. The K concentration may be as high in the NH_4^+-nourished plants as it is in those supplied with NO_3^-. Apparently, additional K is needed for proper utilization of high levels of NH_4^+.

Tillage

Most fertilizer recommendations originally were based on a plow or tillage depth of 6⅔ in.; however, deep tillage (i.e., chiseling 12–15 in.) is becoming more extensive in some regions. When tillage is increased from 6⅔ to 10 in.,

TABLE 7.6   Effect of K Uptake and Growth on Young Corn Plants in a High-Lime Iowa Soil

Soil Condition	Uptake (meq/100 g of Dry Matter)			Weight of Corn Plants (g)
	K	Ca	Mg	
High lime	23	55	101	1.2
Normal	107	32	39	12.0

SOURCE: Pierre and Bower, *Soil Sci.*, **55**:23 (1943).

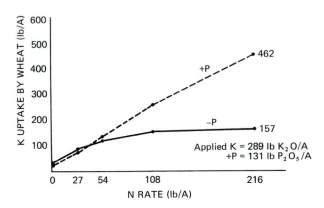

FIGURE 7.14   K uptake by wheat increases with N and P rates. *Schwartz and Kafkafi,* Agron. J., ***70****:227, 1978.*

50% more soil is involved. Where the practice of soil K buildup is used, the K requirement will be increased about 50%, depending on the K status of soil deeper in the tilled layer.

Tillage practices can influence K availability by modifying soil aeration, temperature, moisture, and positional availability of applied K. Crop residues left on the soil surface may lower soil temperature, reduce evaporation, and enhance moisture movement into soil, thus changing nutrient–moisture relations.

Conservation tillage is increasing, and there is evidence that K availability is reduced under this system because of increased compaction, less aeration, lower temperature, and positional availability of K applied on the soil surface. In a number of studies comparing the effects of tilled and till-planted operations on corn yields and K uptake, reducing tillage decreased K availability and crop yields. For example, %K in corn leaves was significantly less under no till compared to full tillage (Table 7.8). Higher corn yields are usually associated with leaf concentrations near 2.0%. Fertilizer K has helped to counteract the reduced K availability under no-till.

Loss of K by Leaching

In most soils, except those that are quite sandy or subject to flooding, K leaching losses are small. Research in Illinois indicates that losses from the silt

TABLE 7.7   Effect of K on Yields When the N Supply
for Corn Is Increased

	Yield of Corn (bu/a)		
	K_2O (lb/a)		Yield Change
N (lb/a)	0	96	(bu/a)
0	78	71	− 7
40	99	94	− 5
80	127	130	+ 3
160	139	143	+ 4
320	144	157	+ 13

SOURCE: Schulte, *Better Crops Plant Food,* **63**:19 (Spring 1979).

TABLE 7.8 Effect of Added K on the Uptake of K by Corn and the Yield Loss from No Till on a Medium K Soil in Wisconsin

K_2O Applied Annually, 1972–1976 (lb/a)	K in Leaves at Silking (%)		Yield Loss from Not Plowing (bu/a)
	Plowed	No-till	
0	0.73	0.59	51
80	1.40	1.04	29
160	1.71	1.42	18

SOURCE: Schulte, *Better Crops Plant Food*, **63:**25 (Fall 1979).

loam soil studied were equivalent to only about 1.4 lb K/a/yr. Other studies, however, showed that large amounts of K would be lost from coarse, sandy soils with low CEC. Passage of 16 in. of water through such soils under experimental conditions resulted in removal of 126 lb K/a, or over 93% of the total supply of K.

The wide differences illustrated by these examples can be explained on the basis of the CEC of these soils and their relative degrees of K saturation. The silt loam soil had a much higher CEC than the sandy soil and was able to hold a larger amount of K. Additions of fertilizer K would result in adsorption of a higher proportion of the added K than in the sandy soil. The unadsorbed K^+ remains in the soil solution and is removed in leaching waters, which explains why a large quantity was lost from the sandy soil with the low CEC.

While organic soils, such as mucks, have high CEC, the bonding strength for K is not great, and the exchangeable K level tends to vary somewhat with the intensity of rainfall. Thus emphasis should be placed on annual applications rather than on buildup of soil K. Since crops grown on organic soils characteristically have a high K need, it is important to monitor the fertility level with soil tests.

In the humid tropics, leaching is recognized as a major factor in limiting productivity. Under natural vegetation leaching is low, in the range of 0 to 5 lb/a/yr. On cleared land after fertilizer application, 35% of the K may be leached with cropping and much higher losses occur on bare land. In these soils emphasis should be placed on annual or split applications rather than on buildup of soil K. Monitoring with soil tests is crucial. In the final analysis, leaching losses of K are important only in coarse-textured, organic, or humid tropic soils in areas of high rainfall.

Plant Factors Affecting K Availability

CEC of Roots

One explanation for the varying abilities of crops to utilize soil K is related to the CEC properties of plant roots. Grasses and cereals with low root CEC respond least to applications of fertilizer K, in contrast to plants such as clovers, which have higher root CEC. Although the CEC of roots may be important in determining the ability of plants to absorb the more slowly available forms of soil K, it is only a minor factor involved in the K supply to roots.

Root System and Crop

The plant factors that have considerable influence on K uptake are as follows:

1. Ion flux (movement) and concentration.
2. Root radius.
3. Rate of water uptake.
4. Root length.
5. Rate of root growth.

Root type and density are two of the major characteristics affecting K availability to crops. The quantity of water transpired will also influence crop uptake of K. The K absorption properties of three crops are compared in Table 7.9. Not shown in this table is root density, which was much lower for onions than for corn. Because of differences in root morphology and K influx, corn removed 47% of the exchangeable and solution K, whereas onions removed only 6% from the same soil volume.

Most grasses have a fibrous root system with many lateral branches, whereas alfalfa, especially under low-K conditions, is tap rooted, with less root renewal. High availability of K actually enhances root development, producing more branching and lateral roots in both alfalfa (Fig. 7.15) and grasses such as small grains and corn (Fig. 7.16).

Differences in rate of K uptake among crop species can result in competition for K. A stable, productive association between white clover and ryegrass can be difficult to maintain because of the ability of the ryegrass to absorb K two to five times faster than white clover. Figure 7.17 shows that the rate of absorption for ryegrass is twice the rate for red clover. This is in addition to the advantage provided by the more extensive root system of ryegrass.

Variety or Hybrid

Ion absorption by plants is partially under genetic control, and considerable differences exist both between and within genera. For example, corn hybrids have different capacities to take up K, ranging from 20 to 40 lb K/a under low fertility and from 160 to 280 lb K/a under high fertility. Similar differences exist among soybean varieties (Fig. 7.18). Dare is more responsive to additions of fertilizer K.

TABLE 7.9   K Absorption Properties of Corn, Onion, and Wheat

Parameter	Maize (16 days)	Onion (20 days)	Wheat (15 days)
Maximum influx, I_{max} [pmol/(cm · sec)]	4.07	3.97	2.93
Average influx, I_n [pmol/(cm · sec)]	1.42	0.53	0.33
Absorption cylinder (mm^2)	2.00	0.18	1.50
Depletion zone (mm)	11.00	7.00	7.00

SOURCE: Baligar and Barber, *11th Congr. Int. Soc. Soil Sci. Abstr.*, **1**:309 (1978).

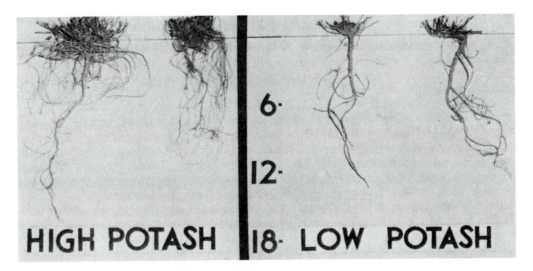

FIGURE 7.15   High available K encourages root development and density of taprooted legumes such as alfalfa.   *Munson, in* Potassium for Agriculture—A Situation Analysis. *Atlanta: Potash & Phosphate Institute, 1980.*

FIGURE 7.16   Adequate K on corn increases root branching and density, size of stalks, and size of ear.   *Courtesy of Potash & Phosphate Institute, Atlanta.*

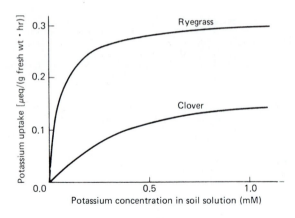

FIGURE 7.17   K uptake by ryegrass and white clover plants growing under a K stress.   *Dunlop et al.*, New Phytol., *83:365, 1979.*

Growth of five barley cultivars grown at five different levels of available K reveals a consistent pattern of superior performance by Herta and Fergus and generally poor performance by Conquest (Table 7.10). The varietal differences were explained by the observation of a strong correlation between K uptake and H^+ efflux from the barley roots. This relationship for 24 barley varieties is illustrated in Figure 7.19.

Plant Population and Spacing

In several of the principal crops, there is a tendency to use higher plant populations or closer row spacing. In many regions, narrower row spacing of 3 to 4 in. has been shown to be advantageous for small grains such as wheat and barley. Higher plant populations and closer row spacings affect nutrient needs, including K. Numerous studies have demonstrated that increasing the population increased K removal, which would imply that high populations may require higher K fertilization rates. For example, as the plant population of two corn hybrids in Florida was raised from 19,000 to 39,000 plants/a, the K uptake increased from 150 to 228 lb K/a for one hybrid and from 128 to 170 lb K/a for another hybrid.

Where plant populations are increased without appropriate increases in K fertilization or availability, there can be greater K uptake, but the yield may

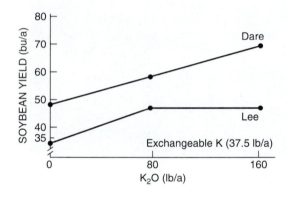

FIGURE 7.18   Soybean varieties have different capacities to respond to soil and fertilizer K.   *Terman*, Agron. J., *69:234, 1977.*

TABLE 7.10   Capacities of Barley Cultivars to Respond to K

	Fresh Weight in Grams of Plant with a Nutrient Solution of K Concentration (μM):				
Cultivar	0.1	1.0	10	100	500
Herta	0.256	0.294	0.287	0.474	1.121
Fergus	0.227	0.278	0.256	0.494	1.116
Carlsberg	0.236	0.209	0.227	0.379	0.987
Olli	0.130	0.189	0.208	0.547	0.946
Conquest	0.143	0.139	0.180	0.408	0.820

SOURCE: Glass and Perley, *Plant Physiol.*, **65:**160 (1980).

actually be reduced. Increasing the corn population from 12,146 to 36,438 plants/a on an irrigated, low-K soil in Florida decreased the yield from 110 to 56 bu/a but increased the K uptake from 117 to 160 lb/a. Inclusion of 240 lb/a of K_2O with these same plant populations boosted the yield from 110 to 231 bu/a and the uptake from 117 to 259 lb/a.

Yield Level

As yield level increases, the amount of K available usually must be increased to compensate for the greater K uptake. For example, 250 bu/a of corn may remove 300 lb/a of K, compared to 136 lb/a with 145 bu/a corn yield. The yield goal or level of production desired is considered by most soil test laboratories in making recommendations for fertilizer K, as shown in Figure 7.9. With higher yields, it is important to adjust maintenance rates of K in accordance with optimum soil K levels and the yield attained.

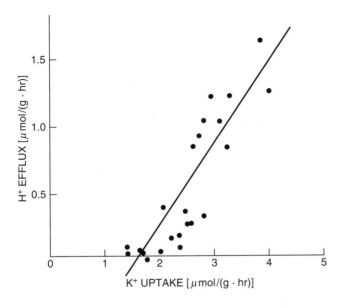

FIGURE 7.19   Plot of net H^+ efflux against K^+ uptake for 24 barley varieties. Plants, previously grown in a balanced inorganic medium containing 60 μM K^+ were exposed to 1 mM K_2SO_4 plus 0.5 mM $CaSO_4$ for 24 hours. Correlation coefficient = 0.88. *Glass et al.*, Plant Physiol., *68:457, 1981.*

Time Factor

Cropping intensity also can influence the management of soil K. Initially, some soils are well supplied with available K, and they may show little response to K. However, in soils that do not have a large BC for K, K levels may decrease rapidly and large crop responses to applied K will be obtained after several cropping seasons.

Figure 7.20 illustrates the long-term development of a K requirement for rice grown on a soil unresponsive to K. Exchangeable soil K was 356 lb/a initially. By the end of the third year of cropping, addition of K was beneficial and the response continued to increase with time. At the end of 8 years, soil test K on the NP-only plots had declined to 102 lb/a, a decrease of 254 lb/a or 32 lb/a/yr.

Another consideration is the maximum K demand as plant growth progresses. Crop requirements for K obviously vary at different growth stages (Fig. 7.21). The young corn seedling does not need much K, but the rate of uptake rises rapidly to a maximum at 3 weeks prior to tasseling. Thus, for satisfactory growth, the rate of K supply from soil must at least match crop needs during critical periods of maximum uptake.

The actual amounts of K measured during the growing season by corn and soybean yielding 308 and 101 bu/a, respectively, are shown in Tables 7.11 and 7.12. These values clearly indicate the changing demands for K supplies as well as N and P.

Where multiple cropping is practiced, both total yield and K removal per unit of time are increased. This intensifies the demand for soil K, accelerating the rate of K drawdown and increasing the K needed for soil test K maintenance.

K Soil Testing

The exchangeable plus soil solution K is usually extracted with a neutral salt solution. Ammonium acetate is a common extractant, but many others are also used. The objective of a soil test is to characterize the nutrient-supplying power of soils. This is done by trying to simulate the feeding action of plant

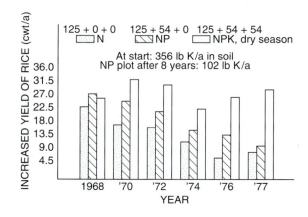

FIGURE 7.20   K fertilization becomes beneficial after several years of rice production on a soil initially unresponsive to K.   *von Vexküll in G. S. Sekhon, Ed., Potassium in Soils and Crops, pp. 241–259. New Delhi: Potash Research Institute of India, 1978.*

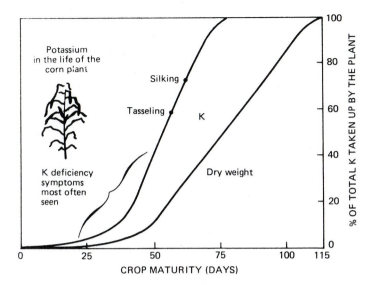

FIGURE 7.21   K is taken up rapidly early in the life of the corn plant.   *Aldrich et al.,* Modern Corn Production, *2nd ed. A. & L. Publications, Champaign, Ill., 1975.*

roots and to determine the fraction of the nutrients in the soil that is available to the plant during the growing season. This is a challenge for K.

First, K^+ moves to the plant root through diffusion and mass flow. The many factors affecting such movement were mentioned earlier. Second, the BC and the mechanisms of release and adsorption complicate the measurement. As discussed earlier, clay and OM content and clay mineralogy are factors.

Laboratories vary in analytical approaches. Some adjust for CEC, removal in harvested crop, and/or yield goal. In addition to estimating the K status and needs for the next crop, the soil test is helpful in detecting whether the

TABLE 7.11   Nutrient Uptake per Day for the Various Growth Periods of Corn—Yield 308 bu/a

Sampling Stage	Days in Period	Nutrient Uptake Per Day (lb)		
		N	P_2O_5	K_2O
4-leaf	32	.38	.08	.58
8-leaf	12	1.63	.35	3.35
12-leaf	15	3.43	.90	3.37
Early tassel	13	11.05	2.85	15.32
Silk	12	−1.43	.88	2.63
Blister	18	1.00	.70	.68
Early dent	31	3.71	1.44	1.41
Mature	13	0.16	1.16	−1.65

SOURCE: Flannery, *Better Crops with Plant Food* (Fall) 4–5 (1986).

TABLE 7.12   Nutrient Uptake per Day
for the Various Growth Periods
of Soybeans—Yield 101 bu/a

Sampling Stage	Days in Period	Nutrient Uptake Per Day (lb)		
		N	P_2O_5	K_2O
3 tri-leaf	40	.75	.25	.68
6 tri-leaf	11	1.45	.55	2.72
Full bloom	16	7.81	1.75	5.75
Pod dev.	15	9.13	2.27	9.60
Soft seed	21	11.43	2.76	2.43
Near mature	16	−3.38	−1.25	−2.25

SOURCE: Flannery, *Better Crops with Plant Food*, 6–7 (1986).

fertilizer program is increasing or decreasing exchangeable K levels. Depending on the soil, about 3 to 6 lb K_2O/a is required to increase the K soil test 1 lb/a.

In spite of the difficulties, exchangeable K^+ has been used for 50 years to provide a measure of the need for K on a wide range of soils. This practice will be continued, but the need for more precise tests does exist. This need will be met by further research in the laboratory, greenhouse, and field.

K Fertilizers

Extensive deposits of soluble K salts are found in many areas of the world, most of them well beneath the surface of the earth but some in the brines of dying lakes and seas. Many of these deposits and brines have high purity and lend themselves to mining operations for the production of agricultural and industrial K salts, usually termed *potash* by the fertilizer industry. In North America, K production is located in New Mexico, California, and Utah in the United States and in Saskatchewan and New Brunswick in Canada (Fig. 7.22).

The Saskatchewan deposit is the world's largest known high-grade potash deposit. It extends in a broad belt 450 mi long and up to 150 mi wide into Manitoba on the east and southward into North Dakota and Montana. The depth of this deposit increases from 3,000 ft at the northern edge to 7,000 ft at the international boundary.

As with N and P, there have been large increases in world K consumption in recent years. As more knowledge of crop production is acquired, made urgent by the rise in world population and the need for higher and more profitable yields, usage not only of K but of all fertilizers will continue to increase.

Like P, the K content of fertilizers is presently guaranteed in terms of its K oxide (K_2O) equivalent. This is determined analytically by measuring the amount of K salt that is soluble in an aqueous solution of ammonium oxalate. Converting %K to %K_2O, and the reverse, can be accomplished by the two following expressions:

$$\%K = \%K_2O \div 1.2$$

$$\%K_2O = \%K \times 1.2$$

Practically all of the K fertilizers are water soluble. They consist essentially of K in combination with Cl^-, SO_4^{2-}, NO_3^-, PO_4^{3-}, and $P_2O_4^-$. Some double salts exist, such as potassium-magnesium sulfate.

Potassium Chloride (KCl)

This salt is sold under the commercial term *muriate of potash*. The term *muriate* is derived from muriatic acid, a common name for hydrochloric acid (HCl). Fertilizer-grade KCl contains 50 to 52% K (60 to 63% K_2O) and varies in color from pink or red to brown or white, depending on the mining and recovery process used. There is no agronomic difference among the products. The white soluble grade is popular in the fluid fertilizer market.

Muriate of potash is by far the most widely used K fertilizer. It is used for direct application to the soil and for the manufacture of N-P-K fertilizers. When added to the soil, it readily dissolves in the soil water.

FIGURE 7.22   Major potash production areas in North America.

Potassium Sulfate (K_2SO_4)

Sulfate of potash is the term usually applied to this salt by the fertilizer industry. It is a white material containing 42 to 44% K (50 to 53% K_2O) and 17% S. It is produced by different processes, some of which involve reactions of other salts with KCl and some of which involve reaction with S or H_2SO_4.

K_2SO_4 finds its greatest use on potatoes and tobacco, which are sensitive to large applications of Cl. Its behavior in the soil is essentially the same as that of KCl, but it has the advantage of supplying S.

Potassium Magnesium Sulfate (K_2SO_4, $MgSO_4$)

This is a double salt with a small amount of NaCl, which is largely removed in processing. The material contains 18% K (22% K_2O), 11% Mg, and 22% S. It has the advantage of supplying both Mg and S and is frequently included in mixed fertilizers for that purpose on soils deficient in these two elements. It reacts as it would any other neutral salt when applied to the soil.

Potassium Nitrate (KNO_3)

KNO_3 contains 13% N and 37% K (44% K_2O). Agronomically, it is an excellent source of fertilizer N and K. KNO_3 is being marketed largely for use on fruit trees and on crops such as cotton and vegetables. If production costs can be lowered, it might compete with other sources of N and K for use on crops of a lower value.

Potassium Phosphates (KPO_3, $K_4P_2O_7$, KH_2PO_4, K_2HPO_4)

Several K phosphates have been produced and marketed on a limited basis. Their advantages are as follows:

1. High analysis.
2. Low salt index.
3. Adapted to preparation of clear fluid fertilizers high in K_2O.
4. Formulation of polyphosphates with controlled solubility.
5. Fluorine and chlorine free, making them well suited for use on tobacco, potatoes, and other crops sensitive to excessive amounts of Cl^-.

Potassium Carbonate (K_2CO_3), Potassium Bicarbonate ($KHCO_3$), and Potassium Hydroxide (KOH)

These salts are used primarily for the production of high-purity fertilizers for foliar application or other specialty uses. The high cost of manufacture has precluded their widespread use as commercial fertilizers. Studies on the agronomic suitability of $KHCO_3$ suggest that its use on acid soils will reduce loss of cations by leaching. There are also indications that this material will increase the effectiveness of P fertilizers. The composition of important K fertilizers and other K salts is shown in Table 7.13.

TABLE 7.13 Plant Nutrient Content of Common K Fertilizers and Other Sources

Material	N (%)	P_2O_5 (%)	K_2O (%)	S (%)	Mg (%)
Potassium chloride	—	—	60–62	—	—
Potassium sulfate	—	—	50–52	17	—
Potassium magnesium sulfate	—	—	22	22	11
Potassium nitrate	13	—	44	—	—
Potassium and sodium nitrate	15	—	14	—	—
Manure salts	—	—	22–27	—	—
Potassium hydroxide	—	—	83	—	—
Potassium carbonate	—	—	< 68	—	—
Potassium orthophosphates	—	30–60	30–50	—	—
Potassium polyphosphates	—	40–60	22–48	—	—
Potassium metaphosphate	—	55–57	38	—	—
Potassium calcium pyrophosphate	—	39–54	25–26	—	—
Potassium thiosulfate	—	—	25	17	—
Potassium polysulfide	—	—	22	23	—

Potassium Thiosulfate ($K_2S_2O_3$) and Potassium Polysulfide (KS_x)

Analysis of these new liquid fertilizer products is (0-0-25-17) and (0-0-22-23), respectively. $K_2S_2O_3$ is compatible with most liquid ferilizers and is well suited for foliar application and drip irrigation.

Agronomic Value of K Fertilizers

K fertilizers have been compared in numerous field and greenhouse trials. The following list summarizes the results.

1. In general, if the material is being used for its K content alone, one material is as good as another. The selection should be based on the cost per unit of K applied to the soil.
2. Sources such as KNO_3 and $K_4P_2O_7$, which contain the other major nutrients, are as effective as KCl and must be evaluated based on the economics of supplying K as well as N and P. These K sources may be completely absorbed by plants, and no anions such as Cl^- or SO_4^{2-} will remain. Their use in greenhouse culture permits the maintenance of adequate N, P, and K without danger of accumulation of excess salts. The same principle would apply, at least in part, under some field conditions in which heavy fertilization is required.
3. Accompanying elements, such as S, Mg, Cl, and Na, are agronomically important on some soils. The value of the accompanying nutrient must be considered in a K source.
4. Tobacco is a crop that is extremely sensitive to excessive amounts of Cl^-. Although rates of up to 20 lb/a are beneficial, quantities in excess of 30 to

40 lb/a will impair the burning quality. In some potato, sweet potato, and citrus areas, high quantities of Cl^- are avoided. Instead, sources such as K_2SO_4 or KNO_3 may furnish the major portion of the K.

Summary

1. The K content of soils is variable, but many soils contain large amounts. The total amount present, however, is no criterion of the amount available to plants.

2. The availability of K to plants is governed by the equilibria in the soil system among four forms of K that vary in their availability. These fractions, together with estimates of the approximate amounts in each soil, are as follows: mineral or structural, 5,000 to 25,000 ppm; nonexchangeable (fixed or difficultly available), 50 to 750 ppm; exchangeable, 40 to 600 ppm; and solution, 1 to 10 ppm.

3. Fixed or difficultly available K is nonexchangeable, and it occurs mainly within the lattice of clay minerals such as illites or hydrous mica, vermiculite, chlorite, and interstratified minerals, in which two or more of the preceding types occur in more or less random arrangement in the same particle.

4. K present in the soil solution and held in an exchangeable state by soil colloids is classified as readily available. About 0.1 to 2% of the total K in soils will normally be accounted for by solution and exchangeable forms.

5. The activity ratio (AR_e^K) is useful for estimating the effectiveness of soil solution K in the presence of competing ions such as Ca^{2+} and Mg^{2+}. It may also be desirable to take into account the influence of Al^{3+} in very acid soils and that of Na^+ in salt-affected soils. This ratio is a measure of the intensity of labile K in the soil or the amount of K that is immediately available to plant roots.

6. Transport of K from soil to absorbing plant roots is achieved mainly by mass flow and diffusion, both of which occur in the soil solution. The former is the less important mechanism for satisfying K requirements of crops, except in soils naturally high in solution K or those in which fertilizer K has elevated K levels in the soil solution.

7. A number of soil and environmental conditions will influence the rate of K diffusion. Some of these are moisture content, temperature, clay content, salt content, and K concentration, and gradient.

8. Exchangeable K is held around negatively charged soil colloids by electrostatic attraction. K held in this manner is easily displaced or exchanged when the soil is placed in contact with neutral salt solutions.

9. Exchangeable K is bound at three types of exchange sites: planar, edge, and inner positions. The planar position on the outside surfaces of clay minerals is rather unspecific for K, while the edge and inner binding positions have high specificity for K.

10. Additions of fertilizer K, as well as K in the soil solution, can be fixed in some soils, especially those high in 2:1 clays and illite. K fixation capacities are affected by a number of factors, including type and amount of clay minerals, soil reaction, K concentration, wetting and drying, and freezing and thawing.

11. K fixation is not completely unfavorable since it helps to conserve the element, which might otherwise be lost from soils where leaching is a problem. Also, fixed K tends to become available over a long period of time. In most soils, K fixation is not a serious crop production problem.

12. Plant uptake of available soil K can be impaired significantly by various soil and environmental conditions such as low moisture, low temperature, restricted aeration, compaction and unsatisfactory tilth, low pH, high concentrations of basic cations such as Ca^{2+} and Mg^{2+}, and other interfering cations such as Al^{3+} in very acidic soils.

13. Liming acidic soils will improve K availability by increasing the effective CEC, thus increasing the ability of soils to retain exchangeable K. The competitive effect of Al^{3+} on plant uptake of K^+ is reduced when insoluble Al is formed following the addition of lime.

14. Crops vary in their ability to utilize soil K. They may also have different internal requirements to satisfy the metabolic and osmotic needs of the crop.

15. Root type and density are major factors influencing availability among crops. Grasses tend to have a more fibrous root system, and legumes tend to be tap rooted. High availability of K enhances root development.

16. When rates of such nutrients as N and P are more adequate and as yields increase, the demand for available K increases.

17. With today's intensive agriculture, which demands the production of high-yielding crops, considerable K is required to fulfill the needs of these crops. Under such conditions, the K released from slowly available forms in the soil will be insufficient in most soils.

18. Leaching losses of K are of importance only on coarse, sandy, organic, or humid tropic soils in areas of high rainfall. In such situations annual or split applications are effective.

19. Exchangeable K^+ in the soil, extracted by a neutral salt solution, is used as a measure of the need for additional K in crop production. Much more research is needed to evaluate more completely the effects of diffusion, mass flow, and buffer capacity on the supply of K to the plant roots.

20. The principal fertilizer compounds containing K are KCl, K_2SO_4, $K_2SO_4 \cdot MgSO_4$, and KNO_3.

21. Agronomic effectiveness of the principal K fertilizers is about equal in cropping situations where only supplemental K is needed. Obviously, there will be differences among sources if there is also a need for other plant nutrients such as N, P, Cl^-, SO_4^{2-}, or Mg^{2+}. The important factor governing selection of a K fertilizer source where only K is needed is the cost per unit of K applied to the land.

Questions

1. Why are soils with a high clay content generally more fertile than sandy soils? Are they always more productive? Why?

2. Under what soil conditions is there most likely to be reversion of available or added K to less available forms?

3. A soil was found to contain 1.5 meq/100 g of exchangeable K. To how many lb KCl/a does this correspond?

4. What effect will the liming of an acid soil have on the retention of K?

5. Why does the addition of gypsum to an acid soil not result in an increased conservation of K?

6. What is the original source of soil K? Is finely ground granite dust a suitable source of K? Why?

7. Is K released more readily from feldspar than from the K-bearing micas?

8. Describe the general nature of the micas.

9. Do members of the mica group have similar abilities to supply K?

10. In which soil particle-size fractions are feldspar and micas usually found?

11. Describe the changes and end results that occur when mica minerals weather in soils.

12. Is soil solution K important in the nourishment of crops? Summarize its role.

13. What is the activity ratio, and what does it measure?

14. Is exchangeable K available to crops? Are some fractions of nonexchangeable K available?

15. Which forms of K are not readily available to plants?

16. How does CEC affect the amount of K in solution at a given level of exchangeable K?

17. Are there different binding sites for exchangeable K? Name the mechanism responsible for K retention by soil colloids.

18. What mechanisms are thought to account for K fixation in soil?

19. Does fixed K tend to become available to plants?

20. From your knowledge of CEC in soils, would you predict that the addition of Na^+ would tend to deplete or conserve the supply of soil K? Why?

21. By what processes is K transported to the plant root surface? What factors govern this movement?

22. What factors control the amount of K present in the soil solution?

23. Under what conditions (soil and environmental) would you expect to obtain the least and greatest crop responses to surface-applied top dressings of a soluble K fertilizer?

24. Why does applied K increase the rate of diffusion to the plant root?

25. Low moisture, low temperature, or more clay in a soil decreases the rate of diffusion of K to the root. Why does applied K increase the rate of diffusion?

26. Under what conditions is band-placed K likely to be superior to broadcast applications?

27. Can plant uptake of available soil K be impaired by soil and environmental factors? If so, list the principal factors.

28. Do crops vary in their ability to use soil K? If there are differences, explain why they occur.

29. Are there differences in the ability of crop varieties and hybrids to absorb K? Summarize any examples.

30. What are the two major root characteristics that cause differences among crops in uptake of K?

31. List important plant factors that can influence the availability of and need for K.

32. Why might continuous cropping at high yield levels deplete available K over time and increase the probability of a response to K?

33. What is *luxury consumption* of K? Is it a serious problem under most soil and cropping conditions? How can it be minimized?

34. Name the common sources of fertilizer K.

35. A fertilizer is guaranteed to contain 30% K_2O. To what percentage of K does this correspond?

36. Under what soil conditions might you prefer to use $K_2SO_4 \cdot MgSO_4$ rather than KCl and dolomite or KCl alone?

37. Are there situations where KCl will be more effective than K_2SO_4 or KNO_3?

38. How many pounds of K_2SO_4 would be required to supply the same amount of K in 350 lb of 60% KCl?

39. In what form is K absorbed by plants?

Selected References

BARBER, S. A. 1984. *Soil Nutrient Bioavailability: A Mechanistic Approach.* John Wiley & Sons, New York.

BARBER, S. A., R. D. MUNSON, and W. B. DANCY. 1985. Production, marketing, and use of potassium fertilizers, pp. 377–410. In O. P. Englestad (Ed.), *Fertilizer Technology and Use.* Soil Science Society of America, Madison, Wisc.

FOLLETT, R. H., L. S. MURPHY, and R. L. DONAHUE. 1981. *Fertilizers and Soil Amendments.* Prentice-Hall, Inc., Englewood Cliffs, N.J.

MUNSON, R. D. (ed.). 1985. *Potassium in Agriculture.* Soil Science Society of America, Madison, Wisc.

CHAPTER 8

Soil and Fertilizer Sulfur, Calcium, and Magnesium

Sulfur (S), Ca, and Mg are macronutrients required in relatively large amounts for good crop growth. S and Mg are needed by plants in about the same quantities as P, whereas for many plant species, the Ca requirement is greater than that for P. S reactions in soil are very similar to those of N, which are dominated by the organic or microbial fraction in the soil (see Chapter 5). In contrast, Ca^{2+} and Mg^{2+} are associated with the soil colloidal fraction and behave like K^+ (see Chapter 7).

Sulfur

S Sources in Soils

S is the 13th most abundant element in the earth's crust, averaging between 0.06 and 0.10%. The main S-bearing minerals in rocks and soils are gypsum ($CaSO_4 \cdot 2H_2O$), epsomite ($MgSO_4 \cdot 7H_2O$), mirabilite ($Na_2SO_4 \cdot 10H_2O$), pyrite (FeS_2), sphalerite (ZnS), chalcopyrite ($CuFeS_2$), and cobaltite ($CoAsS$). Other important sulfides, including pyrrhotite ($Fe_{11}S_{12}$), galena (PbS), arsenopyrite ($FeS_2 \cdot FeAs_2$), and pentlandite ($Fe,Ni)_9S_8$, are found throughout the world.

Silicate minerals generally contain less than 0.01% S; however, S can be much more abundant in biotites, chlorites, and layer-type clay minerals. The S content of igneous rocks usually ranges from 0.02 to 0.07%. The S content in sedimentary rocks varies from 0.02 to 0.22%; thus, they are an important source of S in soils.

Elemental S^0 occurs in deposits over salt domes, in volcanic deposits, and in deposits associated with calcite and gypsum. Minor amounts of S occur as gases released during volcanic activity and as hydrogen sulfide (H_2S) of volcanic, hydrothermal, and biological origin. Hydrogen sulfide, an important commercial source of S, is a contaminant in many natural gas fields. Organic compounds containing S occur in crude oil, coal, and tar sands.

The original source of soil S was doubtless the metal sulfides in rocks. As these rocks were exposed to weathering, the minerals decomposed and S^{2-} was oxidized to SO_4^{2-}. The SO_4^{2-} was then precipitated as soluble and insoluble SO_4^{2-} salts in arid or semiarid climates, absorbed by living organisms, or reduced by other organisms to S^{2-} or S^0 under anaerobic conditions. Some of the SO_4^{2-} formed during mineral weathering found its way to the sea in drainage waters. Oceans contain approximately 2,700 ppm SO_4^{2-}. In other natural waters SO_4^{2-} ranges from 0.5 to 50 ppm but may reach 60,000 ppm (6%) in highly saline lakes and sediments.

Another source of soil S is the atmosphere. In regions where coal and other S-containing products are burned, SO_2 is released into the air and is later brought back to earth in precipitation. Plants may also absorb SO_2 by diffusion into the leaves, which is metabolized by the plant. However, exposure to as little as 0.5 ppm of SO_2 for 3 hr can cause visible injury to the foliage of sensitive vegetation.

Liberation of SO_2 to the atmosphere by industrial nations was approximately 100 million tons in 1990. Most of this resulted from the combustion of fossil fuels, but industrial processes such as ore smelting, petroleum refining, and others contributed about 20% of the total amount emitted. The amount of S brought down in rainfall in the United States ranges from about 1 lb/a per year in rural areas to 100 lb/a near industrial areas. S emissions are partly responsible for the acid rainfall and snowfall in industrialized nations. Strong acids including H_2SO_4 have lowered the pH of precipitation in much of these areas to between 4 and 5. Direct adsorption is an important way that SO_2 enters soils.

Because of the growing concern over air pollution, legislation may ultimately be enacted to require the cleaning and scrubbing of all waste gases. Many industrial companies are presently required to clean effluent gases, and it is likely that this practice will spread and, in turn, will reduce the S content in precipitation and removed from the atmosphere directly by plants and soil. While the emission of S compounds into the atmosphere by industrial activity has received unfavorable publicity, the fact remains that nearly 70% of the S compounds in the atmosphere are due to natural processes. Volatile S compounds are released in large quantities from volcanic activity, from tidal marshes, from decaying OM, and from other sources.

Forms of S in Soils

S is present in the soil in both organic and inorganic forms, although nearly 90% of the total S in most noncalcareous surface soils exists in organic forms. The inorganic forms are solution SO_4^{2-}, adsorbed SO_4^{2-}, insoluble SO_4^{2-}, and reduced inorganic S compounds. Solution plus adsorbed SO_4^{2-} represents the readily available fraction of S utilized by plants.

S cycling in the soil–plant–atmosphere continuum is shown in Figure 8.1. There are similarities between the N and S cycles in that both have gaseous components and their occurrence in soils is associated with OM (see Chapter 5).

SOLUTION SO_4^{2-} S is absorbed by roots as SO_4^{2-} ions that reach roots by diffusion and mass flow. In soils containing 5 ppm or more SO_4^{2-}, all of the

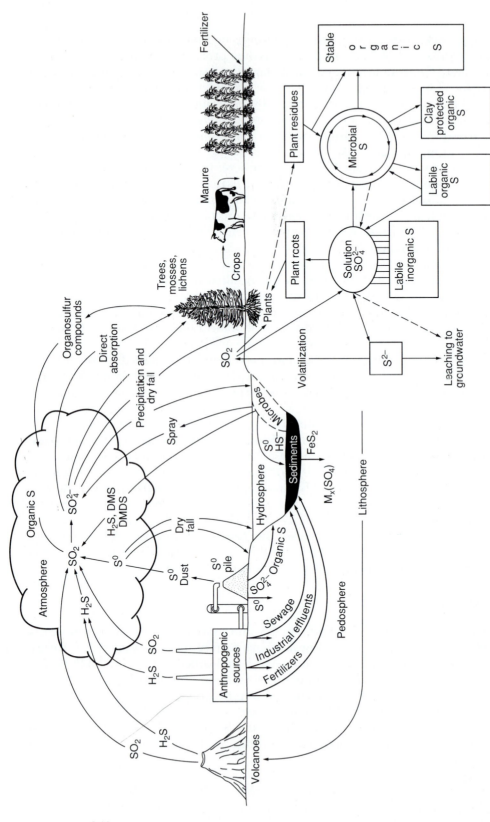

FIGURE 8.1 Simplified version of the overall S cycle in nature. *Krouse and McCready, in P. A. Trudinger and D. J. Swaine, Eds., Studies in Environmental Science, Vol. 3: Biogeochemical Cycling of Mineral-Forming Elements. Amsterdam: Elsevier Scientific, 1979; Stewart and Bettany, Proc. Alberta Soil Sci. Workshop, Edmonton, Alberta, p. 184, February 23–24, 1982.*

requirement of most crops can be supplied by mass flow. Concentrations of 3 to 5 ppm of SO_4^{2-} in the soil solution are adequate for the growth of many plant species, although some crops, like rape and alfalfa, require higher concentrations. Concentrations of 5 to 20 ppm of readily soluble SO_4^{2-} are common in North American soils. Sandy S-deficient soils often contain less than 5 ppm. Except for soils in dry areas that may have accumulations of SO_4^{2-} salts, most soils contain less than 10% of their total S as inorganic SO_4^{2-}.

Levels of readily soluble SO_4^{2-} vary greatly with depth of soil and fluctuate considerably within the profile. Soluble SO_4^{2-} may sometimes reach a maximum in subsurface horizons but it can also be very low in sandy subsoils. Accumulations of soluble SO_4^{2-} commonly occur in calcareous or gypsiferous horizons and in impervious horizons where moisture movement and leaching are restricted.

Large seasonal and year-to-year fluctuations in soluble SO_4^{2-} in the soil can occur (Table 8.1). This variability is caused by the interaction of environmental and seasonal conditions on the mineralization of organic S, either downward or upward movement of SO_4^{2-} in soil water, and SO_4^{2-} uptake by plants.

Sulfate content of soils is also affected by the application of S-containing fertilizers and by the SO_4^{2-} present in precipitation and irrigation waters. In localized areas near centers of industrial activity, the SO_4^{2-} content of soils can be increased by direct adsorption of SO_2 and the fallout of dry particulates.

Because of its anionic nature and the solubility of SO_4^{2-} salts, SO_4^{2-}, like NO_3^-, can be readily leached from surface soil. However, its tendency to disappear from soils varies widely. The relation between the amount of percolating water and the SO_4^{2-} leaching can be determined with radioactive ^{35}S (Fig. 8.2). The greater the amount of added water, the greater the net downward movement of the SO_4^{2-}.

Another factor influencing the loss of SO_4^{2-} is the nature of the cation in the soil solution. Leaching losses of SO_4^{2-} are greatest when monovalent ions

TABLE 8.1 Sulfate Levels in Two Soil Depths at Locations in Southern Alberta

Site and Year	Previous Crop*	Growing Season Precipitation (mm)	SO₄-S (kg/ha) Soil Depth (cm) 0–15	0–30
Barons				
1976	Smf	204	6	9
1977	St	116	3	7
1978	St	160	6	11
Milk River				
1977	St	81	2	39
1978	Smf	292	4	11
Jefferson				
1977	Smf	110	5	24
Welling				
1976	St	132	14	23

*Smf, summer fallow; St, stubble (recrop).

SOURCE: Bole and Pittman, in *Effective Use of Nutrient Resources in Crop Production*, p. 335, Proc. Alberta Soil Sci. Workshop (1979).

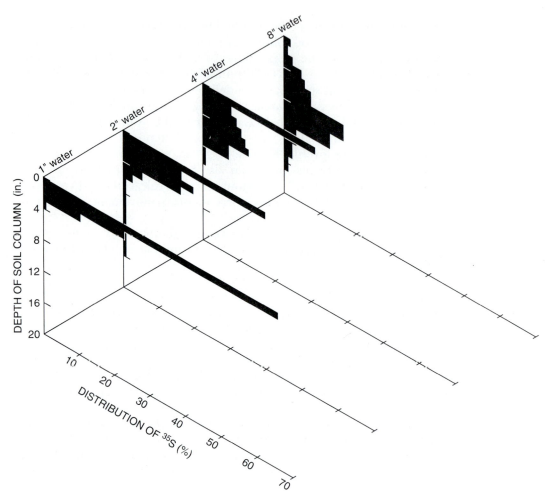

FIGURE 8.2   Distribution of S throughout columns of a Willamette soil as a function of the amount of added water.   *Chao et al., SSSA Proc., 26:27, 1962.*

such as K^+ and Na^+ predominate; next in order are the divalent Ca^{2+} and Mg^{2+} ions; leaching losses are least in acid soils with appreciable amounts of exchangeable Al^{3+}.

ADSORBED SO_4^{2-} Adsorbed SO_4^{2-} is an important fraction in soils containing large amounts of Al and Fe oxides. Many Ultisol (Red–Yellow Podzol) and Oxisol (Latosol) soils contain appreciable amounts of adsorbed SO_4^{2-}. These soils are highly weathered and occur in regions of high rainfall. For example, the amounts of adsorbed SO_4^{2-} in various southeastern U.S. soils are shown in Table 8.2. Adsorbed SO_4^{2-} in highly weathered soils can contribute significantly to the S needs of plants because it is usually readily available. Adsorbed SO_4^{2-} may not be as rapidly available as soluble SO_4^{2-}, but it can be released over longer periods of time.

Reserves of adsorbed SO_4^{2-} in subsoils are the result of eluviation or leaching of SO_4^{2-} from the upper part of soil profiles followed by their retention at lower depths. Adsorbed SO_4^{2-} can account for up to one-third of the total

S in subsoils. In surface soils it usually represents less than 10% of the total S present.

Although crops can utilize adsorbed SO_4^{2-} in subsoils, they may experience S deficiency in the early growth stages until root development is sufficient to reach the subsoil. Deep-rooted crops such as alfalfa and lespedeza are unlikely to have temporary shortages of available S.

Sulfate adsorption is readily reversible and is influenced by the following soil properties:

1. *Clay content and type of clay mineral.* Adsorption of SO_4^{2-} increases with the clay content in soils. In general, SO_4^{2-} adsorption on H-saturated clays follows the order kaolinite > illite > montmorillonite. When saturated with Al^{3+}, adsorption is about the same for kaolinite and illite but is much lower for montmorillonite.
2. *Hydrous oxides.* Al and Fe oxides are responsible for most of the SO_4^{2-} adsorption in many soils.
3. *Soil horizon or depth.* Capacity for SO_4^{2-} adsorption is often greater in subsoils due to the presence of more clay and Fe and Al oxides.
4. *Effect of pH.* Adsorption of SO_4^{2-} in soil is favored by strongly acid conditions, and it becomes negligible at pH > 6.5. Anion exchange capacity (AEC) increases with decreasing pH.
5. *SO_4^{2-} concentration.* Adsorbed SO_4^{2-} is in equilibrium with SO_4^{2-} in solution; therefore, increased solution SO_4^{2-} will increase adsorbed SO_4^{2-}.
6. *Effect of time.* Sulfate adsorbtion increases with the length of time SO_4^{2-} is in contact with the adsorbing surfaces.
7. *Presence of other anions.* Sulfate is considered to be weakly held with the strength of adsorption decreasing in the order $OH^- > H_2PO_4^- > SO_4^{2-} > C_2H_3O_2^- > NO_3^- = Cl^-$. Phosphate will displace SO_4^{2-}, but SO_4^{2-} has little effect on $H_2PO_4^-$. There is little effect of Cl^- on SO_4^{2-} adsorption.
8. *Effect of cations.* The amount of SO_4^{2-} retained is affected by the associated cation or by the exchangeable cation and follows the lyotropic series: $H^+ > Ca^{2+} > Mg^{2+} > K^+ = NH_4^+ > Na^+$. Both the cation and the SO_4^{2-} from a salt may be retained, but the strength of adsorption of each will likely differ.
9. *Organic matter.* In some soils, OM may contribute to SO_4^{2-} adsorption.

TABLE 8.2   Concentration of Adsorbed SO_4^{2-} in Selected Soils of the Southeastern United States

Soil Depth (in.)	Concentration of SO_4^{2-} Extracted with Sodium Acetate (ppm)				
	Norfolk Sandy Loam A	*Norfolk Sandy Loam B*	*Chewacla Silt Loam*	*Fannin Clay Loam*	*Magnolia Fine Sandy Loam*
0–6	0	2	1	3	1
6–12	1	7	1	83	54
12–18	63	114	1	100	95
18–24	93	131	6	58	106
24–30	97	135	24	44	146
30–36	110	127	47	28	87

SOURCE: Ensminger, *Alabama Agr. Exp. Sta. Bull. 312* (1958); and Jordan, *U.S. Dept. Agr. Tech. Bull. 1297* (1964).

Of all these factors, the amount and type of soil colloids, pH, SO_4^{2-} concentration, and the presence of other ions in the equilibrium solution influence SO_4^{2-} adsorption most significantly.

Possible mechanisms of SO_4^{2-} adsorption are as follows:

1. Anion exchange caused by positive charges on Fe and Al oxides or on the broken edges of clays, especially kaolinite, at low pH values (see Chapter 4).
2. Adsorption of SO_4^{2-} by hydroxyaluminum complexes.
3. Salt adsorption resulting from attraction between the surface of soil colloids and the salt.
4. Amphoteric properties of soil OM develop positive charges under certain conditions.

SO_4^{2-} COPRECIPITATED WITH $CaCO_3$ S occurs as a coprecipitated or co-crystallized impurity with $CaCO_3$ and is an important fraction of the total S in calcareous soils. Solubility and availability of SO_4^{2-} coprecipitated with $CaCO_3$ are influenced by several factors, including particle size of the $CaCO_3$, soil moisture content, common ion effects, and ionic strength. The common ion effect also plays an important role in the formation of this fraction.

The grinding that is normally done in soil sample preparations will render the SO_4^{2-} in this fraction accessible to chemical extraction. Consequently, more S will be extracted by a particular soil-test procedure than is available under field conditions.

REDUCED INORGANIC S (S^{2-} and S^0)   Sulfides do not exist in well-drained upland soils. Under anaerobic conditions in waterlogged soils, there may be accumulations of H_2S formed by the decay of OM. Also, SO_4^{2-} present in the soil serves as an electron acceptor for SO_4^{2-}-reducing bacteria, and it is usually reduced to H_2S. Little or no S^{2-} accumulates in oxidized soil (> -150 mV) or with a pH outside the range 6.5 to 8.5. Sulfide accumulation is limited primarily to coastal regions influenced by seawater. In normal submerged soils well supplied with Fe, the H_2S liberated from OM and from SO_4^{2-} is almost completely removed from solution by reaction with Fe^{2+} to form amorphous FeS, which undergoes conversion to pyrite (FeS_2). The deep color of the shore of the Black Sea is caused by the accumulation of FeS_2.

Pyrite may persist in soils, with appreciable amounts still present in previously waterlogged land many years after drainage and cultivation. In contrast, oxidation of amorphous FeS precipitates may be complete after only a few hours of exposure to the atmosphere.

Sulfates added to waterlogged soils are reduced to H_2S. If H_2S is not subsequently precipitated by Fe and other metals, it escapes to the atmosphere. The effect of waterlogging on the production of H_2S in a rice paddy soil increases with both time and added OM (Table 8.3).

In some tidal marsh lands, large quantities of reduced S compounds accumulate, which increase soil pH. When the areas are drained, the S compounds are oxidized to SO_4^{2-}, considerably reducing soil pH. The general reaction for FeS_2 oxidation in soils is

$$FeS_2 + H_2O + \tfrac{7}{2}O_2 \longrightarrow Fe^{2+} + 2SO_4^{2-} + 2H^+$$

TABLE 8.3 Effect of Time and Added OM on the
Production of H_2S in a Waterlogged Soil

Time After Submergence (days)	Concentration of H_2S in Soil for Three Treatments		
	Control (ppm)	Green Manure (ppm)	Straw (ppm)
0	Nil	Nil	Nil
7	Nil	12.5	10.5
14	3.5	23.8	18.5
21	5.0	32.9	28.8
28	6.8	52.8	40.5
38	8.2	57.6	54.8
51	10.5	60.3	62.5
63	13.0	62.5	65.6
78	15.0	65.0	66.0

SOURCE: Mandal, *Soil Sci.*, **91:**121 (1961). Reprinted with the
permission of The Williams & Wilkins Co., Baltimore.

A classic example is found in Kerala State on the southwestern coast of India. During the monsoon season these soils are underwater, and the pH is about 7.0. With restoration of adequate drainage after the monsoon, the S compounds are oxidized and the pH drops to 3.5. This cycle is repeated annually.

Occurrence of pyrite is widespread in coal and allied materials. It may be finely dispersed and invisible to the naked eye or may occur in large aggregates. When pyritic spoil is exposed to the atmosphere, H_2SO_4 is formed, resulting in acidification of mine waters and mine spoil areas, which can create problems in revegetation and use of the affected areas.

Elemental S^0 is not a direct product of SO_4^{2-} reduction in reduced soils but is an intermediate formed during chemical oxidation of S^{2-}. Accumulation of S^0 may occur, however, in soils where oxidation of reduced forms of S is interrupted by periodic flooding.

Factors Affecting S^0 Oxidation in Soils. Elemental S^0, S^{2-}, and other inorganic S compounds can be oxidized in the soil by purely chemical means, but these are usually much slower and therefore of less importance than microbial oxidation. The rate of biological S^0 oxidation depends on the interaction of three factors: (1) the microbial population in soil, (2) characteristics of the S source, and (3) soil environmental conditions.

SOIL MICROFLORA. Three classes of bacteria are involved in S oxidation. These are chemolithotrophic S bacteria that utilize energy released from the oxidation of inorganic S for the fixation of CO_2 in OM. Their activity is described in the following general equation:

$$CO_2 + S^0 + \tfrac{1}{2}O_2 + 2H_2O \rightarrow [CH_2O] + SO_4^{2-} + 2H^+$$

Thiobacilli are typical chemolithotrophic bacteria. Many of them are strict autotrophic aerobes, but some are facultative autotrophs. These S^0 oxidizers were once believed to dominate S^0 oxidation in soils entirely, and they are very active in certain environments.

The second class of S^0 oxidizers is photolithotrophic S bacteria, which carry out photosynthetic C fixation using S^{2-} and other S compounds as "oxidant sinks." The following equation summarizes their behavior:

$$CO_2 + 2H_2S \xrightarrow{\text{light}} [CH_2O] + H_2O + S^0$$

In this class of microorganisms, there are two principal groups: the *Chlorobium*, or green bacteria, and *Chromatium*, or purple bacteria. They are obligate anaerobes, and their habitat is usually H_2S-containing muds and stagnant waters exposed to light.

A third class of S^0 oxidizers occurs as the general population of heterotrophic organisms. Heterotrophs are the most abundant S^0 oxidizers in some soils, where 3 to 37% of the total heterotrophic population are capable of converting S^0 to thiosulfate. S^0 oxidation is greater in the plant rhizosphere, where there are larger and more diverse populations of S^0-oxidizing heterotrophs than in nonrhizosphere soil.

The best-known, and usually considered to be the most important, group of S-oxidizing organisms are the autotrophic bacteria belonging to the genus *Thiobacillus*. Considerable variability in S^0 oxidation rates among soils exists due to differences in the number of *Thiobacillus* (Table 8.4). The initial rate of S oxidation in laboratory studies can be greatly increased by inoculation, usually with thiobacilli, but also with heterotrophs. Figure 8.3 demonstrates the stimulatory effect of inoculation with two species of *Thiobacillus* on oxidation rate of S^0.

These favorable effects of inoculation are frequently short-lived under most conditions, and less benefit is usually obtained in field trials. Addition of S^0 to soil will encourage the growth of S^0-oxidizing microorganisms. This enrichment of the S^0-oxidizing capacity and the overriding influence of environmental factors could be the reason for limited success from field-scale inoculations.

Where inoculation is being attempted, addition of the inoculum to the soil will probably be more successful than direct inoculation of S fertilizers because of poor microbial survival.

TABLE 8.4 S^0 Oxidation Rates in Soils

Location and Number of Soils	Percent S^0 Oxidation in Various Incubation Periods
Australia	
51	0– 0.4
64	0.5– 4.0
58	4.1–12.9
100	13.0–61.6
Nebraska (1)	14.0–56.0
Oregon (1)	2.0–50.0
Saskatchewan (4)	20.0–44.0
Wisconsin (54)	3.0–73.0

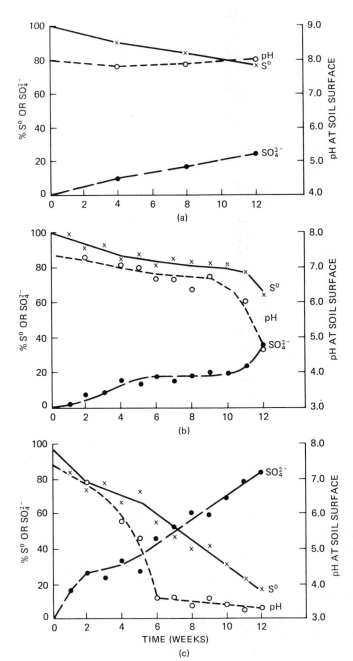

FIGURE 8.3 Changes in S^0 and SO_4^{2-} content and pH during the oxidation of S^0 in Solonetzic soils: (a) uninoculated; (b) inoculated with *Thiobacillus thiooxidans*; (c) inoculated with *T. thioparus*. McCready and Krouse, Can. J. Soil Sci., **62**:105, 1982.

SOIL TEMPERATURE. An increase in temperature increases the S^0 oxidation rate in the soil (Fig. 8.4). The data indicate an increasing rate of oxidation up to 30°C. At temperatures above 55° to 60°C, the S^0-oxidizing organisms are killed. Optimum temperatures will naturally vary for the different S^0-oxidizing organisms, but temperatures between 25° and 40°C will be ideal for most of them.

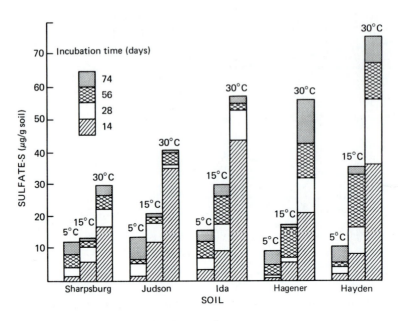

FIGURE 8.4   Effects of temperature and time of incubation on oxidation of S^0 (100 μg of S^0 per gram of soil) in soils. *Nor and Tabatabai, Soil Sci. Soc. Am. J., 41:739, 1979.*

SOIL MOISTURE AND AERATION. S^0-oxidizing bacteria are mostly aerobic, and their activity will decline if O_2 is lacking due to waterlogging. S^0 oxidation is favored by soil moisture levels near field moisture capacity (FMC) (Fig. 8.5). Also evident is the decline in oxidizing activity when the soils are either excessively wet or dry. Dry soils retain their ability to oxidize S^0, but there can be a lag period following rewetting before they regain full capacity.

SOIL pH. Generally, microbial oxidation of S^0 occurs over a wide range in soil pH, although with some species, optimum pH can be 4.0 or lower. For example, as the initial soil pH dropped during the incubation study referred to in Figure 8.3, oxidation of S^0 was enhanced.

SOIL TYPE AND PROPERTIES. The large differences observed in the initial rate of S^0 oxidation by soils are unrelated to soil type or to various soil properties, including texture, OM content, field moisture capacity, pH, initial SO_4^{2-}, or degree of S deficiency of the soil. Liming effects on this process are variable.

Microorganisms responsible for S^0 oxidation require most of the same nutrients needed by plants plus a few others. They can compete with plants for nutrient supply, and temporary N deficiencies have been reported in plants following stimulation of S^0-oxidizing activity by the addition of S^0. More rapid S^0 oxidation has been observed in fertilized soil than in one low in both P and K.

Thiobacilli require NH_4^+, whereas NO_3^- can be injurious. These organisms can withstand high concentrations of NH_3 in anhydrous NH_3 injection zones. High Cl^- concentrations will cause some depression of S oxidation, but the

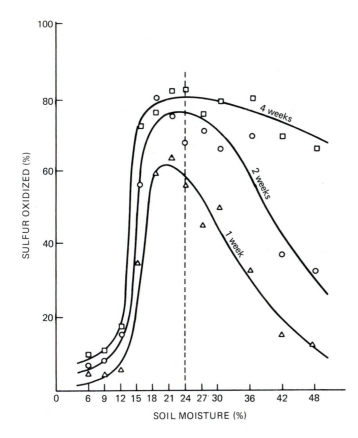

FIGURE 8.5 Percentage of added S^0 oxidized in a Miami silt loam incubated at various moisture levels after 1-, 2-, and 4-week periods. Dashed line, field moisture capacity. *Kittams and Attoe,* Agron. J., *57:331, 1965.*

effect is less than for most other biological transformations in the soil. OM is not essential for activity of autotrophic S bacteria, but heterotrophs require OM as a source of energy.

ORGANIC S   Most of the S in surface horizons of well-drained agricultural soils is present in organic forms, which account for over 90% of the total S in most noncalcareous surface soils. The proportion of total S existing in organic forms varies considerably according to soil type and depth in the soil profile.

As described in Chapter 5, there is a close relationship between organic C, total N, and total S in soils. The C/N/S ratio in most well-drained, noncalcareous soils is approximately 120/10/1.4. Generally more variability exists in C/S ratio in soils than in N:S ratio. Differences in the C/N/S ratios among and within types of soils are related to variations in parent material and other soil-forming factors, such as climate, vegetation, leaching intensity, and drainage. The N/S ratio in most soils falls within the narrow range 6 to 8:1.

The nature and properties of the organic S fraction in soils are important since they govern the release of plant-available S (Fig. 8.1). While much of the organic S in soils remains uncharacterized, three broad groups of S compounds are recognized. These are HI-reducible S, C-bonded S, and residual or inert S. The relative importance of these three categories is shown in Table 8.5.

TABLE 8.5   Fractionation of Organic S in Surface Soils

Location*	HI-Reducible S as Percent of Total		C-Bonded S[†] as Percent of Total		Residual, Inert S as Percent of Total	
	Range	Mean	Range	Mean	Range	Mean
Quebec, Canada (3)	44–78	65	12–32	24	0–44	11
Alberta, Canada (15)	25–71	49	12–32	21	7–45	30
Saskatchewan, Canada (54)	28–59	45	n.d.‡		n.d.	
Australia (21)	—	52	n.d.		n.d.	
Australia (15)	32–63	47	22–54	30	3–31	23
Iowa, U.S. (24)	36–66	52	5–20	11	21–53	37
Brazil (6)	36–70	51	5–12	7	24–59	42

*Figures in parentheses refer to number of samples.

†Determined by reduction with Raney nickel.

‡n.d., not determined.

SOURCE: Biederbeck, in M. Schnitzer and S. U. Khan, Eds., *Soil Organic Matter*, chap.6. New York: Elsevier, 1978.

HI-Reducible S. This fraction is composed of organic S that is reduced to H_2S by hydriodic acid. Its S is largely in the form of SO_4^{2-} esters and ethers with C—O—S linkages. Examples of substances in this grouping include arylsulfates, alkylsulfates, phenolic sulfates, sulfated polysaccharides, and sulfated lipids. About 50% of the organic S occurs in this fraction, but it can range from about 27 to 59%. Values as high as 94.5% have been reported for subsoils.

Carbon-Bonded S. S directly bonded to C distinguishes this group and is determined by reduction to S^{2-} with nickel.

The S-containing amino acids, cystine and methionine, are principal components of this fraction, which accounts for about 10 to 20% of the total organic S. More oxidized S forms, including sulfoxide, sulfones, sulfenic, sulfinic, and sulfonic acids, are also included in this fraction.

Inert or Residual S. Organic S that is not reduced by either hydriodic acid or nickel is considered to be inert or residual. This unidentified fraction generally represents approximately 30 to 40% of the total organic S.

S Transformations in Soils

Numerous transformations of S in soil occur as it converts back and forth from inorganic to organic forms due to the presence of microorganisms (Fig. 8.1). Biological cycling among the major S pools is shown in Figure 8.6. The principal classes of organisms responsible for these reactions are: dissimilatory reducers (*Desulfovibrio, Desulfotomaculum*)—reaction 2; *Desulfuromonas*—reaction 5; assimilatory reducers (bacteria, fungi, algae, plants)—reactions 1 and 3; chemolithotrophs (*Thiobacillus, Beggiatoa*)—reactions 4, 6, and 8; photolithotrophs (*Chlorobium* and *Chromatium*)—reactions 4, 6, and 8; and heterotro-

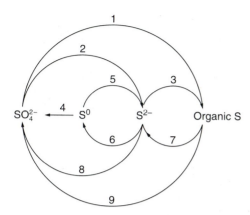

FIGURE 8.6   Biological cycling of the major S pools.   *Trudinger, "The biological sulfur cycle." in P. A. Trudinger and D. J. Swaine, Eds.,* Studies in Environmental Science, Vol. 3: Biogeochemical Cycling of Mineral-Forming Elements, *Chapter 6.1. Amsterdam: Elsevier Scientific, 1979.*

phic microorganisms—reactions 7 and 9, with some involved in reactions 4, 6, and 8.

When plant and animal residues are returned to the soil, they are digested by microorganisms, releasing some of the S as SO_4^{2-}; however, most of the S remains in organic form and eventually becomes part of the soil humus. In contrast to the relatively rapid decomposition of fresh organic residues in soil, degradation and release of S from the large humus fraction are limited and slow.

The S supply to plants is largely dependent on the SO_4^{2-} released from the organic soil fraction and from plant and animal residues. Approximately 4 to 13 lbs/a of S as SO_4^{2-} is mineralized each year from the organic fraction, which in most surface soils may typically contain several hundred lbs/a.

S MINERALIZATION AND IMMOBILIZATION   Mineralization of S is the conversion of organic S to inorganic SO_4^{2-} and is similar to mineralization of organic N (see Chapter 5). Similarly, immobilization is the conversion of SO_4^{2-} to organic S. Any factor that affects the growth of microorganisms is expected to alter the mineralization and immobilization of S.

Factors Affecting S Mineralization and Immobilization
    s CONTENT OF OM.   Mineralization of S depends on the S content of the decomposing material in much the same way that N mineralization depends on the N content (Fig. 8.7). Smaller amounts of SO_4^{2-} are liberated from low-S-containing residue, which is similar to N mineralization. In residue containing less than 0.15% S, there is a reduction in SO_4^{2-} at the end of the incubation period, which may suggest immobilization of S.

S may be immobilized in soils in which either the C/S or N/S ratio is too large. At or below a C/S ratio of approximately 200/1, only mineralization of S occurs. Above this ratio, immobilization or tie-up of SO_4^{2-} in various organic forms is favored, particularly if the ratio is greater than 400/1. The immobilized S is bound in soil humus, in microbial cells, and in by-products of microbial synthesis. Immobilization occurs during the mineralization of organic fractions with large C/S ratios because of conversion of a larger portion of C into microbial biomass, with a resultant higher need for S than when the C/S ratio is low. Fresh organic residues commonly have C/S ratios of about 50/1.

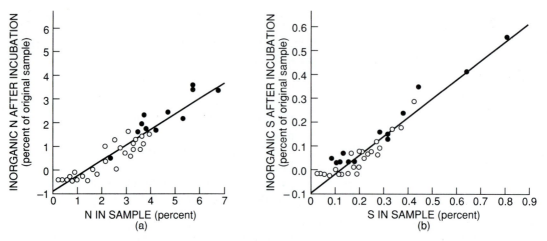

FIGURE 8.7   Relationships between (a) the N mineralized and the N content of the original material; (b) the inorganic S present after incubation and the S content of the original material.   *Barrow*, Australian J. Agr. Res., *11:960, 1960.*

In practical farming operations where large amounts of straw, stover, or other OM is returned to the soil, adequate N and S availability is necessary to promote rapid decomposition of the straw. Otherwise, a temporary N or S deficiency may be induced in the following crop.

SOURCE OF MINERALIZABLE S.   Because of the contribution of S mineralization to plant-available SO_4^{2-}, there is interest in characterizing the labile S reserve in soil OM. Most of the available S removed by plants probably comes from the ester SO_4^{2-} fraction, although other organic S compounds are also involved.

Canadian studies showed that the contributions of six soil OM fractions to S mineralization during 65 years of cultivation were, in decreasing order, clay-associated humic acid (36%), conventional humic acid (26%), humin <2 μm (18%), conventional fulvic acid (14%), clay-associated fulvic acid (4%), and humin >2 μm (3%). The ester SO_4^{2-} fraction was much lower in the cultivated soil than in the pasture soil (Table 8.6). During the 65 years of cultivation, some fractions lost a greater percentage of this form of S than others.

SOIL TEMPERATURE.   Mineralization of S is severely impeded at 10°C, increases with increasing temperatures from 20° to 40°C, and decreases at temperatures >40°C. In samples representing 12 major soil series, more S was released during incubation at 35°C than at 20°C (Fig. 8.8). An average Q_{10} of S mineralization of 1.9 occurred in these soils. This temperature effect on S mineralization is consistent with the relatively greater S content of soils formed in northern climates.

SOIL MOISTURE.   Soil moisture affects the activity of sulfatases, the rate of S mineralization, the form S released from organic matter, and the movement of SO_4^{2-} in soil. Mineralization of S in soils incubated at low (<15%) and

TABLE 8.6 Distribution of HI-Reducible S Among Soil OM Fractions of a Well-Drained Black Chernozemic Soil (Udic Haploboroll)

	HI-Reducible S ($\mu g/g$ soil) in:		
	Pasture	*Cultivated Soil*	*Percent Difference†*
Humic acid A	28	12	
Humic acid B	71	42	
Fulvic acid A	85	76	57
Fulvic acid B	11	10	41
Humin $> 2\mu m$	12	8	11
Humin $< 2\mu m$	37	18	9
Total fractions	244	166	33
Percent recovery	98.1	96.9	51

†The concentration of HI-S of the cultivated soil was subtracted from that of the pasture soil and expressed as a percentage of the pasture HI-S soil fractions.
SOURCE: Bettany et al., *Soil Sci. Soc. Am. J.*, **44**:70 (1980).

high (>40%) moisture levels is reduced compared to the optimum moisture content of 60% of field moisture-holding capacity.

Gradual moisture changes in the range between field capacity and wilting point have little influence on S mineralization. However, drastic differences in soil moisture conditions can produce a flush of S mineralization in some soils. Increased availability of S due to soil wetting and drying may explain the observations of increased plant growth after dry periods in S-deficient soils.

Part of the soil organic S appears to be very unstable and is readily converted to SO_4^{2-} by physical treatments such as heating, air drying, or grinding. Soils prepared for laboratory and greenhouse studies usually receive some or all of these treatments. Air drying releases considerable SO_4^{2-}, where increases of 20 to 80% in extractable SO_4^{2-} have been found upon air drying of soils at room temperature. The SO_4^{2-} liberated by air drying is readily available to plants. Increased SO_4^{2-} availability from air drying is probably not significant under field conditions, but it may contribute to the seasonal and year-to-year fluctuations in soil SO_4^{2-} levels.

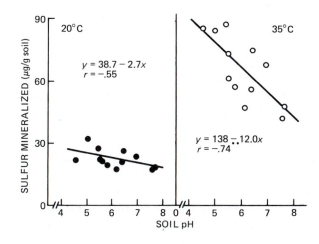

FIGURE 8.8 Relationship between total S mineralized and pH of soils incubated at 20 or 35°C for 26 weeks. *Tabatabai and Al-Khafaji, Soil Sci. Soc. Am. J., **44**:1000, 1980.*

SOIL pH. The effect of pH on S mineralization is not clear. Rates of S mineralization in 12 Iowa soils were found to be negatively correlated with soil pH (Fig. 8.8). In soils from other regions, the amount of S released is directly proportional to pH up to a value of 7.5. At pH values above 7.5 in these soils, mineralization increases more rapidly, suggesting that an additional factor such as chemical hydrolysis might be involved. Formation of SO_4^{2-} in some soils has been found to be proportional to the amount of $CaCO_3$ applied and not to the resulting pH. Near-neutral soil pH is normally expected to encourage microbial activity and S mineralization.

PRESENCE OR ABSENCE OF PLANTS. Soils generally mineralize more S in the presence of growing plants than in their absence. This phenomenon has been explained by stimulation of microbial activity in the "rhizosphere" brought about by the excretion of amino acids and sugars by plant roots. Appreciable immobilization of added SO_4^{2-} has been observed in uncropped soils. This suggests that SO_4^{2-} applied to fallowed soils could be immobilized.

TIME AND CULTIVATION. As with N, when soil is first cultivated, its S content declines rapidly. With time an equilibrium level is reached that is characteristic of the climate, cultural practices, and soil type. Before reaching this point, the rate of S mineralization gradually diminishes and becomes inadequate to meet plant needs. In western Canada, the C/N/S ratios of virgin soils are larger than those of the corresponding cultivated surface soils. Reduction of this ratio on cultivation suggests that S is relatively more resistant to mineralization than C and N or that the losses of organic C and N are proportionately greater than those of S.

SULFATASE ACTIVITY. As much as 50% of the total S in surface soils may be present as organic SO_4^{2-} esters. Sulfatase enzymes that hydrolyze these esters and release SO_4^{2-} may be important in the mineralization process. The general action of sulfatases is

$$R \cdot O \cdot SO_3^- + H_2O \overset{\text{sulfatase}}{\rightleftharpoons} R \cdot OH + HSO_4^-$$

The HI-reducible ester sulfates are considered to be the natural substrates for sulfatase enzymes in soil.

Sulfur Volatilization   Volatile S compounds are produced through microbial transformations under both aerobic and anaerobic conditions. Where volatilization occurs, the volatile S compounds are dimethyl sulfide (CH_3SCH_3), carbon disulfide (CS_2), methyl mercaptan (CH_3SH), and/or dimethyl disulfide (CH_3SSCH_3). CH_3SCH_3 can account for 55 to 100% of all S volatilized.

In low-OM soils, S volatilization is negligible and generally increases with increasing OM content. The actual amount of S volatilized is very small and represents <0.05% of the total S present in soil. Such losses are probably insignificant under field conditions.

Volatilization of S has also occurred from soils treated with residues of a variety of cruciferous crops. Such compounds may represent 30% of the total organic S in these crops but not over 5% in alfalfa, clover, and beets. Volatile

S compounds also evolve from intact plants. Losses in plants can range from about 0.3 to 6.0% of the total S content of crops.

Even though potential losses of soil and plant S by volatilization are probably only of minor consequence, the various volatile compounds may have important side effects. Those substances released from soil during the decomposition of cruciferous crop residues have been reported to control root rot in peas, beans, and sesame. CS_2 is a potent inhibitor of nitrification, and CH_3SH, CH_3SCH_3, and CH_3SSCH_3 are also capable of retarding it. The offensive and sometimes toxic odors from decomposing animal manures are caused by volatile S compounds such as H_2S, CH_3SH, CH_3SCH_3, CH_3SSCH_3, COS, and CS_2.

Volatile S compounds released by intact plants may affect the palatability and acceptability of forage plants to grazing animals. S losses from forage species when they are dried in haymaking or pelleting might also influence quality and palatability.

PRACTICAL ASPECTS OF S TRANSFORMATIONS   Crops grown on coarse-textured soils are generally more susceptible to S deficiency, because these soils often have low OM contents and are subject to SO_4^{2-} leaching. Leaching losses of SO_4^{2-} can be especially high on coarse-textured soils under high rainfall. Under such conditions, SO_4^{2-}-containing fertilizers may have to be applied more frequently than on fine-textured soils and under lighter rainfall. In some areas, a fertilizer containing both SO_4^{2-} and S^0 may be required to extend the period of S availability to crops. This practice is followed in some of the more humid regions.

Immobilization of added S can occur in some soils, particularly those that have a high C/S or N/S ratio. In contrast, S mineralization is favored in soils with a low C/S or N/S ratio. Again, S availability will generally increase with increasing OM content. Crops grown on soils that have <1.2–1.5% OM often require S fertilization.

There are substantial differences in the S fertilizer requirement of crops (Table 8.7). The actual amount of S needed will depend on the balance between all soil additions of S by precipitation, air, irrigation water, crop residues, fertilizers, and other agricultural chemicals and all soil losses through crop removal, leaching, and erosion. In addition to the observed differences in S needs among crop species, there are varietal or cultivar differences. For example, differences in the response of annual clovers to S are as great among varieties of a given species as between species. Similar results have been reported for alfalfa and corn.

Grasses are better able to utilize SO_4^{2-} than legumes. In grass-legume meadows the grasses can absorb available SO_4^{2-} at a faster rate than the legumes. Unless adequate S availability is maintained, the legumes will be forced out of the mixture, because S is required for N fixation by the *Rhizobia*.

Withdrawal of S from soils is accelerated by numerous management factors, including high-yielding varieties, multiple cropping, irrigation, and heavier rates of other plant nutrients. Prolonged use of S-free fertilizers will eventually induce an S deficiency in crops.

S requirements are closely related to the amounts of N applied to crops. Because both N and S are involved in protein synthesis, the full benefit from the addition of one is dependent on an ample supply of the other (Fig. 8.9).

TABLE 8.7   Tentative Classification of Crops According to Their S Fertilizer Requirement

Crop	Fertilizer Required in Deficient Areas* (kg S/ha)	Crop	Fertilizer Required in Deficient Areas* (kg S/ha)
Group I (high)		Group III (low)	
Cruciferous forages	40–80	Sugar beet	15–25
Lucerne	30–70	Cereal forages	10–20
Rapeseed	20–60	Cereal grains	5–20
Group II (moderate)		Peanuts	5–10
Coconuts	–50		
Sugarcane	20–40		
Clovers and grasses	10–40		
Coffee	20–40		
Cotton	10–30		

*Figures cited for the high end of the range apply where the potential yield is high, rainfall is low, the soil is low in available S, and there is considerable loss in effectiveness of applied S. Figures cited for the low end refer to the opposite situation.

For perennials, a further consideration is whether the requirement refers to a corrective fertilizer dressing or to an annual maintenance dressing. The former is typically about four times the latter.

SOURCE: Spencer, in K. D. McLachlan, Ed., *Sulphur in Australasian Agriculture.* Sydney: Sydney University Press (1975).

Here the magnitude of response to S increases with the rate of S applied. Table 8.8 lists the most common S-containing fertilizer materials and gives their typical content of S and other plant nutrients.

S Soil Testing

S soil tests generally have not been successful in predicting S fertilization requirements for crops. Like NO_3^-, SO_4^{2-} is very mobile in the soil; thus, in humid regions, extractable SO_4^{2-} has not been a reliable measure of the S status of the soil. However, a number of laboratories determine the water-, $Ca(H_2PO_4)_2$-, or $CaCl_2$-extractable SO_4^{2-}. In some soils, especially in regions of low rainfall, these soil tests have been somewhat reliable. As previously discussed, most of the plant-available SO_4^{2-} is supplied by mineralization of organic S during the growing season; however, soil tests to quantify potential mineralizable S also have not been successful.

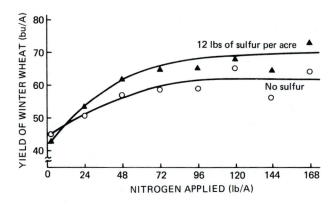

FIGURE 8.9   S increases the response of winter wheat to N on a low-S Oregon soil 0–24 in. = 2 ppm, 24–48 in. = 1.3 ppm, and 48–72 in. = 1.4 ppm). *Rasmussen*, Proc. Pendleton–Walla Walla Fert. Dealers Conf., *Walla Walla, Wash. (January 6, 1976).*

TABLE 8.8 S-Containing Fertilizer Materials

Material	Formula	Plant Nutrient Content (%)					S Content (lb/ton)
		Nitrogen	P_2O_5	K_2O	Sulfur	Other	
Ammonium bisulfite	NH_4HSO_3	14.1	0	0	32.3		646
Ammonium bisulfite solution	$NH_4HSO_3 + H_2O$	8.5	0	0	17		340
Ammonium nitrate-sulfate	$NH_4NO_3 \cdot (NH_4)_2\,SO_4$	30	0	0	5		100
Ammonium phosphate	MAP (crude)	11	48	0	2.2		44
Ammonium phosphate-sulfate	$MAP, DAP + (NH_4)_2SO_4$	16.5	20.5	0	15.5		310
		13	39	0	7		140
Ammonium polysulfide	NH_4S_x	20.5	0	0	45		900
Ammonium polysulfide solution	NH_4S_x	20	0	0	40		800
Ammonium sulfate	$(NH_4)_2SO_4$	21	0	0	24.2		484
Ammonium thiosulfate	$(NH_4)_2S_2O_3$	18.9	0	0	43.3		866
Ammonium thiosulfate solution	$(NH_4)_2S_2O_3 + H_2O$	12	0	0	26		520
Ferrous sulfate	$FeSO_4 \cdot H_2O$	0	0	0	18.8	32.8 (Fe)	376
Gypsum (hydrated)	$CaSO_4 \cdot 2H_2O$	0	0	0	18.6	32.6 (CaO)	372
Magnesium sulfate (Epsom salt)	$MgSO_4 \cdot 7H_2O$	0	0	0	13	9.8 (Mg)	260
Potassium sulfate	K_2SO_4	0	0	50	17.6		352
Pyrites	FeS_2	0	0	0	53.5	46.5 (Fe)	1,070
Potassium-magnesium sulfate	$K_2SO_4 \cdot 2MgSO_4$	0	0	22	22	11 (Mg)	440
Potassium thiosulfate		0	0	25	17		
Potassium polysulfide		0	0	22	23		
Sulfuric acid (100%)	H_2SO_4	0	0	0	32.7		654
Sulfur	S	0	0	0	100		2,000
Sulfur (granular with additives)		0–7	0	0	68–95		1,000
Sulfur dioxide	SO_2	0	0	0	50		1,000
Superphosphate, single	$Ca(H_2PO_4)_2 + CaSO_4 \cdot 2H_2O$	0	20	0	13.9		278
Superphosphate, triple	$Ca(H_2PO_4)_2 + CaSO_4 \cdot 2H_2O$	0	46	0	1.5		30
Urea-sulfur	$CO(NH_2)_2 + S$	36–40	0	0	10–20		200–400
Urea-sulfuric acid	$CO(NH_2)_2 \cdot H_2SO_4$	10–28			9–18		
Zinc sulfate	$ZnSO_4 \cdot H_2O$	0	0	0	17.8	36.4 (Zn)	356

SOURCE: Bixby and Beaton, *Tech. Bull. 17.* Washington, D.C.: The Sulphur Institute (1970).

Probably the best way to assess the potential for S deficiency is to evaluate the soil texture and OM content. Numerous studies have demonstrated that a crop response to S is more likely on coarse-textured, low-OM soils. Other factors to consider are (1) the crop requirement for S, (2) the crop history, (3) use of manures, (4) proximity to industrial S emissions, and (5) S content of irrigation water.

S Fertilizers

Sulfate materials applied to the soil surface and moved into the profile with rainfall or irrigation are immediately plant available unless immobilized by microbes degrading high C/S or N/S residues. Studies comparing the effectiveness of SO_4^{2-} sources (Table 8.8) suggest that one source of SO_4^{2-} is generally equal to any other (provided that the accompanying cation is not Zn, Cu, or Mn, which must be applied sparingly) and that the factor determining the selection should be the cost per unit of S applied.

ELEMENTAL S^0 Elemental S^0 is a yellow, inert, water-insoluble crystalline solid. Commercially it is stored in the open, where it remains unaltered by moisture and temperature changes. When S^0 is finely ground and mixed with soil, however, it is oxidized to SO_4^{2-} by soil microorganisms (see the discussion on pp. 273–276). The effectiveness of S^0 in supplying S to plants compared with SO_4^{2-} depends on several factors, including particle size, rate, method, and time of application; S^0-oxidizing characteristics of the soil; and environmental conditions. S^0 oxidation rates increase as particle size is reduced. As a general rule, 100% of the S^0 material must pass through a 16-mesh screen, and 50% of that should, in turn, pass through a 100-mesh screen. The finer the S^0 particle size, the greater the surface area and the faster the SO_4^{2-} formation. Because of the inverse relationship between surface area and particle diameter, the oxidation rate increases exponentially with decreasing particle diameter. Thus, increasing the S^0 surface area results in increased SO_4^{2-} availability to crops (Fig. 8.10).

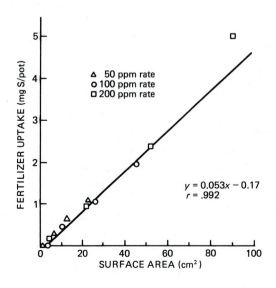

$$y = 0.053x - 0.17$$
$$r = .992$$

FIGURE 8.10   Influence of surface area of applied S^0 on the uptake of S by canola. *Janzen et al.*, Proc. Alberta Soil Sci. Workshop, *p. 229, Edmonton, Alberta (February 23–24, 1982).*

When S^0 is finely ground and mixed with soil possessing a high oxidizing capacity, it is usually just as effective as other sources. Time of application is especially important with S^0 products. Finely divided S^0 should be worked into the soil as far ahead of planting as possible.

Application of heavier rates of S^0 will increase the surface area exposed to S^0-oxidizing organisms, which should result in a linear increase in the release of plant-available S. In poorly buffered soils the reverse can occur, with the percentage of S^0 oxidized declining with higher rates of application.

Placement of S^0 can often affect its oxidation rate, with broadcasting, followed by incorporation, being superior to banding. Uniform distribution of S^0 particles throughout the soil will (1) provide greater exposure of S^0 particles to oxidizing microorganisms, (2) minimize any potential problems caused by excessive acidity, and (3) provide more favorable moisture relationships.

If S^0 is placed on the soil surface and compared with SO_4^{2-} placed similarly, the SO_4^{2-} may give initially better responses. Because of its solubility, it can move into the root zone with percolating waters. The S^0 must first be oxidized to SO_4^{2-}. This is not a rapid process, particularly when left on the soil surface. However, in the case of granular S^0 products, a period of exposure on the soil surface to wet-dry and freeze-thaw cycles is required to disrupt the granules and disperse the S^0. This dispersion process prior to soil incorporation is essential for satisfactory conversion of S^0 to plant-available SO_4^{2-}. With the exception of perennial crops, topdressing of S^0 sources is not normally recommended, and in all cases, S^0 should be topdressed well in advance of crop need.

S^0-BENTONITE   A variety of S^0-bentonite fertilizers have been produced containing about 90% S^0 and 10% bentonite. Particles of S^0-bentonite are sized for blending with solid N, P, and K materials. When S^0-bentonite is applied to soil, the bentonite component imbibes moisture, causing granules to disintegrate, forming more finely divided S^0, which is more rapidly converted to SO_4^{2-}. This material has gained wide acceptance as a source of plant nutrient S for high-analysis, bulk-blend formulations; however, dust problems have occurred with it due to breakdown of its irregular, nonspherical particles.

Because of the uncertainty of adequate availability of S^0-bentonite during the first growing season after application, it should be incorporated into soil at least 4 or 5 months prior to planting. When it is applied just before seeding and on severely S-deficient soils, some SO_4^{2-} should also be provided. The dispersion of S-bentonite products on the soil surface, referred to previously, before soil incorporation is critical for their effectiveness as S sources.

S^0 SUSPENSIONS   The addition of finely ground S^0 to water containing 2 to 3% attapulgite clay results in a suspension containing 40 to 60% S. These suspensions can be applied directly to the soil or they can be combined with suspension fertilizers. They are easy to handle and have the added advantage of being nondusty.

AMMONIUM THIOSULFATE [$(NH_4)_2S_2O_3$ or ATS]   ATS is a clear liquid containing 12% N and 26% S and is the most popular S-containing product used in the fluid fertilizer industry. ATS is compatible with N solutions and complete (N-P-K) liquid mixes, which are neutral to slightly acid in reaction.

It cannot be used with acidic (pH $\leqslant 5.8$) materials. In addition to its wide adaptability for clear liquid mixtures, it is commonly used in suspensions. It is essentially noncorrosive and may be stored in mild steel or aluminum containers.

ATS can be applied to the soil directly, in mixtures, or to both sprinkler and open-ditch irrigation systems. When applied to the soil, ATS forms colloidal S and $(NH_4)_2SO_4$. The SO_4^{2-} is immediately available, whereas the S^0 must be oxidized to SO_4^{2-}, thus extending the availability to the crop. Potassium thiosulfate (KTS) behaves similarly as ATS and was mentioned in Chapter 7.

AMMONIUM POLYSULFIDE (NH_4S_x)   Ammonium polysulfide is a red to brown to black solution having a H_2S odor. It contains approximately 20% N and 45% S. In addition to use as a fertilizer, it is used for reclaiming high-pH soils and for treatment of irrigation water to improve water penetration into the soil.

Ammonium polysulfide is recommended for mixing with anhydrous NH_3, aqua NH_3, and UAN solutions. For stability with UAN solutions, not less than 10% by volume of ammonium polysulfide should be used. The simultaneous application of ammonium polysulfide and anhydrous NH_3 is a common way of providing both N and S in the small grain-producing areas of eastern Oregon, eastern Washington, and northern Idaho. Normally, it is considered incompatible with phosphate-containing liquids. This material has a low vapor pressure, and it should be stored at a pressure of 0.5 psi to prevent loss of NH_3 and subsequent precipitation of S^0. Potassium polysulfide (0-0-22-23) has been used on a limited basis in sprinkler and flood irrigation systems for salt removal and to supply K (see Chapter 7).

UREA-SULFURIC ACID   A simple, rapid, and economic batch process has been developed for mixing urea and sulfuric acid to produce this concentrated liquid N-S fertilizer. Two typical grades used as acidifying amendments, as well as sources of N, contain 10% N and 18% S and 28% N and 9% S, respectively. They can be applied directly to the soil or added through sprinkler systems.

Because these urea-sulfuric acid formulations have pH values between 0.5 and 1.0, the equipment used must be made from stainless steel and other noncorrosive materials. Workers must be safety conscious and wear protective clothing. Monitoring the pH of irrigation water applied through aluminum equipment is necessary to avoid acidification below pH 5.0.

S^0-N/P FERTILIZERS S^0 can be readily incorporated into N/P fertilizer materials to provide 5–20% S. S^0 as an integral part of N-P fertilizer materials may be oxidized more rapidly than when it is added separately. S^0 in granulated triple superphosphate and diammonium phosphate fertilizers oxidize faster than S alone in both acid and calcareous soils. There are several possible reasons for this enhancement, including the effects of N, P, or Ca on S^0 oxidizing microorganisms and the existence of more favorable moisture conditions around the fertilizer granule. The temporary low pH resulting from dissolution of certain fertilizer materials such as triple superphosphate might also improve the growth of S^0-oxidizing microorganisms.

S responses from urea-S^0 (36-0-0-20(S)) in the Canadian Prairie Provinces have been variable, perhaps due to differences in the population and activity of oxidizing organisms in these soils; however, at several locations in the United States, S^0 in this product becomes available sufficiently fast to meet crop needs.

For satisfactory conversion of S^0 in N-P sources to SO_4^{2-}, it is necessary to use the same procedure of permitting granule disintegration and S^0 dispersion at the soil surface followed by adequate soil incorporation, that was described previously for granular S^0 products.

S^0-fortified normal superphosphates are popular in some countries, such as Australia and New Zealand. Ordinary superphosphate is enriched with S^0 to make mixtures containing 18 to 35% S^0. The added S^0 is superior in its residual effect to the $CaSO_4$ already in the ordinary superphosphate.

Fertilizer Use Guidelines

For purposes of convenience, recommendations for the use and proper application of common S-containing fertilizers are summarized in Table 8.9.

Calcium

Form Utilized by Plants

Calcium (Ca) is absorbed by plants as Ca^{2+} from the soil solution and is supplied to the root surface by mass flow and root interception. Ca deficiency is uncommon but can occur in highly leached and unlimed acid soils. In soils abundant in Ca^{2+}, excessive accumulation in the vicinity of roots can occur.

Ca in the soil solution of temperate region soils ranges from 30 to 300 ppm. In soils of higher-rainfall areas, soil solution Ca^{2+} concentrations will vary from 8 to 45 ppm and usually averages about 33 ppm. A level of 15 ppm Ca^{2+} in the soil solution is adequate for high corn yields.

Ca concentrations in the soil higher than necessary for proper plant growth will normally have little effect on Ca^{2+} uptake, because the Ca^{2+} uptake is largely genetically controlled. Although the Ca^{2+} concentration of the soil solution is about 10 times greater than that of K^+, its uptake is usually lower than that of K^+. Capacity of plants for Ca^{2+} uptake is limited because it can be absorbed only by young root tips in which the cell walls of the endodermis are still unsuberized.

Conditions impairing the growth of new roots will reduce access of plant roots to Ca^{2+} and induce deficiency. Problems related to inadequate Ca^{2+} uptake are more likely to occur with plants that have small root systems than with those possessing more highly developed rooting systems.

Special attention must be given to the Ca^{2+} requirements of certain crops, including peanuts, tomatoes, and celery, which are often unable to obtain sufficient Ca^{2+} from soils supplying adequate Ca^{2+} for most other crops. Proper Ca^{2+} supply is important for crops such as alfalfa, cabbage, potatoes, and sugar beets, which are known to have high requirements for Ca^{2+}.

TABLE 8.9 Recommendations for Use of Fertilizers Containing S

Solid Materials	Recommended Use	Remarks
Ammonium phosphate–S^0; Urea–S^0; Water-degradable S^0; flake S and porous granular S	For direct application and bulk blends, apply materials several months before beginning of the growing season; fall applications should be encouraged and allowances made for dispersion prior to incorporation of broadcast applications	If used in starter fertilizer or shortly before beginning of growing season, some readily available SO_4^{2-} should be included (15 to 20% of the total S applied). Dispersion of water degradeable granular S products at the soil surface before incorporation will greatly improve their agronomic effectiveness. Where feasible, incorporate into soil 4 or 5 months prior to planting; when applied just in advance of planting or on severely S-deficient soils, some readily available SO_4^{2-} should be included.
Ammonium sulfate	For direct application and to some extent for bulk blending; should be effective at almost any time	Tends to segregate in bulk blends unless physical properties are improved by granulation; where significant leaching losses are expected, apply shortly before planting
Ammonium nitrate–sulfate; ammonium phosphate-sulfate; ordinary superphosphate; potassium sulfate; potassium magnesium–sulfate	For direct application and bulk blends; should be effective at almost any time	Where significant leaching losses of SO_4^{2-} are expected, apply shortly before planting or the beginning of the growing season
Calcium sulfate (gypsum)	For direct application; should be effective at almost any time	Difficulties may be encountered in application (dustiness, clogging)
Fluid Materials	Recommended Use	Remarks
Ammonium thiosulfate Potassium thiosulfate	For direct application and blending with most fluid fertilizer products; can be broadcast prior to planting or applied in starter fertilizers; can be topdressed on certain growing crops; can also be added through open-ditch and sprinkler irrigation systems.	Can be blended with all neutral fluid phosphate products now available, all nitrogen solutions (except anhydrous ammonia) and most micronutrient solutions.

Ammonium bisulfite	For direct application and blending with most other N and N-phosphate fluids; can be sprayed or dribbled on prior to planting; can also be added through open-ditch and sprinkler irrigation systems; should be effective at almost any time	Can be blended with all neutral fluid phosphate products now available and all N solutions except anhydrous NH_3; can be applied simultaneously with anhydrous NH_3
Ammonium polysulfide Potassium polysulfide	For direct application and blending with other N solutions; frequently injected into soil; broadcast spray applications are possible following dilution with water; single preplant applications are effective; repeated applications at low rates are often made to growing crops through open-ditch irrigation systems. For mixing with wet process ammonium polyphosphate and anhydrous ammonia in the preparation of clear liquid blends.	Ammonium polysulfide is generally not considered suitable for mixing with phosphate-containing fluids
Sulfuric acid		Sulfuric acid has been applied directly to crops such as onions and garlics for weed control purposes
Suspensions containing elemental S^0	For direct application and for simultaneous application with other fertilizers, the suspensions should be applied several months before beginning of the growing season	If used in starter fertilizer or shortly before beginning of growing season, readily available SO_4 should be included (15 to 20% of the total S applied)
Suspensions containing sulfate salts	Should be effective at almost any time	Where significant leaching losses are expected, apply shortly before planting or the beginning of the growing season

SOURCE: Bixby and Beaton, *Tech. Bull. 17*. Washington, D.C.: The Sulphur Institute (1970).

Source of Soil Ca

The Ca concentration of the earth's crust is about 3.64%; however, the Ca^{2+} content in soils varies widely. Sandy soils of humid regions contain very low amounts of Ca^{2+}, while in noncalcareous soils of humid temperate regions, Ca^{2+} normally ranges from 0.7 to 1.5%; however, highly weathered soils of the humid tropics may contain as little as 0.1 to 0.3% Ca. Ca levels in calcareous soils vary from less than 1% to more than 25%. Values of more than approximately 3% indicate the presence of $CaCO_3$.

Ca in soils originated in the rocks and minerals from which the soil was formed. The plagioclase mineral anorthite ($CaAl_2Si_2O_8$) is the most important primary source of Ca, although pyroxenes (augite) and amphiboles (hornblende) also are fairly common in soils. Small amounts of Ca may also originate from biotite, apatite, and certain borosilicates.

The Ca content of arid region soils is generally high, regardless of texture, as a result of low rainfall and little leaching. Calcite ($CaCO_3$) is often the dominant source of Ca in soils of semiarid and arid regions. Dolomite [$CaMg(CO_3)_2$] may also be present in association with $CaCO_3$. In some arid-region soils, calcium sulfate or gypsum ($CaSO_4 \cdot 2H_2O$) may be present.

The fate of solution Ca^{2+} is less complex than that of K^+. It may be (1) lost in drainage waters, (2) absorbed by organisms, (3) adsorbed onto the CEC, or (4) reprecipitated as a secondary Ca compound, particularly in arid climates (Fig. 8.11).

As a general rule, coarse-textured, humid-region soils formed from rocks low in Ca minerals are low in Ca. Soils that are fine-textured and formed from rocks high in the Ca minerals are much higher in both exchangeable and total Ca. However, in humid regions, even soils formed from limestones are frequently acid in the surface layers because of the removal of Ca and other cations by excessive leaching. As water containing dissolved CO_2 percolates through the soil, the H^+ formed displaces Ca^{2+} (and other basic cations) on the exchange complex.[1] If there is considerable percolation of such water through the soil profile, soils gradually become acid.

Ca deficiencies are rare in agricultural soils, since most acid soils usually contain sufficient Ca^{2+} for plant growth. More common are indirect deficiencies caused by low Ca^{2+} in fruit and storage organs that grow rapidly but with restricted internal Ca^{2+} supplies.

Behavior of Ca in Soil

Ca in acid, humid-region soils occurs largely in the exchangeable form and as primary minerals. In most of these soils, Ca^{2+}, Al^{3+}, and H^+ ions dominate the exchange complex. As with any other cation, the exchangeable and solution forms are in dynamic equilibrium (Fig. 8.11). If the activity of solution Ca^{2+} is decreased by leaching or plant removal, Ca^{2+} will desorb to resupply solution Ca^{2+}. Other cations, like H^+ and/or Al^{3+}, will occupy the exchange sites left by the desorbed Ca^{2+}. Conversely, if solution Ca^{2+} is increased, the equilibrium shifts in the opposite direction, with subsequent adsorption of some of the Ca^{2+} by the exchange complex.

[1] $CO_2 + H_2O \rightleftarrows H^+ + HCO_3^-$.

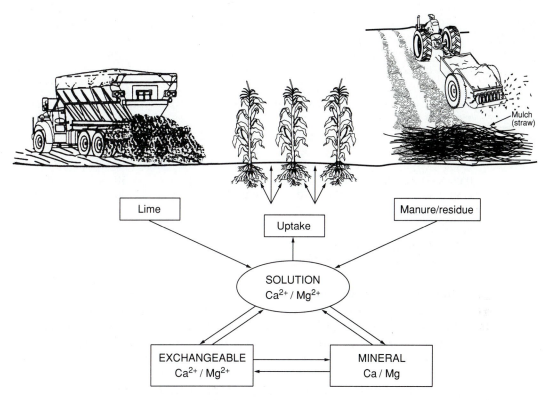

FIGURE 8.11   Simple representation of Ca and Mg equilibrium and transformations in soil.

In soils not containing $CaCO_3$, $CaMg(CO_3)_2$, or $CaSO_4$, the amount of soil solution Ca^{2+} depends on the amount of exchangeable Ca^{2+}. Soil factors of the greatest importance in determining the Ca^{2+} availability to plants are the following:

1. Total Ca supply.
2. Soil pH.
3. CEC.
4. Percent % Ca^{2+} saturation on CEC.
5. Type of soil colloid.
6. Ratio of Ca^{2+} to other cations in solution.

Total Ca in very sandy, acid soils with low CEC can be too low to provide sufficient available Ca^{2+} to crops. On such soils supplemental Ca may be needed to supply Ca^{2+}, as well as to correct the acidity. High H^+ activity (low soil pH) will impede Ca^{2+} uptake. For example, much higher Ca^{2+} concentrations are required for soybean root growth as the pH is lowered from 5.6 to 4.0 (Table 8.10).

In acid soils, Ca is not readily available to plants at low saturation. For example, a low-CEC soil having only 1,000 ppm exchangeable Ca^{2+} but representing a high $\%Ca^{2+}$ saturation might well supply plants with more Ca^{2+} compared to 2000 ppm exchangeable Ca^{2+} with a low %Ca saturation on a high-CEC soil. In other words, as the $\%Ca^{2+}$ saturation decreases in proportion to the total CEC, the amount of Ca^{2+} absorbed by plants decreases.

TABLE 8.10   Effect of Ca Concentration and pH in Subsurface Nutrient Solution on Soybean Taproot Elongation in the Nutrient Solution

	Experiment 2				Experiment 3		
pH	Ca Concentration Added (ppm)	Taproot Elongation Rate* (mm/hr)	Taproot Harvest Length† (mm)	Oven Dry Wt/mm (mg)	pH	Ca Concentration Added (ppm)	Taproot Elongation Rate (mm/hr)
5.6	0.05	2.66	461	0.20	4.75	0.05	0.11
	0.50	2.87	453	0.23		0.50	0.91
	2.50	2.70	455	0.32			
4.5	0.05	0.04	24	0.54	4.0	2.50	0.44
	0.50	1.36	270	0.26		5.00	1.26
	2.50	2.38	422	0.31			

*Elongation rate during first 4 hr in solution.
†Harvested 7½ days after entering the solution.
SOURCE: Lund, *Soil Sci. Soc. Am. J.*, **34:**457 (1970).

High Ca^{2+} saturation indicates a favorable pH for plant growth and microbial activity. Also, a prominence of Ca will usually mean low concentrations of undesirable exchangeable cations such as Al^{3+} in acidic soils and Na^+ in sodic soils. Many crops will respond to Ca applications when the %Ca^{2+} saturation falls below 25%. Ca^{2+} saturation <40 to 60% and Al^{3+} saturation >40 to 60% have lowered cotton yields. Soybeans are reported to suffer Ca deficiency at <20% Ca^{2+} saturation and >65% Al^{3+} saturation. However, normal growth of sugarcane in Hawaii is possible with 12% Ca^{2+} saturation in volcanic soils.

The type of clay influences Ca^{2+} availability; 2:1 clays require higher Ca^{2+} saturation than 1:1 clays. Specifically, montmorillonitic clays require a >70% Ca^{2+} saturation for adequate Ca availability, whereas kaolinitic clays are able to supply sufficient Ca^{2+} at 40 to 50% Ca^{2+} saturation.

Increasing the Al^{3+} concentration in the soil solution reduces Ca^{2+} uptake by corn, cotton, soybeans, and wheat. The depressing action of Al^{3+} on rate of Ca^{2+} uptake by wheat is evident in Figure 8.12. As with Ca^{2+}, the concentration of Al^{3+} in the solution of mineral soils is determined by the percent saturation of the CEC. With organic soils, the amount of exchangeable CEC rather than the percent saturation determines the quantity of Al^{3+} in the soil solution.

Ca availability and uptake by plants are also influenced by the ratios between Ca^{2+} and other cations in the soil solution. A Ca/total cation ratio of 0.10 to 0.15 is desirable for an adequate Ca^{2+} supply to most crops.

While Ca^{2+} uptake is depressed by NH_4^+, K^+, Mg^{2+}, Mn^{2+}, and Al^{3+}, its absorption is increased when plants are supplied with NO_3^--N. A high level of NO_3^- nutrition stimulates organic anion synthesis and the resultant accumulation of cations, particularly Ca^{2+}.

Losses of Ca

Where leaching occurs, Na^+ is lost more readily then Ca^{2+} (see the lyotropic series in Chapter 4); however, since exchangeable and solution Ca^{2+} is much greater than Na^+ in most soils, the quantity of Ca^{2+} lost is much greater. Ca is often the dominant cation in drainage waters, springs, streams, and lakes.

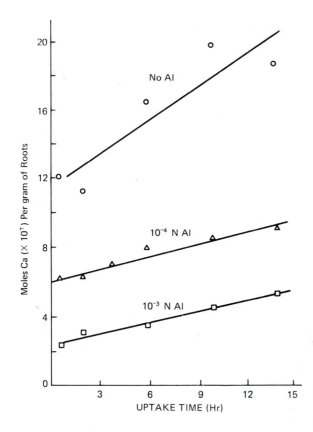

FIGURE 8.12   Influence of $AlCl_3$ on the rate of Ca uptake from 10^{-3} N $CaCl_2$ by excised wheat roots as determined by ^{45}Ca measurements. *Johnson and Jackson,* Soil Sci. Soc. Am. J., *28:381, 1964.*

Leaching of Ca^{2+} ranges from 75 to 200 lb/a per year. Since Ca^{2+} is adsorbed on the CEC, losses by erosion may be considerable in some soils.

Ca Fertilizers

Ca is not usually formulated into mixed fertilizers but rather is present as a component of the materials supplying other nutrients, particularly P. Single superphosphate (SSP) and triple superphosphate (TSP) contain 18 to 21 and 12 to 14% Ca, respectively. The Ca concentration in $Ca(NO_3)_2$ is about 19%. Synthetic chelates such as CaEDTA contain approximately 3 to 5% Ca, while some of the natural complexing substances used as micronutrient carriers contain 4 to 12% Ca. Chelated Ca can also be foliarly applied to crops. Phosphate rocks contain about 35% Ca, and when applied at high rates to acid tropical soils, substantial amounts of Ca are supplied.

The primary sources are liming materials such as $CaCO_3$, $CaMg(CO_3)_2$, and others that are applied to neutralize soil acidity. In situations where Ca is required without the need for correcting soil acidity, gypsum is used.

Gypsum ($CaSO_4 \cdot 2H_2O$) deposits are found at several locations in the United States and Canada and in many other areas of the world. Large amounts of by-product gypsum are produced in the manufacture of phosphoric acid (see Chapter 6).

Gypsum is a common source of Ca^{2+} for peanuts in the United States and is applied directly to the plant in early bloom. In several African countries, the CaSO$_4$ contained in the superphosphate applied to peanut plants is valuable for both its S and Ca content. Gypsum has little effect on soil reaction; hence it may have some value on crops that demand an acid soil yet need considerable Ca. It is widely used on the sodic soils in arid climates. The Ca^{2+} replaces Na$^+$ on the exchange complex, and the Na$^+$ is then leached out below the root zone. Reducing % Na saturation will flocculate the soil and make it more permeable to water.

Magnesium

Form Utilized by Plants

Magnesium (Mg) is absorbed by plants as Mg^{2+} from the soil solution and, like Ca^{2+}, is supplied to plant roots by mass flow and diffusion. Root interception contributes much less Mg^{2+} to uptake than Ca^{2+}. The quantity of Mg^{2+} taken up by plants is usually less than that of Ca^{2+} or K$^+$. The Mg concentration of soil solutions is typically 5 to 50 ppm in temperate region soils, although Mg^{2+} concentrations between 120 and 2400 ppm have been observed.

Plant species and varieties differ in their Mg requirement. Pastures, corn, potatoes, oil palm, cotton, citrus, tobacco, and sugar beets are highly Mg responsive. Mg deficiencies often become apparent on several apple varieties in high yield years.

Corn inbreds and hybrids differ in their uptake of and response to Mg (Fig. 8.13). At Mg^{2+} concentrations below 1.2 mM (29 ppm) inbred B57 produced higher leaf and root mass than did Oh40B; however, at higher Mg^{2+} levels, the reverse occurred.

Seasonal and environmental conditions will interact with plant varieties to produce Mg deficiency. For example, an Mg deficiency in a cool and wet year was evident in all corn varieties except Pioneer 3965 (Table 8.11). Mg deficiency symptoms and reduced growth in certain corn varieties have occurred in previous years at this location when growing conditions were similar to those that prevailed in 1976.

Source of Soil Mg

Mg constitutes 1.93% of the earth's crust; however, the Mg^{2+} content of soils is variable, ranging from 0.1% in coarse, sandy soils in humid regions to 4% in fine-textured, arid, or semiarid soils formed from high-Mg parent materials. Mg in the soil originates from the weathering of rocks containing primary minerals such as biotite, dolomite, hornblende, olivene, and serpentine. It is also found in the secondary clay minerals chlorite, illite, montmorillonite, and vermiculite. Substantial amounts of epsomite (MgSO$_4 \cdot$ 7H$_2$O) and bloedite [Na$_2$Mg(SO$_4$)$_3 \cdot$ 4H$_2$O] may occur in arid or semiarid soils. During mineral weathering Mg is released into the soil solution, where it may be (1) lost in percolating waters, (2) absorbed by living organisms, (3) adsorbed on the CEC, or (4) reprecipitated as a secondary mineral, predominately in arid climates (Fig. 8.11).

Mg in clay minerals is slowly weathered out by leaching and exhaustive

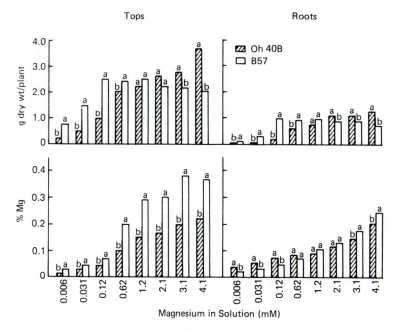

FIGURE 8.13   Dry matter yields and Mg concentrations of
Oh40B and B57 tops and roots grown at 8 levels of Mg. The
values of bars with different letters differ significantly at the 0.05
level.   *Clark,* Soil Sci. Soc. Am. J., ***38:488, 1975.***

TABLE 8.11   Mg Deficiency Rating, Aboveground Yield, and Mg, N, K Composition for
Selected Corn Varieties

| | *Symptom Rating in 1976* * | *Dry Matter Yield*† *(tons/ha)* ** | | | *Nutrient Concentration (% in Dry Matter)* | | | | | |
| | | | | | *Mg* | | *N* | | *K* | |
		1975	*1976*†	*1977*	*1976*	*1977*	*1976*	*1977*	*1976*	*1977*
Idahybrid										
BX110	3	17.8	7.4	18.3	0.32	0.56	3.27	2.94	2.50	1.77
Pride										
R097	3	—	8.1	18.8	0.22	0.52	2.41	2.37	2.10	1.90
Pioneer										
3965	6	19.7	18.6	17.5	0.32	0.37	3.13	2.66	2.77	1.90
Pioneer										
3960	2	13.9	3.6	—	0.18	—	2.40	—	2.30	—
Buckerfields										
BX4331	3	16.4	8.5	—	0.18	—	2.69	—	2.93	—

*Ranking: 1, very severe deficiency; 6, no deficiency.

**Tons per hectare × 0.446 = tons per acre.

†May to August 1976—cool, wet, and cloudy.

SOURCE: Broersma and van Ryswyk, in *Research 1977 Roundup.* Kamloops, B.C.: Range Research Station,
Agriculture Canada (1977).

cropping. Vermiculite has a high Mg content, and it can be a significant source of Mg in soils. Conditions where Mg is likely to be deficient include acid, sandy, highly leached soils with low CEC; calcareous soils with inherently low Mg levels; acid soils receiving high rates of liming materials low in Mg; high rates of NH_4^+ or K^+ fertilization; and crops with a high Mg demand.

Excess Mg can occur in certain situations where soils are formed on serpentine bedrock or are influenced by groundwaters high in Mg. Normal Ca nutrition can be disrupted when exchangeable Mg^{2+} exceeds Ca^{2+}.

Behavior of Mg in the Soil

Mg occurs predominately as exchangeable and solution Mg^{2+} (Fig. 8.11). The absorption of Mg by plants depends on the amount of solution Mg^{2+}, soil pH, the %Mg saturation on the CEC, the quantity of other exchangeable ions, and the type of clay.

Like K, but to a lesser extent, Mg occurs in soils in a slowly available form, which is in equilibrium with exchangeable Mg^{2+}. The formation of these relatively unavailable forms in acid soils is favored by the presence of soluble Mg compounds and a 2:1 clay. Presumably Mg^{2+} could be trapped in the interlayer of expanding and contracting minerals.

Large changes in exchangeable Mg^{2+} can occur following the addition of an Mg liming material. At first, Mg^{2+} levels increase, but as the pII approaches neutrality they decrease. This reduction is attributed to Mg^{2+} fixation through reactions with soluble silica or aluminum chlorite and to coprecipitation with $Al(OH)_3$.

Coarse-textured soils in humid regions exhibit the greatest potential for Mg deficiency. These soils normally contain small amounts of total and exchangeable Mg^{2+}. Soils are probably deficient when they contain less than 25 to 50 ppm exchangeable Mg^{2+}.

Exchangeable Mg normally accounts for 4 to 20% of the CEC of soils, but in soils derived from serpentine rock, exchangeable Mg^{2+} can exceed Ca^{2+}. The critical Mg saturation for optimum plant growth coincides closely with this range, but in most instances, %Mg saturation should not be less than 10%.

Reduced Mg^{2+} uptake in many strongly acid soils is caused by high levels of exchangeable Al^{3+}. Al saturation of 65 to 70% is often associated with Mg deficiency. The availability of Mg^{2+} can also be adversely affected by high H^+ activity in acid organic soils where exchangeable Al^{3+} is not a major cause of the acidity. Mg deficiencies also can occur in soils with high ratios of exchangeable Ca/Mg, where this ratio should not exceed 10/1 to 15/1. On many humid-region, coarse-textured soils the continued use of high-calcic liming materials may increase the Ca/Mg ratio and induce Mg deficiency on certain crops.

High levels of exchangeable K can interfere with Mg uptake by crops. Generally, the recommended K/Mg ratios are <5/1 for field crops, 3/1 for vegetables and sugar beets, and 2/1 for fruit and greenhouse crops.

Competition between NH_4^+ and Mg^{2+} also can lower the Mg^{2+} availability to crops. Ammonium-induced Mg^{2+} stress is greatest when high rates of NH_4^+ fertilizers are applied to low exchangeable Mg^{2+} soils. This interaction may contribute to grass tetany problems. The mechanism of this interaction probably involves the H^+ released when NH_4^+ is absorbed by roots, as well as the direct effect of NH_4^+.

Losses of Mg

Mg^{2+}, like Ca^{2+}, can be leached from soils, and losses of 5 to 60 lb Mg/a have been observed. The amounts lost depend on the interaction of several factors, including the Mg content of soil, rate of weathering, intensity of leaching, and uptake by plants. Leaching of Mg^{2+} is often a severe problem in sandy soils, particularly following the addition of fertilizers such as KCl and K_2SO_4 (Table 8.12). Very little Mg displacement occurs when equivalent amounts of K are applied as either CO_3^{2-}, HCO_3^-, or $H_2PO_4^-$. Apparently, Mg^{2+} desorption and leaching in coarse-textured soils are enhanced by the presence of soluble Cl^- and SO_4^{2-}. As with Ca^{2+}, erosion losses can be considerable in some soils.

Mg Fertilizers

In contrast to Ca, few primary nutrient fertilizers contain Mg, with the exception of $K_2SO_4 \cdot MgSO_4$ (see Chapter 7). Dolomite is commonly applied to low-Mg acid soils. $K_2SO_4 \cdot MgSO_4$ and $MgSO_4$ (Epsom salts) are the most widely used materials in dry fertilizer formulations (Table 8.8). Other materials containing Mg are magnesia (MgO, 55% Mg), magnesium nitrate [$Mg(NO_3)_2$, 16% Mg], magnesium silicate (basic slag, 3 to 4% Mg; serpentine, 26% Mg), magnesium chloride solution ($MgCl_2 \cdot 10H_2O$, 8 to 9% Mg), synthetic chelates (2 to 4% Mg), and natural organic complexing substances (4 to 9% Mg).

$MgSO_4$, $MgCl_2$, $Mg(NO_3)_2$, and synthetic and natural Mg chelates are well suited for application in clear liquids and foliar sprays. Mg deficiency of citrus trees in California is frequently corrected by foliar applications of $Mg(NO_3)_2$. In some tree-fruit growing areas, $MgSO_4$ solutions are foliar applied to maintain levels, and in seriously deficient orchards several annual applications are necessary.

$K_2SO_4 \cdot MgSO_4$ is the most widely used Mg additive in suspensions. A special suspension grade, 100% passing through a 20-mesh screen, of this material is available commercially.

Grass Tetany

Low Mg content of forage crops, particularly grass forages, can be a problem in some areas. Cattle consuming low-Mg forages may suffer from hypomagnesemia, or grass tetany, which is an abnormally low level of blood Mg. High rates of NH_4^+ or K^+ fertilizers may depress the Mg^{2+} level in plant tissue.

TABLE 8.12 Percentage of Exchangeable Mg Displaced by Various K Salts in the First Week of Leaching

	Soil Type and Initial Exchangeable Mg (meq %)		
	Taupo Sandy Silt (0.56 meq %)	Te Kopuru Sand (0.59 meq %)	Patea Sand (0.22 meq %)
	% displaced		
KCl (0.06 g)	12.1	6.8	31.4
K_2SO_4	4.3	6.1	30.8
$KHCO_3$ or K_2CO_3	< 0.1	< 0.1	1.6
KH_2PO_4	< 0.1	< 0.1	2.6

SOURCE: Hogg., *New Zealand J. Sci.*, **5**:64 (1962).

For example, the Mg content of young corn plants is markedly reduced when NH_4^+ rather than NO_3^- is applied. As grass tetany often occurs in the spring, the N may still be in the NH_4^+ form, particularly if cool weather has prevailed. In addition, the high protein content of ingested forages (and other feeds) will depress the absorption of Mg by the animal.

Levels of soil Mg may be increased through the use of dolomitic limestone, if liming is advisable, or through the use of Mg-containing fertilizers. Also, the inclusion of legumes in the forage program is advisable as these plants have a higher content of Mg than do grasses. Cattle can also be fed an Mg salt to help prevent grass tetany.

Although some workers maintain that hypomagnesemia is the result of excessive N or K fertilization, it seems more reasonable to class it as an Mg deficiency and to treat it accordingly.

Summary—S

1. In most humid region soils, a large proportion of the S occurs in organic combination. With soils containing appreciable amounts of hydrous oxides, some adsorbed sulfate S will be found. In arid regions, of course, soluble sulfate salts accumulate in the soil profile.

2. Plants require concentrations of about 3 to 5 ppm of SO_4^{2-} in the soil solution. Sulfate, which is the form absorbed by plants, reaches plant roots mainly by mass flow, especially when solution concentrations are 5 ppm or greater.

3. In S-deficient soils, the level of soluble SO_4^{2-} is usually less than 5 to 10 ppm.

4. There are large seasonal and year-to-year fluctuations in SO_4^{2-} concentrations in surface soils. This variability is due to the interaction of environmental and seasonal conditions on the release of SO_4^{2-} from organic forms, movement of SO_4^{2-} in soil moisture, and uptake by plants.

5. Sulfate is subject to leaching, but appreciable quantities can be retained by adsorption in soils high in Fe and Al oxides and in 1:1 clays.

6. The mobilization and immobilization of soil S depend on the supply of organic C, N, and P, as well as on the activity of soil microorganisms.

7. Factors affecting the growth of microorganisms will alter the mineralization of organic S. Some of the most important factors are the mineral content of OM, temperature, pH, presence or absence of plants, time and cultivation, and availability of energy supply.

8. Although not all of the organic S in soils has been characterized, three broad groups of S compounds are recognized. These are (a) HI-reducible or S esters and ethers, (b) C-bonded S compounds such as cystine and methionine, and (c) residual or inert S, which is very stable since it resists degradation by drastic chemical treatment.

9. Sulfides, polysulfides, and S^0 are converted to SO_4^{2-} by soil microorganisms. The speed with which this conversion takes place depends on the temperature, the moisture, organisms present, and the soil pH. Of great importance, particularly with S^0, are the fineness of the material, the temperature, and the population and activity of S^0-oxidizing organisms.

10. The best-known group of S-oxidizing organisms are the autotrophic bacteria belonging to the genus *Thiobacillus*, but the general population of heterotrophs also can be important.

11. Additions of S^0 and its compounds to soil will encourage greater populations of S^0-oxidizing microorganisms and improved oxidizing power of the soil. Inoculation of soil will improve S^0 oxidation rates in the laboratory or greenhouse.

12. Sulfatase enzymes have been detected in soil. These enzymes hydrolyze organic sulfate esters releasing inorganic SO_4^{2-}. Activity of the sulfatases is dependent on the organic C content of soils, and it decreases with soil depth.

13. There are substantial differences in the S requirements of crops. The actual amount of S fertilizer needed depends on the balance between all soil additions of this nutrient from precipitation, air, irrigation water, crop residues, fertilizers and other agricultural chemicals and all soil losses through crop removal, leaching, and erosion.

14. In addition to differences in S needs among crop species, there are varietal or cultivar differences.

15. The need for S fertilizers on low-S soils is closely related to the amount of N being applied to crops.

16. The effectiveness of S^0 fertilizers depends on several factors. These are particle size, rate, method, time of application, S-oxidizing capacity of the soil, and environmental conditions, particularly those that influence dispersion of water degradeable S^0 materials purposely left unincorporated on the soil surface.

17. When granular S^0 fertilizers are applied shortly before a crop is to be grown on severely S-deficient soils, some readily available SO_4^{2-} should also be provided. The same guideline for SO_4^{2-} supplementation should be followed when the S-oxidizing ability of soil is low or uncertain.

Summary—Ca and Mg

1. Ca is absorbed by plants as the ion Ca^{2+} and Mg is taken up as the ion Mg^{2+}. Both are readily transported to root surfaces by the mechanism of mass flow. Root interception can also provide these two nutrients in some soils.

2. In spite of ample supplies of Ca in the soil solution, plant uptake of this nutrient is limited because it can be absorbed only by young root tips where the cell walls of the endodermis are unsuberized.

3. Ca deficiencies causing reductions in crop growth are rare in agricultural soils. More common are disorders related to Ca shortage in fruit and storage organs caused by restrictions in its movement within plant structures.

4. Ca and Mg, which are somewhat similar in their behavior in soils, are held as exchangeable ions by electrostatic attraction around negatively charged soil colloids. Mg may, under some soil conditions, become fixed in the lattice structure of certain clay minerals and by other soil constituents. Ca is not subject to fixation reactions.

5. A high degree of Ca saturation of soil exchange complexes is desirable because of the resultant favorable pH for the growth of most plants and for microbial activity. Also, when Ca is prevalent there will usually be low

concentrations of such undesirable exchangeable cations as Al^{3+} in acidic soils and Na^+ in sodic soils.

6. Mg deficiencies are not widespread, and the conditions in which they are most likely to occur include acid, sandy, highly leached soils with low CEC; calcareous soils with inherently low Mg levels; acid soils receiving high rates of NH_4^+ and K^+ fertilizers; and cropping with plants that have high-Mg requirements.

7. There are important genetic differences in the Mg requirement and efficiency of plants.

8. The occurrence of Mg deficiency can be brought on by climatic factors (e.g., cool, wet, cloudy weather). Lack of Ca can result from conditions such as low temperatures, drought, poor aeration, and so on, which impair the growth of new roots and reduce the access of plants to Ca.

9. Major sources of Mg are $K_2SO_4 \cdot MgSO_4$, $MgSO_4$, MgO, $Mg(NO_3)_2$, and $MgCl_2$. Ca is supplied by both dolomitic and calcitic limestone, as well as gypsum. Some Ca is supplied incidentally in $Ca(NO_3)_2$, in Ca phosphate fertilizers, and in heavy rates of phosphate rock.

Questions—S

1. What are the forms of S found in soils?
2. Which of the S forms is of special importance in plant nutrition?
3. How does the plant-available form of S reach plant roots? What is the approximate concentration of this form needed in the soil solution?
4. What factors influence the release of inorganic SO_4^{2-} from OM?
5. What is the effect of drying the soil on the availability of organic S?
6. What is the importance of the C/N/P/S ratio to the availability of soil S?
7. What are the soil conditions under which losses of S by leaching would be expected?
8. What are the soil conditions under which little loss of S by leaching would be expected?
9. Describe the soil and climatic conditions under which S deficiencies in the field are most likely to occur.
10. Discuss SO_4^{2-} adsorption by soils, with emphasis on the factors affecting this phenomenon.
11. What are the three broad groups of organic S compounds?
12. Which of these three groups seems to be an active reserve of mineralizable S?
13. What are the factors affecting the oxidation of S^0 in soils?
14. What soil microorganisms are responsible for oxidation of S^0? Which group is the best known and most thoroughly investigated?
15. Are there active populations of S^0-oxidizing organisms in all soils?
16. Will the addition of S^0 to soil influence the population of S^0-oxidizing microorganisms?
17. Does inoculation with S^0-oxidizing organisms improve S^0-oxidation rates? What is the preferable method of inoculation?
18. What is sulfatase? What is its importance?
19. Is S volatilized from soil, crop residues, or animal manures?
20. Can S volatilize from intact plants and during oven drying of top and root samples?
21. What side effects in soil and changes in plants are possible because of S volatilization?

22. What factors influence the amount of S fertilizer needed?
23. Are there differences in S needs among crop species and varieties?
24. When using granular S^0 fertilizer, are there conditions making it desirable to apply some fertilizer SO_4^{2-} as well? What are the conditions?
25. What are the important measures for ensuring that granular S^0-containing products are satisfactory S sources?

Questions—Ca and Mg

1. In what forms are Ca and Mg absorbed by plants? What is the primary transport mechanism of these forms to the root surface?
2. Are deficiencies of Ca and Mg common? What conditions are conducive to shortages of these two nutrients in soils?
3. Is Mg fixed in soils?
4. Why is it desirable to have a high degree of Ca saturation of the CEC of soils?
5. Why is soil acidity usually associated with impaired uptake of Ca and Mg?
6. Does Ca influence Mg uptake by plants? Does the reverse take place?
7. What other fertilizer nutrients influence Mg availability?
8. By what common agricultural practice is Mg most easily added to soils?
9. Under what soil conditions might you prefer to use $K_2SO_4 \cdot MgSO_4$ rather than KCl and dolomite or KCl alone?
10. What are some incidental sources of plant-nutrient Ca?
11. How is a low-Ca soil condition usually corrected?
12. Are there genetic differences in the requirement of plants for Mg and in their efficiency of using it?
13. Which crops have a high Ca requirement?
14. Can climate influence availability and uptake of Ca and Mg?
15. What are some common sources of fertilizer Mg?

Selected References

BARBER, S. A. 1984. *Soil Nutrient Bioavailability: A Mechanistic Approach.* John Wiley & Sons, New York.

BEATON J.D., R. L. FOX, and M. B. JONES. 1985. Production, marketing, and use of sulfur products, pp. 411–454. *In* O. P. Englestad (Ed.), *Fertilizer Technology and Use.* Soil Science Society of America, Madison, Wisc.

FOLLETT, R. H., L. S. MURPHY, and R. L. DONAHUE. 1981. *Fertilizers and Soil Amendments.* Prentice-Hall, Inc., Englewood Cliffs, N.J.

MORTVEDT, J. J., and F. R. FOX. 1985. Production, marketing, and use of calcium, magnesium, and micronutrient fertilizers, pp. 455–482. *In* O.P. Englestad (Ed.), *Fertilizer Technology and Use.* Soil Science Society of America, Madison, Wisc.

TABATABAI, M. A. (Ed.). 1986. *Sulfur in Agriculture.* No. 27. ASA, CSSA, Soil Science Society of America, Madison, Wisc.

CHAPTER 9

Micronutrients and Other Beneficial Elements in Soils and Fertilizers

General Relationships of Micronutrients in Soils

As with any plant nutrient, many soil factors influence the availability of micronutrients to plants. Among the most important factors are soil solution pH and OM. The various steps and processes involved between the weathering of soil minerals or decomposition of organic residues and plant uptake of micronutrient cations are represented in Figure 9.1. This diagram implies that soluble complexing or chelating compounds secreted from roots and produced during microbial degradation of residues increase the transport of micronutrients to roots. Knowledge is limited about the stage in the uptake process where breakdown of these complexes releases the micronutrient. In addition, mass flow and diffusion are the main processes responsible for movement of micronutrient cations from the soil solution to plant roots.

Iron

Iron (Fe) comprises about 5% of the earth's crust and is the fourth most abundant element in the lithosphere. Common primary and secondary Fe minerals are olivene [$(Mg,Fe)_2SiO_4$], pyrite (FeS), siderite ($FeCO_3$), hematite (Fe_2O_3), goethite (FeOOH), magnetite (Fe_3O_4), and limonite [$FeO(OH) \cdot nH_2O + Fe_2O_3 \cdot nH_2O$]. Iron can be either concentrated or depleted during soil development; thus, Fe concentration in soil varies widely, from 0.7 to 55%. Most of this soil Fe is found in primary minerals, clays, oxides, and hydroxides.

Fe deficiency has been reported in more than 12 million acres in 22 western states in the United States. Iron deficiency of Fe-susceptible soybeans also is common on high-pH soils in the upper Midwest. Fe deficiencies in western states and Canada occur principally in ornamental shrubs and fruit trees

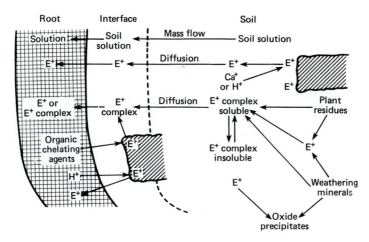

FIGURE 9.1 Principal processes operative in the transfer of trace elements from the soil to the plant root. The general symbol E^+ is used to represent any trace-element cation that can participate in the particular process illustrated. A complementary diagram could be produced to illustrate anionic processes. *Mitchell,* Geol. Soc. Am. Bull., *83:1069, 1972. Reprinted with permission of the Macaulay Institute for Soil Research, Aberdeen, Scotland.*

grown on basic soils which are calcareous, sandy, or organic. Fe deficiencies induced by high levels of Mn have also been reported on certain soils.

Forms of Soil Fe

Fe occurs in four major forms in soil: (1) primary and secondary mineral Fe, (2) adsorbed Fe, (3) organic Fe, and (4) solution Fe (Fig. 9.2). The Fe "cycle" in soils is very similar to other micronutrient cycles in soil and will be referred to throughout this chapter. Understanding the relationships and dynamics among these forms is essential for eliminating Fe stress in plants grown on Fe-deficient soils.

SOIL SOLUTION FE   Fe in solution occurs primarily as $Fe(OH)_2^+$, although other hydrolysis species contribute to total Fe in solution (Fig. 9.3). Compared to other cations in soils, the Fe^{3+} concentration in solution is very low. In well-drained, oxidized soils, the solution Fe^{2+} concentration is less than that of the dominant Fe^{3+} species in solution. The following equation describes the pH-dependent relationship for Fe^{3+}:

$$Fe(OH)_3(soil) + 3H^+ \longleftrightarrow Fe^{3+} + 3H_2O$$

For every unit increase in pH, Fe^{3+} concentration decreases 1,000-fold. In contrast, Fe^{2+} decreases 100-fold for each unit increase in pH, which is similar to the behavior of other divalent metal cations (Fig. 9.3).

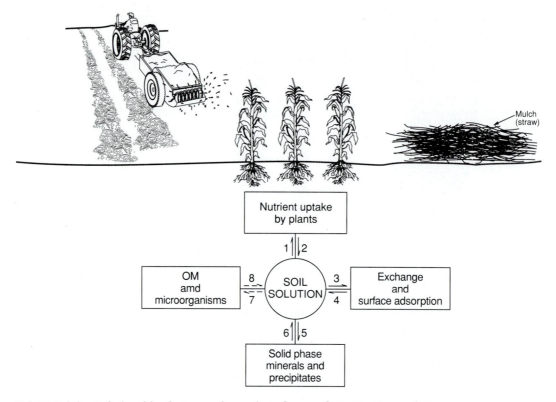

FIGURE 9.2   Relationships between the various forms of Fe, Zn, Cu, and Mn micronutrients in soils. Reactions 1 and 2 represent plant absorption and exudation, respectively; reactions 3 and 4 represent adsorption and desorption, respectively; reactions 5 and 6 represent precipitation and dissolution, respectively; and reactions 7 and 8 represent immobilization and mineralization, respectively. All these processes interact to control concentration of Fe, Zn, Cu, and Mn in soil solution.   *Lindsay*, Micronutrients in Agric., *1972, ASA, p. 42.*

The solubility of the common Fe minerals in soil is very low, only 10^{-6} to $10^{-24} M$ Fe^{3+} in solution, depending on pH (Fig. 9.4). The mineral denoted by "Soil Fe" represents an amorphous $Fe(OH)_3$ precipitate, which appears to control the solution Fe^{3+} concentration in most soils.

Oxidation-reduction reactions, normally the result of changes in O_2 partial pressure, exert considerable influence on the amount of soluble Fe in the soil solution. The insoluble Fe^{3+} form predominates in well-drained soils, while levels of soluble Fe^{2+} increase significantly when soils become waterlogged. In general, lowering redox increases Fe^{2+} solubility 10-fold for each unit decrease in pe + pH, a term used to quantify the redox state in a soil (Fig. 9.5).

Over the normal pH range in soils, total solution Fe is not sufficient to meet plant requirements for Fe, even in acid soils, where Fe deficiencies occur less frequently than in high pH and calcareous soils (Fig. 9.6). Obviously, another mechanism that increases Fe availability to plants exists; otherwise, crops grown on almost all soils would be Fe deficient.

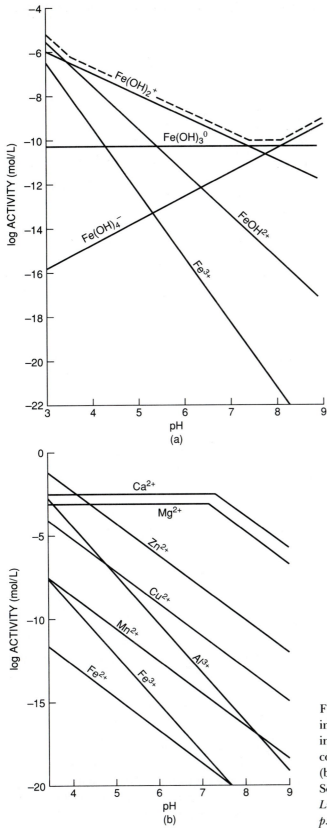

FIGURE 9.3 The solution Fe species in equilibrium with soil Fe (a) and the influence of pH on solution Fe^{3+} concentration relative to other cations (b). *(a) Lindsay,* Chemical Equilibria in Soils, *John Wiley & Sons, 1979; (b) Lindsay,* Chemistry in Soil Environment, *p. 189. Madison, Wisc.: ASA.*

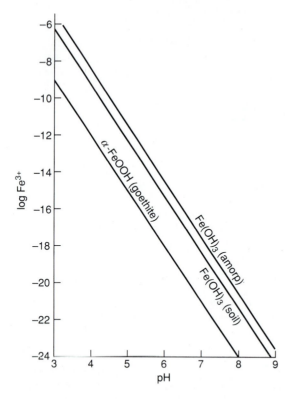

FIGURE 9.4   The common Fe minerals that control solution Fe in soils.
Lindsay, Chemical Equilibria in Soils, *Wiley, New York, 1979.*

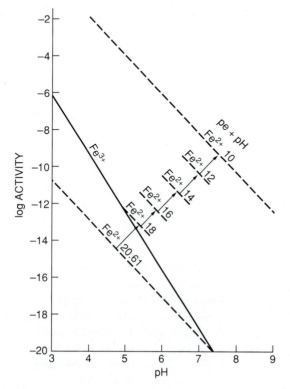

FIGURE 9.5   Effect of redox and pH on the equilibrium of Fe^{2+} with soil Fe.
Lindsay, Chemical Equilibria in Soils, *Wiley, New York, 1979.*

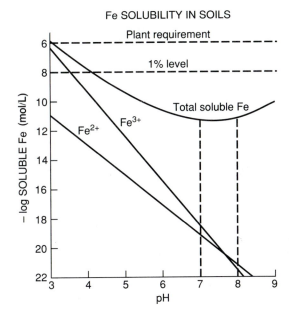

Fe SOLUBILITY IN SOILS

FIGURE 9.6   The influence of pH total solution Fe concentration and its relationship to and Fe required by plants.   *Lindsay,* Plant Root and Its Environment, *p. 508, Univ. Press of Virginia, 1974.*

CHELATES

Chelate Dynamics.   Numerous natural organic compounds in soil, or synthetic compounds added to soils, are able to complex or *chelate* Fe^{3+} and other micronutrients. The concentration of Fe in solution and the quantity of Fe transported to the root by mass flow and diffusion can be greatly increased through *complexation* of Fe with natural organic chelating compounds in the soil (Fig. 9.7). Diffusion of Fe to sorghum roots was encouraged by the higher concentration of soluble Fe. An example of increased Zn diffusion by chelation is shown in Figure 9.18.

Chelate is a term derived from a Greek word meaning "claw." Chelates are soluble organic compounds that bond with metals such as Fe, Zn, Cu, and Mn, increasing their solubility and their supply to plant roots. Natural organic chelates in soils are products of microbial activity and degradation of soil OM and plant residues. Root exudates also are capable of complexing micronutrients.

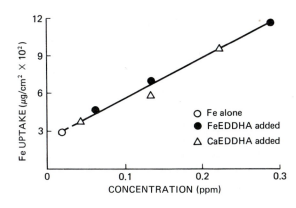

O Fe alone
● FeEDDHA added
△ CaEDDHA added

FIGURE 9.7   Uptake of Fe at the end of 18 hours by sorghum roots as a function of the concentration of Fe in the soil solution and chelate treatment.   *O'Connor et al.,* Soil Sci. Soc. Am. J., *35:407, 1971.*

Substantial quantities of organic-complexed Fe can be cycled through crop residue, which remain available for the succeeding crop. Many of the natural organic chelates have not been identified; however, compounds like citric and oxalic acids have chelating properties (Table 9.1). Examples of the molecular structure of several synthetic chelates are shown in Figure 9.8.

The dynamics of chelation in increasing solubility and transport of micronutrients is illustrated in Figure 9.9. During active plant uptake, the concentration of chelated Fe or other micronutrients is greater in the bulk solution than at the root surface; thus, chelated Fe diffuses to the root surface in response to the concentration gradient. At the root surface the Fe^{3+} "unhooks" or dissociates from the chelate by a mechanism not well understood. After Fe^{3+} dissociates from the chelate, the "free" chelate will diffuse away from the root back to the "bulk" solution, again because of a concentration gradient (free chelate concentration near the root > free chelate in bulk solution). The free chelate subsequently complexes another Fe^{3+} ion from solution. As the unchelated Fe^{3+} concentration decreases in solution because of chelation, additional Fe is desorbed from mineral surfaces or Fe minerals dissolve to resupply solution Fe. The chelate–micronutrient "cycling" is an extremely important mechanism in soils that greatly contributes to plant available Fe and other micronutrients.

Synthetic Chelate Stability. In soils, synthetic chelates behave similarly to natural organic chelates, and have been used as micronutrient fertilizers and for micronutrient soil testing. Therefore, understanding chelate chemistry is important for management of micronutrients in soils. The agriculturally important chelates are listed in Table 9.1. The choice of which chelate to use as a fertilizer or for soil testing depends on (1) the specific micronutrient and (2) the stability of the chelate in the soil.

When synthetic or natural chelates are added to soils, they readily complex cations in soil solution. For example, citric and oxalic acids, two natural chelates, complex Al^{3+} at low pH, but when pH increases above 5 or 6, Ca^{2+} and/or Mg^{2+} are more readily complexed (Fig. 9.10). Notice that citric acid is not effective in complexing Fe in solution. In contrast, DTPA and EDTA readily

TABLE 9.1   Chemical Formula for Several Common Synthetic
and Natural Chelates

Name	Formula	Abbreviation
Ethylenediaminetetraacetic acid	$C_{10}H_{16}O_8N_2$	EDTA
Diethylenetriaminepentaacetic acid	$C_{14}H_{23}O_{10}N_3$	DTPA
Cyclohexanediaminetetraacetic acid	$C_{14}H_{22}O_8N_2$	CDTA
Ethylenediaminedi-*o*-hydroxyphenlyacetic acid	$C_{18}H_{20}O_6N_2$	EDDHA
Hydroxyethylethylenediaminetriacetic acid	$C_{10}H_{18}O_7N_2$	HEDTA
Nitrilotriacetic acid	$C_6H_9O_6N$	NTA
Ethylene glycol bis(2-aminoethyl ether)tetraacetic acid	$C_{14}H_{24}O_{10}N_2$	EGTA
Citric acid	$C_6H_8O_7$	CIT
Oxalic acid	$C_2H_2O_4$	OX
Pyrophosphoric acid	$H_4P_2O_7$	P_2O_7
Triphosphoric acid	$H_5P_3O_{10}$	P_3O_{10}

SOURCE: Mortvedt et al., Eds., *Micronutrients in Agriculture*, p. 116. Madison, Wisc.: ASA, 1972.

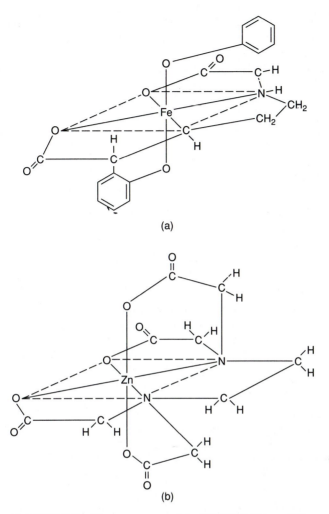

FIGURE 9.8   The structure of Fe-EDDHA (A) and Zn-EDTA (B). *Follett et al.*, Fertilizers and Soil Amendments, *Prentice-Hall, Englewood Cliffs, N.J., 1981.*

chelate Fe at pH <7 and pH <6.5, respectively, while above these pH values, Ca replaces Fe in both chelates (Fig. 9.11). Thus, in neutral and calcareous soils, Ca will dominate most synthetic chelates except for EDDHA (Fig. 9.12). The chelate EDDHA will strongly complex Fe and is stable over the entire pH range. As a result, Fe-EDDHA is commonly used as an Fe fertilizer because of its affininty for Fe. For example, when Fe-EDTA, Fe-DTPA, and Fe-EDDHA were added to a soil, the EDDHA chelate provided more plant-available Fe than the other chelates but only partially corrected the Fe deficiency in sorghum (Fig. 9.13).

Factors Affecting Fe Availability

Low total soil Fe is seldom related to plant deficiencies. As previously discussed, low Fe mineral solubility results in very low solution Fe concentration.

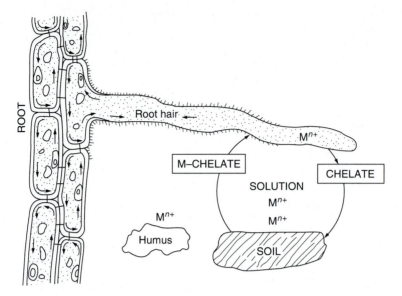

FIGURE 9.9   Cycling of chelated micronutrients (M) in soils. *After Lindsay,* Plant Root and Its Environment, *p. 517. Univ. Press of Virginia, 1974.*

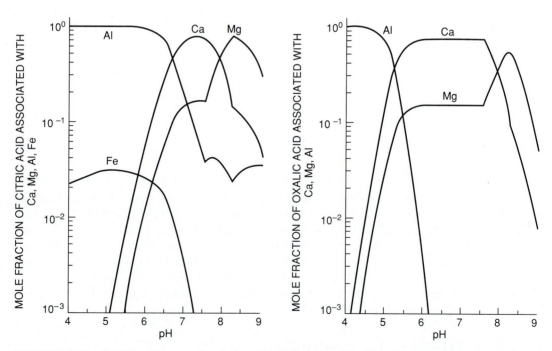

FIGURE 9.10   Stability diagram for citric and oxalic acids with Fe, Al, Ca, and Mg in soils.   *Norvell,* Micronutrients in Agriculture, *p. 125. Madison, Wisc.: ASA, 1972.*

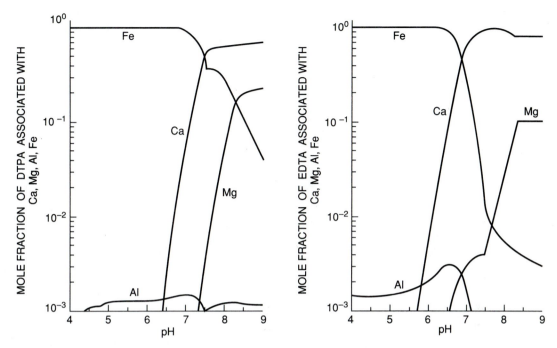

FIGURE 9.11   Stability diagram for DTPA and EDTA with Fe, Al, Ca, and Mg in soils. *Norvell,* Micronutrients in Agriculture, *pp. 121–122. Madison, Wisc.: ASA, 1972.*

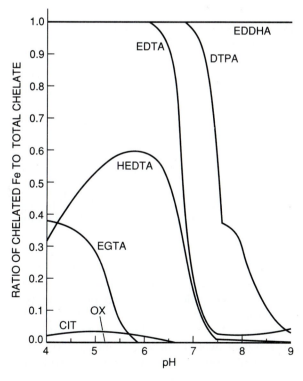

FIGURE 9.12   Stability of selected synthetic chelates with Fe in soils. *Norvell,* Micronutrients in Agriculture, *p. 126, 1972.*

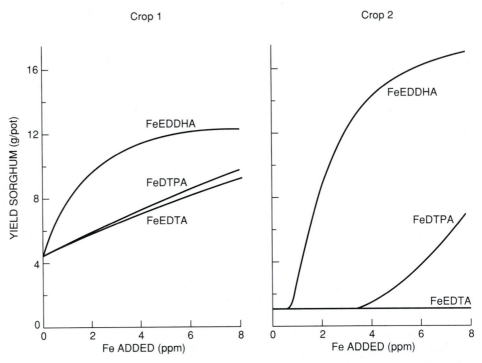

FIGURE 9.13 Effectiveness of selected synthetic Fe chelates in supplying Fe to Fe-deficient sorghum. *Lindsay,* Plant Root and Its Environment, *p. 511. University Press of Virginia, 1974.*

SOIL pH, BICARBONATE, AND CARBONATES Fe deficiency is most often observed on high-pH and calcareous soils in arid regions, but it may also occur on acid soils that are very low in total Fe. Irrigation waters and soils high in bicarbonate (HCO_3^-) may aggravate Fe deficiencies, probably because of the high pH levels associated with HCO_3^- accumulation. The pH of most soils containing $CaCO_3$ falls in the range 7.3 to 8.5, which coincides with the greatest incidence of Fe deficiency and the lowest solubility of soil Fe. Bicarbonate ion can be formed in calcareous soils by the following reaction:

$$CaCO_3 + CO_2 + H_2O \rightarrow Ca^{2+} + 2HCO_3^-$$

Although the presence of lime alone does not necessarily induce Fe deficiency, its interaction with certain soil environmental conditions is related to Fe deficiency.

EXCESSIVE WATER AND POOR AERATION The preceding reaction is promoted by the accumulation of CO_2 in excessively wet and poorly drained soils. Consequently, any compact, heavy-textured, calcareous soil is potentially Fe deficient. Iron chlorosis is often associated with cool, rainy weather when soil moisture is high and soil aeration poor. Also, root development and nutrient absorption are reduced under these cool, wet conditions, which contributes to Fe stress. Lime-induced chlorosis will often disappear when these soils are allowed to dry out.

Flooding and submergence of soils in which HCO_3^- formation is of no concern can improve Fe availability by increasing Fe^{2+} concentrations (Fig. 9.6). There can actually be a buildup of toxic concentrations of Fe^{2+} in the soil solution of latosols.

Although lime-induced Fe deficiency occurs in wet soils, semiarid region calcareous soils that are low in OM are often low in plant-available Fe. This occurs especially on eroded portions of the field where the OM-rich topsoil has been removed, exposing calcareous subsoils. Land leveling for irrigation can also expose calcareous, low-Fe subsoils.

ORGANIC MATTER   Additions of OM to well-drained soils can improve Fe availability. Organic materials such as manure may supply chelating agents that aid in maintaining the solubility of micronutrients. Improved structure of fine-textured soils resulting from applications of organic manures also should increase Fe availability because of better soil aeration. However, this ameliorating effect may be negated by higher levels of CO_2 and associated HCO_3^- produced as a result of greater microbial activity.

The presence of OM can increase Fe^{2+} solubility in waterlogged soils (Table 9.2). Fe reduction is greatly accelerated by the addition of OM, and the longer the period of submergence, the greater the amount of soluble plus exchangeable Fe present.

INTERACTIONS WITH OTHER NUTRIENTS   Metal cations can interact with Fe to induce Fe stress in plants. Fe deficiencies can result from an accumulation of Cu after extended periods of Cu fertilization. Pineapples in Hawaii exhibited Fe chlorosis when grown on soils high in Mn, and other plants growing on soils developed from serpentine exhibited Fe deficiency because of excess Ni. Fe deficiencies on soybeans can occur because of a low Fe/(Cu + Mn) ratio in plants. In addition to Fe deficiencies caused by excess Cu, Mn, Zn, and Mo, Fe–P interactions have been observed in some plants, probably related to precipitation of Fe-P minerals. In addition, plants can be more tolerant of low Fe when P is also low.

TABLE 9.2   Effect of Time and Added Organic Matter on the Reduction of Fe^{3+} in a Submerged Soil

Time of Submergence (days)	Fe^{2+} in Solution (ppm)	Exchangeable Fe^{2+} (ppm)	Fe^{2+} in Solution (ppm)	Exchangeable Fe^{2+} (ppm)
	——Control——		——Straw——	
0	Nil	Nil	Nil	Nil
7	Nil	Nil	Nil	26.0
14	Nil	Trace	30.0	108.0
21	Nil	10	90.4	162.0
28	Nil	20	132.0	200.0
38	Nil	31	192.0	446.0
51	Trace	40	184.0	450.0
63	2.0	46	128.0	482.0
78	4.0	50	104.0	523.0

SOURCE: Mandal. *Soil Sci.*, **91**:121 (1961). Reprinted with permission of the Williams & Wilkins Co., Baltimore.

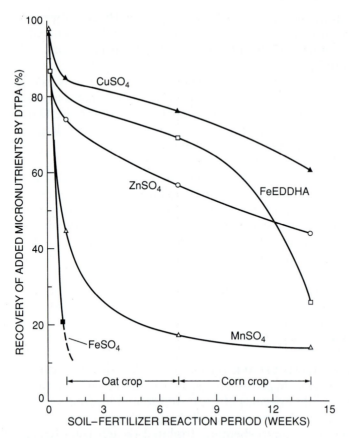

FIGURE 9.14 Recovery of micronutrients in soils fertilized with various inorganic micronutrient fertilizers. *Follett and Lindsay,* Soil Sci. Soc. Am. J., *35:600–602, 1971.*

Plants receiving NO_3^- are more likely to develop Fe stress than those nourished with NH_4^+. When a strong acid anion (NO_3^-) is absorbed and replaced with a weak acid (HCO_3^-), the pH of the root zone increases, particularly in low buffered systems, which decreases Fe availability. Thus, Fe solubility and availability are favored by the acidity that develops when NH_4^+ is utilized by plants.

Deficiencies of K or Zn can reduce Fe translocation within plants. Lack of either of these two nutrients causes Fe to accumulate in the stem nodes of corn; this method has been used to diagnose deficiencies.

PLANT FACTORS   Although diffusion of both Fe^{3+} and Fe^{2+} to the root occurs, the Fe^{3+} is reduced to Fe^{2+} prior to absorption. Plant genotypes differ in their ability to take up Fe. Table 9.3 rates plants according to their sensitivity or tolerance to low levels of available Fe. Some crops appear in more than one category because of variations in soil, growing conditions, and differential response of varieties of a given crop. Fe-efficient varieties should be selected for conditions where Fe deficiencies are likely to occur.

The ability of plants to absorb and translocate Fe appears to be a genetically

TABLE 9.3 Sensitivity of Crops to Low Levels of
Available Fe in Soil*

Sensitive	Moderately Tolerant	Tolerant
Berries	Alfalfa	Alfalfa
Citrus	Barley	Barley
Field beans	Corn	Corn
Flax	Cotton	Cotton
Forage sorghum	Field beans	Flax
Fruit trees	Field peas	Grasses
Grain sorghum	Flax	Millet
Grapes	Forage legumes	Oats
Mint	Fruit trees	Potatoes
Ornamentals	Grain sorghum	Rice
Peanuts	Grasses	Soybeans
Soybeans	Oats	Sugar beets
Sundangrass	Orchard grass	Vegetables
Vegetables	Ornamentals	Wheat
Walnuts	Rice	
	Soybeans	
	Vegetables	
	Wheat	

*Some crops are listed under two or three categories because of
variations in soil, growing conditions, and differential response
varieties of a given crop.
SOURCE: Mortvedt, *Farm Chem.*, **143**:42 (1980).

controlled adaptive process that responds to Fe deficiency or stress. Roots of
Fe-efficient plants alter their environment to improve the availability and
uptake of Fe. Some of the biochemical reactions and changes enabling Fe-
efficient plants to tolerate and adapt to Fe stress are as follows:

1. Excretion of H^+ ions from roots.
2. Excretion of various reducing or chelating compounds from roots.
3. Rate of reduction (Fe^{3+} to Fe^{2+}) increases at the root.
4. Increase in organic acids, particularly citrate, in the root sap.
5. Adequate transport of Fe from roots to tops.
6. Less accumulation of P in roots and shoots, even in the presence of relatively high P in the growth medium.

Fe Soil Testing

Chelate–micronutrient relationships and stability in soils are utilized in soil
testing for micronutrients. Figure 9.12 shows that when EDDHA is added to
soil, it is 100% complexed with Fe over the pH range in soil. Therefore,
EDDHA might make a good extractant for Fe; however, the Fe-EDDHA is so
stable that very few other micronutrient cations would be complexed with
EDDHA. Although Fe-DTPA is not as stable at high pH as Fe-EDDHA (Fig.
9.12), the other micronutrients (i.e., Zn and Cu) exhibit considerable stability
with DTPA, especially at pH >7 (Fig. 9.15).

Knowledge of chelate stability in soil, as shown in Figures 9.12 and 9.15,

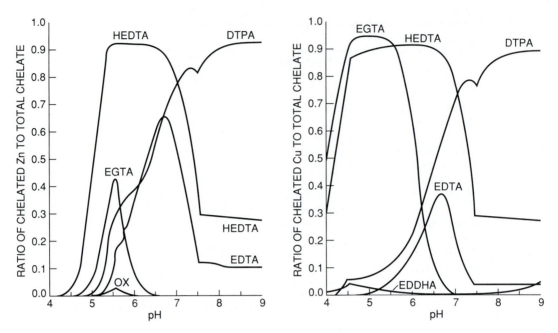

FIGURE 9.15 Stability of selected synthetic chelates with Zn and Cu in soils. *Norvell,* Micronutrients in Agriculture, *pp. 130–133. Madison, Wisc.: ASA, 1972.*

provides the basis for developing the DTPA soil test for Fe, Zn, Cu, and Mn, which is used in most soil testing laboratories. The DTPA soil test has been calibrated for most crops, and the general interpretation for DTPA extractable micronutrients is shown in Table 9.4. Before chelate relationships were developed, the most common micronutrient soil test was based on an acid extraction, usually HCl. Although some laboratories still use acid-extractable micronutrient soil tests, the DTPA test is preferred.

Fe Fertilizers and Fertilization

Fe chlorosis is one of the most difficult micronutrient deficiencies to correct in the field. Table 9.5 lists the Fe-containing materials that are commonly used to treat Fe deficiencies. In general, soil applications of inorganic Fe sources are not effective in correcting Fe deficiency because of the rapid precipitation of very insoluble $Fe(OH)_3$ (Fig. 9.4). For example, when $FeSO_4 \cdot 7H_2O$ and Fe-EDDHA were added to soil, only 20% of the $FeSO_4 \cdot 7H_2O$ was DTPA extractable after just 1 week, compared to 70% FeEDDA after 7 weeks and 26% after 14 weeks (Fig. 9.14).

TABLE 9.4 DTPA-Extractable Fe, Zn, Cu, and Mn for Deficient, Marginal, and Sufficient Soils

Category	Fe	Zn	Mn	Cu
	------------------- ppm -------------------			
Low (deficient)	0–2.5	0–0.5	<1.0	<0.2
Marginal	2.6–4.5	0.6–1.0	—	—
High (sufficient)	>4.5	>1.0	>1.0	>0.2

TABLE 9.5   Some Sources of Fertilizer Fe

Source	Formula	Percent Fe (Approx.)
Ferrous sulfate	$FeSO_4 \cdot 7H_2O$	19
Ferric sulfate	$Fe_2(SO_4)_3 \cdot 4H_2O$	23
Ferrous oxide	FeO	77
Ferric oxide	Fe_2O_3	69
Ferrous ammonium phosphate	$Fe(NH_4)PO_4 \cdot H_2O$	29
Ferrous ammonium sulfate	$(NH_4)_2SO_4 \cdot FeSO_4 \cdot 6H_2O$	14
Iron ammonium polyphosphate	$Fe(NH_4)HP_2O_7$	22
Iron chelates	NaFeEDTA	5–14
	NaFeHEDTA	5–9
	NaFeEDDHA	6
	NaFeDTPA	10
Natural organic materials	—	5–10

SOURCE: Mortvedt et al., Eds., *Micronutrients in Agriculture*, p. 357. Madison, Wisc.: Soil Science Society of America, 1972.

Correction of Fe deficiencies is done mainly with foliar application of Fe. One application of a 2% $FeSO_4$ solution at a rate of 15 to 30 gal/a is usually sufficient to alleviate mild chlorosis. However, several applications 7 to 14 days apart may be needed to remedy more severe Fe deficiencies. Injections of Fe salts directly into trunks and limbs of fruit-tree species such as pears and plums have been very effective in controlling Fe chlorosis. Treatments in California orchards typically consist of pressure injection at 200 psi of between 1 to 2 pints to 1 to 2 quarts per tree of 1–2% $FeSO_4$ solutions.

With the exception of $FeSO_4$, perhaps the most widely used Fe sources are the synthetic chelates (Table 9.5). These materials are water soluble and can be applied to the soil or foliage. Chelated Fe is protected from the usual soil reactions, which result in formation of insoluble $Fe(OH)_3$.

The stability of Fe chelates in soils was previously discussed. Since Fe-EDDHA is the most stable Fe chelate, it is the preferred chelate fertilizer source, although Fe-DTPA has also been used, especially on acid soils (Fig. 9.12). Unfortunately, Fe chelates are very expensive for soil application, except on high-value crops or plants.

Local acidification of small portions of the root zone can be effective in correcting Fe deficiencies in calcareous and high-pH soils. Several S products, such as S^0, ammonium thiosulfate, sulfuric acid, ammonium bisulfite, sulfur dioxide, and ammonium polysulfide, will lower soil pH and increase solution Fe concentration.

Complexing with polyphosphate fertilizers also increases the plant availability of both SO_4^{2-} and chelated Fe, but Fe-EDDHA is more effective than $FeSO_4$ at the same Fe rates.

Zinc

Zinc (Zn) content of the lithosphere is about 80 ppm, and Zn in soils ranges from 10 to 300 ppm and averages approximately 50 ppm. The igneous rocks contain about 70 ppm, while sedimentary rocks (shale) contain more Zn (95 ppm) than limestone (20 ppm) or sandstone (16 ppm). Franklenite ($ZnFe_2O_4$), smithsonite ($ZnCO_3$), and willemite (Zn_2SiO_4) are common Zn-containing minerals (Fig. 9.16). Zn mineral solubility in soils often resembles the solubility represented by "soil Zn."

Zn deficiencies are widespread in the United States and throughout the world, especially in the rice cropland of Asia. Soil conditions most associated with Zn deficiencies are acid sandy soils low in Zn; neutral, basic, or calcareous soils; fine-textured soils; soils high in available P; some organic soils; and subsoils exposed by land leveling or by wind and water erosion.

Forms of Soil Zn

The forms of Zn in soils are: solution Zn^{2+}; adsorbed Zn^{2+} on clay surfaces, OM, carbonates, and oxide minerals; organically complexed Zn^{2+}; and Zn^{2+} substituted for Mg^{2+} in the crystal lattices of clay minerals; and Zn in primary and secondary minerals. Identification of these various phases has been difficult because of the small amounts of Zn involved; however, the relationships and cycling between these Zn fractions in soils are very similar to those described for Fe (Fig. 9.2).

SOIL SOLUTION ZN   Zinc in the soil solution is very low, ranging between 2 and 70 ppb, with over half of the Zn^{2+} in solution complexed by OM. Several Zn hydrolysis species exist in solution with Zn^{2+} predominating below pH 7.7. Above this pH, $ZnOH^+$ becomes the most abundant species, with small amounts of $Zn(OH)_2^0$ above pH 9.1 (Fig. 9.17). Zinc solubility is highly pH

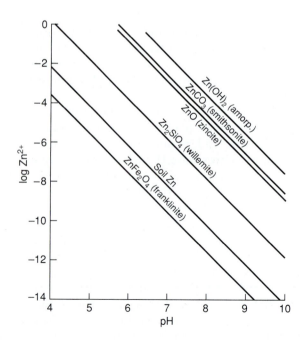

FIGURE 9.16   Solubilities of the common Zn minerals in soils.   *Lindsay. Chemical Equilibria in Soils, Wiley, New York, 1979.*

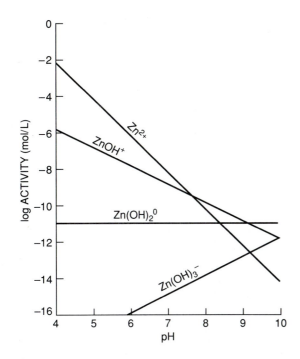

FIGURE 9.17   The common Zn species in soil solution as influenced by pH. *Lindsay*. Chemical Equilibria in Soils, *Wiley, New York, 1979.*

dependent and decreases 100-fold for each unit increase in pH. This relationship is represented by the following formula:

$$Soil-Zn \ + \ 2H^+ \rightleftharpoons Zn^{2+}$$

Thirtyfold reductions of Zn concentration in soil solutions for every unit of pH increase in the pH range 5 to 7 typically have been observed.

Diffusion is the dominant mechanism for transporting Zn^{2+} to plant roots. Complexing agents or chelates from root exudates or from decomposing organic residues facilitate the diffusion of Zn^{2+} to a root, as previously described (Fig. 9.9). Diffusion of chelated Zn^{2+} can be significantly greater than that of unchelated Zn^{2+} (Fig. 9.18).

Factors Affecting Zn Availability

Plant-available Zn^{2+} is determined by pH, adsorption on clay surfaces, OM, carbonates, and oxide minerals; complexation by OM; interactions with other nutrients; and climatic conditions.

SOIL pH   The availability of Zn^{2+} decreases with increased soil pH, as demonstrated earlier (Fig. 9.17). Most pH-induced Zn deficiencies occur in neutral and calcareous soils, although not all these soils exhibit Zn deficiency because of increased availability from chelation of Zn^{2+} (Fig. 9.9).

At high pH, Zn precipitates as insoluble amorphous soil Zn, $ZnFe_2O_4$, and/or $ZnSiO_4$, which reduces Zn^{2+} in soils (Fig. 9.16). Liming acid soils, especially ones low in Zn, will reduce uptake of Zn^{2+}, which is related to the pH effect

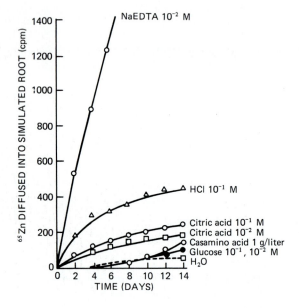

FIGURE 9.18   Effect of various complexing agents and acids on the accumulative diffusion of Zn into a simulated root over a 14-day period. *Elgawhary et al.,* Soil Sci. Soc. Am. J., *34:211, 1970.*

on Zn^{2+} solubility. Zn adsorption on the surface of $CaCO_3$ also could reduce solution Zn^{2+}. Adsorption of Zn^{2+} by clay minerals, Al and Fe oxides, OM, and $CaCO_3$ increases with increasing pH.

Zn Adsorption   The mechanism of Zn^{2+} adsorption on oxide surfaces is likely that shown below:

$$
\begin{array}{ccc}
\text{OH} & & \text{OH} \\
\diagdown \diagup \; \text{H} & & \diagdown \diagup \; \text{H} \\
\text{Fe} \longleftarrow \text{OH} & & \text{Fe} \longleftarrow \text{O} \\
\diagup & & \diagup \qquad \diagdown \\
\text{O} \qquad + Zn^{2+} & \rightleftharpoons & \text{O} \qquad Zn + 2H^+ \\
\diagdown & & \diagdown \qquad \diagup \\
\text{Fe} \longleftarrow \text{OH} & & \text{Fe} \longleftarrow \text{O} \\
\diagup \diagdown \; \text{H} & & \diagup \diagdown \; \text{H} \\
\text{OH} & & \text{OH}
\end{array}
$$

Such adsorption is considered an extension of the oxide surface resulting in specific or irreversible retention of Zn^{2+}. Nonspecific adsorption of Zn^{2+} also occurs where the retained Zn^{2+} is less firmly held and can be replaced by other cations, such as Ca^{2+} and Mg^{2+}.

Adsorption of Zn^{2+} by bentonite, illite, and kaolinite clay minerals is directly related to the CEC of clays. Zn reversibly bound by clay minerals is exchangeable and, thus, can be desorbed to solution.

Zn adsorption by carbonates is partly responsible for reduced Zn^{2+} availability in calcareous soils. Analysis of the Zn^{2+} adsorption curves in Figure 9.19 reveals that $CaCO_3$ content was the principal factor contributing to Zn adsorption. Although $ZnCO_3$ precipitated at higher Zn^{2+} concentrations, Zn availability to plants will probably not be seriously impaired because the solubility of this compound is too high for it to persist in soils (Fig. 9.16).

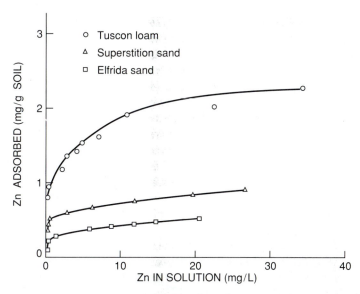

FIGURE 9.19   Adsorption of Zn by calcareous Arizona soils.   *Udo et al., Soil Sci. Soc. Am. J., **34***:405, 1970.*

Zinc is strongly adsorbed by magnesite ($MgCO_3$), to an intermediate degree by dolomite [$CaMg(CO_3)_2$], and least of all by calcite ($CaCO_3$). In magnesite and dolomite, it appears that Zn is actually adsorbed into the crystal surfaces at sites in the lattice normally occupied by Mg atoms.

ORGANIC MATTER   As previously discussed, Zn^{2+} forms stable complexes with soil OM-components. The humic and fulvic acid fractions are prominent in Zn adsorption. Three classes of reactions of OM with Zn and other micronutrients have been distinguished:

1. Immobilization by high molecular weight organic substances such as lignin.
2. Solubilization and mobilization by short-chain organic acids and bases.
3. Complexation by initially soluble organic substances that then form insoluble salts.

The action of OM on Zn^{2+} can be expected to vary depending on the characteristics and amounts of the organic materials involved. When reactions 1 and/or 3 prevail, availability of Zn will be reduced; this occurs in Zn-deficient peats and humic gley soils. On the other hand, formation of soluble chelated Zn compounds will enhance availability by keeping Zn^{2+} in solution. Substances present in or derived from freshly applied organic materials have the capacity to chelate Zn^{2+}; however, the increased solution Zn^{2+} is not always reflected in enhanced Zn uptake by plants.

INTERACTION WITH OTHER NUTRIENTS   Other metal cations, including Cu^{2+}, Fe^{2+}, and Mn^{2+}, inhibit Zn^{2+} uptake, possibly because of competition for the same carrier site. The antagonistic effect of several cations, especially Cu^{2+} and Fe^{2+}, on Zn uptake by rice is clearly demonstrated in Figure 9.20.

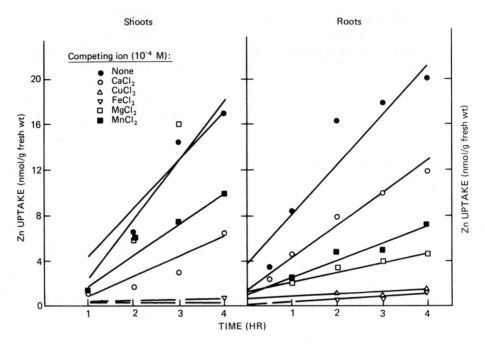

FIGURE 9.20   Effect of competing ions on uptake of Zn in shoots and roots of rice seedlings immersed in $5 \times 10^{-3}\ M\ ^{65}ZnCl_2$.   *Giordano et al.*, Plant Soil, *41*:637, *1974.*

Phosphorus. High P availability can induce Zn deficiency, commonly in soils that are marginally Zn deficient. When plants are Zn deficient, their ability to regulate P accumulation is severely impaired. As a consequence, P is absorbed by roots and transported to plant tops in such excess that it becomes toxic and produces symptoms resembling Zn deficiency, in spite of adequate Zn concentrations in plant tops.

TABLE 9.6   Sensitivity of Crops to Low Levels of Available Zn

Very Sensitive	Mildly Sensitive	Insensitive
Beans, lima and pea	Alfalfa	Asparagus
Castor beans	Barley	Carrots
Citrus	Clovers	Forage grasses
Corn	Cotton	Mustard and other crucifers
Flax	Potatoes	Oats
Fruit trees (deciduous)	Sorghum	Peas
Grapes	Sugar beets	Peppermint
Hops	Tomatoes	Rye
Onions	Wheat	Safflower
Pecans		
Pine		
Rice		
Soybeans		
Sudangrass		

SOURCE: Adapted from *Zinc in Crop Nutrition.* New York: International Lead Zinc Research Organization, Inc., and Zinc Institute, Inc., 1974; and Mortvedt, *Farm Chem.*, **143**(11):56 (1980).

Serious concerns have been expressed about inducing Zn deficiencies, particularly in sensitive crops such as corn, flax, and beans, as a result of systematically building up available P in soil. The problem can be corrected or avoided by ensuring that Zn supplies are also adequate.

The influence of high P in soil on mycorrhizal uptake of Zn might also be a factor in Zn deficiency of crops. Mycorrhizae increase the micronutrient uptake by many plants. P fertilization can suppress mycorrhizal uptake of Zn in corn and soybeans.

The popularly held belief that P–Zn reactions in soil, such as the formation of insoluble $Zn_3(PO_4)_2 \cdot 4H_2O$, are responsible for P-induced Zn deficiency should be discounted. Solubility of this compound is sufficiently high that it will readily provide Zn to plants.

Sulfate and Nitrogen. The highly mobile $ZnSO_4^0$ complex is an important species in soils and contributes significantly to total Zn in solution. For example, $CaSO_4$ added to a slightly acid soil can increase the Zn and Fe concentrations in plants.

Application of N fertilizer can stimulate plant growth and increase Zn requirements. Acid-forming N fertilizers will increase the uptake of both native and supplemental Zn. On the other hand, products with a neutral to basic effect are known to depress Zn uptake.

FLOODING   Zn deficiency in flooded rice soils was once thought to be related to high pH or the presence $CaCO_3$. However, Zn deficiency also occurs in acid soils. When soils are submerged, the concentration of many nutrients increases, but not that of Zn. In acid soils, Zn deficiency may be attributed to the increase in pH under reducing conditions and subsequent precipitation of franklinite ($ZnFe_2O_4$) or sphalerite (ZnS). Decreasing pH in submerged, calcareous soils would usually increase Zn solubility. However, the higher the soil pH and the poorer the aeration, the greater the Zn deficiency.

CLIMATIC CONDITIONS   Zn deficiencies are generally more pronounced during cool, wet seasons and often disappear in warmer weather. Climatic conditions during early spring contribute to Zn deficiency by poor light, as well as low temperature and excessive moisture. Increasing soil temperature increases the availability of Zn to crops by increasing solubility and diffusion of Zn^{2+}.

PLANT FACTORS   Species and varieties of plants differ in their susceptibility to Zn deficiency (Table 9.6). Corn and beans are very susceptible to low Zn. Fruit trees in general, and citrus and peach in particular, are also sensitive.

Cultivars differ in their ability to take up Zn, which may be caused by differences in Zn translocation and utilization, differential accumulations of nutrients that interact with Zn, and differences in plant roots to exploit for soil Zn.

Zn Soil Testing

See section in Fe soil testing (pp. 317–318).

Zn Fertilizers and Fertilization

Zinc sulfate ($ZnSO_4$), containing about 35% Zn, is the most common Zn fertilizer source, although use of synthetic Zn chelates has increased (Table 9.7). Inorganic Zn sources are satisfactory fertilizers because they are very soluble in soils. Zinc phosphate, $Zn_3(PO_4)_2$, is less soluble than the oxides, hydroxides, or carbonates, but in soils it provides plants available Zn over extended periods of time.

Fertilizer Zn rates depend on the crop, Zn source, method of application, and severity of Zn deficiency. Rates usually range from 3 to 10 lb/a with inorganic Zn and from 0.5 to 2.0 lb/a with a chelate or an organic Zn source. For most field and vegetable crops, 10 lb/a is recommended in clay and loam soils and 3 to 5 lb/a in sandy soils. In most cropping situations, applications of 10 lb/a of Zn can be effective for 3 to 5 years.

Because of limited Zn mobility in soils, broadcast Zn should be thoroughly incorporated into the soil; however, band application may be more effective, especially in fine-textured and very low Zn soils. The efficiency of band-applied Zn can be improved by the presence of acid-forming N fertilizers.

In the case of long-term perennials such as hops, grapes, and tree fruits, preplant soil applications of Zn are effective. Zinc rates for hops and grapes are 20 lb/a and 100 lb/a for tree fruits. Soil applications are only of limited value after these crops have been established.

Foliar applications are used primarily for tree crops. Sprays containing 10 to 15 lb/a of Zn are usually applied to dormant orchards, whereas 2–3 lb/a can be foliar applied to growing crops. Damage to foliage can be prevented by adding lime to the solution or by using less soluble materials like ZnO or $ZnCO_3$. Other methods include seed coatings, root dips, and tree injections. The former treatment may not supply enough Zn for small seeded crops, but dipping potato seed pieces in a 2% ZnO suspension is satisfactory.

Foliar applications of chelates and natural organics are particularly suitable for quick recovery of Zn-deficient seedlings. These Zn sources can be used in high-analysis liquid fertilizers because of their high solubility and compatability. Chelates such as ZnEDTA are mobile and can be soil applied; however, high cost usually limits their use.

TABLE 9.7   Some Sources of Fertilizer Zn

Source	*Formula*	*Percent Zn (Approx.)*
Zinc sulfate monohydrate	$ZnSO_4 \cdot H_2O$	35
Zinc oxide	ZnO	78
Zinc carbonate	$ZnCO_3$	52
Zinc phosphate	$Zn_3(PO_4)_2$	51
Zinc chelates	$Na_2ZnEDTA$	14
	NaZnNTA	13
	NaZnHEDTA	9
Natural organics	—	5–10

SOURCE: Mortvedt et al., Eds., *Micronutrients in Agriculture*, p. 371. Madison, Wisc.: Soil Science Society of America, 1972.

Copper

Copper (Cu) concentration in the earth's crust averages about 55 to 70 ppm. Igneous rocks contain 10 to 100 ppm Cu, while sedimentary rocks contain between 4 and 45 ppm Cu. Cu concentration in soils ranges from 1 to 40 ppm and averages about 9 ppm. Total soil Cu may be 1 or 2 ppm in deficient soils.

Chalcopyrite ($CuFeS_2$), chalcocite (Cu_2S), and bornite ($CuFeS_4$) are the important Cu-containing primary minerals. Secondary Cu minerals include oxides, carbonates, silicates, sulfates, and chlorides, but most of them are too soluble to persist.

Cu deficiencies are less common than deficiencies of other micronutrients. The geographical pattern of deficiency is usually localized and often associated with plants grown on peats and mucks (Histosols). Cu deficiency is often the first nutrient problem to appear in plants grown on newly reclaimed acid Histosols; this condition is often referred to as *reclamation disease*. Cu deficiency also is common when virgin organic soils are brought into production.

Cu deficiencies are common in Florida, Wisconsin, Michigan, New York, and Manitoba, where high-value speciality crops are intensively grown on peats and mucks. Small grains have frequently shown Cu deficiency on peat soils in several Canadian provinces. There are also instances of inadequate Cu in well-drained sandy and calcareous sandy soils in Manitoba, and in sandy soils in Saskatchewan and Alberta. Approximately 8.5 million acres of cultivated land in the three Canadian Provinces are Cu deficient and often reduce productivity in small grains.

Use of Cu-containing fungicides in the past probably prevented or corrected Cu deficiencies; however, excessive use has created some Cu toxicity problems. Cu toxicity has also been encountered in soils affected by mine wastes.

Forms of Soil Cu

In addition to the Cu in primary and secondary minerals, it exists in the following forms:

1. In the soil solution—ionic and complexed.
2. On cation exchange sites of clays and OM.
3. Occluded and coprecipitated in soil oxide material.
4. On specific adsorption sites.
5. In OM and living organisms.

The relationships and cycling between these Cu fractions in soils are very similar to those described for Fe (Fig. 9.2). Most of the Cu in soils is very insoluble and can only be extracted by strong chemical treatments that dissolve various mineral structures or solubilize OM. There is, however, a significant "pool" of organically complexed Cu, which is in equilibrium with soil solution Cu and can contribute to Cu^{2+} diffusion to plant roots.

SOIL SOLUTION CU   The Cu concentration in soil solution is usually very low, ranging between 10^{-8} and $10^{-6} M$ (Fig. 9.21). The dominant solution

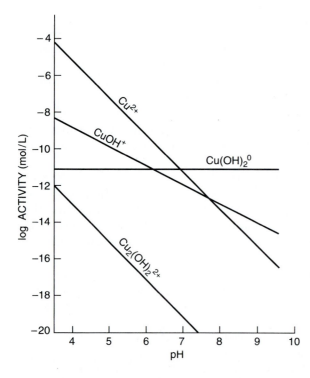

FIGURE 9.21   The common Cu species in soil solution as influenced by pH. *Lindsay,* Chemical Equilibria in Soils, *Wiley, New York, 1979.*

species are Cu^{2+} at pH <7 and $Cu(OH)_2^0$ at pH >7. Hydrolysis reactions of Cu ions are shown in the following equations:

$$Cu^{2+} + H_2O \rightleftharpoons CuOH^+ + H^+$$

$$CuOH^+ + H_2O \rightleftharpoons Cu(OH)_2^0 + H^+$$

The $CuSO_4^0$ and $CuCO_3^0$ complexes are also important forms of Cu. Solubility of Cu is pH dependent and it increases 100-fold for each unit decrease in pH (Figs. 9.21, 9.22). The line indicated by "soil Cu" represents the solubility of Cu in most soils and is very close to that of cupric ferrite ($CuFe_2O_4$) (Fig. 9.22).

Cu is most likely supplied to plant roots by diffusion of organically bound, chelated Cu, similar to chelated Fe diffusion in soil (Fig. 9.9). Organic compounds in the soil solution are capable of chelating solution Cu^{2+}, which increases the solution Cu^{2+} concentration above that predicted by Cu mineral solubility (Fig. 9.22).

ADSORBED CU   The Cu^{2+} ion is specifically or chemically adsorbed by layer silicate clays, OM, and Fe, Al, or Mn oxides. With the exception of Pb^{2+}, Cu^{2+} is the most strongly adsorbed of all the divalent metals on Fe and Al oxides. The mechanism of adsorption by oxides, unlike the electrostatic attraction of Cu^{2+} on the CEC of clay particles, involves formation of Cu-O-Al or Cu-O-Fe surface bonds (Fig. 9.23). This chemisorption process is thus controlled by the quantity of surface OH^- groups.

Cu adsorption increases with increasing pH due to (1) increased pH-dependent sites on clay and OM, (2) reduced competition with H^+, and (3) a change in the hydrolysis state of Cu in solution. As the pH is raised, hydrolysis

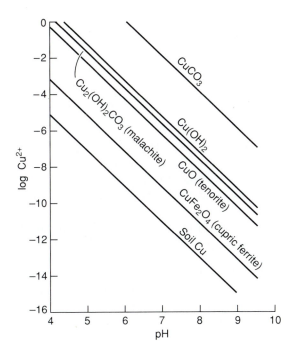

FIGURE 9.22   Solubilities of the common Cu minerals in soils.   *Lindsay.* Chemical Equilibria in Soils, *Wiley, New York, 1979.*

of Cu^{2+} adsorbed on the CEC decreases exchangeable Cu^{2+} and increases chemisorbed Cu (i.e., decreasing H^+ shifts equilibrium to the right in Fig. 9.23).

OCCLUDED AND COPRECIPITATED CU   A significant fraction of soil Cu is occluded or buried in various mineral structures, such as clay minerals and Fe and Mn oxides. Cu is capable of isomorphous substitution in octahedral positions of crystalline silicate clays. It is present as an impurity within $CaCO_3$ and $MgCO_3$ minerals in arid soils, and within $Al(OH)_3$ and $Fe(OH)_3$ in acid soils.

ORGANIC CU   Most of the soluble Cu in surface soils is organically complexed and is more strongly bound to organic matter than any other micronutrient. The Cu^{2+} ion is directly bonded to two or more organic functional groups, chiefly carboxyl or phenol (Fig. 9.24). Humic and fulvic acids contain

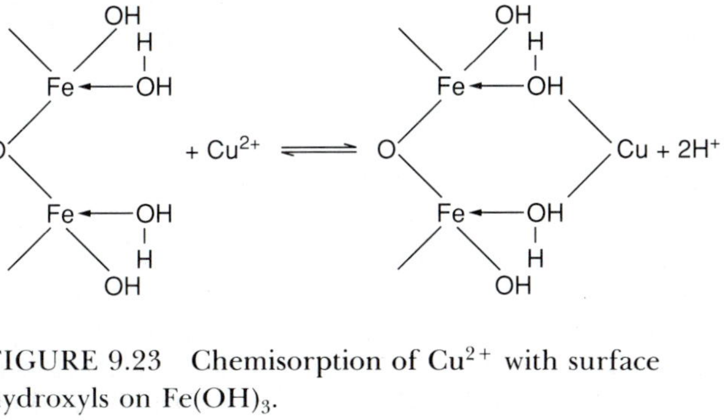

FIGURE 9.23   Chemisorption of Cu^{2+} with surface hydroxyls on $Fe(OH)_3$.

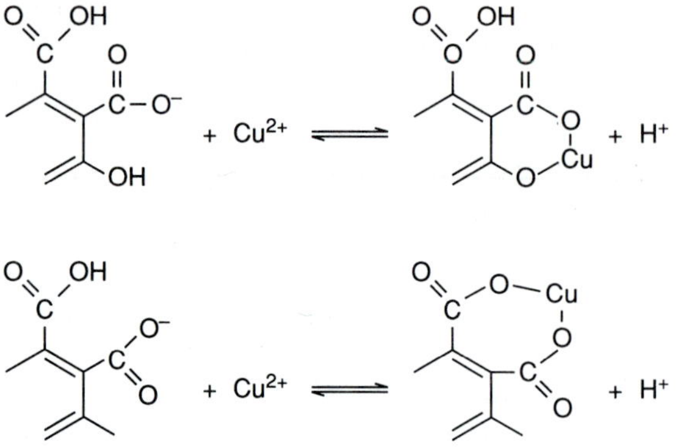

FIGURE 9.24   Mechanism of Cu complexed by organic matter. *Stevenson and Ardakani, 1972, p. 90, Micronutrients in Agric., ASA.*

multiple binding sites, primarily carboxyl groups, for Cu. In most mineral soils, OM is intimately associated with clay, probably as a clay-metal-organic complex (Fig. 9.25).

At soil OM levels up to 8%, both organic and mineral surfaces are involved in Cu adsorption, while at higher concentrations of OM, binding of Cu takes place mostly on organic surfaces. For soils having similar clay and OM contents, the contribution of OM to the complexing of Cu will be highest when the predominant clay mineral is kaolinite and lowest with montmorillonite.

Factors Affecting Cu Availability

The availability and movement of Cu are influenced by soil texture, pH, CEC, OM content, and hydrous oxides.

TEXTURE   Cu content in the soil solution is usually lower in excessively leached podzolic sands and calcareous sands than in other soil types.

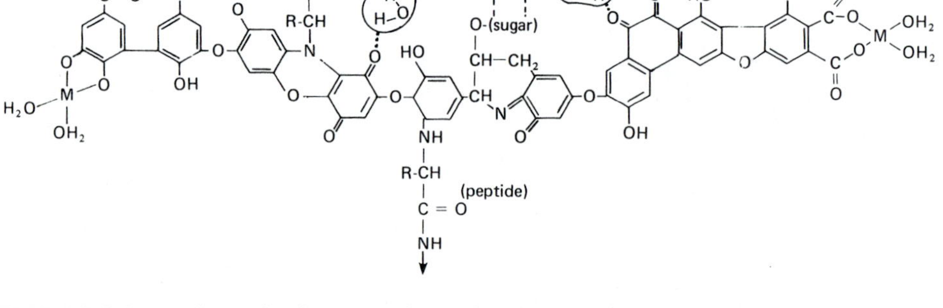

FIGURE 9.25   Schematic diagram of the clay-OM-metal (M) complex.   *Stevenson and Fitch, in J. F. Loneragan, Ed.,* Copper in Soils and Plants, *p. 70. Academic Press, New York, 1981.*

SOIL PH   The concentration of soil solution Cu decreases with increasing pH, and its supply to plants is reduced because of decreased solubility and increased adsorption.

INTERACTIONS WITH OTHER NUTRIENTS   There are numerous interactions involving Cu and other nutrients. Applications of N-P-K fertilizers can induce Cu deficiencies. Cu deficiencies following the use of acid-forming N fertilizers may be related to increased Al^{3+} levels in the soil solution. Furthermore, increased growth resulting from the application of N or other nutrients may be proportionally greater than Cu uptake, which dilutes Cu concentration in plants. Increasing the N supply to crops can reduce mobility of Cu in plants, since high N in plants impedes translocation of Cu from older leaves to new growth. High concentrations of Zn, Fe, and P in soil solution also can depress Cu absorption by plant roots and may intensity Cu deficiency.

PLANT FACTORS   Crops that are highly responsive to Cu include wheat, rice, alfalfa, carrots, lettuce, spinach, table beets, sudangrass, citrus, and onions, while crops with the greatest tolerance to low Cu are beans, peas, potatoes, asparagus, rye, pasture grasses, *Lotus* spp., soybeans, lupine, rape, and pines. Among small-grain species, rye has exceptional tolerance to low levels of soil Cu and will be healthy where wheat fails completely without the application of Cu. Rye can extract up to twice as much Cu as wheat under the same conditions. This genetic advantage over wheat is inherited by the wheat-rye hybrid triticale. The usual order of sensitivity of the small grains to Cu deficiency in the field is wheat > barley > oats > rye. Varietal differences in tolerance to low Cu are important, and sometimes they can be as large as those among crop species.

The genotypic differences in the Cu nutrition of plants are related to (1) differences in the rate of Cu absorption by roots, (2) better exploration of soil through greater root length per plant or per unit area, (3) better contact with soil through longer root hairs, (4) modification of Cu availability in soil adjacent to roots by root exudation, (5) acidification or change in redox potential, (6) more efficient transport of Cu from roots to shoots, and/or (7) lower tissue requirement for Cu.

Plowing unharvested *Brassica* root crops into soil aggravates Cu deficiency in the following crop, which is probably related to the large amounts of S released during decomposition of these residues.

Severe Cu deficiency in crops planted in high C/N residues is related to (1) reactions of Cu with organic compounds originating from decomposing straw, (2) competition for available Cu by stimulated microbial populations, and (3) inhibition of root development and the ability to absorb Cu. If the soil-available Cu is low, manure added to a field may accentuate the problems. OM from manure, straw, or hay can tie up Cu, making it unavailable to plants.

Cu Toxicity

Toxicity symptoms include reduced shoot vigor, poorly developed and discolored root systems, and leaf chlorosis. The chlorotic condition in shoots superficially resembles Fe deficiency. Toxicities are uncommon, occurring in limited areas of high Cu availability; after additions of high-Cu materials such as sewage sludge, municipal composts, pig and poultry manures, and mine wastes; and from repeated use of Cu-containing pesticides.

Cu Soil Testing

See Fe soil testing on pp. 317–318.

Cu Fertilizers and Fertilization

The usual Cu source is $CuSO_4 \cdot 5H_2O$, although CuO, mixtures of $CuSO_4$ and $Cu(OH)_2$ and Cu chelates are also used (Table 9.8). Copper sulfate contains 25.5% Cu, is soluble in water, and is compatible with most fertilizer materials. Copper ammonium phosphate can be either soil or foliar applied. It is only slightly soluble in water but can be suspended and sprayed on the plants. It contains 30% Cu and, like the other metal ammonium phosphates, is slowly available.

Soil and foliar applications are both effective, but soil applications are more common, with Cu rates of 0.6 to 20 lb/a needed to correct deficiencies. Effectiveness is increased by thoroughly mixing Cu fertilizers into the root zone or by banding them in the seed row. When band applied, rates are reduced to prevent possible root injury. Additions of Cu can be ineffective when root activity is restricted by excessively wet or dry soil, root pathogens, toxicities, and deficiencies of other nutrients. Residual Cu fertilizer availability from as little as several pounds per acre can persist for 2 or more years, depending on the soil, crop, and Cu rate.

Application of Cu in foliar sprays is confined mainly to emergency treatment of deficiencies identified after planting. In some areas, however, Cu is included in regular foliar spraying programs. Cu chelates (CuEDTA) can be used as a foliar Cu fertilizer; however, soil application is often too costly.

TABLE 9.8   Cu Compounds Used as Fertilizers

Source	Formula	Percent Cu	H_2O Solubility
Copper sulfate	$CuSO_4 \cdot 5H_2O$	25	Soluble
Copper sulfate monohydrate	$CuSO_4 \cdot H_2O$	35	Soluble
Copper nitrate	$Cu(NO_3)_2 \cdot 3H_2O$		Soluble
Copper acetate	$Cu(C_2H_3O_2)_2 \cdot H_2O$	32	Slightly
Copper ammonium phosphate	$Cu(NH_4)PO_4 \cdot H_2O$	32	Insoluble
Copper chelates	Na_2Cu EDTA	13	Soluble
Copper chelates	NaCu HEDTA	9	Soluble
Copper polyflavanoids	—	5–7	Soluble

SOURCE: Gilkes, in J. F. Loneragan et al., Eds., *Copper in Soils and Plants*, p. 98. New York: Academic Press, 1981.

Manganese

Manganese (Mn) concentration in the earth's crust averages 1000 ppm, and Mn is found in most Fe-Mg rocks. Mn, when released through weathering of primary rocks, will combine with O_2, CO_3^{2-}, and SiO_2 to form a number of secondary minerals, including pyrolusite (MnO_2), hausmannite (Mn_3O_4), manganite ($MnOOH$), rhodochrosite ($MnCO_3$), and rhodonite ($MnSiO_3$). Pyrolusite and manganite are the most abundant.

Total Mn in soils generally ranges between 20 and 3000 ppm and averages about 600 ppm. Mn in soils occurs as various oxides and hydroxides coated on soil particles, deposited in cracks and veins, and mixed with Fe oxides and other soil constituents.

It has been estimated that 13 million acres in 30 states of the United States may be low in Mn, predominately in humid regions of the East, especially around the Great Lakes. Neutral or basic soils are generally low in Mn but may not be deficient. Mn deficiency has been identified in Canada's Prairie Provinces on neutral or alkaline soils of abundant OM and on sandy and peat soils underlain with calcareous subsoils. It also occurs in British Columbia in a few orchards situated on alluvial fans. Low Mn is the most common micronutrient deficiency in both soybeans and small grains in Ontario.

The soil conditions commonly identified with Mn deficiency are the following:

1. Thin, peaty soils overlying calcareous subsoils.
2. Alluvial silt and clay soils and marsh soils derived from calcareous materials.
3. Poorly drained calcareous soils high in OM.
4. Calcareous black sands and reclaimed acid soils.
5. Calcareous soils freshly broken up from old grassland.
6. Old black garden soils where manure and lime have been applied regularly.
7. Very sandy acid mineral soils that are low in native Mn content and where available Mn may have been leached.

Forms of Soil Mn

Mn exists as solution Mn^{2+}, exchangeable Mn^{2+}, organically bound Mn, and as various Mn minerals. The equilibrium among these forms determines Mn availability to plants (Fig. 9.2). For satisfactory Mn nutrition of crops, solution and exchangeable Mn should be 2 to 3 ppm and 0.2 to 5 ppm, respectively.

The major processes in this cycle are Mn oxidation-reduction and complexing solution Mn^{2+} with natural organic chelates. The continuous cycling of OM significantly contributes to soluble Mn. Factors influencing the solubility of soil Mn include pH, redox, and complexation. Soil moisture, aeration, and microbial activity influence redox, while complexation is affected by OM and microbial activity.

SOIL SOLUTION MN   The principal species in solution is Mn^{2+}, which decreases 100-fold for each unit increase in pH, similar to the behavior of other divalent metal cations (Fig. 9.26). The concentration of Mn^{2+} in solution is predominantly controlled by MnO_2 (Fig. 9.27). Concentration of Mn^{2+} in the soil solution of acid and neutral soils is commonly in the range 0.01 to 1 ppm, with organically complexed Mn^{2+} comprising about 90% of solution Mn^{2+}. Plants take up Mn^{2+}, which moves to the root surface by diffusion of principally chelated Mn^{2+}, as described for Fe (Fig. 9.9).

Mn in the soil solution is greatly increased under acid, low-redox conditions. In extremely acid soils, Mn^{2+} solubility can be sufficiently great to cause

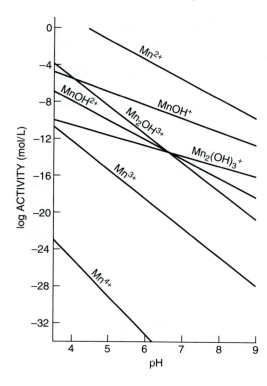

FIGURE 9.26   The common Mn species in soil solution as influenced by pH.   *Lindsay.* Chemical Equilibria in Soils, *Wiley, New York, 1979.*

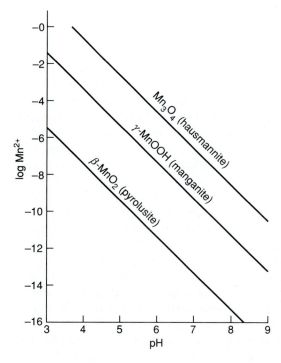

FIGURE 9.27   Solubilities of the common Mn minerals in well-aerated soils. Reducing conditions will increase Mn solubility.   *Lindsay.* Chemical Equilibria in Soils, *Wiley, New York, 1979.*

toxicity problems in sensitive plant species (Fig. 9.26). On a loamy sand below pH 5.0 in Indiana, extensive foliar damage on muskmelon was due to Mn toxicity. Liming increased yields and reduced Mn levels in plant tissue.

Because of the mobility of Mn^{2+}, it can leach from soils, particularly from acid podzols. The frequent occurrence of Mn deficiency in poorly drained mineral and organic soils is often attributed to low Mn levels resulting from leaching of soluble Mn^{2+}.

Factors Affecting Mn Availability

Many soil, environmental, and management factors influence Mn availability to crops.

SOIL pH AND CARBONATES   Since Mn^{2+} in solution varies with pH, management practices that change soil pH will also greatly influence Mn^{2+} availability and uptake. Liming very acid soils decreases solution and exchangeable Mn^{2+}, due to Mn^{2+} precipitation as MnO_2 (Fig. 9.27). On the other hand, low Mn availability in high pH and calcareous soils and in overlimed, poorly buffered, coarse-textured soils can be largely overcome by acidification through the use of acid-forming N or S materials. High pH also favors the formation of less available organic complexes of Mn. Activity of the soil microorganisms that oxidize soluble Mn to unavailable forms reaches a maximum near pH 7.

EXCESSIVE WATER AND POOR AERATION   Soil waterlogging will reduce O_2 and lower redox potential, which increases soluble Mn^{2+}, especially in acid soils. Mn availability can be increased by poor aeration in compact soils and by local accumulations of CO_2 around roots and other soil microsites. The resulting low-redox conditions will render Mn more available without appreciably affecting the redox potential or pH of the bulk soil.

ORGANIC MATTER   Availability of Mn^{2+} can be strongly influenced by reactions with OM. The low availability of Mn in high-OM basic soils is attributed to the formation of unavailable chelated Mn^{2+} compounds. It may also be held in unavailable organic complexes in peats or muck soils. Additions of natural organic materials such as peat moss, compost, and wheat and clover straw have increased the solution and exchangeable Mn.

INTERACTION WITH OTHER NUTRIENTS   High levels of Cu, Fe, or Zn can reduce Mn uptake by plants. Addition of acid-forming NH_4^+ to soil will enhance Mn uptake. Neutral KCl, NaCl, and $CaCl_2$ applied to acid soils also can increase the Mn availability to and concentration in plants. The relative order of the salt effect on increasing available Mn is $KCl > KNO_3 > K_2SO_4$. The effect of KCl on Mn uptake can be so strong that it produces toxicity symptoms in sensitive crops.

CLIMATIC EFFECTS   Very pronounced seasonal variations in Mn availability have been observed. Wet weather favors the presence of Mn^{2+}, whereas warm, dry conditions encourage the formation of less available oxidized forms of Mn. Dry weather either induces or aggravates Mn deficiency, particularly in fruit trees. Wet weather is one of the conditions usually associated with a high incidence of gray speck of oats, a Mn deficiency disorder. Increasing soil temperature during the growing season improves Mn uptake, presumably because of greater plant growth and root activity.

SOIL MICROORGANISMS   Mn deficiencies caused by soil organisms oxidizing Mn^{2+} to Mn^{4+} have been reported. A number of bacterial and fungal genera are recognized as being capable of Mn oxidation. Mn also may contribute to disease resistance in some crops. Practices that reduce the activity of Mn oxidizers, such as NH_4-N sources, and inhibition of nitrification increase Mn solubility and help suppress certain diseases.

PLANT FACTORS   Several plant species, exhibit differences in sensitivity to Mn deficiency (Table 9.9). Although it is difficult to generalize, small grains, soybeans, some vegetables, and tree fruits are usually considered to be most affected by inadequate Mn nutrition. Lack of Mn is one of the most common micronutrient problems in soybean.

These differences in the response of Mn-efficient and Mn-inefficient plants are due to internal factors rather than to the effects of the plants on the soil. Reductive capacity at the root may be the factor restricting Mn uptake and translocation. There may also be significant differences in the amounts and properties of root exudates generated by plants, which can influence Mn^{2+} availability. It is possible that plant characteristics possessed by Fe-efficient plants may similarly influence Mn uptake in plants tolerant of Mn stress.

Mn Soil Testing

See Fe soil testing on pp. 317–318.

Mn Fertilizers and Fertilization

Manganese sulfate is widely used for correction of Mn deficiency and may be soil or foliar applied (Table 9.10). In addition to inorganic Mn fertilizers, natural organic complexes and chelated Mn are available and are usually foliar applied.

TABLE 9.9   Sensitivity of Crops to Low Levels of Available Mn in Soil*

Sensitive		Moderately Tolerant		Tolerant	
Alfalfa	Soybeans	Barley	Potatoes	Barley	Rye
Citrus	Sugar beets	Corn	Rice	Corn	Soybeans
Fruit trees	Wheat	Cotton	Rye	Cotton	Vegetables
Oats		Field beans	Soybeans	Field beans	Wheat
Onions		Fruit trees	Vegetables	Fruit trees	
Potatoes		Oats	Wheat	Rice	

*Some crops are listed under two or three categories because of variation in soil, growing conditions, and differential response of varieties of a given crop.

SOURCE: Mortvedt, *Farm Chem.*, **143**:42 (1980).

TABLE 9.10   Sources of Mn Used for Fertilizer

Source	Formula	*Percent Mn (Approx.)*
Manganese sulfate	$MnSO_4 \cdot 4H_2O$	26–28
Manganous oxide	MnO	41–68
Manganese chloride	$MnCl_2$	17
Natural organic complexes	—	5–9
Synthetic chelates	MnEDTA	5–12

SOURCE: Mortvedt et al., Eds., *Micronutrients in Agriculture*, p. 363. Madison, Wisc.: Soil Science Society of America, 1972; and Mortvedt, *Farm Chem.*, **143**:42 (1980).

Manganese oxide is only slightly water soluble, but it is usually a satisfactory source of Mn. The particle size of this material is important, and it must be finely ground to be effective. Rates of Mn application range from 1 to 25 lb/a; higher rates are recommended for broadcast application, while lower rates are foliar applied. Band-applied Mn is generally more effective than broadcast Mn, and band treatments are usually about one-half the broadcast rates. Oxidation to less available forms of Mn is apparently delayed with band-applied Mn. Applications at the higher rates may be required on organic soils. Band application of Mn in combination with N-P-K fertilizers is commonly practiced.

Broadcast application of Mn chelates and natural organic complexes is not normally advised because soil Ca or Fe can replace Mn in these chelates and the freed Mn is usually converted to unavailable forms. Meanwhile, the more available complexed Ca or Fe probably accentuates the Mn deficiency. Lime or high-pH-induced Mn deficiencies can be rectified by acidification resulting from the use of S or other acid-forming materials.

Boron

Boron (B) is the only nonmetal among the micronutrients. It occurs in low concentrations in the earth's crust and in most igneous rocks ($\sim$ 10 ppm). Among sedimentary rocks, shales have the highest B concentrations (up to 100 ppm), present mainly in the clay minerals. The total concentration of B in soils varies between 2 and 200 ppm and frequently ranges from 7 to 80 ppm. Less than 5% of the total soil B is available to plants.

Tourmaline, a borosilicate, is the main B-containing mineral found in soils. It is insoluble and resistant to weathering; consequently, release of B is quite slow. Increasing frequency of B deficiencies suggests that it is incapable of supplying plant requirements under prolonged heavy cropping. B in soils of arid climates is usually sufficient because alkaline earth borates are plentiful.

In North America, B deficiency occurs mainly in the eastern and northwestern U.S. and in the adjoining regions of Canada (Fig. 9.28). B toxicity to plants is uncommon in most arable soils unless it has been added in excessive amounts in fertilizers. In arid regions, however, B toxicity may occur naturally or may develop because of a high B content in irrigation waters.

FIGURE 9.28   Areas of B deficiency in the United States.   *Turner,* Boron in Agriculture, *U.S. Borax, New York, 1976.*

Forms of Soil B

B exists in four major forms in soil: in rocks and minerals, adsorbed on clay surfaces and Fe and Al oxides, combined with OM, and as boric acid ($H_3BO_3^0$) and $B(OH)_4^-$ in the soil solution (Fig. 9.29). Understanding B cycling between the solid and solution phases is very important because of the narrow range in solution concentration that separates deficiency and toxicity in crops.

SOIL SOLUTION B   Undissociated $H_3BO_3^0$ is the predominant species expected in soil solution at pH values ranging from 5 to 9. At pH > 9.2 $H_2BO_3^-$ can hydrolyze to $H_4BO_4^-$. B can be transported from the soil solution to absorbing plant roots by both mass flow and diffusion.

ADSORBED B   B adsorption and desorption can buffer solution B, which helps to reduce B leaching losses. It is a major form of B in alkaline, high-B soils. The main B adsorption sites are (1) broken Si—O and Al—O bonds at the edges of clay minerals, (2) amorphous hydroxide structures, and (3) Fe and Al oxy and hydroxy compounds. Increasing pH, clay content, and OM and the presence of Al compounds favor $H_4BO_4^-$ adsorption.

ORGANICALLY COMPLEXED B   OM represents a large potential source of plant-available B in soils, which increases with increasing OM. The B–OM complexes are probably

$$=C-O \diagdown B-OH \quad \text{or} \quad H^+ \left| \begin{array}{c} =C-O \diagdown \diagup O-C= \\ \diagup B \diagdown \\ =C-O \diagup \diagdown O-C= \end{array} \right|^-$$

B in Minerals   B can substitute for Al^{3+} and/or Si^{4+} ions in silicate minerals. Following its adsorption on clay surfaces, B will slowly diffuse into interlayer positions.

Factors Affecting B Availability

The factors that influence the availability and movement of B are soil texture, amount and type of clay, pH and liming, OM, interrelationships with other elements, and soil moisture. Extraction with hot water is the most common soil test of soil B availability. In addition to soil test B, evaluating these factors is important in assessing the potential crop response to B.

Soil Texture   Coarse-textured, well-drained, sandy soils are low in B, and crops with a high requirement, such as alfalfa, respond to B applications of 3 or more lb/a. Sandy soils with fine-textured subsoils generally do not respond to B the same way as those with coarse-textured subsoils. B added to soils remains soluble where up to 85% can be leached in low-OM, sandy soils. Finer-textured soils retain added B for longer periods than coarse-textured soils

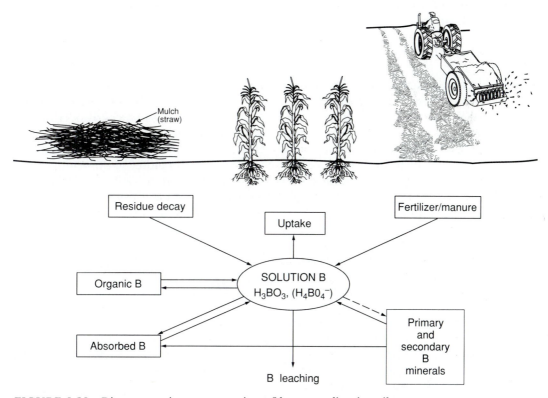

FIGURE 9.29   Diagrammatic representation of boron cycling in soils.

because of greater B adsorption in clays. The fact that clays retain B more effectively than sands does not imply that plants will absorb B from clays in greater quantities than from sands. Plants can take up much larger quantities of B from sandy soils than from fine-textured soils at equal concentrations of water-soluble B.

TYPE OF CLAY   B-adsorption capacities generally follow the order mica > montmorillonite > kaolinite.

SOIL pH AND LIMING   B normally becomes less available to plants with increasing soil pH, decreasing dramatically above pH 6.3 to 6.5 (Table 9.11). Liming strongly acid soils frequently causes a temporary B deficiency in susceptible plants. The severity of the deficiency depends on the moisture status of the soil, the crop, and the time elapsed following liming.

The reduction in B availability following liming is caused mainly by B adsorption on freshly precipitated $Al(OH)_3$, with maximum adsorption at pH 7. Alternatively, moderate liming can be used to depress B availability and plant uptake on soils high in B. It should be noted that heavy liming does not always lead to greater B adsorption and reduced plant uptake. Higher pH resulting from liming of soils high in OM may encourage OM decomposition and release of B.

ORGANIC MATTER   The greater availability of B in surface soils compared with subsurface soils is related to the greater quantities of OM in surface soil. Applications of OM to soils can increase the B concentration in plants and even cause phytotoxicity.

INTERACTIONS WITH OTHER ELEMENTS   Low tolerance to B occurs when plants have a low Ca supply. When Ca availability is high, there is a greater requirement for B. The occurrence of Ca^{2+} in alkaline and recently overlimed soils will restrict B availability; thus, high solution Ca^{2+} can protect crops from excess B. The Ca/B ratio in leaf tissues has been used to assess the B status of crops. To avoid misleading results, it is essential that neither of these nutrients is deficient or toxic. B deficiency is indicated by ratios greater than 1370:1 in barley sampled at the boot stage; 3300:1 in rutabaga leaf tissue; and 1200 to 1500:1 in tobacco.

TABLE 9.11   B Uptake and Percentage Recovery of Added B by Five Harvests of Tall Fescue at Five Soil pH Levels

| | *Amount of Added B (mg/pot)* | | | | | | |
| | *0* | *4.5* | *8.9* | *17.8* | *4.5* | *8.9* | *17.8* |
Soil pH	*B Uptake (mg/pot)*				*% Recovery*		
4.7	0.47	1.85	4.15	9.40	30.7	41.3	50.2
5.3	0.45	1.92	4.45	9.51	32.7	44.9	50.9
5.8	0.44	1.98	4.14	9.10	34.2	41.6	48.7
6.3	0.45	1.98	4.03	9.37	34.0	40.2	50.1
7.4	0.22	0.80	1.40	3.76	12.9	13.3	19.9

SOURCE: Peterson and Newman, *Soil Sci. Soc. Am. J.*, **40**:280 (1976).

At low levels of B, increased rates of applied K may accentuate B-deficiency symptoms. B deficiency in alfalfa and other crops such as oil palm can be aggravated by K fertilization to the extent that B addition is needed to prevent yield loss. The effect of K may be related to its influence on Ca absorption. In contrast, increased K rates may accentuate B toxicity at high levels of B supply.

SOIL MOISTURE   B deficiency is often associated with dry weather and low soil moisture conditions. This behavior is related to restricted release of B from OM and to reduced B uptake due to lack of moisture in the root zone. Although B levels in soil may be high, low soil moisture impairs B diffusion and mass flow to absorbing root surfaces.

PLANT FACTORS   The sensitivity to B deficiency sufficiency varies between crops (Table 9.12). Genetic variability contributes to differences in B uptake. Investigations with tomatoes revealed that susceptibility to B deficiency is controlled by a single recessive gene. Tomato variety T3238 is B-inefficient, while the variety Rutgers is B-efficient. Corn hybrids have similarly exhibited genetic variability related to B uptake.

Because of the narrow range between sufficient and toxic levels of available soil B, the sensitivity of crops to excess B is important. Sensitive crops to B toxicity are peach, grapes, kidney beans, soybeans, peanuts, and figs. Semitolerant plants include barley, peas, corn, potato, alfalfa, tobacco, and tomato. The most tolerant crops are turnips, sugar beet, and cotton.

B Fertilizers and Fertilization

B is one of the most widely applied micronutrients. Sodium tetraborate, $NA_2B_4O_7 \cdot 5H_2O$, is the most commonly used B source and contains about 15% B (Table 9.13). Solubor is a highly concentrated, completely soluble

TABLE 9.12   Relative Sensitivity of Selected Crops to B Deficiency

High Sensitivity		Low Sensitivity	
Alfalfa	Peanut	Asparagus	Pea
Cauliflower	Sugar beet	Barley	Peppermint
Celery	Table beet	Bean	Potato
Rapeseed	Turnip	Blueberry	Rye
Conifers		Cucumber	Sorghum
		Corn	Spearmint
Medium Sensitivity		Grasses	Soybean
		Oat	Sudangrass
Apple	Cotton	Onion	Sweet corn
Broccoli	Lettuce		Wheat
Cabbage	Parsnip		
Carrot	Radish		
Clovers	Spinach		
	Tomato		

SOURCE: Robertson et al., *Mich. Coop. Ext Bul* E-1037 (1976).

TABLE 9.13 Principal B Fertilizers and Their Formulas and B Percentages

Source	Formula	Percent B (Approx.)
Borax	$Na_2B_4O_7 \cdot 10H_2O$	11
Boric acid	H_3BO_3	17
Colemanite	$Ca_2B_6O_{11} \cdot 5H_2O$	10–16
Sodium pentaborate	$Na_2B_{10}O_{16} \cdot 10H_2O$	18
Sodium tetraborate	$Na_2B_4O_7 \cdot 5H_2O$	14–15
Solubor	$Na_2B_4O_7 \cdot 5H_2O +$ $Na_2B_{10}O_{16} \cdot 10H_2O$	20–21

SOURCE: Fleming, in B. E. Davies, Ed., *Applied Soil Trace Elements*, p. 171. New York: Wiley, 1980; Gupta, *Adv. Agron.*, **31**:273 (1979); and Turner, *Boron in Agriculture*. New York: U.S. Borax, 1970.

source of B that can be applied as a spray or dust directly to foliage of fruit trees, vegetables, and other crops. It is also used in liquid and suspension fertilizer formulations. Solubor is preferred to borax because it dissolves more readily and causes minimum changes in crystallization temperatures. The Ca borate mineral, colemanite, is often used on sandy soils because it is less soluble and less subject to leaching than the sodium borates.

The most common methods of B application are broadcast, banded, or applied as a foliar spray or dust. In the first two methods, the B fertilizer source is usually mixed with N-P-K-S products and applied to soil. B salts can also be coated on dry fertilizer materials.

B fertilizers should be applied uniformly to soil because of the narrow range between deficiency and toxicity. Segregation of granular B sources in dry fertilizer blends must be avoided. Application of B with fluid fertilizers eliminates the segregation problem.

Foliar application of B is practiced for perennial tree fruit crops, often in combination with pesticides other than those formulated in oils and emulsions. B may also be included in sprays of chelate, Mg, Mn, and urea. Foliar applications of B with insecticides are also used in cotton. B may be used with herbicides for peanuts.

Rates of B fertilization depend on plant species, soil cultural practices, rainfall, liming, and soil OM, as well as other factors. Application rates of 0.5 to 3 lb/a are generally recommended. The amounts of B recommended depends on the method of application. For example, the B rate for vegetable crops is 0.4 to 2.7 lb/a broadcast, 0.4 to 0.9 lb/a banded, and 0.09 to 0.4 lb/a foliar.

Chloride

Chloride (Cl) concentration is 0.02 to 0.05% in the earth's crust and it occurs primarily in igneous and metamorphic rocks. Most of the soil Cl^- commonly exists as soluble salts such as NaCl, $CaCl_2$, and $MgCl_2$. Cl^- is sometimes the principal anion in extracts of saline soils. The quantity of Cl^- in soil solutions may range from 0.5 ppm or less to over 6000 ppm.

The majority of Cl^- in soils originates from salts trapped in parent material, from marine aerosols, and from volcanic emissions. Nearly all of the soil Cl^- has been in the oceans at least once, being returned to the land surface either by uplift and subsequent leaching of marine sediments or by oceanic salt spray carried in rain or snow. Annual depositions of 12 to 35 lb Cl^-/a in precipitation are common and may increase to over 100 lb/a in coastal areas.

Levels of 2 ppm Cl^- are typical in the precipitation near seacoasts. The actual quantities depend on the amount of sea spray, which is related to temperature; the foam formation on tops of waves; the strength and frequency of winds sweeping inland from the sea; the topography of the coastal region; and the amount, frequency, and intensity of precipitation. Salty droplets or dry salt dust may be whirled to very great heights by strong air currents and carried over great distances. Concentration of Cl^- in precipitation drops off rapidly inland, with areas about 500 miles inland averaging about 0.2 ppm. In the Midwest and Great Plains regions of the United States, the values are uniform over large areas and range from 0.1 to 0.2 ppm.

Behavior in Soil

The Cl^- anion is very soluble in most soils; however, appreciable exchangeable Cl^- can occur in acid, kaolinitic soils which have significant pH-dependent positive charge (AEC).

Because of the mobility of Cl^- in all but extremely acid soils, it can be rapidly cycled through soil systems. Cl will accumulate where the internal drainage of soils is restricted and in shallow groundwater where Cl^- can be moved by capillarity into the root zone and deposited at or near the soil surface.

Problems of excess Cl^- occur in some irrigated areas and are usually the result of interactions of two or more of the following factors:

1. Significant amounts of Cl^- in the irrigation water.
2. Failure to apply sufficient water to leach out Cl^- accumulations adequately.
3. Unsatisfactory physical properties and drainage conditions for proper leaching.
4. High water table and capillary movement of Cl^- into the root zone.

Environmental damage in localized areas from high concentrations of Cl^- has resulted from road deicing, water softening, saltwater spills associated with the extraction of oil and natural gas deposits, and disposal of feedlot wastes and various industrial brines.

The principal effect of too much Cl^- is to increase the osmotic pressure of soil water and thereby lower the availability of water to plants. In addition, some woody plants, including most fruit trees, berry and vine crops, conifers, and ornamental shrubs, are specifically sensitive to Cl^- and develop leaf-burn symptoms when the Cl^- concentration reaches about 0.50% on a dry matter basis. Leaves of tobacco and tomatoes thicken and begin to roll with excessive Cl^-.

Plant Responses

Depression of Cl^- uptake by high concentrations of NO_3^- and SO_4^{2-} has been observed in a number of plants, including potatoes, *Nitella*, beans, tomatoes, buckwheat, sugar beets, and perennial ryegrass. Lowering of NO_3^- uptake by increasing levels of Cl^- has also been observed in barley, corn, and wheat. The negative interaction between Cl^- and NO_3^- has been attributed to competition for carrier sites at root surfaces.

Cl^- is beneficial for some "salt-loving" plants like beet, spinach, cabbage, and celery. Growth responses of coconut to KCl are closely related to the Cl^- content of leaves and negatively correlated with K^+ levels, which explains the observed positive effect of NaCl or even seawater on coconuts, oil palm, and kiwi fruit.

Another example of the favorable effect of increasing crop yield and Cl^- concentration appears in Figure 9.30. Here potato yields increase as the Cl^- level in petioles increases from 1.1% to 6.9% and that of NO_3^- decreases. Although beneficial effects of Cl^- on plant growth are not fully understood, improved plant–water relationships and inhibition of plant diseases are two important factors.

The effect of Cl^- fertilization on root and leaf disease suppression has been observed on about 10 different crops (Table 9.14). Several mechanisms have been suggested and include (1) increased NH_4^+ uptake through inhibition of nitrification by Cl^-, which reduces take-all root disease by decreased rhizo-

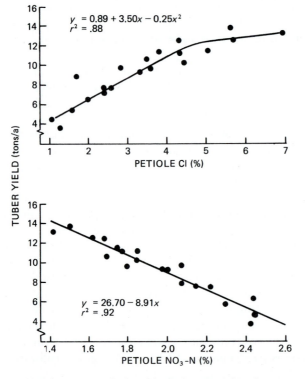

FIGURE 9.30   Relationship between yield of potatoes and Cl^- and NO_3^- concentrations in petiole samples taken on August 5, 1980. *Jackson et al., unpublished data, Oregon State Univ., 1981.*

TABLE 9.14   Diseases Suppressed by Cl Fertilization

Location	Crop	Suppressed Disease
Oregon	Winter wheat	Take-all
	Winter wheat	Septoria
	Potatoes	Hollow heart
	Potatoes	Brown center
North Dakota	Winter wheat	Tanspot
	Spring wheat	Common root rot
	Barley	Common root rot
	Barley	Spot blotch
	Durum wheat	Common root rot
South Dakota	Spring wheat	Leaf rust
	Spring wheat	Tanspot
	Spring wheat	Septoria
New York	Corn	Stalk rot
California	Celery	Fusarium yellows
Saskatchewan	Spring wheat	Common root rot
	Barley	Common root rot
Manitoba	Spring wheat	Take-all
Alberta	Barley	Common root rot
	Barley	Net blotch
Germany	Winter wheat	Take-all
Great Britain	Winter wheat	Stripe rust
India	Pearl millet	Downy mildew
Indonesia	Rice	Stem rot
	Rice	Sheath blight
Philippines	Coconut palm	Gray leaf spot

SOURCE: Fixen, *2nd National Wheat Res. Conf.*, 1987.

sphere pH or (2) competition between Cl^- and NO_3^- for uptake. Plants with a low NO_3^- level are less susceptible to root rot diseases.

In some regions, the Cl^- response in some crops has not been related to disease suppression. For example, in South Dakota, Cl^- deficiency is caused by low soil Cl^-, with the probability of a response to Cl^- fertilization increasing with decreasing water-extractable soil Cl^- (Table 9.15).

TABLE 9.15   Frequency of Spring Wheat to Cl Fertilization as a Function of Soil Cl Levels

Category	Soil Cl Content	Yield Response Frequency
	lb/a-2 ft	%
Low	0–30	69
Medium	31–60	31
High	>60	0

SOURCE: Fixen, *J. Fert. Issues. 4:95 (1979)*.

Cl Fertilizers and Fertilization

When additional Cl^- is desirable, it can be supplied by the following sources:

Ammonium chloride (NH_4Cl)	66% Cl
Calcium chloride ($CaCl_2$)	65% Cl
Magnesium chloride ($MgCl_2$)	74% Cl
Potassium chloride (KCl)	47% Cl
Sodium chloride ($NaCl$)	60% Cl

Rates of Cl^- will vary, depending on a number of conditions, including the crop, method of application, and purpose of addition (i.e., for correction of nutrient deficiency, for disease suppression, or for improved plant water status). Where take-all root rot of winter wheat is suspected, banding 35 to 40 lb/a of Cl^- with or near the seed at planting is recommended. Broadcasting 75 to 125 lb/a of Cl^- has effectively reduced crop stress from take-all and by leaf and head diseases (i.e., stripe rust and septoria).

Molybdenum

The average concentration of molybdenum (Mo) in the earth's crust is about 2 ppm, and in soils it typically ranges from 0.2 to 5 ppm. The main forms of Mo in soil include nonexchangeable Mo in primary and secondary minerals; exchangeable Mo held by Fe and Al oxides; Mo in the soil solution; and organically bound Mo. Although Mo is an anion in solution, the relationships between these fractions is similar to those of other metal cations (Fig. 9.2).

Deficiencies of Mo occur largely on acid sandy soils of the Atlantic and Gulf coasts in the United States. Large soil areas in New Zealand, Australia, and eastern Canada are also deficient in Mo. This nutrient may often be lacking in highly podzolized soils low in total Mo, in acid soils high in hydrous Fe and Al oxides, or in soils derived from calcareous parent materials.

Soil Solution Mo

Mo in solution occurs predominantly as MoO_4^{2-}, $HMoO_4^-$, and $H_2MoO_4^0$. Concentration of MoO_4^{2-} and $HMoO_4^-$ increases dramatically with increasing soil pH (Fig. 9.31). The solubility of Mo in soils is mainly controlled by soil Mo, which is very close to the solubility of $PbMoO_4$ or wulfenite (Fig. 9.32).

Plants absorb Mo as MoO_4^{2-}. The extremely low concentrations of Mo in soil solution is reflected in the low Mo content of plant material (~ 1 ppm Mo). At concentrations above 4 ppb in the soil solution, Mo is transported to plant roots by mass flow, while Mo diffusion to plant roots occurs at levels <4 ppb.

Factors Affecting Mo Availability

The most important factors affecting Mo availability are soil pH and the amount of Al and Fe oxides.

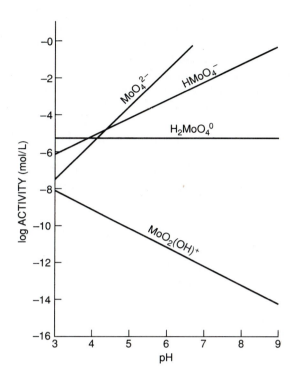

FIGURE 9.31 The common Mo species in soil solution as influenced by pH. *Lindsay,* Chemical Equilibria in Soils, *Wiley, New York, 1979.*

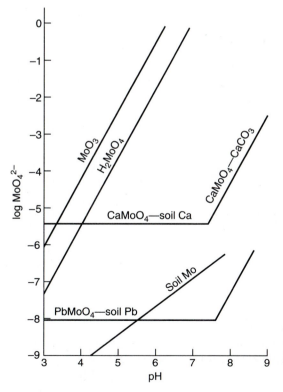

FIGURE 9.32 Solubilities of the common Mo minerals, which increase with increasing pH. *Lindsay,* Chemical Equilibria in Soils, *Wiley, New York, 1979.*

SOIL pH AND LIMING   Mo availability, unlike that of other micronutrients, increases with increasing pH (Fig. 9.33). This diagram shows about a 10-fold increase in MoO_4^{2-} activity per unit increase in soil pH.

Liming to correct soil acidity will increase Mo availability and prevent Mo deficiency. Alternatively, Mo availability is decreased by application of acid-forming fertilizers such as $(NH_4)_2SO_4$ to a coarse soil.

REACTION WITH FE AND AL   Mo is strongly adsorbed by Fe and Al oxides, a portion of which becomes unavailable to the plant. As plant roots remove Mo from solution, more Mo is desorbed into the solution by simple mass action. Because of adsorption reactions, soils that are high in Fe, especially amorphous Fe on clay surfaces, tend to be low in available Mo.

INTERACTIONS WITH OTHER NUTRIENTS   P enhances Mo absorption by plants, probably due to exchange of adsorbed MoO_4^{2-}. In contrast, high levels of SO_4^{2-} in the solution depress Mo uptake by plants (Table 9.16). On soils with marginal Mo deficiencies, the application of heavy rates of SO_4^{2-}-containing fertilizers may induce an Mo deficiency in plants, and inclusion of Mo in the fertilizer may be advisable.

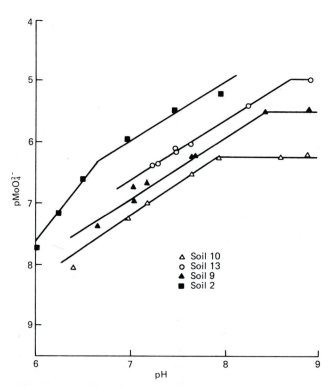

FIGURE 9.33   pH dependence of Mo solubility in four Colorado soils.   *Vlek and Lindsay,* Soil Sci. Soc. Am. J., *41*:42, 1977.

TABLE 9.16   Effect of S and Mo on the Yield and S and
Mo Concentration of Brussels Sprouts

Treatment	Yield (g/pot)	Aboveground Tissue	
		Mo (ppm)	S (%)
Sulfur[†]			
No S added	12.6	5.09	0.25
50 ppm S	13.8	0.88	0.60
100 ppm S	13.8	0.50	0.70
Molybdenum			
No Mo added	12.7	0.08	0.53
Seed treated with Mo	13.6	0.16	0.49
2.5 ppm Mo	13.9	6.23	0.51

[†]S treatments did not alter soil pH or the exchangeable Mo content.
SOURCE: Gupta, *Sulphur Inst. J.,* **5**(1):4 (1969).

Both Cu and Mn also can reduce Mo uptake; however, Mg has the opposite effect and will encourage Mo absorption by plants.

Nitrate N encourages Mo uptake, while NH_4–N sources reduce Mo uptake. This beneficial effect of NO_3^- nutrition is perhaps related to the release of OH^- ions and an accompanying increase in solubility of soil Mo.

CLIMATIC EFFECTS   Mo deficiency is more severe under dry soil conditions, probably due to reduced mass flow or diffusion under low soil moisture content.

PLANT FACTORS   Soil Mo is very low in most soils; thus, plants contain very little Mo. Crops that are very sensitive to low solution Mo are legumes, crucifers (broccoli, brussels sprouts, cauliflower, rapeseed, etc.), and citrus. Other crops that are also sensitive to a low Mo supply are beet, cotton, lettuce, spinach, sweet corn, sweet potatoes, and tomatoes. Subterranean clover is one of the legumes most sensitive to a lack of Mo. Small grains and some row crops tend to be more tolerant to low levels of available Mo in soil.

Mo-efficient and Mo-inefficient varieties of alfalfa, cauliflower, corn, and kale have been identified. The differential susceptibility of cauliflower varieties to Mo deficiency is related to their ability to extract soil Mo.

Mo Toxicities

Excessive amounts of Mo are toxic, especially to grazing cattle or sheep. Areas of Mo toxicity occur in high-pH soils in the western regions of North America. Mo toxicity also exists in certain areas of Australia. The soil conditions on which high-Mo forage may be found are wet soils mostly neutral to alkaline in reaction, often with either a thick A_1 horizon or an A_1 horizon capped by a thin surface layer of peat or muck. Pockets of peats may also be present in these problem areas.

Molybdenosis, a disease in cattle, is caused by an imbalance of Mo and Cu in the diet, when the Mo content of the forage is greater than 5 ppm. Mo toxicity causes stunted growth and bone deformation in the animal and can be corrected by oral feeding of Cu, injections of Cu, or the application of $CuSO_4$ to the soil. Other practices used to decrease Mo toxicity are application of S or Mn and improvement of soil drainage.

Mo Fertilizers and Fertilization

Sources of Mo used for fertilizers are listed in Table 9.17. Rates of Mo application are very low, only 0.5 to 5 oz/a, and the solution may be applied to soil, sprayed on foliage, or put on seed prior to planting. The optimum Mo rate depends on the application method, with lower rates used in the latter two methods. Seed treatments involving soaking seed in a solution of sodium molybdate before seeding are widely used because of the low application rates needed. Seed treatments with a slurry or dust are also effective. To obtain satisfactory distribution of the small quantities of Mo applied to soil, Mo sources are sometimes combined with N-P-K fertilizers. Foliar spray applications with NH_4 or Na molybdate also are effective in correcting deficiencies.

Application of Mo to clovers will in some cases increase the yield equivalent to that achieved with the addition of limestone. Since liming can be more expensive, Mo fertilization is often preferred.

Cobalt

Cobalt (Co) has been established as essential for symbiotic N_2 fixation in some microrganisms and in the synthesis of vitamin B_{12} in ruminant animals, but the requirement for higher plants has not been documented. Soil is an important source of plant Co for animals.

The average total Co concentration in the earth's crust is 40 ppm. Acidic rocks, including granites, containing large amounts of Fe-rich ferromagnesian minerals are low in Co, with levels ranging from 1 to 10 ppm. Much higher levels (100 to 300 ppm Co) may be present in Mg-rich ferromagnesian minerals. Sandstones and shales are normally low in Co, with concentrations frequently below 5 ppm.

Total Co content of soils typically ranges from 1 to 70 ppm and averages about 8 ppm. Co deficiencies in ruminants are often associated with forages produced on soils containing less than 5 ppm of total Co. Co in soil is primarily related to the presence of Mg and, to a lesser extent, that of Ni and Mn, primary minerals.

TABLE 9.17   Sources of Mo Used for Fertilizer

Sources	Formula	Percent Mo (Approx.)
Ammonium molybdate	$(NH_4)_6Mo_7O_{24} \cdot 2H_2O$	54
Sodium molybdate	$Na_2MoO_4 \cdot 2H_2O$	39
Molybdenum trioxide	MoO_3	66
Molybdenum frits	Fritted glass	1–30

SOURCE: Mortvedt, *Farm Chem.*, **143**:42 (1980).

Soils in which Co deficiency can occur are (1) acidic, highly leached, sandy soils with low total Co; (2) some highly calcareous soils; and (3) some peaty soils. Extremely low Co soils are found in the lower Atlantic coastal plains, which contain 1 ppm or less of total Co. Soils formed on granitic glacial drift also are generally low in total Co. Spodosols formed in coarse-textured deposits and organic (Histosol) soils are implicated in Co deficiency. Low-Co soils also occur in Australia, Canada, Ireland, Kenya, New Zealand, Norway, and Scotland.

Factors Affecting Co Availability

Co is adsorbed on the exchange complex and occurs as clay–OM complexes similar to those of the other metal cations (Fig. 9.2). The order of adsorption is muscovite > hematite > bentonite = kaolin. Co concentration in the soil solution is very low, often <0.5 ppm HCl extractable Co. Because Co behaves similarly to Fe or Zn, excess Co produces visual symptoms similar to Fe and Mn deficiencies.

Among the several factors that influence Co availability is the presence of crystalline Mn oxide minerals. These minerals have a high adsorption capacity for Co. They are capable of retaining almost all of the Co in soil, leading to fixation of soil-applied Co fertilizer. Co appears to replace Mn in the surface layers of these minerals. Co availability is favored by increasing acidity and waterlogging conditions, which solubilize Mn oxide; therefore, liming and drainage are practices that reduce Co availability.

Co Fertilizers and Fertilization

Co deficiency of ruminants can be corrected by (1) adding it to feed, salt licks, or drinking water; (2) drenching; (3) using Co bullets; and (4) fertilizing forage crops with small amounts of Co. Co fertilization with 1.5 to 3 oz/a as $CoSO_4$ is recommended.

Some soils in Australia are too low in available Co for satisfactory nodulation and N_2 fixation by clover and alfalfa. Applications of 0.5 to 2 oz/a of Co, as $CoSO_4$, can correct this condition. Superphosphate, with small amounts of $CoSO_4$, also has been used to increase the Co concentration in subterranean clover.

Sodium

Sodium (Na) content in the earth's crust is about 2.8%, while soils contain 0.1 to 1%. Low Na in soils indicates weathering of Na from Na-containing minerals. Very little exchangeable and mineral Na occurs in humid regions soils, whereas Na is common in most arid and semiarid regions soils.

Forms of Na in Soil

Na exists in soils as solution and exchangeable Na and in silicate minerals. Sodium in highly leached soils occurs in high-albite plagioclases and in small amounts in micas, pyroxenes, and amphiboles, which exist mainly in the fine sand and silt fractions. In arid and semiarid soils, Na exists in silicates, as well as $NaCl$, Na_2SO_4, and Na_2CO_3.

Sodium is absorbed by plants as Na^+ ion from the soil solution, which contains between 0.5 and 5 ppm Na^+ in temperate-region soils. Solution and exchangeable Na^+ vary greatly among soils. In humid-region soils the proportion of exchangeable Na^+ to other cations is $Ca^{2+} > Mg^{2+} > K^+ = Na^+$. Exchangeable Na^+ can be utilized by crops. Sugar beets respond to fertilization when exchangeable Na^+ in soil is < 0.05 meq/100 g.

In arid regions and if soils are irrigated with sodic waters, exchangeable Na^+ levels generally exceed those of K^+. Sodium salts accumulating in poorly drained soils of the arid and semiarid regions will be contributors to soil salinity and sodicity (see Chapter 10).

Effect of Na on Soil Properties

The dispersing action of Na^+ on clay and organic matter reduces soil aggregation, permeability to air and water, germination, and root growth, depending on the soil. Soil dispersion occurs when exchangeable Na exceeds 10 to 20% of the CEC. With fine-textured soils, 10% exchangeable Na^+ can be tolerated, whereas in sandy soils the upper limit is 30%. Sodic soils are commonly found in arid/semiarid regions.

Growth of most crop plants is severely reduced on sodic soils (Table 9.18). In addition, the associated high pH can cause micronutrient deficiencies, although the main impediment to growth is the loss of soil permeability.

Effect of Na on Plant Growth

On the positive side, Na is beneficial for the growth of some plants (see Chapter 3). Na is recognized as essential for some C-4 plants. The beneficial effects of Na on plant growth are often observed in low-K soils, as Na^+ can partially substitute for K^+. Crops have been categorized according to their potential for Na uptake (Table 9.19). Growth of those crops with high and medium ratings will be favorably influenced by Na^+.

Na Fertilizers and Fertilization

Responses to Na have been observed in crops with a high uptake potential (Table 9.19). The Na demand of these crops appears to be independent of,

TABLE 9.18   Reductions in Crop Yields at Various Exchangeable Na Percentages (ESP)

Type of Soil	ESP (%)	Average Decrease in Crop Yield (%)
Slightly sodic	7–15	20–40
Moderately sodic	15–20	40–60
Very sodic	20–30	60–80
Extremely sodic	> 30	> 80

SOURCE: Velasco, *Sulphur Agri.*, **5**:2 (1981).

TABLE 9.19   Na Uptake Potential of Various Crops

High	Medium	Low	Very Low
Fodder beet	Cabbage	Barley	Buckwheat
Sugar beet	Coconut	Flax	Maize
Mangold	Cotton	Millet	Rye
Spinach	Lupins	Rape	Soya
Swiss chard	Oats	Wheat	Swede
Table beet	Potato		
	Rubber		
	Turnips		

SOURCE: Marschner, *Proc. 8th Colloq. Int. Potash Inst.*, pp. 50–63 (1971); cited by Mengel and Kirkby, *Principles of Plant Nutrition*, 3rd ed. Bern: International Potash Institute, 1982.

and perhaps even greater than, their K demand. In some parts of Europe, fertilization of grass forage crops with Na is considered desirable. Na concentrations of between 1 and 2% in pasture grasses will improve palatability and will provide part of the animals' Na requirements.

The important Na-containing fertilizers are the following:

· K fertilizers with various NaCl contents.
· Sodium nitrate (about 25% Na).
· Rhenania phosphate (about 12% Na).
· Multiple-nutrient fertilizers with Na.

Silicon

Silicon (Si) is the second most abundant element in the earth's crust, averaging 27.6%, whereas Si in soils ranges between 23 and 35%. Unweathered sandy soils can contain as much as 40% Si, compared with as little as 9% Si in highly weathered tropical soils. Tropical soils consisting largely of Al and Fe oxides are left after Si is leached under the intense weathering conditions of soils. Si is solubilized during weathering, and reprecipitation as secondary minerals is important to soil development. Major sources of Si include primary and secondary silicate minerals, and quartz (SiO_2). Quartz is the most common mineral in soils, comprising 90 to 95% of all sand and silt fractions.

Low-Si soils exist in intensively weathered, high-rainfall regions. Properties of the Si-deficient soils include low total Si, high Al, low base saturation, and low pH. Further, they all have extremely high P-fixing capacity due to their high AEC and Al and Fe oxide content. Plant-available Fe^{2+} and Mn^{2+} may also be high in these soils.

Soil Solution Si

Silicic acid ($H_4SiO_4^0$) is the principal Si species in solution (Fig. 9.34). At high concentrations of Si in solution, $H_4SiO_4^0$ polymerizes to form precipitates of amorphous silica SiO_2. The solubility of Si in water is unaffected by pH in the range 2 to 9 (Fig. 9.35). Si concentrations of less than 0.9 to 2 ppm in soil

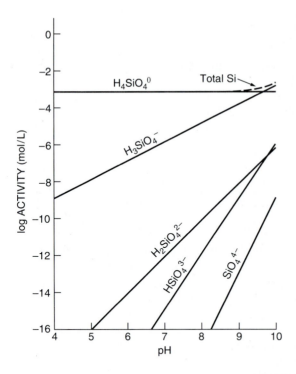

FIGURE 9.34   The common Si species in soil solution as influenced by pH. *Lindsay,* Chemical Equilibria in Soils, *Wiley, New York, 1979.*

solution are insufficient for proper nutrition of sugarcane. By comparison, levels of 3 to 37 ppm Si in solution have been reported for a wide range of normal soils. Figure 9.36 illustrates the large differences in soluble Si observed in several Hawaiian soils. Levels of NaOAc-extractable Si rated as adequate for rice production are >130 ppm in Japan and Korea and >90 ppm in Taiwan.

The concentration of $H_4SiO_4^0$ in soil solutions is largely controlled by a pH-dependent adsorption reaction. Si is adsorbed on the surfaces of Fe and Al oxides. The adsorptive capacity of Al oxides declines markedly with greater degree of crystallinity.

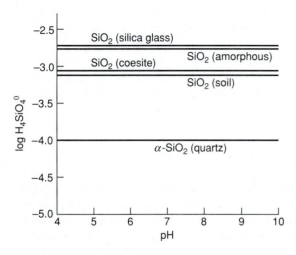

FIGURE 9.35   Solubilities of the common Si minerals, which are not influenced by pH. *Lindsay,* Chemical Equilibria in Soils, *Wiley, New York, 1979.*

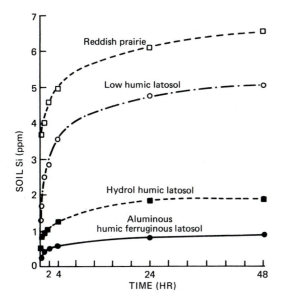

FIGURE 9.36 Concentration of water soluble Si in four Hawaii soils. The Reddish Prairie and Hydrol Humic Latosol are developed from ash with low and high rainfall, respectively. The Low Humic Latosol and Humic Ferruginous Latosol are from basalt, with low and high rainfall, respectively. *Fox et al., Soil Sci. Soc. Am. J., 31:775, 1967.*

Factors Affecting Si Availability

The soil and management conditions influencing Si uptake are Fe and Al oxides, liming, flooding, and nutrient supply. Si removal in highly weathered soils will reduce solution Si and Si uptake. Liming can decrease Si uptake in some plants, including barley, oats, red clover, rice, ryegrass, sorghum, and sugarcane. Conversely, acidification increases Si uptake by oats.

High soil water content encourages Si uptake, particularly by crops such as rice. Solution concentration of Si increases with time of submergence, probably due to increased concentration of H_4SiO_4 hydrolysis species (Fig. 9.34).

Heavy applications of N render rice plants more susceptible to fungal attack because of decreases in Si concentration in the straw. To correct this problem, Si-bearing materials are added when high rates of N fertilizer are used.

Si uptake differs with plant species. Gramineae contain 10 to 20 times the Si concentration normally found in legumes and other dicotyledons. Lowland or paddy rice commonly contains between 4.6 and 7.0% Si in the straw.

Si Fertilizers and Fertilization

The primary Si fertilizers include the following:

Calcium silicate slag ($CaAl_2Si_2O_8$)	18 to 21% Si
Calcium silicate ($CaSiO_3$)	31% Si
Sodium metasilicate ($NaSiO_3$)	23% Si

Minimum rates of at least 5000 lb/a of $CaSiO_3$ are broadcast applied and incorporated before planting sugarcane. Annual $CaSiO_3$ applications of between 500 and 1000 lb/a applied in the row have also improved sugarcane yields. The beneficial effect of increasing rates of slag on sugar yields in Hawaii

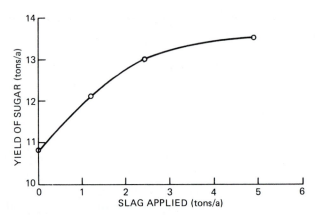

FIGURE 9.37   Effect of electric furnace slag on yield of
sugar from sugarcane grown on an aluminous humic
ferruginous latosol in Hawaii. Mean results for plant and
ratoon crops combined.   *Ayres, Soil Sci., **101***:216 (1966).*

is evident in Figure 9.37. Additions of lime that increase Ca levels and decrease
soil acidity do not produce similar dramatic improvements in the growth of
sugarcane. Rates of 1.5 to 2.0 t/ha of silicate slag usually provide sufficient Si
for rice produced on low-Si soils.

Selenium

Selenium (Se) is not needed by plants, but it must be present in forage since
it is essential for animals. Se occurs in very small amounts in nearly all materials
of the earth's crust. It averages only 0.09 ppm in rocks and is found mainly
in sedimentary minerals. Se is similar in behavior to S; however, it has five
important oxidation states: -2, 0, $+2$, $+4$, and $+6$.

The total Se concentration in most soils is between 0.1 and 2 ppm and
averages about 0.3 ppm. There are extensive areas of high-Se soils in western
North America and in other semiarid regions, which produce vegetation toxic
to livestock. The parent materials of these soils are predominantly sedimentary
shale deposits. High-pH, calcareous soil in regions of low rainfall (<20 in.
total precipitation) are usually high in Se.

Several areas in the United States and Canada have been identified where
the crop plants contain low concentrations of Se. Many of these low-Se soils
are acid, mostly Spodosols, Inceptisols, and contiguous areas of Alfisols. Insufficient plant uptake of Se is usually caused by one or both of the following soil
factors: low total Se in the soil parent material or low availability of Se in acid
and poorly drained soils.

Soil Se

The forms of Se generally considered to be present in soil are selenides (Se^{2-}),
elemental Se^0, selenites (Se^{+4}), selenates (Se^{+6}), and organic Se compounds

(Fig. 9.38). Se species in soils and sediments are closely related to redox potential, pH, and solubility.

SELENIDES (SE^{2-})   Selenides are largely insoluble, and they are associated with S^{2-} in soils of semiarid regions where weathering is limited. They contribute little to Se uptake because of their insolubility.

ELEMENTAL SE (SE^0)   Se^0 is present in small amounts in some soils. Significant amounts of Se^0 may be oxidized to selenites and selenates by microorganisms in neutral and basic soils.

SELENITES (SEO$_3{}^{2-}$)   A large fraction of Se in acid soils may occur as stable complexes of selenites with hydrous iron oxides. The low solubility of Fe–selenite complexes is apparently responsible for the nontoxic levels of Se in plants growing on acid soils having very high total Se contents. Plants will absorb selenite, but generally to a lesser extent than selenate.

SELENATES (SEO$_4{}^{2-}$)   Selenates are frequently associated with SO$_4{}^{2-}$ in arid-region soils and are stable in many well-aerated, semiarid seleniferous soils. Other forms of Se will be oxidized to selenates under these conditions. Only limited quantities of selenate occur in acid and neutral soils. Selenates are highly soluble and readily available to plants, and, thus, are largely responsible for toxic accumulations in plants grown on high-pH soils. Most of the water-soluble Se in soils probably occurs as selenates.

ORGANIC SE   Organically complexed Se can be an important fraction, since up to 40% of the total Se in some soils is present in humus. Soluble organic

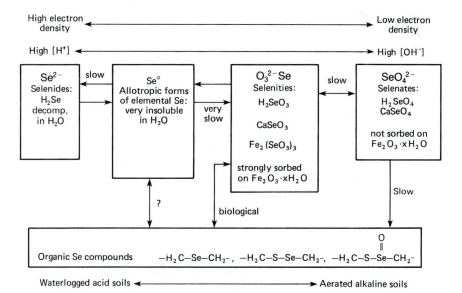

FIGURE 9.38   Generalized chemistry of Se in soils and weathering sediments. *Allaway, in D. E. Hemphill, Ed.,* Trace Substances in Environmental Health, II. *University of Missouri, 1968.*

Se compounds are liberated through the decay of seleniferous plants. Such substances derived from accumulator or indicator plants are readily taken up by other plants. Se in residue is stable in semiarid areas, and much of it remains available in soil. Organic Se is more soluble under basic than acidic soils, which would enhance availability to plants in semiarid region soils.

Factors Affecting Se Availability

Plant uptake of Se is generally greater in high-pH soil than in acid soils. Se in the soil solution is lowest at slightly acid to neutral pH, and increases under both more acidic and basic soil pH. High soil pH will facilitate the oxidation of selenites to the more readily available selenates.

Increased yields with N and S fertilization may lower Se concentrations in crops through dilution. There has been some concern about increased incidence and severity of Se deficiencies in cattle due to the negative interaction of SO_4^{2-} on SeO_4^{2-} uptake by crops.

A greater frequency of livestock nutritional disorders caused by low Se has been observed after cold, rainy summers than after hot, dry ones. High summer temperatures are amenable to increased Se concentration in feedstuffs.

Plant species differ in Se uptake. Certain species of *Astragalus* absorb many times more Se than do other plants growing in the same soil, because they utilize Se in an amino acid peculiar to the species. Plants such as the cruciferae (cabbage, mustard, etc.) and onions, which require large amounts of S, absorb intermediate amounts of Se, while grasses and grain crops absorb low to moderate amounts.

Se Fertilizers and Fertilization

Although Se deficiency disorders such as muscular dystrophy or white muscle disease in cattle and sheep can be corrected by therapeutic measures, there is interest in Se fertilization to produce forages adequate in Se for grazing animals rather than to satisfy any particular plant requirements. Se fertilization is acceptable if proper precautions are taken:

1. At no stage should herbage become toxic to grazing animals; topdressing of growing plants must be avoided.
2. High levels of Se in edible animal tissue should not result.
3. Protection against Se deficiency should be provided for at least one grazing season following application during the dormant season.

Fertilization with selenites is preferred because they are slower acting and thus less likely to produce excessive levels of Se in plants than the rapidly available selenates, which are effective if rapid Se uptake is desired. The addition of Na selenite at rates of 1 oz of Se/a is satisfactory for forages. Foliar application of Na selenite at 6 g Se/a is an efficient way to increase Se in field corn.

Se is present in phosphate rocks and in superphosphate produced from them. Superphosphate containing 20 ppm or more Se may provide sufficient Se to the plants in Se-deficient areas to protect livestock from Se deficiency disorders.

Summary

1. The solubility and availability of the micronutrient cations are affected by the presence of complexing and chelating agents, as well as by the oxidation-reduction (redox) potential in the soil.

2. Plant nutrient deficiencies of Fe have been reported in 25 states in the United States and in western Canada. The solubility and availability of Fe in soils are controlled by pH, chelation, redox potential of the soil, and the presence of carbonates and bicarbonates.

3. Fe is taken up by plants principally as the ion Fe^{2+} and is transported to the roots as Fe chelates. Because of low solubility, only small amounts of Fe^{3+} are available to plant roots.

4. Genotypes differ widely in their ability to utilize Fe from the growth medium. Under field conditions, Fe chlorosis is one of the most difficult micronutrient element deficiencies to correct. Its correction is accomplished primarily with foliar sprays and, to a lesser extent, by band application of some acidifying agent in the root zone.

5. Zn deficiencies have been reported in most of the states in the United States and in at least five Canadian provinces. Deficiencies of Zn are found most frequently in sandy soils low in total Zn and neutral or basic soils, especially those that are calcareous, and in exposed subsoils from land leveling and wind and water erosion.

6. Zn availability is highly pH dependent, decreasing 100-fold for each unit increase in pH. The Zn^{2+} ion is the form absorbed, and diffusion is the main process of transporting this ion to plant root surfaces. The formation of Zn–organic complexes facilitates this diffusion.

7. Plant nutrient Zn can be supplied from several fertilizer sources directly to the soil. Tree injections and foliar applications of Zn have also been successfully utilized. The effectiveness of banded applications of Zn in the soil has been increased by the presence of acid-forming N fertilizers and certain acid-forming S compounds.

8. Cu deficiencies are most frequently found in organic soils that have just been brought into production and with high-value specialty crops that are grown intensively in peat and muck soils. Cu is strongly bound to soil OM, and its solubility in the soil solution decreases 100-fold for each unit increase in soil pH.

9. In soil solutions with pH values below 6.9, the divalent ion Cu^{2+} is the dominant form of this element. Above this pH value, $Cu(OH)_2^0$ is the major species. In addition to soil texture, pH, and OM, the plant availability of Cu is influenced by its interaction with other micronutrients.

10. Cu deficiencies can be corrected by the application of a wide variety of commercially available fertilizer materials applied at rates ranging from 0.6 to 20 lb/a.

11. Deficiencies of Mn have been reported in a number of states in the United States and in the three Prairie Provinces of Canada.

12. The solubility of soil Mn is influenced by the soil pH, the redox potential, and the formation of ion complexes with organic matter. Mn^{2+} is the principal ion species in the soil solution, and diffusion is the principal mechanism for its movement to plant root surfaces. The formation of Mn organic

complexes enhances its movement to root surfaces. As with Fe, plant susceptibility to Mn deficiency varies greatly with its genetic makeup.

13. Mn deficiencies can be corrected by foliar sprays of simple inorganic salts or chelated Mn compounds. Soil applications of Mn compounds, as well as lowering of the soil pH, are effective in correcting Mn deficiencies.

14. The availability and movement of B in the soil are influenced by a number of factors, the most important of which are climatic conditions, soil texture, pH, and liming. B reaches plant roots primarily by mass flow and, to a lesser extent, by diffusion. B deficiencies can be overcome through a number of B-containing fertilizers.

15. Cl^- is a minor constituent of the lithosphere, and the variable amounts of this element in soils are thought to come from marine aerosols, volcanic emissions, fertilization, crop protection programs, and irrigation waters. The mobility of the Cl^- ion is great in most soils, and the strength with which it is held is very weak, its adsorption decreasing with increases in soil pH.

16. While Cl^- has been found to be beneficial for many plants, large amounts of this ion increase the osmotic pressure of the soil solution and thereby lower the availability of water to plants.

17. Unlike most of the micronutrients, deficiencies of Mo increase with a decrease in soil pH.

18. Plants absorb Mo from the soil as the ion MoO_4^{2-}, and its movement to plant roots is largely through mass flow. At low concentrations, however, diffusion becomes the main transfer mechanism.

19. Deficiencies of Mo can be prevented or overcome through seed treatment prior to planting, by applying Mo directly to the soil in the fertilizer, or by spraying it on the foliage. Rates of application are extremely low, 0.5 to 5.0 oz/a of Mo.

20. Co is needed to improve legume nodulation and for the synthesis of vitamin B_{12} in ruminant animals. Soil areas low in Co have been identified in the United States, Canada, and other countries.

21. Na is an element that is beneficial to the growth of some plants, and it is essential for certain halophytic species. It is detrimental, however, to the growth of many other plants. The benefits of Na generally tend to be greatest when K nutrition is poorest.

22. There is evidence that Na can partially substitute for K in plant nutrition, which in turn is closely related to the plants' ability to absorb Na. Harmful levels of Na in plants are usually accompanied by low concentrations of K and Ca.

23. Some plants, notably rice and sugarcane, have responded to applications of silicate fertilizers. The major sources of Si in soils are primary Si minerals, secondary aluminosilicates, and several forms of Si. Soils low in Si are usually found in the tropics under highly weathered conditions. Plant uptake of Si is influenced by soil content of Fe and Al, liming, flooding, and nutrient supply.

24. Se is not an essential plant nutrient, but it is essential for animals. A deficiency of it results in white muscle disease or muscular dystrophy.

25. There are large areas of soils in the United States and Canada that are low in Se, and there are also significant soil areas in each where the Se level is high and plants absorb amounts sufficient to be toxic to livestock.

26. Se deficiencies in animals can be overcome through Se fertilization of the forages ingested by these animals. Several precautions must be strictly observed when fertilizing forages with this element to ensure that toxicities to the animals do not result.

27. Chelates are soluble organic compounds that bond with metals such as Zn, Cu, Fe, and Mn, keeping them soluble and mobile in the soil solution.

Questions

1. What plant nutrient elements are classified as micronutrients?
2. Name other nutrients discussed in this chapter that may also be beneficial for some plants. Are they micronutrients?
3. What micronutrient elements exist in soil solution as cations? Which are present in anionic form?
4. Give the principal form(s) in which each element discussed in this chapter occurs in the soil solution.
5. Report the primary mechanism(s) responsible for movement from the soil to root surfaces of each nutrient reviewed in this chapter.
6. Indicate how the solubility of each nutrient is influenced by pH.
7. The theoretical solubility of the heavy metal micronutrients in water seems in most cases to be too low for satisfactory crop nutrition. In spite of this, soil solution concentrations are often much higher and plant nutrition is better than expected. Discuss the mechanism(s) responsible for higher solubility and improved plant uptake.
8. Availabilities of the plant nutrient elements discussed in this chapter are altered by liming. Which ones are decreased by liming? Which ones are increased by liming?
9. What elements in the micronutrient grouping react in various ways with carbonate minerals? Is their availability altered by these reactions?
10. Which of the various micronutrients and other beneficial elements react with oxides of Fe and Al? Is their availability affected by these reactions?
11. The solubility and availability of certain of the micronutrients and other beneficial elements are influenced by flooding and submergence. Which elements are affected by these conditions? What happens to their solubility and availability?
12. Describe at least three major roles that OM additions can have on micronutrient supplies to plants.
13. Can microbial activity influence the solubility and availability of micronutrients? Explain the reasons for your answer.
14. Why are nutrient interrelationships important in plant uptake of micronutrients and the other beneficial elements? Give examples of strong interactions (a) among heavy metal cations, (b) between N and at least five other elements, (c) between P and at least five other elements, (d) between K and at least two other elements, (e) between SO_4^{2-}-containing substances and at least two other elements, and (f) between Mo and one other micronutrient.
15. What is meant by a mutually antagonistic interaction? Give an example of such an interaction.
16. Mn availability can be greatly increased by the addition of which neutral fertilizer salts?
17. Describe briefly the nature of the

latest findings related to the interaction of P and Zn.

18. Is the sensitivity of plants to deficiencies of micronutrients and the other beneficial elements related in any way to plant genetic differences? Explain your answer.

19. What is meant by Fe-efficient and Fe-inefficient plants? Give examples of plants in these two categories.

20. Outline the biochemical reactions and changes associated with Fe-efficient plants. Do these plant characteristics have any bearing on the efficiency of plant uptake of other heavy metal micronutrients?

21. Availability and plant absorption of micronutrients and other beneficial elements can be influenced by climatic and environmental conditions. Describe the effect of climatic factors on the occurrence of B, Mn, Mo, and Zn deficiencies.

22. Why is Cu deficiency often referred to as *reclamation disease*?

23. The deficiency of which micronutrient is generally considered one of the most difficult to correct in the field?

24. What nutrient deficiencies are best controlled by foliar treatments of micronutrient fertilizers?

25. Correction of what nutrient deficiencies is best accomplished by soil applications of micronutrient fertilizers?

26. Acidification of high-pH and calcareous soils in localized zones such as fertilizer bands can be helpful in the treatment of what micronutrient deficiencies?

27. In the treatment of certain micronutrient deficiencies, it is necessary to thoroughly incorporate the fertilizer additives deeply into the soil. This is true for what micronutrients? Why?

28. Name the important external sources of Cl$^-$ reaching soils. Why are some contributions typically much higher near seacoasts?

29. How is the availability of Cu affected by the level of soil OM and soil pH?

30. How is a Co deficiency in animals commonly remedied?

31. What are some common Cu fertilizers?

32. What are the reasons that have been suggested to explain Fe chlorosis of plants?

33. What are the ways commonly employed to overcome Fe chlorosis?

34. In what forms is Mn believed to exist in soils?

35. Under what soil conditions has Mn toxicity been observed?

36. What are some common Mn fertilizer materials?

37. In what way is the behavior of Mo in soils different from the behavior of the other microelements?

38. What soil factors influence the availability of Zn in soils?

39. In what ways is Se important in soils?

40. What are chelates? Which of the microelements are frequently applied as chelates?

41. What soil form of Se is readily available to plants?

42. What soil conditions favor the existence of the Se form readily available to plants?

43. Why doesn't vegetation grown on seleniferous soils in Hawaii and Puerto Rico contain toxic levels of Se?

44. Are there practical ways of increasing the Se concentration of plants? List them.

Selected References

BARBER, S. A. 1984. *Soil Nutrient Bioavailability: A Mechanistic Approach.* John Wiley & Sons, New York.

FOLLETT, R. H., L. S. MURPHY, and R. L. DONAHUE. 1981. Fertilizers and Soil Amendments. Prentice-Hall, Inc. Englewood Cliffs, N.J.

MORTVEDT, J. J., and F. R. COX. 1985. Production, marketing, and use of calcium, magnesium, and micronutrient fertilizers, pp. 455–482. *In* O. P. Engelstad (Ed.), *Fertilizer Technology and Use.* Soil Science Society of America, Madison, Wisc.

MORTVEDT, J. J., et al. (Eds.) 1991. *Micronutrients in Agriculture.* No. 4. Soil Science Society of America, Madison, Wisc.

Soil Acidity and Basicity

Although briefly discussed in Chapter 4, soil pH and associated properties strongly influence plant nutrient availability and soil productivity. Therefore, understanding soil acidity and basicity is essential for proper management of soil pH and optimum soil and crop productivity.

General Concepts

In aqueous systems, an acid is a substance that donates H^+ to some other substance. Conversely, a base is any substance that accepts H^+. An acid, when mixed with water, ionizes into H^+ and the accompanying anions, as represented by the the dissociation of acetic acid (CH_3COOH) or hydrochloric acid (HCl):

$$CH_3COOH \rightleftharpoons CH_3COO^- + H^+$$

$$HCl \rightleftharpoons Cl^- + H^+$$

Dissociation of H^+ in a strong acid like HCl is 100%, whereas only 1% H^+ dissociation occurs in a weak acid like CH_3COOH.

The H^+ ions, or *active* acidity, increase with the strength of the the acid. The undissociated acid is considered *potential* acidity. The total acidity of a solution is the sum of the active and potential acid concentrations. For example, suppose that the active and potential acidity are 0.099 M and 0.001 M, respectively. The total acid concentration is 0.100 M, and since the H^+ activity is nearly equal to the total acidity, this would be a strong acid.

With weak acids, the H^+ activity is much less than the potential acidity. For example, a 0.100 M weak acid that is 1% dissociated means that the H^+ activity is $0.1 \times 0.01 = 0.001$ M.

Pure water undergoes slight self-ionization:

$$H_2O \leftrightarrow H^+ + OH^-$$

The H^+ actually attaches to another H_2O molecule to give

$$H_2O + H^+ \leftrightarrow H_3O^+$$

Since both H^+ and OH^- are produced, H_2O is both a weak acid and a weak base. The concentration of H^+ (or H_3O^+) and OH^- in pure H_2O, not in equilibrium with atmospheric CO_2, is 10^{-7} M. The product of H^+ and OH^- concentration, as shown in the following equation, is the dissociation constant for water or K_w.

$$[H^+][OH^-] = [10^{-7}][10^{-7}] = 10^{-14} = K_w$$

The pH of H_2O in equilibrium with CO_2 in our atmosphere is about 5.7 because of the following reaction:

$$H_2O + CO_2 \xleftarrow{\quad H_2CO_3 \quad} H^+ + HCO_3^-$$

Adding an acid to H_2O will increase $[H^+]$, but $[OH^-]$ would decrease because K_w is a constant 10^{-14}. For example, in a 0.1 M HCl solution, the $[H^+]$ is 10^{-1} M; thus the $[OH^-]$ is

$$K_w = [H^+][OH^-] = 10^{-14}$$
$$[10^{-1}][OH^-] = 10^{-14}$$
$$[OH^-] = 10^{-13} \ M$$

The $[H^+]$ in solution can be conveniently expressed using pH and is defined as follows:

$$pH = \log 1/[H^+] = -\log [H^+]$$

Each unit increase in pH represents a 10-fold decrease in $[H^+]$ (Table 10.1). A solution with $[H^+] = 10^{-5}$ M will have a pH of 5.0.

Solutions with pH <7 are acidic, those with pH >7 are basic, and those with pH = 7 are neutral. The pH represents the H^+ concentration in solution and does not measure the undissociated or potential acidity. For example, the pH of completely dissociated 0.1 M HCl is 1.0, while the pH of 0.1 M CH_3COOH,

TABLE 10.1   Relationship Between pH and $[H^+]$
Concentration

Conc. of H_3O^+ (*mol/l*)	*pH*	*Conc. of H_3O^+* (*mol/l*)	*pH*
10^{-1}	1	10^{-8}	8
10^{-2}	2	10^{-9}	9
10^{-3}	3	10^{-10}	10
10^{-4}	4	10^{-11}	11
10^{-5}	5	10^{-12}	12
10^{-6}	6	10^{-13}	13
10^{-7}	7	10^{-14}	14

a weak acid, is 3.0. Similarly, the pH of 0.1 M NaOH, a strong base, is 13.0, while the pH of 0.1 M NH₄OH, a weak base, is 11.0.

When acids and bases are combined, both are neutralized, forming a salt and water:

$$HCl \quad + \quad NaOH \longleftrightarrow H_2O + Na^+ + Cl^-$$

$$H^+ \quad Cl^- \qquad Na^+ \qquad OH^-$$

If a given quantity of acid is titrated with a base and the pH of the solution is determined at intervals during the titration, a curve is obtained by plotting pH against the amounts of base added (Fig. 10.1). Titration curves for strong and weak acids differ markedly. The neutralization reaction of HCl with NaOH is given in the previous equation, and that of CH_3COOH with NaOH is as follows:

$$CH_3COOH \quad + \quad NaOH \longleftrightarrow H_2O + CH_3COO^- + Na^+$$

$$CH_3COO^- + \quad H^+ \quad Na^+ \quad OH^-$$

Buffers

Buffers or buffer systems can maintain the pH of a solution within a narrow range when small amounts of acid or base are added. *Buffering* defines the resistance to a change in pH (see Chapter 4). An example of a buffer system is CH_3COOH and CH_3COONa:

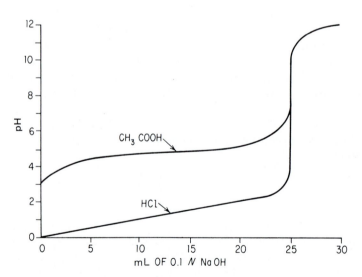

FIGURE 10.1   Titration of 0.10 N CH₃COOH and 0.10 N HCl with 0.10 N NaOH.

$$CH_3COOH \rightleftharpoons H^+ + CH_3COO^-$$

$$CH_3COONa \rightleftharpoons Na^+ + CH_3COO^-$$

For example, a solution containing 1 M CH_3COOH and 1 M CH_3COONa has a pH of 4.6, compared to a pH 2 for CH_3COOH alone. Adding the highly dissociated CH_3COONa to CH_3COOH increases the CH_3COO^- concentration, which shifts the equilibrium to form the undissociated CH_3COOH. The pH remains at 4.6 even with a 10-fold dilution with H_2O; however, dilution of 1 M CH_3COOH would raise the pH to 3.0. Thus, adding CH_3COONa to CH_3COOH buffers the solution pH.

If 10 ml of 1 M HCl is added to the CH_3COOH/CH_3COONa buffer solution, the pH will decrease to only 4.5, because the additional H^+ will shift the $CH_3COOH \leftrightarrow H^+ + CH_3COO^-$ equilibrium to the left and the decreased CH_3COO^- will be replaced by the CH_3COO^- from the dissociation of CH_3COONa (equilibrium shift to the right in $CH_3COONa \leftrightarrow CH_3COO^- + Na^+$).

Conversely, if 10 ml of 1 M NaOH is added, the OH^- neutralizes H^+ to form water. Because of the large supply of undissociated CH_3COOH, the equilibrium shifts to the right, replacing the neutralized H^+; thus, the pH increases to only 4.7.

Soils behave like buffered weak acids with the CEC of humus and clay minerals providing the buffer for soil solution pH.

Soil Acidity

The sources of soil acidity include OM, clay minerals, Fe and Al oxides, exchangeable Al^{3+}, soluble salts, and CO_2.

Humus

Soil OM or humus contains reactive carboxylic and phenolic groups that behave as weak acids (see Chapter 4). They will dissociate, releasing H^+. The soil OM content varies with the environment, vegetation, and soil; thus, its contribution to soil acidity will vary accordingly. In peat and muck soils and in mineral soils containing large amounts of OM, organic acids contribute significantly to soil acidity.

Clay Minerals

Clay minerals in soils typified by kaolinite (1:1) and montmorillonite (2:1) can buffer soil pH. The dissociation of H^+ from "broken edges" of clay minerals, Al and Fe oxide surfaces, and OM contributes to pH buffering in soil. High clay and/or high-OM soils will exhibit greater buffering capacity (BC) than coarse-textured and/or low-OM soils. The pH-dependent charge and BC associated with clay minerals, Al and Fe oxides, and OM are as follows:

On Fe and Al oxides:

$$\begin{bmatrix} \text{Fe, Al} \end{bmatrix}\begin{matrix} \diagup \text{OHH}^+ \\ \diagdown \text{OHH}^+ \end{matrix} \quad \xleftarrow{\text{H}^+} \quad \begin{bmatrix} \text{Fe, Al} \end{bmatrix}\begin{matrix} \diagup \text{OH} \\ \diagdown \text{OH} \end{matrix} \quad \xrightarrow{\text{OH}^-} \quad \begin{bmatrix} \text{Fe, Al} \end{bmatrix}\begin{matrix} \diagup \text{O}^- \\ \diagdown \text{O}^- \end{matrix}$$

acid   neutral   alkaline

On clay minerals:

acid   neutral   alkaline

(Si–O and Al–O clay structures shown with HH⁺ (acid), H (neutral), and O⁻ (alkaline) forms, transformed by H⁺ and OH⁻.)

On organic matter:

$$\text{R—COOH} \xrightarrow{\text{OH}^-} \text{R—COO}^-$$
acid   alkaline

$$\text{R—OH} \xrightarrow{\text{OH}^-} \text{R—O}^-$$
acid   alkaline

Al and Fe Polymers

The Al^{3+} ions displaced from the CEC are hydrolyzed to hydroxyaluminum complexes. Hydrolysis of Al^{3+} liberates H^+ and lowers pH unless there is a source of OH^- to neutralize the H^+. Each successive step occurs at a higher pH. Figure 10.2 illustrates the range in pH wherein the various Al hydrolysis species predominate. The following reactions illustrate the equilibria between Al species.

$$Al^{3+} + H_2O \rightarrow Al(OH)^{2+} + H^+$$

$$Al(OH)^{2+} H_2O \rightarrow Al(OH)_2^+ + H^+$$

$$Al(OH)_2^+ + H_2O \rightarrow Al(OH)_3^0 + H^+$$

$$AL(OH)_3^0 + H_2O \rightarrow Al(OH)_4^- + H^+$$

If a base is added to a soil, H^+ will be neutralized first. When more of the base is added, the Al^{3+} hydrolyzes, with the production of H^+ in amounts equivalent to the Al^{3+} present. It should be noted that insoluble $Al(OH)_3$ will precipitate at pH > 6.5 whenever the $Al(OH)_3$ solubility product is exceeded.

Like H_2O, Al^{3+} can function as either an acid or a base, as illustrated in the following equations:

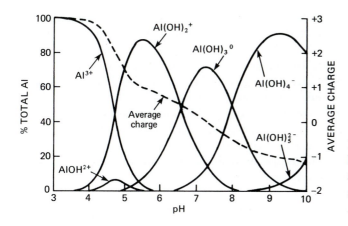

FIGURE 10.2 Relationship between pH and the distribution and average charge of soluble Al species. *Marion et al.*, Soil Sci., *121*:76, 1976.

Al as a base:

$$Al(OH)_3 + H^+ \rightleftharpoons Al(OH)_2^+ + H_2O$$
$$Al(OH)_2^+ + H^+ \rightleftharpoons Al(OH)^{2+} + H_2O$$
$$Al(OH)_2^{2+} + H^+ \rightleftharpoons Al^{3+} + H_2O$$
$$Al(OH)_3 + 3H^+ \rightleftharpoons Al^{3+} + 3H_2O$$

Al as an acid:

$$Al^{3+} + OH^- \rightleftharpoons Al(OH)^{2+}$$
$$Al(OH)^{2+} + OH^- \rightleftharpoons Al(OH)_2^+$$
$$Al(OH)_2^+ + OH^- \rightleftharpoons Al(OH)_3$$

Al as an anion:

$$Al(OH)_3 + OH^- \rightleftharpoons Al(OH)_4^-$$

The hydroxy Al ions combine to form large, multicharged polymers. Polymerization is favored in the presence of clay surfaces. The mechanism for polymer formation is the sharing of OH^- groups by adjacent Al^{3+} ions, as represented in the following equations:

$$\left[(H_2O)Al \begin{matrix} OH \\ \\ H_2O \end{matrix} \right]^{2+} + \left[\begin{matrix} H_2O \\ \\ HO \end{matrix} Al(H_2O)_4 \right]^{2+} \rightarrow \left[(H_2O)_4Al \begin{matrix} OH \\ \\ OH \end{matrix} Al(H_2O)_4 \right]^{4+} + 2H_2O$$

$$\downarrow$$

$$\left[(H_2O)_4Al \begin{matrix} O \\ \\ O \end{matrix} Al(H_2O)_4 \right]^{2+} + 2H^+$$

The Al polymers have high positive charge and are essentially nonexchange-able. CECs of soil can be affected by the adsorption of the positively charged Al polymers. At high pH, the CEC is increased because $Al(OH)_3$ precipitates and $(+)$ charged Al polymer formation is reduced or is nonexistent [see the increasing $(-)$ charge with increasing pH in Fig. 10.2]. Decreasing the soil pH will increase $(+)$ charged Al polymer formation, and adsorption to clay surfaces will decrease CEC [see the increasing $(+)$ charge in Fig. 10.2].

Fe hydrolysis, as shown in the following equation, is similar to that of Al. Although this reaction is more acidic than Al hydrolysis, the acidity is buffered by Al hydrolysis reactions. Thus, Fe hydrolysis will have little effect on soil pH until most of the soil Al has reacted:

$$Fe^{3+} + H_2O \leftrightharpoons Fe(OH)^{2+} + H^+$$

Al and Fe polymers can occur as amorphous or crystalline colloids, which coat the clay and other mineral surfaces. They are also held between the lattices of expanding soil minerals, preventing collapse of these lattices as water is removed during drying or freezing.

Soluble Salts

Acid, neutral, or basic salts in the soil solution originate from mineral weather-ing, OM decomposition, or addition as fertilizers and manures. The cations of these salts will displace adsorbed Al^{3+} in acid soils and thus decrease soil solution pH. Divalent cations have a greater effect on lowering soil pH than monovalent metal cations (see the lyotropic series in Chapter 4).

Band-applied fertilizer will result in a high soluble-salt concentration in the affected soil zone, which will decrease pH through Al hydrolysis. With high rates of band-applied fertilizer in soils with pH $<5.0-5.5$, this acidification may be detrimental to plant growth.

Carbon Dioxide

In calcareous soils the pH is influenced by the partial pressure of CO_2 in the soil atmosphere. The pH of a soil containing free $CaCO_3$ in equilibrium with atmospheric CO_2 is 8.5; however, increased CO_2 in soil air causes the pH to decrease to 7.3–7.5.

Decomposition of organic residues and root respiration increase CO_2 in soil air, which combines with water to provide a source of H^+ to lower pH ac-cording to the following:

$$H_2O + CO_2 \leftrightarrow H_2CO_3$$

$$H_2CO_3 \leftrightarrow H^+ + HCO_3^-$$

Factors Affecting Soil Acidity

Increasing soil acidity in crop production systems is caused by (1) use of commercial fertilizers, especially NH_4-N sources that produce H^+ during nitrification; (2) crop removal of cations in exchange for H^+; (3) leaching of these cations being replaced first by H^+ and subsequently by Al^{3+}; and (4) decomposition of organic residues. Natural acidification of soils is enhanced with increasing rainfall since rain has a pH of 5.7 or less, depending on pollutants like SO_2, NO_2, and others.

ACIDITY AND BASICITY OF FERTILIZERS   Fertilizer materials vary in their soil reaction pH. Nitrate sources carrying a basic cation should be less acid-forming than NH_4^+ sources. Compared to P fertilizers, those containing or forming NH_4-N exhibit greater effect on soil pH (Table 10.2).

Phosphoric acid released from dissolving P fertilizers such as TSP (monocalcium phosphate) and MAP can temporarily acidify localized zones at the site of application. The former material will reduce pH to as low as 1.5, while the latter will decrease pH to approximately 3.5; however, the quantity of H^+ produced is very small and has little long-term effect on pH. Diammonium phosphate (DAP) will initially raise the pH of the soil to about 8 (see Chapter 6). Acidity produced by the nitrification of the NH_4^+ in DAP will offset this initial pH rise.

The Association of Official Agricultural Chemists (AOAC) approved method for determining the acidity or basicity of fertilizers is based on the following assumptions:

1. The acidifying effect of fertilizer is caused by all of the contained S and Cl^-, one-third of the P, and one-half of the N.
2. The presence of Ca, Mg, K, and Na in the fertilizer will slightly increase or cause no change in soil pH.
3. Half of the fertilizer N is taken up as NO_3^-, and accompanied by equivalent amounts of K^+, Ca^{2+}, Mg^{2+}, or Na^+. Uptake of the other half is associated with H^+ as the counterion or exchanged for HCO_3^- from plant roots.

This method assumes that plant growth reduces the potential or theoretical amount of acidity produced by nitrification of NH_4^+ fertilizers because of unequal absorption of cations and anions by the crop.

Significant amounts of NH_4^+ utilized directly by plants, with resultant acidification of soil close to roots, is not considered. Accordingly, 1.8 lb of pure $CaCO_3$ would be required to neutralize the acidity resulting from the addition of each pound of NH_4-N (Table 10.2). When cation/anion uptake effects are not considered, 3.57 lb of $CaCO_3$ is needed to neutralize the acidity produced per pound of NH_4-N nitrified to NO_3^-.

REMOVAL OF BASIC CATIONS   Since solutions containing salts must be electrically neutral [equal (+) and (−) charges], water leaching below the root zone will contain both cations and anions. Every pound of N as NO_3^- leached from the soil takes with it 3.57 lb of $CaCO_3$ or its equivalent in cations.

Crop uptake of cations can either reduce or increase the soil acidity produced by nitrification of NH_4^+ from fertilizers, crop and animal wastes, or in OM. These variations are explained by differences in N and excess bases taken up by the plant. Excess bases (EB) are defined as total cation (Ca^{2+}, Mg^{2+}, K^+, and Na^+) uptake minus total anion (Cl^-, SO_4^{2-}, NO_3^-, and $H_2PO_4^-$)

TABLE 10.2 Soil Acidity Produced by N Fertilizers

N Source	Nitrification Reaction	Residual Soil Acidity				Official Value*
		Maximum		Minimum		
		Acid Residue	CaCO$_3$ Equiv.	Acid Residue	CaCO$_3$ Equiv.	
		kg CaCO$_3$/kg of N		— kg CaCO$_3$/kg of N —		— kg CaCO$_3$/kg of N —
Anhydrous ammonia	$NH_3(g) + 2O_2 \rightarrow H^+ + NO_3^- + H_2O$	H^+ NO_3^-	$50/14 = 3.6$	None	0	1.8
Urea	$(NH_2)_2CO + 4O_2 \rightarrow$ $2H^+ + 2NO_3^- + CO_2 + H_2O$	$2H^+$ $2NO_3^-$	$100/28 = 3.6$	None	0	1.8
Ammonium nitrate	$NH_4NO_3 + 2O_2 \rightarrow$ $2H^+ + 2NO_3^- + H_2O$	$2H^+$ $2NO_3^-$	$100/28 = 3.6$	None	0	1.8
Ammonium sulfate	$(NH_4)_2SO_4 + 4O_2 \rightarrow$ $4H^+ + 2NO_3^- + SO_4^{2-} + 2H_2O$	$4H^+$ $2NO_3^-$ SO_4^{2-}	$200/28 = 7.2$	$2H^+$ SO_4^{2-}	$100/28 = 3.6$	5.4
Monoammonium phosphate	$NH_4H_2PO_4 + O_2 \rightarrow$ $2H^+ + NO_3^- + H_2PO_4^- + H_2O$	$2H^+$ NO_3^- $H_2PO_4^-$	$100/14 = 7.2$	H^+ $H_2PO_4^-$	$50/14 = 3.6$	5.4
Diammonium phosphate	$(NH_4)_2HPO_4 + O_2 \rightarrow$ $3H^+ + 2NO_3^- + H_2PO_4^-$ $+ H_2O$	$3H^+$ $2NO_3^-$ $H_2PO_4^-$	$150/28 = 5.4$	H^+ $H_2PO_4^-$	$50/28 = 1.8$	3.6

*Value adopted by the Association of Official Analytical Chemists (Pierre, 1934).

SOURCE: Adams, *Soil Acidity and Liming*, no. 12, p.234. Madison, Wisc.: ASA, 1984.

uptake. Plants with an EB/N ratio below 1.0 decrease the acidity formed by nitrification, whereas those with a ratio above this value increase acidity. Only a few crops (e.g. buckwheat and spinach) have values slightly above 1.0. Cereal and grass crops have average ratios of 0.43 and 0.47, respectively, meaning that only 43 and 47%, respectively, of N uptake is acid-forming.

LONG-TERM EFFECT   Soil acidity does not develop in a year or two, but rapidity varies among soils. A problem might develop in 5 years on a sandy soil or in 10 years on a silt loam soil, but might take 15 years or more on a clay loam. Soil acidity is an increasing problem because lime use is decreasing, while N use and yields have been increasing. An example of long-term use of $NaNO_3$ and $(NH_4)_2SO_4$ with and without lime is shown in Table 10.3. With $(NH_4)_2SO_4$ soil pH in the surface 15 cm decreased 1.6–1.8 units after 32 years compared to $(NH_4)_2SO_4$ plus lime. With $NaNO_3$ pH declined only 0.7–1.0 pH units, in contrast with $NaNO_3$ plus lime.

The Soil as a Buffer

Soil behaves like a weak acid that will buffer the pH accordingly. In acid soils the adsorbed Al^{3+} will maintain an equilibrium with Al^{3+} in the soil solution, which hydrolyzes to produce H^+.

$$\text{clay} \begin{bmatrix} Al^{3+} \\ Ca^{2+} \\ Mg^{2+} \\ K^+ \\ Al^{3+} \end{bmatrix} \rightleftharpoons Al^{3+} + H_2O \rightleftharpoons AlOH^{2+} + H^+$$

If the H^+ is neutralized by adding a base and the Al^{3+} in solution precipitates as $Al(OH)_3$, more exchangeable Al^{3+} will desorb to resupply solution Al^{3+}. Thus, the pH remains the same or is buffered. As more base is added, the

TABLE 10.3   Effect of N Sources and Lime on Soil Profile Acidity in Two 2-Yr Cotton-Corn Rotation Experiments in Alabama After 32 Yr (1930–1962)

N and Lime Treatment[†]	Soil pH by Solid Depth After 32 Yr*							
	Dothan ls				Lucedale sl			
	0–15 cm	15–30 cm	30–45 cm	45–60 cm	0–15 cm	15–30 cm	30–45 cm	45–60 cm
$(NH_4)_2SO_4$	4.8	4.8	4.3	4.3	4.8	4.8	4.9	4.9
$(NH_4)_2SO_4$ + lime	6.2	5.4	4.7	4.6	6.0	5.5	5.3	5.1
$NaNO_3$	5.7	5.5	5.1	5.0	5.8	5.5	5.3	5.1
$NaNO_3$ + lime	6.4	6.4	6.3	6.3	6.8	6.6	5.8	5.2

* Initial surface soil pH was 6.0 at both sites.

[†] Nitrogen rate was 40 kg/ha during 1930 to 1945 and 53 kg/ha during 1946 to 1962.

Basic slag was lime source; it was applied annually at a rate of 500 kg/ha during 1930 to 1945 and at a rate of 800 kg/ha during 1946 to 1962.

SOURCE: Adams, *Soil Acidity and Liming.* No. 12, p. 238. Madison, Wisc.: ASA, 1984.

preceding reaction continues, with more adsorbed Al^{3+} neutralized and re-placed on the CEC with the cation of the added base. As a result, there is a gradual increase in soil pH.

The reverse of the preceding reaction also occurs. As acid is continually added to a neutral soil, OH^- in the soil solution is neutralized. Gradually, the $Al(OH)_3$ dissolves, which increases Al^{3+} in the solution and subsequently on the CEC. As the reaction continues, there is a continual but slow decrease in soil pH as the Al^{3+} replaces adsorbed basic cations.

The total amount of clay and OM in a soil and the nature of the clay minerals will determine the extent to which soils are buffered. Soils containing large amounts of clay and OM are highly buffered and require larger amounts of lime to increase the pH than soils with a lower BC. Sandy soils with small amounts of clay and OM are poorly buffered and require only small amounts of lime to effect a given change in pH. Soils containing mostly 1:1 clays (Ultisols and Oxisols) are generally less buffered than soils with principally 2:1 clay minerals (Alfisols and Mollisols). For example, an increasing lime requirement with an increasing clay content and CEC is shown in Figure 10.3.

Determination of Active and Potential Acidity in Soils

Active Acidity

Currently, the most accurate and most widely used method involves measuring pH in a saturated paste or a more diluted soil:water mixture with a pH meter and glass electrode. Soil pH is a useful indicator of the presence of exchangeable Al^{3+} and H^+. Exchangeable H^+ is present at pH <4, while exchangeable Al^{3+} predominantly occurs at pH 4 to 5.5. Al polymers occur in the pH range 5.5 to 7.0.

Increasing the dilution of the soil from saturation to 1:1 to 1:10 soil:water will generally increase the measured pH compared to the pH of a saturated paste. To minimize differences in salt concentration among soils, some labora-tories dilute the soil with 0.01 $CaCl_2$ instead of water. Adding Ca^{2+} will de-crease the pH compared to soil diluted with water. Changes in measured pH with dilution and added salt are generally small, ranging between 0.1 and 0.5 pH unit.

Potential Acidity

Soil pH measurements are excellent indicators of soil acidity or basicity; how-ever, as an indicator of active acidity, soil pH does not measure potential acidity. Quantifying potential soil acidity requires titrating the soil with a base, which can be used to determine the lime requirement or quantity of $CaCO_3$ needed to increase the pH to a desired level. Thus, the lime requirement of a soil is related not only to the soil pH but also to its BC or CEC (Fig. 10.3). High clay and/or high-OM soils have higher BCs and will have a high lime

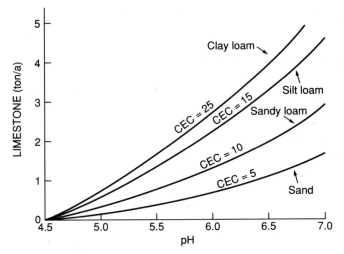

FIGURE 10.3   Approximate tons of limestone required to raise the pH of a 17-cm (7-in.) layer of soil of four textural classes with typical cation exchange capacities (CEC) in milliequivalents per 100 g of soil.   Sources: *Modified from D. R. Christenson and E. C. Doll, Ext. Bul. 471, 1979; and R. G. Hanson,* Science and Technology Guide 9102, Agronomy, *11, 1977.*

requirement, while coarse-textured soils with little or no clay and OM have a low BC and a low lime requirement.

An example liming calculation for two soils with CEC = 20 meq/100 g and 10 meq/100 g is shown below.

Clay Soil: CEC = 20 meq/100 g
Initial pH = 5.0 and % B.S. = 50%
Final pH = 6.5 and % B.S. = 80%

Need to neutralize 30% of acids on CEC (80–50%); thus

$$(0.30) \left(\frac{20 \text{ meq/CEC}}{100 \text{ g Soil}} \right) = \frac{6.0 \text{ meq Acids}}{100 \text{ g Soil}} = \frac{6.0 \text{ meq CaCO}_3}{100 \text{ g Soil}}$$

therefore, the quantity of pure $CaCO_3$ needed is:

$$\frac{6.0 \text{ meq CaCO}_3}{100 \text{ g Soil}} \left(\frac{50 \text{ mg CaCO}_3}{\text{meq}} \right) = \frac{300 \text{ mg CaCO}_3}{100 \text{ g Soil}}$$

thus,

$$\frac{300 \text{ mg CaCO}_3}{100 \text{ g Soil}} \left(\frac{1000 \text{ g}}{\text{mg}} \right) = \frac{0.30 \text{ g CaCO}_3}{100 \text{ g Soil}} = \frac{0.30 \text{ lb CaCO}_3}{100 \text{ lb Soil}}$$

$$\frac{0.30 \text{ lb CaCO}_3}{100 \text{ lb Soil}} \left(\frac{2 \times 10^4}{2 \times 10^4} \right) = \frac{6000 \text{ lb CaCO}_3}{\text{afs}}$$

Sandy Loam Soil: CEC = 10 meq/100 g

Assuming the same initial and final pH and % B.S., the $CaCO_3$ required would be only 3000 lb $CaCO_3$/afs.

Determining the Lime Requirement of Soils

The lime requirement of a soil can be determined by several different methods. Titrating the soil with an acid or a base will subsequently decrease or increase the soil pH, respectively (Fig. 10.4). The base usually employed is $Ca(OH)_2$ and the acid is HCl. After equilibration, pH is determined and the values are plotted against the meq of acid or base added. From these data it is simple to determine the amount of lime to be added.

For example, increasing pH from 5.7 to 6.5 requires adding 1.0 meq base/100 g soil (Fig. 10.4). Thus, the quantity of pure $CaCO_3$ needed to increase pH from 5.7 to 6.5 would be

$$\frac{1.0 \text{ meq } CaCO_3}{100 \text{ g soil}} \left(\frac{50 \text{ mg } CaCO_3}{\text{meq}} \right) = \frac{50 \text{ mg } CaCO_3}{100 \text{ g soil}}$$
$$= 1000 \text{ lb } CaCO_3/\text{a-6 in.}$$

The methods in common use are based on the change in pH of a buffered solution added to a soil. When an acid soil is added to a buffered solution, the buffer pH is depressed in proportion to the original soil pH and its BC. By calibrating pH changes in the buffered solution, the amount of lime required to increase pH to the desired level, that is, the lime requirement, can be calculated.

The Shoemaker, McLean, and Pratt (SMP) single-buffer method for measurement of the lime requirement of acid soils has been widely adopted by

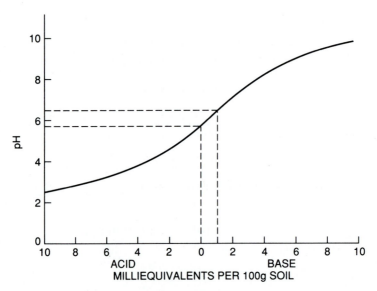

FIGURE 10.4   General lime requirement curve for a sandy loam soil.

U.S. soil-testing laboratories. The buffer solution is a dilute mixture of triethanolamine, paranitrophenol, and K, chromate. The SMP method is especially well suited for soils with the following properties: lime requirement >4 meq/100 g (>4,000 lb $CaCO_3$/a), pH <5.8, OM <10%, and appreciable quantities of soluble (extractable) Al. The SMP method used on soils with low lime requirements will frequently result in overliming.

In very acid soils, the lime requirement may be based on the $CaCO_3$ needed to lower Al on the CEC. For example, in Columbia, it is recommended that 1 ton of $CaCO_3$ be applied for each 1 meq of exchangeable Al^{3+}/100 g soil. In Brazil the amount of lime needed to neutralize exchangeable Al is determined as follows:

$$\text{meq } CaCO_3/100 \text{ g} = 2 \times \text{meq exchangeable Al/100 g}$$

Other studies show that the range in $CaCO_3$ recommendation is from 1000 to 6000 lb $CaCO_3$/meq Al^{3+}/100 g soil.

Soil pH for Crop Production

For many years the optimum soil pH for crop production was considered to be between 6.5 and 7.0; however, lime sufficient to raise pH to about 5.6 or 5.7 and reduce exchangeable Al^{3+} to less than 10% of the CEC will eliminate pH-related crop production problems. Liming Ultisols and Oxisols to pH 7.0 can impair productivity by

1. Decreasing water percolation.
2. Reducing the growth of legumes and nonlegumes.
3. Restricting plant uptake of P and some micronutrients.

Reducing soluble Al^{3+} to <1 ppm is recommended for sensitive crops. On the Alfisols and Mollisols, liming to pH 6.5 to 6.8 is optimum for most crops except alfalfa, where liming to pH 6.8 to 7.0 is recommended.

Crops vary widely in their tolerance to acid soils (Table 10.4).

TABLE 10.4 Optimum Soil pH Levels for Various Crops

Soil pH	Crop	Soil pH	Crop
>6.5	Alfalfa	5.0–5.5	Blueberries
	Sweet clover		Cranberries
	Sugar beets		Potatoes
5.5–6.5	Red clover		Tobacco
	Corn	<5.0	Azalea
	Wheat		Hydrangea
	Barley		Rhododendron
	Bluegrass		

Liming Materials

The materials commonly used for liming soils are Ca and/or Mg oxides, hydroxides, carbonates, and silicates (Table 10.5). The accompanying anion *must* lower H^+ activity and hence Al^{3+} in the soil solution. Gypsum ($CaSO_4 \cdot 2H_2O$) and other neutral salts cannot neutralize H^+, as shown in the following reaction:

$$CaSO_4 \cdot 2H_2O + 2H^+ \leftrightarrow Ca^{2+} + 2H^+ + SO_4^{2-} + 2H_2O$$

In fact, the addition of neutral salts will actually lower soil pH. Their addition, especially in a band, results in replacement of adsorbed Al^{3+} in a localized soil zone, sometimes with a significant lowering of the pH in this region.

Liming reactions begin with the neutralization of H^+ in the soil solution by either OH^- or HCO_3^- originating from the liming material. For example, $CaCO_3$ behaves as follows:

$$CaCO_3 + H_2O \rightarrow Ca^{2+} + HCO_3^- + OH^-$$

The rate of the reaction is directly related to the rate at which the OH^- ions are removed from solution. As long as sufficient H^+ ions are in the soil solution, Ca^{2+} and HCO_3^- will continue to go into solution. When the H^+ ion concentration is lowered, however, formation of the Ca^{2+} and HCO_3^- ions is reduced.

The continued removal of H^+ from the soil solution will ultimately result in the precipitation of Al^{3+} and Fe^{3+} as $Al(OH)_3$ and $Fe(OH)_3$ and their replacement on the CEC with Ca^{2+} and/or Mg^{2+}.

The overall reaction for neutralization of Al-derived soil acidity can be written as follows:

$$\text{clay} \begin{bmatrix} Al^{3+} \\ Ca^{2+} \\ Mg^{2+} \\ K^+ \\ Al^{3+} \end{bmatrix} + 3\ CaCO_3 + 3\ H_2O \rightarrow \text{clay} \begin{bmatrix} K^+ \\ Ca^{2+} \\ Ca^{2+} \\ Mg^{2+} \\ Ca^{2+} \\ Ca^{2+} \end{bmatrix} + 2\ Al(OH)_3 + 3\ CO_2$$

Obviously, as pH increases, the percent base saturation also increases, as discussed in Chapter 4.

TABLE 10.5   Neutralizing Value (CCE) of Pure Forms of Some Liming Materials

Material	Molecular Wt. (g/mole)	Eq. Wt. (g/eq)	Neutralizing Value %
CaO	56	28	179
Ca(OH)$_2$	72	36	136
CaMg(CO$_3$)$_2$	184	46	109
CaCO$_3$	100	50	100
CaSiO$_3$	116	58	86

Calcium Oxide

Calcium oxide (CaO) is the only material to which the term *lime* may be correctly applied. Known also as *unslaked lime, burned lime*, or *quicklime*, CaO is a white powder, shipped in paper bags because of its caustic properties. It is manufactured by roasting $CaCO_3$ in a furnace, driving off the CO_2. CaO is the most effective of all liming materials, with a neutralizing value or calcium carbonate equivalent (CCE) of 179%, compared to pure $CaCO_3$ (Table 10.5). When unusually rapid results are required, either this material or $Ca(OH)_2$ should be selected. Complete mixing of CaO with the soil may be difficult, because immediately after application, absorbed water causes the material to form flakes or granules. These granules may harden due to $CaCO_3$ formation on their surfaces, which can remain in the soil for long periods of time.

Calcium Hydroxide

Calcium hydroxide [$Ca(OH)_2$], referred to as *slaked lime, hydrated lime*, or *builders' lime*, is a white powder and is difficult and unpleasant to handle. Neutralization is rapidly effected. Slaked lime is prepared by hydrating CaO and it has a neutralizing value (CCE) of 136.

Calcium and Calcium-Magnesium Carbonates

Calcium carbonate ($CaCO_3$) or calcite and calcium-magnesium carbonate [$CaMg(CO_3)_2$] or dolomite are common liming materials. Limestone is most often mined by open-pit methods. The quality of crystalline limestones depends on the degree of impurities they contain, such as clay. The neutralizing values range from 65 to a little more than 100%. The neutralizing value of pure $CaCO_3$ has been theoretically established at 100%, while pure dolomite has a neutralizing value of 109%. As a general rule, however, the CCE of most agricultural limestones is between 90 and 98% because of impurities.

Marl

Marls are soft, unconsolidated deposits of $CaCO_3$ frequently mixed with earth and usually quite moist. Marl deposits are generally thin, recovered by dragline or power shovel after the overburden has been removed. The fresh material is stockpiled and allowed to dry before being applied to the land. Marls are almost always low in Mg, and their neutralizing value, between 70 and 90%, depends on their clay content.

Slags

Blast-furnace slag ($CaSiO_3$) is a by-product of the manufacture of pig iron. In the reduction of Fe, the $CaCO_3$ in the charge loses its CO_2 and forms CaO, which combines with the molten Si to produce a slag that is either air-cooled or quenched with water. Silicic acid, formed when slag is added to acid soils, is weakly dissociated; thus, the pH of the soil is raised. The neutralizing value of blast-furnace slags ranges from about 75 to 90%, and they usually contain appreciable amounts of Mg.

Basic slag is a by-product of the open-hearth method of making steel from pig iron, which, in turn, is produced from high-P Fe ores. The impurities in the Fe, including Si and P, are removed with lime. In addition to its P content, basic slag has a neutralizing value of about 60 to 70%. It is generally applied for its P content rather than as a liming material, but because of its neutralizing value it is a good material to use on low-P, acid soils.

Electric-furnace slags are produced from the electric-furnace reduction of phosphate rock in the preparation of elemental P and in the manufacture of pig iron and steel. The slag is formed when the Si and CaO fuse, producing Ca silicate. The electric-furnace slag contains 0.9 to 2.3% P_2O_5, and the neutralizing value ranges from 65 to 80%. The data in Table 10.6 illustrate that slags are effective liming materials.

Miscellaneous Liming Materials

Other materials that are used as liming agents in localized areas close to their source include fly ash from coal-burning power generating plants, sludge from water treatment plants, Cotrell lime or flue dust from cement manufacturing, sugar lime, pulp mill lime, carbide lime, acetylene lime, packinghouse lime, and so on. These by-products contain varying amounts of Ca and Mg.

Neutralizing Value or Calcium Carbonate Equivalent (CCE) of Liming Materials

The value of a liming material depends on the quantity of acid that a unit weight will neutralize, which, in turn, is related to the molecular composition and purity. Pure $CaCO_3$ is the standard against which other liming materials are measured, and its neutralizing value is considered to be 100%. The CCE is defined as the acid-neutralizing capacity of a liming material expressed as a weight percentage of $CaCO_3$.

TABLE 10.6 Response of Red Clover to Kimberley Electric Furnace Iron Slag and Other Liming Materials

| | Yield (g) of Oven-Dry Tissue per Pot | | | | | |
| | Alouette Silt Loam | | | Pitt Silty Clay | | |
Liming Treatment	30% Base Satn.	60% Base Satn.	Means	30% Base Satn.	60% Base Satn.	Means
Agricultural lime	1.55	1.51	1.53	4.29	4.49	4.39
Slag (broadcast)	2.03	2.28	2.16	4.42	4.54	4.48
Slag (incorporated)	2.48	3.19	2.84	4.52	5.02	4.77
Dolomitic limestone	2.25	2.66	2.45	4.04	4.94	4.49
Magnesium carbonate	1.58	1.13	1.35	4.14	3.28	3.71
Marl	1.58	1.66	1.62	4.29	4.64	4.47
Control (no liming)	1.10	1.10	1.10	2.30	2.30	2.30
Mean	1.80	1.93	1.86	4.00	4.17	4.09

SOURCE: Beaton et al., *Can. J. Plant Sci.*, **48**:455 (1968).

Consider the following reactions:

$$CaCO_3 + 2H^+ \rightleftharpoons Ca^{2+} + H_2O + CO_2$$

$$MgCO_3 + 2H^+ \rightleftharpoons Mg^{2+} + H_2O + CO_2$$

In each reaction, 1 mole of CO_3^{2-} will neutralize 2 moles of H^+. The molecular weight of $CaCO_3$ is 100, whereas that of $MgCO_3$ is only 84; thus, 84 g of $MgCO_3$ will neutralize the same amount of acid as 100 g of $CaCO_3$. Therefore, the neutralizing value or CCE of equal weights of the two materials is calculated by

$$\frac{84}{100} = \frac{100}{x}$$

$$x = 119$$

Therefore, $MgCO_3$ will neutralize 1.19 times as much acid as the same weight of $CaCO_3$; hence its CCE is 119%. The same procedure is used to calculate the neutralizing value of other liming materials (Table 10.5).

The composition of liming materials is sometimes expressed in terms of the Ca and Mg content of the pure mineral. For example, pure $CaCO_3$ contains 40% Ca and pure $MgCO_3$ contains 28.6% Mg, calculated by the ratio of molecular weights:

$$\frac{24 \text{g/m Mg}}{84 \text{g/m MgCO}_3} \times 100 = 28.6\%$$

To convert %Ca to CCE, multiply by 100/40 or 2.5; to convert %Mg to $MgCO_3$, multiply by 84/24, or 3.5.

CA AND MG OXIDE CONTENT   The quality of a limestone may also be expressed by its Ca or Mg oxide equivalent. For example, pure $CaCO_3$ contains 40% Ca. CaO has a molecular weight of 56, which means that 16 g of O is combined with 40 g of Ca. Therefore, if the Ca in $CaCO_3$ were expressed as the oxide, it would contain (56/100) $\times$ 100, or 56% CaO equivalent. Thus, to convert %Ca to %CaO, multiply the Ca by 56/40, or 1.4; and to convert %$CaCO_3$ to %CaO, multiply the %$CaCO_3$ by 56/100, or 0.56. Similar figures may be derived for the Mg-containing limestones.

TOTAL CARBONATES   The quality of limestones also can be related to the total CO_3^{2-} and is the sum of the %CO_3 contained in a given liming material. For example, assume that a limestone contains 78% $CaCO_3$ and 12% $MgCO_3$. The total CO_3 content would be 90%.

Conversion factors which make possible an expression of the value of a limestone may be determined in any way desired, provided that the content of one of the constituents is given. A few of these factors are listed in Table 10.7.

TABLE 10.7   Limestone Conversion Factors for Determining Neutralizing Equivalents

%	%		Factor	%	%		Factor
Ca	to CaO	multiply by	1.40 *	Mg	to MgCO$_3$	multiply by	3.50
Ca	to Ca(OH)$_2$	multiply by	1.85	Mg	to Ca	multiply by	1.67
Ca	to CaCO$_3$	multiply by	2.50	Mg	to CaCO$_3$	multiply by	4.17
Mg	to MgO	multiply by	1.67	MgO	to CaCO$_3$	multiply by	2.50
Mg	to Mg(OH)$_2$	multiply by	2.42	MgCO$_3$	to CaCO$_3$	multiply by	1.19

* $\frac{56}{40} = 1.40$

Fineness of Limestone

The effectiveness of agricultural limestones also depends on the degree of fineness, because the reaction rate depends on the surface area in contact with the soil. CaO and Ca(OH)$_2$ are powders, so that no problem of fineness is involved, but limestones need to be crushed to reduce the particle size. When crushed limestone is thoroughly incorporated with the soil, the reaction rate will increase with increasing fineness (Table 10.8). These results also show the increasing effectiveness of the coarser fractions relative to the <100 fraction with increasing exposure time in soil. The relative efficiencies, based on change in soil pH, are influenced by the magnitude of the pH change used for making the comparisons. The coarser fractions show higher efficiencies relative to the <100 fraction the lower the reference pH.

Amounts of a particular limestone fraction needed to produce a given rise in pH two years after application are shown in Figure 10.5. Much less of the finer fractions than of the coarser fractions is needed to achieve a certain pH, particularly at lower reference pH values.

TABLE 10.8   Relative Efficiency of Various Dolomitic Limestone Fractions Affected by Reference pH and Time of Equilibration in Withee Silt Loam Under Field Conditions

Fraction (mesh size)	pH 5.5				pH 6.0			pH 6.5	
	1 Mo	1 Yr	2 Yr	3 Yr	1 Yr	2 Yr	3 Yr	2 Yr	3 Yr
Where <100 fraction at 3 years' equilibration for each pH level = 100									
8–20	3	9	27	—	—	21	54	13	24
20–40	5	29	77	100	11	50	80	28	55
40–60	8	53	83	100	16	72	100	53	73
60–100	14	67	100	100	28	85	100	67	92
<100	40	67	100	100	48	93	100	90	100
Where <100 fraction in each column = 100									
8–20	6	14	27	—	—	22	54	15	24
20–40	11	44	77	100	24	54	80	31	55
40–60	21	79	83	100	33	78	100	59	73
60–100	36	100	100	100	59	91	100	74	92
<100	100	100	100	100	100	100	100	100	100

SOURCE: Love et al., *Trans. 7th Int. Congr. Soil Sci.*, IV.37: 293 (1960).

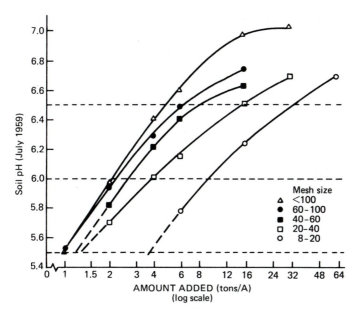

FIGURE 10.5   Effect of various rates of dolomitic limestone fractions on the pH of Withee silt loam after 2 years of equilibration under field conditions.   *Love et al., Trans. 7th Int. Congr. Soil Sci., IV:37:293, 1960.*

Because the cost of limestone increases with its fineness, materials that require minimum grinding, yet contain enough fine material to change pH rapidly, are preferred. Agricultural limestones contain both coarse and fine materials. Many states require that 75 to 100% of the limestone pass an 8- to 10-mesh screen and that 25% pass a 60-mesh screen. This way, there is fairly good distribution of both the coarse and fine particles.

Fineness is quantified by measuring the distribution of particle sizes in a given limestone sample (Table 10.9). The fineness factor is the sum of the percentages of the liming agent in each of the three size fractions multiplied by the appropriate effectiveness factor (Table 10.10). The effective calcium carbonate (ECC) rating of a limestone is the product of its CCE (purity) and the fineness factor.

TABLE 10.9   Fineness Factors for Agricultural Limestone in the United States and Canada

United States		Canada	
Particle Size	*Effectiveness*	*Particle Size*	*Effectiveness*
---- mesh ----	----- % -----	-- sieve no --	----- % -----
>8	0	>10	0
8–60	50	10–60	40
<60	100	<60	100

TABLE 10.10   Fineness Effects on Effective Calcium Carbonate (ECC) Content
of Two Lime Sources

	Solid Agricultural Lime	Suspendable Lime
Percent CaCO$_3$ equivalent	90	98
Sieve analysis		
% on 8-mesh sieve	2	0
% on 60-mesh sieve	21	2
% passing 60-mesh sieve	77	98
Calculated fineness factor		
% on 8-mesh × 0% effectiveness	2 × 0 = 0	0 × 0 = 0
% on 60-mesh × 50% effectiveness	21 × 50% = 11.5	2 × 50% = 1
% through 60-mesh × 100% effectiveness	77 × 100% = 77	98 × 100% = 98
Fineness factor (%)	88.5	99
Percent ECC = purity × fineness factor	90 × 88.5% = 79.6	98 × 99 = 97.0

SOURCE: Murphy et al., In *Situation 1978, TVA Fert. Conf. Bull. Y-131*, Muscle Shoals, Ala.: National Fertilizer
Development Center–Tennessee Valley Authority, 1978.

As a general rule, for the same degree of fineness, the material that costs
the least per unit of neutralizing value applied to the land should be used.
Assume that there are available a calcitic limestone (CCE = 95%), and a
dolomitic limestone (CCE = 105%), both with the same fineness. Assume also
that they both cost $12 per ton applied to the land. Based on the neutralizing
value, the calcitic limestone will cost 105/95 × 12 or $13.26 per ton, compared
with the dolomite at $12 per ton. In addition, the dolomite supplies Mg, which
can be deficient in many humid-region soils.

Use of Lime in Agriculture

Lime is seldom needed in low-rainfall areas where leaching is minimal, such
as parts of the Great Plains states and the arid, irrigated saline, and saline-
alkali soils of the southwestestern, intermountain, and far western states. This
is also true for the majority of soils in the Prairie Provinces of Canada.

Crop responses from the application of lime are usually attributed to decreased toxicity of Al^{3+}, although the plant-nutrient value of the Ca or Mg
also is important.

Direct Benefits

Al toxicity is probably the most important growth-limiting factor in many acid
soils, particularly when pH <5.0 to 5.5. Excess Al interferes with cell division
in plant roots; inhibits nodule initiation; fixes P in less available forms in soils
and in or on plant roots; decreases root respiration; interferes with enzymes
governing the deposition of polysaccharides in cell walls; increases cell wall
rigidity by cross-linking with pectins; and interferes with the uptake, transport,
and use of nutrients and water by plants. At pH 4.5 or less, another benefit
is the removal of H$^+$ toxicity, which damages root membranes and also is
detrimental to the growth of many beneficial bacteria. The greatest single

direct benefit of liming many acid soils is the reduction in the activity or solubility of Al and Mn.

When lime is added to acid soils, the activity of Al^{3+} is reduced by precipitation as $Al(OH)_3$ (Table 10.11). The lime treatment raises soil pH while greatly reducing the level of extractable Al. Not only is Al^{3+} toxic to plants, but increasing Al^{3+} in the soil solution also restricts the plant uptake of Ca and Mg (Fig. 10.6).

Different crops and varieties of the same crop differ widely in their susceptibility to Al^{3+} toxicity. Different varieties of soybeans, wheat, and barley show wide ranges in their tolerance to high concentrations of Al^{3+} in the soil solution (Figure 10.7). Some grasses are quite tolerant to acid mine spoils.

The Al problem is not always economically correctable with conventional liming practices. Differences among genotypes in tolerance to excess Al are in part genetically controlled. A genetic approach has great potential for solving the problem of Al toxicity in acid soils.

Crops also vary in their tolerance to excessive amounts of Mn. For example, rapeseed is very sensitive to Mn toxicity, while barley is more tolerant.

Indirect Benefits

EFFECT ON P AVAILABILITY   At low pH values and on soils high in Al and Fe, P precipitates as insoluble Fe/Al-P compounds (see Chapter 6). Liming acid soils will precipitate Fe and Al as $Fe(OH)_3$ and $Al(OH)_3$, thus increasing plant-available P.

Alternatively, liming soils to pH 6.8–7.0 can reduce P availability because of the precipitation of Ca or Mg phosphates. A liming program should be planned so that the pH can be kept between 5.5 and 6.8 to 7.0 if maximum benefit is to be derived from the applied P.

MICRONUTRIENT AVAILABILITY   With the exception of Mo, the availability of the micronutrients increases with decreased pH (see Chapter 9). This can be detrimental because of the toxic nature of many micronutrients even at relatively low solution concentrations. The addition of adequate lime reduces the solution concentration of many micronutrients, and soil pH values of 5.6

TABLE 10.11   Effects of Lime on Hudson Barley and on the pH and Level of Extractable Al in Tatum Surface Soil

CaCO$_3$ Added (ppm)	Yield of Barley Tops (g/pot)	Soil Properties	
		pH	KCl-Extractable Al (meq/100 g)
0	0.29	4.1	5.75
375	0.91	4.3	4.81
750	2.72	4.5	4.33
1,500	4.29	4.8	2.75
3,000	5.07	5.5	0.37

SOURCE: Foy et al., Agron. J., 57:413 (1965).

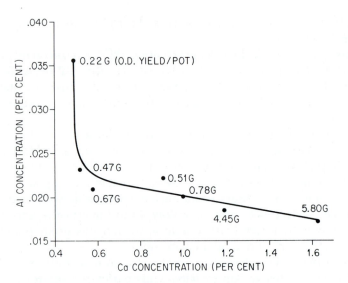

FIGURE 10.6   Relationship between Al and Ca concentration in cotton tops from nonleached subsoil.   *Soileau et al., SSSA Proc., 33:919, 1969.*

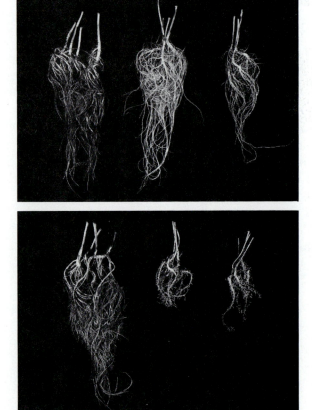

FIGURE 10.7   Differential effects of Al on root growth of Perry (top) and Chief (bottom) soybean varieties grown in ⅕ Steinberg solution containing 2 ppm Ca. *Left to right*: 0, 8, 12 ppm Al added.   *Foy et al., Agron. J., 61:505, 1969*

to 6.0 are usually sufficient to minimize toxicity while maintaining adequate availability of micronutrients.

Mo nutrition of crops is improved by liming, and deficiencies are infrequent in those soils limed to pH values in excess of 7.0. Because of the effect on availability of other micronutrients, liming to this value or above is not normally recommended for most crops in humid areas.

NITRIFICATION   Most of the organisms responsible for the conversion of NH_4^+ to NO_3^- require large amounts of Ca; therefore, nitrification is enhanced by liming to a pH of 5.5 to 6.5 (see Chapter 5). Decomposition of plant residues and breakdown of soil OM are also faster in this pH range than in more acidic soils. The effect of liming on both mineralization of organic N and nitrification is shown in Table 10.12. Application of lime just prior to incubation almost doubled the mineralization of organic N. However, lime added 1 or 2 years before sampling had little or no effect on release of mineral N in two of the soils. Although adding lime at the start of the incubation increased nitrification, earlier applications of lime had an even greater effect.

N FIXATION   Symbiotic and nonsymbiotic N_2 fixation is favored by adequate liming (see Chapter 5). Activity of some *Rhizobia* species is greatly restricted by soil pH levels below 6.0, thus liming will increase the growth of legumes because of increased N_2 fixation. With the nonsymbiotic N_2-fixing organisms, N_2 fixation increases in adequately limed soils, which increases the degradation of crop residues.

SOIL PHYSICAL CONDITION   The structure of fine-textured soils may be improved by liming, as a result of an increase in the OM content and of the flocculation of Ca-saturated clay. Favorable effects of lime on soil structure include reduced soil crusting, better emergence of small-seeded crops, and lower power requirements for tillage operations. However, the overliming of Oxisols and Ultisols can result in the deterioration of soil structure, with the consequent decrease in water percolation through such soils. Ca also improves the physical conditions of sodic soils. Increased electrolyte concentration due

TABLE 10.12   Mineralization of Organic N and Nitrification in Three Acid Soils Incubated for 4 Weeks with and without Lime*

Soil	Treatment	Organic N Mineralized (ppm)	Percent Nitrification
Site 1 (pH 5.5, 0.20% soil N)	No lime	36	8
	Limed at start of incubation	61	66
	Limed 2 yr before in the field	33	94
Site 2 (pH 5.4, 0.13% soil N)	No lime	40	7
	Limed at start of incubation	72	64
	Limed 1 yr before in the field	44	93
Site 3 (pH 5.7, 0.83% soil N)	No lime	90	28
	Limed at start of incubation	177	83
	Limed 1 yr before in the field	134	94

SOURCE: Nyborg and Hoyt, *Can. J. Soil Sci.*, **58:**331 (1978).

to $CaCO_3$ dissolution is responsible for preventing clay dispersion and decreases in hydraulic conductivity of such soils.

DISEASE   Correction of soil acidity by liming may have a significant role in the control of certain plant pathogens. Clubroot is a disease of cole crops that reduces yields and causes the infected roots to enlarge and become distorted. Lime does not directly affect the clubroot organism, but at soil pH greater than 7.0, germination of clubroot spores is inhibited (Table 10.13).

On the other hand, liming will increase the incidence of diseases such as scab in root crops. Severity of take-all infection in wheat, with resultant reductions in yield, is known to be increased by liming soils to near neutral pH.

Application of Liming Materials

Surface applications of lime without some degree of mixing in the soil are not immediately effective in correcting subsoil acidity. In several studies it was observed that 10 to 14 years were required for surface-applied lime without incorporation to raise the soil pH at a depth of 15 cm. For fairly high rates, broadcasting one-half the lime, followed by discing and plowing, and then broadcasting the other half and discing, is a satisfactory method of mixing with the plow layer.

Keeping surface soils at the proper pH over a period of years is a practical way of at least partially overcoming the problem of subsoil acidity (Table 10.14). It is evident that maintenance of surface soil at pH 6.0, 6.5, and 7.2 reduced acidity deeper in the root zone.

Neutralization of subsoil acidity through deep incorporation of surface-applied lime is possible with the tillage equipment now available. The effect of incorporation depth of surface-applied lime on cotton growth showed that the amount and depth of cotton rooting were increased by mixing lime to a depth of 45 cm (Fig. 10.8). Mixing lime even deeper, to a depth of 60 cm, markedly increased corn yields (Fig. 10.9).

TABLE 10.13   Effect of Liming on the Harmful Effects of Clubroot Disease in Cauliflower

Lime (tons/a)	Lime Applied in 1978			Lime Applied in 1979		
	Yield Percent Marketable	Clubroot Rating*	pH at Harvest	Yield Percent Marketable	Clubroot Rating*	pH at Harvest
0	48	3.3	5.6	28	3.8	5.7
2.5	73	1.8	6.6	39	3.8	6.4
5.0	81	1.1	6.9	63	3.4	6.7
10.0	86	0.2	7.1	74	2.5	7.2

*Clubroot rating = $\dfrac{\text{sum of (number of roots at a rating} \times \text{rating)}}{\text{total number of roots}}$.

Rating: 0, no visible clubroot; 1, fewer than 10 galls on the lateral roots; 2, more than 10 galls on the lateral roots, taproot free of clubroot; 3, galls on taproot; 4, severe clubbing on all roots.

SOURCE: Waring, *Proc. 22nd Annu. Lower Mainland Hort. Improvement Assoc. Growers' Short Course*, pp. 95–96. Lower Mainland Horticultural Improvement Association and British Columbia Ministry of Agriculture (1980).

TABLE 10.14   Effect of Maintaining Various Surface pH Levels on Subsoil pH of Wooster Silt Loam Soil

Soil Depth (in.)	pH at Various Depths with Increasing pH in the Surface				
0–7	4.9	5.5	6.0	6.5	7.2
7–14	4.9	5.2	5.9	6.7	7.2
14–21	4.7	4.8	5.2	5.4	6.5

SOURCE: *Ohio Agronomy Guide*, Columbus, Ohio: Cooperative Extension Service, Ohio State Univ., 1985.

FIGURE 10.8   Amount and depth of cotton rooting as affected by depth of lime incorporation. From left to right: unlimed; 0–15 cm (0–6 in.) limed; 0–45 cm (0–18 in.) limed.   *Doss et al., Agron. J., 71:541, 1979.*

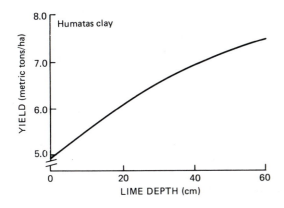

FIGURE 10.9   Effect of depth of lime incorporation on corn yield.   *Bouldin, Cornell Int. Agr. Bull. 74, Ithaca, N.Y.: Cornell Univ., 1979.*

With no-till cropping systems, the surface soil pH can decrease substantially in a few years because of the acidity produced by surface-applied N fertilizers and degradation of crop residues (Table 10.15). Fortunately, the increased acidification is concentrated in the soil surface, where it can be readily corrected by surface liming.

Partial mixing of lime with soil as a means of reducing lime rates may increase soil pH enough to optimize growth (Table 10.16). Maximum yield of wheat dry matter was obtained when lime was confined to 30% of the soil rather than complete mixing.

One of the factors contributing to low yields on acid soils is more difficult weed control (Fig. 10.10). Poor weed control is related to the increased rate of degradation and absorption of simazine at the lower soil pH values. On high-pH soils, residual effects of some herbicides may last for years, which can create problems with sensitive crops.

Equipment

Dry lime application by the supplier who hauls lime to the farm is most efficient. The spinner truck spreader, which throws the lime in a semicircle from the rear of the truck, is often used. Uniform spreading is more difficult with this equipment than with the kind that drops the lime from a covered hopper or conveyor.

Suspending lime in water, frequently referred to as *fluid lime*, is another approach to lime application. Mixing equipment used for conventional suspension fertilizers can be readily utilized to produce fluid lime. Very finely ground lime (100% passing a 100-mesh sieve and 80–90% passing a 200-mesh sieve) with 50% H_2O, along with a suspending agent such as attapulgite clay, may be applied with a fluid fertilizer applicator. Some of the features of fluid lime are as follows:

1. An excellent distribution pattern can be obtained.
2. The finely divided lime reacts rapidly with soil.
3. Only a small amount of liming material is applied at any one time (e.g., 500 to 1000 lb of material per acre).
4. Very-low-pH soils can be corrected quickly.

TABLE 10.15   Soil pH After 7 Years of Continuous Corn Grown on Maury Silt Loam Soil in Kentucky Affected by Tillage Methods, N Fertilization, and Liming

N Treatment	Soil Depth (cm)	Conventional Tillage		No-Tillage	
		Limed	Unlimed	Limed	Unlimed
High N rate	0–5	5.3	4.9	5.5	4.3
(336 kg/ha)	5–15	5.9	5.1	5.3	4.8
	15–30	6.0	5.5	5.8	5.5
Moderate N rate	0–5	5.9	5.2	5.9	4.8
(168 kg/ha)	5–15	6.3	5.6	5.9	5.5
	15–30	6.2	5.7	6.0	5.9

SOURCE: Blevins et al., *Agron. J.*, **40**:322 (1978).

TABLE 10.16   Effect of Mixing Lime with Varying Soil
Volumes on the Growth of Nugaines Winter Wheat in
the Growth Chamber (g/box)

Rate of Lime (tons/ha)	Percent of Soil Limed				
	0	*10*	*30*	*60*	*100*
0	1.27	—	—	—	—
2.24	—	1.95	2.42	2.03	1.83
6.72	—	1.85	3.10	3.02	2.60
11.20	—	1.72	3.50	2.90	2.60

SOURCE: Kauffman and Gardner, *Agron. J.*, **70**:331 (1978).

5. Regular annual applications help to maintain pH.
6. The cost of fluid lime is usually two to four times higher than that of dry
   applications.
7. Fluid fertilizer dealers are supplying fluid lime in areas where conventional
   lime has been difficult to obtain.

Urea-ammonium nitrate (UAN) solutions can be used for suspending the
lime component. Incorporation soon after application of the suspension will
eliminate potential NH_3 loss following urea hydrolysis. N-K-lime suspensions

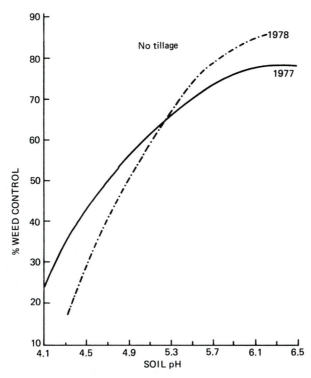

FIGURE 10.10   Effect of pH on the Maury silt loam on
weed control in corn.   *Kells et al.*, Agronomy Notes,
12(2), Univ. of Kentucky, 1979.

are being used successfully. The feasibility of including P sources in lime suspensions has yet to be established. Some lime-herbicide suspensions also may be used.

Regardless of the method employed, care should be taken to ensure uniform application. Nonuniform distribution can result in excesses and deficiencies in different parts of the same field and corresponding nonuniform crop growth.

Factors Determining the Selection of a Liming Program

Intended Crop

Plants differ widely in their sensitivity to soil acidity and thus to added lime (Table 10.4). The type of crop to be grown is the most important factor to consider in developing a lime program.

Soil Texture and OM Content

In a coarse-textured, low-OM soil, the lime requirement will be less than for a fine-textured or high-OM soil (Fig. 10.3). The overliming of coarse-textured soils is not uncommon, but a knowledge of basic soil chemistry can prevent it.

Time and Frequency of Liming Applications

For rotations that include leguminous crops, lime should be applied 3 to 6 months before the time of seeding; this is particularly important on very acid soils. Lime may not have adequate time to react with the soil if applied just before seeding. If clover is to follow fall-seeded wheat, lime is best applied when the wheat is planted. The caustic forms of lime [CaO and $Ca(OH)_2$] should be spread well before planting to prevent injury to germinating seeds.

The frequency of application generally depends on the texture of the soil, N source and rate, crop removal, precipitation patterns, and lime rate. On sandy soils, frequent light applications are preferable, whereas on fine-textured soils, larger amounts may be applied less often. Finely divided lime reacts more quickly, but its effect is maintained over a shorter period than that of coarse materials.

The most satisfactory means of determining reliming needs is by soil tests. Samples should be taken every 3 years.

Depth of Tillage

Lime recommendations are made on the basis of a 6-in. furrow slice. When land is tilled to a depth of 10 in., the lime recommendations should be increased by 50%.

Acidulating the Soil

Acidification may be needed when land is inherently high in carbonates, as in semiarid and arid regions. Land leveling to facilitate irrigation and for other purposes often exposes calcareous and high-pH subsoils that are unfavorable for optimum plant growth. Problems of high soil pH are not confined to arid and semiarid areas. Acidifying paddy soil has increased rice yields, which is often related to increased availability of micronutrients. Farmers in humid regions may overlime or dust from limestone-graveled roads may blow onto field borders, causing localized and excessively high pH. In other areas, moderately acid soils may need further acidification for optimum production of potatoes, blueberries, cranberries, azaleas, rhododendrons, camellias, or conifer seedlings.

The fundamental chemistry of soil acidification is the same as that of liming soils, and several acidic or acid-forming materials can be used.

Elemental S^0

Elemental S^0 is the most effective of the soil acidulents (see Chapter 8). When S^0 is applied to the soil, the following generalized reaction occurs:

$$S^0 + H_2O + 3/2\ O_2 \rightleftharpoons 2H^+ + SO_4^{2-}$$

For every mole of S^0 applied and oxidized, 2 moles of H^+ are produced, which decreases soil pH. In calculating the amount to apply to the soil, reference must be made to the buffer curve of that soil (see Chapter 4).

Finely ground S^0 should be broadcast and incorporated several weeks before planting the crop because the microbial oxidation reaction may be slow, particularly in cold, alkaline soils. Under some conditions, it may be advisable to acidulate a zone near the plant roots to increase water penetration or P and micronutrient availability. Both of these conditions frequently need to be corrected on saline-alkali soils. Elemental S^0 can be applied in bands either as dry ground S^0 or as S^0 suspensions. When S^0 is applied in a band, much smaller amounts are required than with broadcast applications (see Chapter 8 for a discussion of S^0 management).

Soil-applied SO_2 also has been shown to increase nutrient uptake by sorghum (Figure 10.11). In this study, only 25 to 50% neutralization of the basicity was necessary to produce a substantial improvement in nutrient uptake.

Sulfuric Acid

Sulfuric acid (H_2SO_4) has been used for reclaiming Na- or B-affected soils, increasing availability of P and micronutrients, reducing NH_3 volatilization, increasing water penetration, controlling certain weeds and soil-borne pathogens, and enhancing the establishment of range grasses. The favorable influence of H_2SO_4 and other treatments on sorghum yield (Table 10.17) and on rice yield (Table 10.18) is partially related to increased nutrient availability.

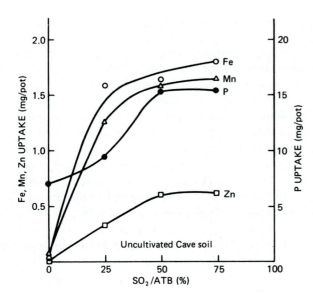

FIGURE 10.11   Effect of SO_2 treatment as percentage of acid-titratable basicity (ATB) on uptake by sorghum of P, Fe, Mn, and Zn. Growth period of approximately 2½ months on an uncultivated calcareous Cave soil.   *Miyamoto et al.*, Sulphur Inst. J., *10(2):14, 1974.*

H_2SO_4 can be added directly to the soil, but it is unpleasant to work with and requires the use of special acid-resistant equipment. It can be dribbled on the surface or applied with a knife applicator, similar to anhydrous NH_3. It can also be applied in ditch irrigation water. H_2SO_4 has the advantage of reacting instantaneously with the soil.

Aluminum Sulfate

Aluminum sulfate $[Al_2(SO_4)_3]$ is a popular material among floriculturists for acidulating the soil for production of azaleas, camellias, and similar acid-tolerant ornamentals. When this material is added to water, it hydrolyzes to produce a very acid solution:

$$Al_2(SO_4)_3 + 6H_2O \rightleftharpoons 2Al(OH)_3 + 6H^+ + 3SO_4^{2-}$$

TABLE 10.17   Effects of H_2SO_4 and $FeSO_4$ on Grain Sorghum Yields on a Calcareous Texas Soil

	Fe (kg/ha)		
H_2SO_4	0	112	560
0	434	1,460	2,275
112	605	1,538	2,274
560	2,169	2,429	2,230
5,600	1,885	1,971	1,810

SOURCE: Mather. In Beaton et al., *Fertilizer Technology and Use*, 3rd ed., chap. 11. Madison, Wisc.: Soil Science Society of America, 1985.

TABLE 10.18 Effect of S, Sulfuric Acid, and Gypsum Soil Ameliorants on the Rice Grain Yield of Bluebonnet 50 and IR661

| | Mean Yield (tons/ha) of: | |
Soil Ameliorant	Bluebonnet 50	IR661
Control	2.69	5.85
Gypsum	2.70	6.00
S	3.24	6.72
H_2SO_4	3.69	6.96

SOURCE: Chapman, *Australian J. Exp. Agr. Anim. Husb.*, **20**:724 (1980).

When $Al_2(SO_4)_3$ is added to the soil, in addition to hydrolysis in the soil solution, the Al^{3+} replaces any exchangeable H^+ and other cations on the CEC and reduces the pH even further:

$$Al_2(SO_4)_3 + clay \begin{bmatrix} 4H^+ \\ Ca^{2+} \end{bmatrix} \longrightarrow clay \begin{bmatrix} Al^{3+} \\ Al^{3+} \end{bmatrix} + Ca^{2+} + 4H^+ + 3SO_4^{2-}$$

$Al_2(SO_4)_3$ is not widely used in general agriculture. Iron sulfate ($FeSO_4$) is also applied to soils for acidification and behaves similarly to $Al_2(SO_4)_3$.

Ammonium Polysulfide (NH_4S_x)

Liquid NH_4S_x is used to lower soil pH and to increase water penetration in irrigated saline-alkali soils. It can be applied in a band 3 or 4 in. to the side of the seed or it can be metered into the ditch irrigation systems. Band application is more effective in correcting micronutrient deficiencies than application through irrigation water. The polysulfide decomposes into ammonium sulfide and colloidal S^0 when applied. The S^0 and S^{2-} are subsequently oxidized to H_2SO_4. Potassium polysulfide was developed for similar purposes.

Acidification in Fertilizer Bands

Because of the high BC of many calcareous and high-pH soils, it is usually too expensive to use enough acidifying material for complete neutralization of soil alkalinity. It is unnecessary to neutralize the alkalinity of the entire soil mass because soil zones more favorable for root growth and nutrient uptake can be created by confining the acid-forming materials to bands and other localized placement.

Band-applied ammonium thiosulfate and ammonium polyphosphate fertilizer solution acidifies the soil in and near the band, which can increase micronutrient availability (Fig. 10.12).

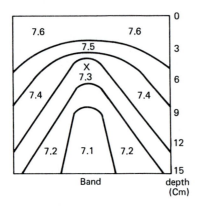

FIGURE 10.12   Application of a P-S fertilizer solution lowers soil pH in the vicinity of fertilizer band (× denotes point of application).   *Leiker, M.Sc. thesis, Kansas State Univ., 1970.*

Saline, Sodic, and Saline/Sodic Soils

In arid and semiarid regions, runoff water collected in depressions evaporates and the salts in the water accumulate. Water also moves upward from artesian sources and shallow water tables. It evaporates, and salts are deposited to form saline, saline-sodic, or sodic soils. These soils are widespread in semiarid and arid regions where the rainfall is not sufficient for adequate leaching, usually less than 15 in./yr. They are particularly prevalent in irrigated areas where improper irrigation and drainage methods are used. The salt marshes of the temperate zones, the mangrove swamps of the subtropics and tropics, and the interior salt marshes adjacent to salt lakes are areas where such soils are found.

There are firm indications that more land is going out of irrigation due to salinity than is being developed with new irrigation. Large areas of the Indian subcontinent have been rendered unproductive by salt accumulation and poor water management. Salinity is a major problem in wetland rice.

Salt buildup is an existing or potential danger on almost all of the irrigated land in semiarid and arid regions of the United States, and salinity on nonirrigated cropland and rangeland in these regions is increasing. The accumulated salts contain the cations Na^+, Ca^{2+}, and Mg^{2+} and the anions Cl^-, SO_4^{2-}, HCO_3^-, and CO_3^{2-}. They can be weathered from minerals and accumulate in areas where the precipitation is too low to provide leaching.

Na is particularly detrimental, both because of its toxic effect on plants and its effect on soil structure. When a high percentage of the CEC is occupied by Na, the soil aggregates disperse. These soils become impermeable to water, develop hard surface crusts, and may keep a water layer or "slick spot" on the surface longer than adjacent soils.

Dispersion problems occur at different exchangeable Na contents. Fine-textured soils with montmorillonitic clays may disperse when about 15% of the exchange complex is saturated with Na. On tropical soils high in Fe and Al oxides and on some kaolinitic soils, 40% Na saturation is required before dispersion is serious. Soils low in clay are also less subject to problems because they are more permeable.

Definitions

SALINE   Saline soils have a saturated *extract conductivity* (EC_{se}) of >4 mmhos/cm, pH <8.5, and have <15% exchangeable Na% (ESP) (Table 10.19). These

TABLE 10.19   Summary of Salt-Affected Soil Classification

Classification	Conductivity (mmhos/cm)	Soil pH	Exchangeable Sodium Percentage	Soil Physical Condition
Saline	>4.0	<8.5	<15	Normal
Sodic	<4.0	>8.5	>15	Poor
Saline/sodic	>4.0	<8.5	>15	Normal

were formerly called *white alkali* because of the deposits of salts on the surface following evaporation. The excess salts, mostly Cl^- and SO_4^{2-} salts of Na^+, Ca^{2+}, and Mg^{2+}, can be leached out, with no appreciable rise in pH. The concentration of soluble salts is sufficient to interfere with plant growth, although salt tolerance varies with plant species.

SODIC   Sodic soils occur when ESP >15%, EC_{se} <4 mmhos/cm, and pH >8.5 (Table 10.19). They were formerly called *black alkali* due to the dissolved organic matter deposited on the surface along with the salts. In sodic soils, the excess Na disperses the soil colloids and the Na creates nutritional disorders in most plants.

SALINE/SODIC   These soils have both the salt concentration (>4 mmhos/cm) to qualify as saline and high exchangeable Na (>15% ESP) to qualify as sodic; however, soil pH is <8.5. In contrast to saline soils, when the salts are leached out, the exchangeable Na hydrolyzes and the pH increases, which results in a sodic soil.

Relationships

Several parameters are commonly used to quantify salt- and Na-affected soils, and many of these parameters are related to each other. By measuring the EC_{se} of a soil, the total quantity of salts in the soil solution can be estimated as follows:

$$EC_{se} \times 10 = \text{total soluble cations (meq/l)}$$

If the soluble cations are measured in the saturated extract, the *sodium adsorption ratio* (SAR) can be calculated as follows:

$$SAR = \frac{Na^+}{\sqrt{\dfrac{(Ca^{2+} + Mg^{2+})}{2}}} \quad \text{(all units in meq/l)}$$

Because of the equilibrium relationships between solution and exchangeable cations in soils, the SAR should be related to the quantity of Na^+ on the CEC, which is expressed as the *exchangeable sodium ratio* (ESR). The ESR is defined as follows:

$$ESR = \frac{\text{exchangeable } Na^+}{\text{exchangeable } (Ca^{2+} + Mg^{2+})} \quad \text{(all units in meq/100 g)}$$

Figure 10.13 illustrates the relationship between solution and exchangeable cations in salt-affected soils. This relationship can be used to estimate the ESR if the quantity of exchangeable cations has not been measured. The following equation represents the linear relationship shown in Figure 10.13:

$$ESR = 0.015(SAR)$$

Subsequently, the ESR is related to the *exchangeable sodium percent* (ESP) previously used to classify Na-affected soils (Table 10.19) and is given by

$$ESP = \frac{100(ESR)}{1 + ESR}$$

These parameters and the relationships between them are extremely valuable in characterizing the solution and exchange chemistry of salt- and Na-affected soils.

Effects on Plant Growth

Growth inhibition in salt-sensitive crops, even at low salinity levels, is caused primarily by toxicity from Na^+ and Cl^-. The high osmotic pressure in the soil solution causes a correspondingly low soil water potential, and when in contact with a plant cell, the solute moves toward the soil solution and the cell collapses (called *plasmolysis*).

Plants differ greatly in their tolerance of saline soils and depend on the stage of growth (Table 10.20). For example, old alfalfa is more tolerant than young alfalfa. Barley and cotton have considerable salt tolerance, but high salt will affect vegetative growth more than grain or bolls of cotton. Cultivar or variety differences also exist. For example, soybean varieties differ in Cl^- exclusion (Table 10.21). Effective excluders of Na^+ and Cl^- may not be very productive because of salt-related water stress. Tolerant crops that do not exclude Na^+ have a capacity to maintain a high K^+/Na^+ ratio in the growing tissue. Conventional breeding and genetic engineering methods are being used to adapt crops to saline environments.

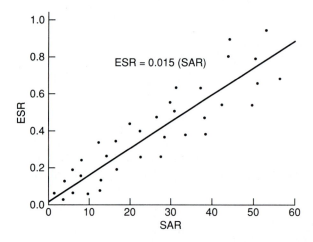

FIGURE 10.13   Relationship between ESR and SAR in salt-affected soils.

TABLE 10.20   Relative Tolerance of Certain Plants to Salty Soils*

Tolerant 8–12 mmhos/cm	Moderately Tolerant 6–8 mmhos/cm	Moderately Sensitive 4–6 mmhos/cm	Sensitive 0–4 mmhos/cm
Barley, grain	Barley, forage	Alfalfa	Apple
Bermuda grass	Beet, garden	Broad bean	Apricot
Bougainvillea	Broccoli	Cabbage	Bean
Cotton	Brome grass	Cauliflower	Blackberry
Date	Clover, berseem	Clover: alsike, Ladino,	Carrot
Mutall alkali grass	Fig	red, strawberry	Celery
Natal plum	Oats	Corn	Grapefruit
Rescue grass	Orchard grass	Cowpea	Lemon
Rosemary	Rye, hay	Cucumber	Onion
Salt grass	Rye grass, perennial	Lettuce	Orange
Sugar beet	Sorghum	Pea	Peach
Wheat grass, crested	Soybean	Peanut	Pear
Wheat grass, fairway	Sudan grass	Potato	Pineapple, guava
Wheat grass, tall	Trefoil, birdsfoot	Rice, paddy	Raspberry
Wild rye, altai	Wheat	Sweet clover	Rose
Wild rye, Russian	Wheat grass, western	Timothy	Strawberry
		Tomato	

*Selected from Carter (1981).

Reclaiming Saline and Sodic Soils for Crop Production

Saline soils are relatively easy to reclaim if adequate amounts of low-salt irrigation waters are available and internal and surface drainage are feasible. Salts must be leached below the root zone and out of contact with subsequent irrigation water.

The quantity of irrigation water needed to leach the salts out of the root zone, or the *leaching requirement* (LR), can be estimated by the following relationship:

$$LR = \frac{EC_{iw}}{EC_{dw}}$$

where LR = leaching requirement
EC_{iw} = EC of irrigation water
EC_{dw} = EC of drainage water

TABLE 10.21   Leaf Scorch Ratings, Yield, and Cl⁻ Concentration in Leaves and Seed of 5 Susceptible and 10 Tolerant Soybean Cultures

	Leaf Scorch*	Yield (bu/a)	Concentration of Cl⁻ Leaves (%)	Seed (ppm)
Cl susceptible	3.4	15	1.67	682
Cl tolerant	1.0	24	0.09	111

*1 = none, 5 = severe (Parker et al., 1986).

LR represents the additional water needed to leach out the salts over that needed to saturate the profile. Although this relationship provides an estimate of the water volume needed reduce the salts in the soil, more sophisticated calculations are generally used to estimate precisely the leaching water needed. The amount of leaching water required will depend on (1) the desired EC_{se}, which depends on the salt tolerance of the intended crop, (2) irrigation water quality (EC_{iw}), (3) rooting or leaching depth, and (4) soil water-holding capacity.

In soils with a high water table, drain installation may be required before leaching. If there is a dense calcareous or gypsiferous layer or the soil is impervious, deep chiseling or plowing may be used to improve infiltration. When only rainfall or limited irrigation is available, surface organic mulches will reduce evaporation and increase drainage.

In sodic and saline-sodic soils, exchangeable Na and/or EC_{se} must be reduced, which can be difficult because the soil clay may be dispersed, preventing infiltration. Exchange of Na^+ is most often accomplished with Ca^{2+} by adding the appropriate rate of gypsum ($CaSO_4 \cdot 2H_2O$). The reaction is

$$\text{clay} \begin{bmatrix} \\ \\ \\ \\ \\ \end{bmatrix} \begin{matrix} Na^+ \\ Na^+ \\ Ca^{2+} \\ Mg^{2+} \\ Na^+ \end{matrix} \quad + Ca^{2+} + SO_4^{2-} \longrightarrow \quad \text{clay} \begin{bmatrix} \\ \\ \\ \\ \end{bmatrix} \begin{matrix} Ca^{2+} \\ Ca^{2+} \\ Mg^{2+} \\ Na^+ \end{matrix} \quad + 2Na^+ \\ \text{leach below} \\ \text{root zone}$$

Estimating the quantity of $CaSO_4 \cdot 2H_2O$ required is similar to the calculation for estimating the $CaCO_3$ required to increase pH. For example, a soil with CEC = 20 meq/100 g contains 15% ESP, and we need to reduce the ESP to 5%; thus, 15% − 5% = 10% reduction in ESP.

$$(0.10) \left(20 \, \frac{\text{meq CEC}}{100 \text{ g}} \right) = 2 \, \frac{\text{meq Na}^+}{100 \text{ g}} = 2 \, \frac{\text{meq CaSO}_4 \cdot 2H_2O}{100 \text{ g}}$$

Thus,

$$2 \, \frac{\text{meq CaSO}_4 \cdot 2H_2O}{100 \text{ g}} \left(86 \, \frac{\text{mg CaSO}_4 \cdot 2H_2O}{\text{meq}} \right) \quad (20)$$

$$= \frac{3440 \text{ lb CaSO}_4 \cdot 2H_2O}{\text{a-6 in.}}$$

Managing Saline Soils

Managing the soil to minimize salt accumulation is essential, especially in semiarid and arid regions. Maintaining the soil near field capacity with frequent watering dilutes salts. Light leaching before planting or light irrigation after planting moves salts below the planting and early rooting zone. If water is available, periodic leaching when crops are not growing will move salts out of the root zone. Much of the salt may precipitate as $CaSO_4 \cdot 2H_2O$ and $CaCO_3$ or $MgCO_3$ during dry periods and will not react as soluble salts, although precipitation of Ca and Mg will increase the proportion of Na^+ present in solution.

Managing soils for improved drainage is essential for controlling soil salinity. When ridge-tillage systems are used, the salt moves upward with capillary H_2O and is deposited on the center of the ridges where the water evaporates. Planting on the shoulders or edge of the ridges helps to avoid problems associated with excess salts.

Summary

1. An acid is a substance that tends to give up protons (hydrogen ions) to some other substance. A base is a substance that tends to accept protons.

2. Acids range from strong to weak. Those that are strong dissociate to a great degree, whereas those that are weak dissociate only slightly. The strength of acids and the methods of expressing it were discussed. The pH concept, which is a convenient means of describing the activity of weak acids, was explained. Neutralization of acids was covered and illustrated with appropriate equations, and the principle of buffering and buffered systems was explained.

3. In many ways, soils are similar to buffered weak acids. This and other current ideas on the fundamental nature of soil acidity were discussed. Soil acidity is affected by the nature and amount of humus and clay colloids present, the amount of hydrous oxides of Al and Fe (especially Al), the content of soluble salts in the soil, and the level of CO_2 in the soil atmosphere. The importance of Al in soil acidity was stressed.

4. Soil acidity in crop production systems is influenced by the use of commercial fertilizers, especially ammoniacal sources, which produce H^+ ions during nitrification; crop removal and leaching of basic cations; and decomposition of organic residues.

5. Some fertilizer materials leave an acidic residue in the soil, others a basic residue, and still others have no influence on soil pH. The effect of various fertilizer materials on soil pH, and the mechanisms responsible for their action were covered.

6. Plant growth reduces the potential or theoretical amount of acidity produced by nitrification of ammoniacal fertilizer materials since plants absorb NH_4-N.

7. Soil pH, or active acidity, is registered by a pH meter on a sample of soil and water. Potential acidity is determined by adding known amounts of base to a series of soil samples and measuring the pH change. The curve plotted from such data is termed a *buffer curve* and is used in calculating the lime requirement of field soils. The lime requirement is generally measured using various strongly buffered solutions.

8. Soil pH determinations are helpful for indicating the presence of exchangeable Al^{3+} and H^+. Exchangeable H^+ is normally present in measurable quantities at pH values below 4, while exchangeable Al^{3+} usually occurs in

significant amounts at pH levels below about 5.5. Multicharged Al polymers are dominant in the pH range 5.5 to 7.0.

9. Several materials, including dolomitic and calcitic limestones, burnt lime, hydrated lime, marl, and slags, are employed commercially in the liming of soils. Their properties and reactions in soil were discussed.

10. The neutralizing value, or CCE, of limestones is a measure of their effectiveness in neutralizing soil acidity. Several methods of determining this property were described.

11. Lime is one of the most important production inputs in the farming system. Its effect on P and microelement availability, nitrification, fixation, soil structure, and disease influences crop production in many ways.

12. Movement and placement of lime were discussed. Developments in liming methods such as fluid lime were described.

13. The selection of a liming program to be followed is determined by the lime requirements of the crop; the pH, texture, and OM content of the soil; the liming material to be used; and the time and frequency of the lime applications.

14. Acidulation of the soil is sometimes necessary. Various S compounds are used, with elemental S^0 being the most effective. The plow layer may be acidified, but sometimes soil zones more favorable for root growth can be created by confining acid-forming materials to bands, furrows, and other localized zones.

15. Saline, sodic, and saline/sodic soils occur in arid and semiarid regions where water, from shallow water tables, artesian sources, and seepage and runoff, evaporates leaving salts. Such soils are common in irrigated areas where drainage is not adequate. The salts are composed largely of the cations Na^+ Ca^{2+}, and Mg^{2+}, and the anions Cl^-, SO_4^{2-}, HCO_3^-, and CO_3^{2-}.

16. Saline soils have a saturation extract conductivity of 4 mmhos/cm or greater and low exchangeable Na. Sodic soils have an exchangeable Na percentage greater than 15% but a low salt content. Saline/sodic soils are a combination of the two.

17. Saline soils can be readily leached, but the Na in sodic soils causes dispersion of the colloids, and permeability is low. Gypsum is widely used for reclaiming saline soils, while S may be employed for sodic soils.

18. Crop species and genotypic differences in salinity tolerance exist and are part of crop management.

Questions

1. What is an acid?
2. How are acids neutralized?
3. What ions are the principal sources of soil acidity?
4. Can pH measurements be used to identify the principal sources of soil acidity? What are the general relationships?
5. For what reasons do soils become acid?
6. What, if any, influence do Al and Fe polymers have on soil acidity? Do they influence other soil properties? Which ones?
7. Is soil pH affected by fertilizer applications?

8. Are there differences in crop uptake of basic cations?

9. The term *agricultural* lime usually refers to what material?

10. Chemically speaking, lime refers to what compound?

11. Distinguish between active acidity and potential acidity. Which of these two forms is measured when a pH determination is made?

12. What is meant by the terms *buffer* and *buffer capacity* (BC)?

13. What soil properties determine its BC?

14. How is the BC of a soil related to the lime requirement of that soil?

15. Define the term *lime requirement*.

16. How is the lime requirement of a soil determined?

17. Calcium chloride ($CaCl_2$) contains 53% Ca. Express this as an equivalent percentage of calcium oxide. Can calcium chloride be used as a liming agent? Why?

18. What three types of slag can be used as effective liming materials?

19. What are marls? What determines their value as liming materials?

20. Define *neutralizing value* or *calcium carbonate equivalent* as it refers to liming materials.

21. What is the calcium carbonate equivalent of sodium carbonate?

22. You analyze a limestone and find that it has a neutralizing value of 85%. How many tons of this limestone would be equivalent to 3 tons of chemically pure calcium carbonate?

23. In addition to its purity and neutralizing value, what other property of crystalline limestones is important with respect to their value as agricultural liming materials?

24. What are the principal direct benefits of adding lime to soil?

25. What are several indirect benefits of adding lime to soil?

26. A solution has a pH value of 6.5. To what hydrogen ion activity does this correspond?

27. Does lime move quickly in soil?

28. Are benefits derived from deep mixing of lime in soil?

29. Can long-term liming of surface soil influence subsoil acidity?

30. What are several factors that will determine the frequency and rate of liming?

31. You have two fields, A and B, that need liming. The characteristics of the soils in each of these fields are the following:

	Field	
Soil Property	A	B
Organic matter (%)	0.8	3.1
Clay (%)	10.0	38.1
Sand (%)	74.0	51.2
Type of clay	1:1	1:2
pH	5.2	5.2

You have lost the liming recommendations sent to you by the soil laboratory, but you do recall that 3 tons/a were recommended for field B. Because the pH is the same in both fields, you apply 3 tons to field A as well. Have you acted wisely? Why?

32. Solution A has a pH of 3.0. Solution B has a pH of 6.0. The active acidity of solution A is how many times greater than that of solution B?

33. What is fluid lime? What are its major advantages and disadvantages?

34. Assume that you have three dolomitic limestones of equal neutralizing value but with the following mechanical analyses:

	Limestone		
	A	B	C
Coarser than 8 mesh	20	5	0
Coarser than 60 mesh	70	30	20
Coarser than 100 mesh	95	60	50

(a) Which limestone would you not buy? (b) Which one would you select for the quickest results? (c) Under what circumstances could you afford to buy limestone B?

35. In what parts of the United States and Canada is liming most needed and a very important practice? Where is it relatively unimportant? In your own particular area, is liming a needed practice? Why?

36. Define the term *effective CEC*. How and why does it differ from the CEC as determined by the ammonium acetate method?

37. With what type of soil would the ammonium acetate method give a fairly good approximation of the effective CEC of soil? On what types of soil would it not give a good estimate? Would the estimate be high or low? Why? What method would give a better estimate of the CEC of such soils?

38. Explain how S acidulates a soil.

39. What are the main cations in saline, sodic, and saline/sodic soils?

40. Explain how sodic soils become rather impermeable.

41. Explain why $CaSO_4 \cdot 2H_2O$ is effective in reclaiming saline soils.

Selected References

ADAMS, F. (Ed.). 1984. *Soil Acidity and Liming*. Soil Science Society of America, Madison, Wisc.

ALLEY, M. M., and L. W. ZELAZNY. 1987. Soil acidity: soil pH and lime needs. *In* J. R. Brown (Ed.), *Soil Testing: Sampling, Correlation, Calibration, and Interpretation*. Special Publication No. 21. Soil Science Society of America, Madison, Wisc.

FOLLETT, R. H., L. S. MURPHY, and R. L. DONAHUE. 1981. *Fertilizers and Soil Amendments*. Prentice-Hall, Inc., Englewood Cliffs, N.J.

KAMPRATH, E. J., and C. D. FOX. 1985. Lime-fertilizer-plant interactions in acid soils. *In* O. P. Engelstad (Ed.), *Fertilizer Technology and Use*. Soil Science Society of America, Madison, Wisc.

MCLEAN, E. O. 1973. Testing soils for pH and lime requirement. *In* L. M. Walsh and J. D. Beaton (Eds.), *Soil Testing and Plant Analysis*. Soil Science Society of America, Madison, Wisc.

CHAPTER 11

Soil Fertility Evaluation

Optimum productivity of any cropping systems depends on an adequate supply of plant nutrients. Although one or more nutrients are commonly applied to most crops, the quantity of nutrients removed in the harvested crop is generally much greater than the quantity added. Continued removal of nutrients, with little or no replacement, will increase the potential for future nutrient-related plant stress and yield loss.

When the soil does not supply sufficient nutrients for normal plant development and optimum productivity, application of supplemental nutrients is required. The proper rate of plant nutrients is determined by knowing the nutrient requirement of the crop and the nutrient-supplying power of the soil. Diagnostic techniques, including identification of deficiency symptoms as well as soil and plant tests, are helpful in determining specific nutrient stresses and the quantity of nutrients needed to eliminate the stress. By the time a plant has shown deficiency symptoms, a considerable reduction in yield potential will already have occurred; thus, the analysis of the nutrient-supplying power or capacity of the soil is essential for quantifying the probability of a crop response to nutrient additions.

The value of soil and plant analysis in quantifying nutrient requirements depends on careful sampling and analysis, and using tests that are calibrated or correlated with plant response. Knowledge of the relationship between test results and crop response is essential for providing the most appropriate nutrient recommendation. Several techniques are commonly employed to assess the fertility status of a soil:

1. Nutrient-deficiency symptoms of plants.
2. Analysis of tissue from plants growing on the soil.
3. Biological tests in which the growth of either higher plants or certain microorganisms is used as a measure of soil fertility.
4. Soil analysis.

Nutrient-Deficiency Symptoms of Plants

Growing plants act as integrators of all growth factors (Fig. 11.1) and are the products in which the grower is interested. Therefore, careful inspection of the growing plant can help identify a specific nutrient stress. If a plant is

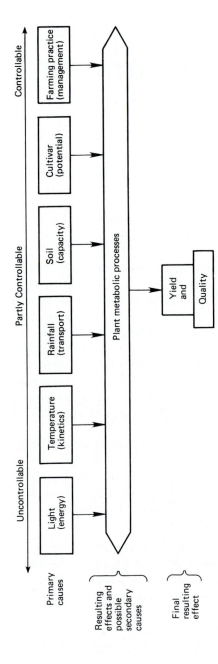

FIGURE 11.1 Schematic representation of the interrelationships between crop yield and quality, metabolic process, and external and genetic factors. *Beaufils, Soil Sci. Bull. 1, Univ. of Natal, Pietermaritzburg, South Africa, 1973.*

lacking in a particular nutrient, characteristic symptoms may appear. Deficiency of a nutrient does not directly produce symptoms. Rather, the normal plant processes are thrown out of balance, with an accumulation of certain intermediate organic compounds and a shortage of others. This leads to the abnormal conditions recognized as symptoms. Visual evaluation of nutrient stress should be used only as a supplement to other diagnostic techniques (i.e., soil and plant analysis). Nutrient-deficiency symptoms may be classified as follows:

1. Complete crop failure at the seedling stage.
2. Severe stunting of plants.
3. Specific leaf symptoms appearing at varying times during the season.
4. Internal abnormalities such as clogged conductive tissues.
5. Delayed or abnormal maturity.
6. Obvious yield differences, with or without leaf symptoms.
7. Poor quality of crops, including differences in protein, oil, or starch content, and storage quality.
8. Yield differences detected only by careful experimental work.

In addition, nutrient deficiencies have a marked effect on the extent and type of root growth (Fig. 11.2). Plant roots have not received much attention because of the difficulty of making observations; however, considering that roots absorb most of the nutrients needed by the plant, inspection of root growth can be an important diagnostic tool.

Each symptom must be related to some function of the nutrient in the plant. A given nutrient may have several functions, which makes it difficult to explain the physiological reason for a particular deficiency symptom. For example,

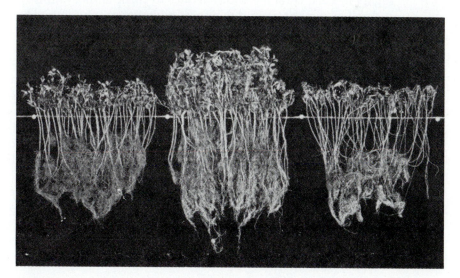

FIGURE 11.2   Omitting P *(left)* or K *(right)* reduced the growth of alfalfa roots as well as tops the spring after seeding. There is evidence of alfalfa heaving above the groundline marked by the string. This soil tested low in P and K.   *Courtesy of the Potash & Phosphate Institute, Atlanta, Ga.*

when N is deficient, the leaves of most plants become pale green or light yellow. When the quantity of N is limiting, chlorophyll production is reduced, and the yellow pigments, carotene and xanthophyll, show through. A number of nutrient deficiencies produce pale green or yellow leaves, and the difficulty must be further related to a particular leaf pattern or location on the plant.

Although discussed in Chapter 3, apparent visual deficiency symptoms can be caused by many factors other than a specific nutrient stress. Precautions in interpreting nutrient-deficiency symptoms include the following:

1. The visual symptom may be caused by more than one nutrient. For example, N-deficiency symptoms may be identified, although S may also be deficient and its symptoms may not be readily apparent. B deficiency is accompanied by a red coloration of the leaves near the growing point when the plant is well supplied with K. On the other hand, when the K content is low, yellowing of alfalfa leaves occurs.
2. Deficiencies are actually relative, and a deficiency of one nutrient may be related to an excessive quantity of another. For example, Mn deficiency may be induced by adding large quantities of Fe, provided that soil Mn is marginally deficient. Also, at a low level of P supply, the plant may not require as much N compared to normal or adequate P. In other words, once the first limiting factor is eliminated, the second limiting factor will appear (Liebig's law of the minimum).
3. It is often difficult to distinguish among the deficiency symptoms in the field, as disease or insect damage can resemble certain micronutrient deficiencies. For example, leaf hopper damage can be confused with B deficiency in alfalfa.
4. A visual symptom may be caused by more than one factor. For example, sugars in corn combine with flavones to form anthocyanins (purple, red, and yellow pigments), and their accumulation may be caused by an insufficient supply of P, low soil temperature, insect damage to the roots, or N deficiency.

Nutrient-deficiency symptoms appear only after the nutrient supply is so low that the plant can no longer function properly. In such cases, it would have been profitable to have applied fertilizer long before the symptoms appeared. If the symptom is observed early, it might be corrected during the growing season. Since the objective is to get the limiting nutrient into the plant as quickly as possible, with some nutrients and under some conditions this may be accomplished with foliar applications or side dressings. Usually the yield is reduced below the quantity that would have been obtained if adequate nutrients had been available at the beginning. However, if the problem is properly diagnosed, the deficiency can be corrected the following year.

Hidden Hunger

Hidden hunger refers to a situation in which a crop needs more of a given nutrient yet has shown no deficiency symptoms (Fig. 11.3). The nutrient content is above the deficiency symptom zone but still considerably below that

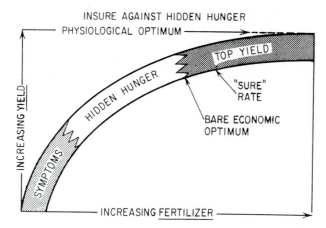

FIGURE 11.3   *Hidden hunger* is a term used to describe a plant that shows no obvious symptoms, yet the nutrient content is not sufficient to give the top profitable yield. Fertilization with the "sure" rate rather than the bare economic optimum for an average year helps to obtain the top profitable yield.   *Courtesy of the Potash & Phosphate Institute, Atlanta, Ga.*

needed for optimum crop production. With most nutrients on most crops, significant responses can be obtained even though no recognizable symptoms have appeared.

The question, then, is how best to eliminate hidden hunger (Fig. 11.4). Testing of plants and soils is helpful for planning or modifying plant-nutrient programs to avoid this problem in subsequent crops. In both approaches, careful consideration must be given to past management practices.

Seasonal Effects

Nutrient shortages in the soil may be caused by abnormal weather conditions. Nutrients may be present in sufficient quantities when conditions are ideal, but in drought, excessive moisture, or unusual temperature conditions the

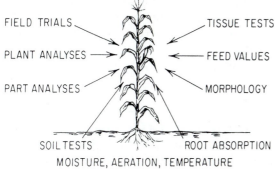

FIGURE 11.4   Detecting hidden hunger in crops is an increasing problem as yield goals rise. With no symptoms to guide us, we must turn to other diagnostic tests to evaluate needs more accurately.   *Courtesy of the Potash & Phosphate Institute, Atlanta, Ga.*

TABLE 11.1 Influence of Applied N, P, and K and
Moisture Stress on Percent N, P, and K in Corn Leaves

Nutrients Applied (kg/ha)			N-P-K Concentration	
N	P	K	No Stress Days	Maximum Stress
			%N	
0	78	47	2.0	1.5
179	78	47	2.9	2.2
			%P	
179	0	47	0.26	0.12
179	78	47	0.32	0.18
			%K	
179	39	0	1.1	0.7
179	39	93	1.6	1.2

SOURCE: Voss, *Proc. 22nd Annu. Fert. Agr. Chem. Dealers' Conf.*,
Iowa State Univ. (1970).

plant may not be able to obtain an adequate supply. For example, with cooler temperatures, nutrient uptake is generally reduced because

1. Mass flow of nutrients is reduced by decreased growth rate and transpiration.
2. Nutrient diffusion rate decreases with declining temperature and a lower concentration gradient.
3. Mineralization of nutrients complexed with OM is reduced.

Likewise, as moisture stress increases, the concentrations of N-P-K in leaves decreases (Table 11.1). Application of these nutrients reduces the effects of moisture stress, but concentrations are still below the optimum in stress years.

Nutrient-deficiency symptoms appearing during early growth may disappear as the growing season progresses, or there may be no measurable yield benefit from supplemental additions of the nutrient(s) in question. For example, fertilizer P may improve the early growth of crops, but at harvest there may be no measurable yield response. Such occurrences are probably related to some of these seasonal effects or to penetration of roots into areas of the soil having higher fertility levels. Disappearance of deficiency symptoms could also be due to slower growth rate, and, thus, less demand for nutrients.

To eliminate plant nutrients as a limiting factor, the nutrient content of the plant must be raised to a level that takes advantage of a good season and prevents climate-related nutrient stress.

Plant Analyses

Two general types of plant analysis have been used. One is the tissue test on fresh tissue in the field, and the other is a tissue analysis performed in the laboratory. Plant analyses are based on the premise that the amount of a given nutrient in a plant is directly related to the availability of the nutrient in the soil. Since a shortage of a nutrient will limit growth, other nutrients may accumulate in the cell sap, regardless of their supply. For example, if corn is low in NO_3^-, the P test may be high. This is no indication, however, that if adequate N were supplied to the corn, the supply of P would be adequate.

Tissue tests and plant analyses are made for the following reasons:

1. To aid in determining the nutrient-supplying power of the soil. They are employed in conjunction with soil tests and management history.
2. To help identify deficiency symptoms and to determine nutrient shortages before they appear as symptoms.
3. To aid in determining the effect of fertility treatment on the nutrient supply in the plant.
4. To study the relationship between the nutrient status of the plant and crop performance.

Tissue Tests

Rapid tests for the determination of nutrients in fresh tissue are important in diagnosing the nutrient needs of growing plants. The plant roots absorb the nutrients from the soil, and these nutrients are transported to other parts of the plant where they are needed. The concentration of the nutrients in the cell sap is usually a good indication of how well the plant is supplied *at the time of testing*. These semiquantitative tests are intended mainly for verifying or predicting deficiencies of N, P, K, S, and several micronutrients. Through the proper application of tissue testing, it is possible to anticipate or forecast certain production problems while still in the field.

GENERAL METHODS   In one test the plant parts may be chopped up and extracted with reagents. The intensity of the color developed is compared with standards and used as a measure of the supply of the nutrient. In another, more rapid test, plant tissue is squeezed with pliers to transfer plant sap to filter paper color-developing reagents are added. The resulting color is compared to a standard chart that indicates very low, low, medium, or high nutrient content. Semiquantitative values for the N, P, and K status of a plant can be obtained in about a minute.

Tissue tests are easy to conduct and interpret, and many tests can be made in a few minutes. These methods have an advantage over those of the laboratory. Because laboratory tests require more time for an answer, there is a tendency to guess rather than send samples to the laboratory. It is important to recognize that application of nutrients to correct a nutrient stress identified with a tissue test may not be feasible because (1) the deficiency may have already caused yield loss, (2) the crop may not respond to the applied nutrient at the specific growth stage tested, (3) the crop may be too large to apply nutrients, and (4) climatic conditions may be unfavorable for fertilization and/or for the crop to benefit from nutrient additions.

PLANT PARTS TO BE TESTED   It is essential to test the part of the plant that will give the best indication of the nutritional status. In general, the conductive tissue of the latest mature leaf is used for testing, while immature leaves at the top of the plant are avoided. For certain plants, the specific part recommended for tissue testing is given in Figure 11.5. The best part to use for testing is generally that showing the greatest range of levels, as the nutrient goes from deficient to adequate levels (Fig. 11.6).

TIME OF TESTING   The growth stage of the plant is important in tissue testing, because the nutrient status will change during the season. In general, the most critical growth stage for tissue testing is at bloom or from bloom to the early fruiting stage. During this period, maximum utilization of nutrients

		Sampling Chart	
Plant	Test	Part to sample	(To avoid hidden hunger) Minimum level
Corn			
	NO$_3$	Midrib, basal leaf	High
Under	PO$_4$	Midrib, basal leaf	Medium
15 in.	K	Midrib, basal leaf	High
15 in. to	NO$_3$	Base of stalk	High
ear showing	PO$_4$	Midrib, first mature leaf*	Medium
	K	Midrib, first mature leaf*	High
Ear to very	NO$_3$	Base of stalk	High
early Dent	PO$_4$	Midrib, leaf below ear	Medium
	K	Midrib, leaf below ear	Medium
Soybeans			
	NO$_3$	Not tested	
Early growth to	PO$_4$	Pulvinus (swollen base of	High
midseason		petiole), first mature leaf*	High
	K	Petiole, first mature leaf	High
Midseason to good	PO$_4$	Pulvinus, first mature leaf	Medium
pod development	K	Petiole, first mature leaf	Medium
Cotton			
	NO$_3$	Petiole, basal leaf*	High
To early bloom	PO$_4$	Petiole, basal leaf*	High
	K	Petiole, basal leaf*	High
	NO$_3$	Petiole, first mature leaf*	High
Boll setting	PO$_4$	Petiole, first mature leaf*	High
to 2/3 maturity	K	Petiole, first mature leaf*	High
	NO$_3$	Petiole, first mature leaf*	Medium
2/3 maturity	PO$_4$	Petiole, first mature leaf*	Medium
to maturity	K	Petiole, first mature leaf*	Medium
Alfalfa			
Before 1st	PO$_4$	Middle 1/3 of stem	High
cutting	K	Middle 1/3 of stem	High
Before other	PO$_4$	Middle 1/3 of stem	Medium
cuttings	K	Middle 1/3 of stem	Medium
Small Grains	NO$_3$	Lower stem	High
Shoot stage to	PO$_4$	Lower stem	Medium
milk stage	K	Lower stem	Medium

*First Mature Leaf—Avoid the immature leaves at the top of the plant. Take the most recently fully matured leaf near the top of the plant.

FIGURE 11.5   Part of the plant used for tissue tests. *Wickstrom et al., Better Crops Plant Food, 47(3):18, 1964.*

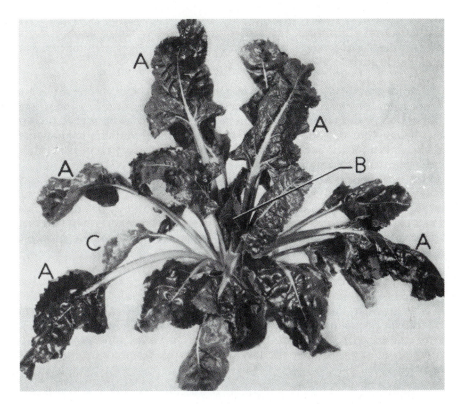

FIGURE 11.6   Selection of sugarbeet leaves for analysis. A leaf stalk from any one of the recently matured, fully expanded leaves marked *A* may be included in the sample. The small leaves in the center or the old leaves should be avoided.
Courtesy of Albert Ulrich, and the Potash & Phosphate Institute, Atlanta, Ga.

occurs and low nutrient levels are likely to be detected. In corn, the leaf opposite and just below the uppermost ear at silking is sampled. At this growth stage, however, it may be difficult to apply nutrients to correct the deficiency. Tissue tests are well suited for forage crops because if deficiencies are found, top-dressings of the required nutrients are effective in later growth.

The time of day can influence the NO_3^- level in plants, for this nutrient is usually higher in the morning than in the afternoon if the supply is short. It accumulates at night and is utilized during the day as carbohydrates are synthesized. Therefore tests should not be made early in the morning or late in the afternoon.

A few points relative to tissue tests are as follows:

1. It is ideal to follow the uptake of nutrients through the season by testing five or six times. Nutrient levels should be higher in the early season when the plant is not under stress.
2. There can be two peak periods of nutrient demand. The first is during maximum vegetative growth, and the second is during the reproductive stage. To determine the adequacy of the fertilization program, these are

the optimum times for tissue testing; however, at the later peak period, it is generally too late for corrective action.
3. Comparison of plants in a field is helpful. Test plants from deficient areas and compare them with plants from normal areas.
4. Plants vary; thus, test 10 to 15 plants and average the results.

INTERPRETATION   The plant diagnostician must be well acquainted with the physiology of the plant being tested. The important factors that are considered before making a decision are as follows:

1. General performance and vigor of the plant.
2. Levels of other nutrients in the plant.
3. Incidence of insects or disease.
4. Soil conditions, such as moisture, aeration, structure, and so on.
5. Climatic conditions.
6. Time of day.
7. Yield goal.

If a plant appears to be discolored or stunted and the tissue tests high for N, P, and K, then some other factor is limiting growth. Only after this condition has been corrected can tissue tests reveal which plant nutrients may be limiting growth. Generally, low to medium test results for N, P, or K in the early part of the growing season mean that the yield will be considerably less than optimum. At bloom stage, a medium to high test result is adequate in most crops.

Total Analysis

Total analysis is performed on the whole plant or on plant parts. After sampling, the plant material is dried, ground, and the nutrient content determined following digesting or ashing of the plant material. With total analysis, the content of all elements, essential and nonessential, can be determined. As in tissue tests, the plant part selected is important, where the most recently matured leaf is preferred (Fig. 11.6). Plant sampling guidelines for selected crops are shown in Table 11.2.

INTERPRETATION   The critical nutrient concentration (CNC) is commonly used in interpreting plant analysis results and diagnosing nutritional problems (Fig. 11.7; also see Fig. 3.1). The CNC is located in that portion of the curve where the plant-nutrient concentration changes from deficient to adequate; therefore, the CNC is the level of a nutrient below which crop yield, quality, or performance is unsatisfactory. For example, CNCs in corn are about 3% N, 0.3% P, and 2% K in the leaf opposite and below the uppermost ear at silking time. For crops such as sugar beets or malting barley, where excessive concentrations of N seriously affect quality, the CNC for N is a maximum rather than a minimum. It is difficult to identify a specific concentration because they vary greatly among crops and regions or climates.

In addition, it is difficult to determine an exact CNC experimentally since considerable variation exists in the points plotted in the transition zone between deficient and adequate nutrient concentrations. Consequently, it is more

TABLE 11.2   Plant Sampling Guidelines for Selected Crops

Crop	When to Sample	Part of Plant to Sample	Number of Plants to Sample
Alfalfa clover	At ¹⁄₁₀ bloom stage or before	Mature leaf blades about ⅓ of the way down the plant	45–55
Cereal grains (including rice)	Seedling stage or	All the aboveground portion	50–75
	Prior to heading	Four uppermost blades from top of plant	30–40
Corn	Seedling stage or	All the aboveground portion	25–30
	Prior to tasseling or	The first fully developed leaves from the top	15–20
	From tasseling to silking	The leaves below and opposite the ear	15–20
Cotton	Prior to or at first bloom or when first squares appear	The youngest fully mature leaves on the main stem	30–35
Hay, forage, or pasture grasses	Before seed head emergence or at the stage for best quality	The four uppermost leaf blades	50–60
Milo-sorghum	Before or at heading	The second leaf from the top of the plant	20–25
Peanuts	Before or at bloom stage	Fully developed leaves from the top of the plant	45–50
Soybeans	Seedling stage or	All the aboveground portion	20–30
	Prior to or during initial flowering	The first fully developed leaves from the top	20–30
Sugar beets	Mid-season	Fully mature leaves midway between the younger center leaves and the oldest leaf whorl on the outside	30–35
Sugarcane	Up to 4 months old	Fourth fully developed leaf from the top	25–30
Beans or Peas	Seedling stage or	Entire aboveground portion	25–30
	Prior to or during initial flowering	Two or three mature leaves at the top of the plant	25–30
Celery	Mid-growth cycle	Petiole of the youngest mature leaf	20–30
Cucumber	Before fruit set	Mature leaves near the base of the main stem	20–25
Leaf crops (lettuce, spinach, etc.)	Mid-growth	Youngest mature leaf	30–50
Melons	Prior to fruit set	Mature leaves near the base of the main stem	20–30

TABLE 11.2   *Continued*

Crop	*When to Sample*	*Part of Plant to Sample*	*Number of Plants to Sample*
Potato	Before or during early bloom	Third to sixth leaf from the growing tip	20–30
Root crops (carrots, beets, onions, etc.)	Before root or bulb enlargement	Center mature leaves	25–35
Apple, apricot, peach, pear	Mid-season	Leaves near the base of the current year's growth	75–100
Grapes	End of bloom period	Petioles from the leaves adjacent to fruit clusters	75–100
Lemon, lime, orange	Mid-season	Mature leaves from the last flush of growth on nonfruiting terminals	30–40
Raspberry, strawberry	Mid-season	Youngest mature leaves	25–40

SOURCE: A&L Agricultural Laboratories.

realistic to use the critical nutrient range (CNR), which is defined as that range of nutrient concentration at a specified growth stage above which the crop is amply supplied and below which the crop is deficient. Figure 11.8 is a generalized illustration of the CNR and how it varies during the growing season. An example of CNRs for macro- and micronutrients in corn is shown in Table 11.3. Critical nutrient ranges have been developed for most of the essential nutrients in many crops.

As in tissue tests, the percentage of plant nutrients often decreases as the crop matures (Figs. 11.8 and Table 11.3). Hence the stage of growth for sampling must be carefully selected and identified.

INCREASE IN YIELD WITH INCREASE IN NUTRIENT CONTENT   When a nutrient is deficient, increasing its availability will increase its content in the plant and the crop yield until the CNR is exceeded (Fig. 11.7). For example, applied N increased the %N in wheat (Fig. 11.9) and in corn (Fig. 11.12). Above the CNR, %N increases with no yield advantage.

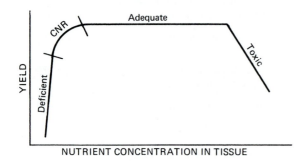

FIGURE 11.7   Relationship between nutrient concentration in plant tissue and crop yield, showing the proposed critical nutrient range (CNR). *Dow and Roberts, Agron: J., 74:401, 1982.*

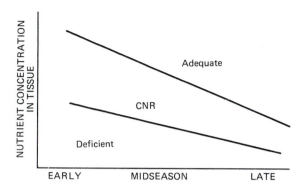

FIGURE 11.8   Generalized interpretive guide based on the concept of critical nutrient range (CNR) for tissue sampled at different times through the season.   *Dow and Roberts,* Agron. J., *74:401, 1982.*

The relationship between corn grain yield and percent K in the leaf is shown in Figure 11.10. It also suggests that rather than referring to a critical level, it would be better to refer to a critical range above 2% K.

Plant analysis interpretations based on the CNR and sufficiency range concepts have limitations. Stage of growth greatly influences nutrient concentrations, and unless the crop sample is taken at the proper time, the analytical results will be of little value. Also, considerable skill on the part of the diagnostician is needed to interpret the crop analysis results in terms of the overall production conditions.

BALANCE OF NUTRIENTS   One of the problems in the interpretation of plant analyses is that of balance among nutrients. Ratios of nutrients in plant tissue are frequently used to study nutrient balance in crops. For example, N/S, K/Mg, K/Ca, Ca + Mg/K, N/P, and other ratios are commonly used.

When a nutrient ratio has an optimal value, optimum yield occurs unless some other limiting factor reduces the yield. When a ratio is too low, a response

TABLE 11.3   Critical Nutrient Range for Macro- and Micronutrients in Corn

Nutrient	Whole Plant 24–45 Days[*]	3rd Leaf, 45–80 Days[†]	Earleaf Green Silks[‡]	Earleaf Brown Silks[§]
N, %	4.0–5.0	3.5–4.5	3.0–4.0	2.8–3.5
P, %	.40–.60	.35–.50	.30–.45	.25–.40
K, %	3.0–5.0	2.0–3.5	2.0–3.0	1.8–2.5
Ca, %	.51–1.6	.20–.80	.20–1.0	.20–1.2
Mg, %	.30–.60	.20–.60	.20–.80	.20–.80
S, %	.18–.40	.18–.40	.18–.40	.18–.35
B, ppm	6–25	6–25	5–25	5–25
Cu, ppm	6–20	6–20	5–20	5–20
Fe, ppm	40–500	25–250	30–250	30–250
Mn, ppm	40–160	20–150	20–150	20–150
Zn, ppm	25–60	20–60	20–70	20–70

[*] Seedlings 6 to 16 in. tall; 24 to 45 days after planting.

[†] Third leaf from top; plants over 12 in. tall; before silking.

[‡] 70 to 90 days after planting.

[§] Grain in developing stage up to "roasting ear."

SOURCE: Schulte and Kelling, *National Corn Handbook, NCH-46*, Purdue Univ. Coop. Ext. Service.

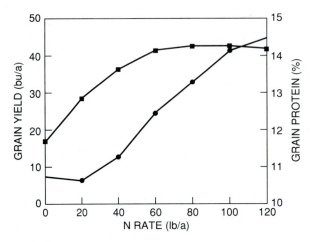

FIGURE 11.9   Influence of N rate on dryland wheat
yield (■) and grain protein content (●).   *Halvorson et al.,*
N.D. Farm Res., **33**:*3–9, 1976.*

to the nutrient in the numerator will be obtained if it is limiting. If the nutrient
in the denominator is excessive, a yield response may or may not occur,
depending on the level of other yield factors. When the ratio is too high, the
reverse is true. These conclusions are supported by the following examples
based on the assumption of an optimum range for the N/S ratio in a particular
plant part within which the crop yield is maximized.

   When the N/S ratio is in this optimum range or balanced, it will be identified
by a horizontal arrow (→). Ratios above the optimum will be recognized by
an upward vertical arrow (↑), and those below it will be assigned a downward
vertical arrow (↓).

   In situations with N/S = → or in its optimal range, three possibilities exist:

$$\frac{N \rightarrow}{S \rightarrow} \qquad or \qquad \frac{N \uparrow}{S \uparrow} \qquad or \qquad \frac{N \downarrow}{S \downarrow}$$

Both numerator	Both numerator	Both numerator
and denominator	and denominator	and denominator
optimal	excessive	insufficient

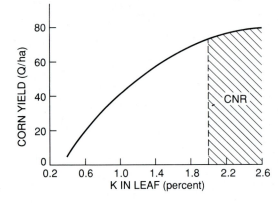

FIGURE 11.10   Corn grain yields
increased with increasing levels of K in
the leaf sampled at silking.   *Loué.*
Fertilite, **20**, *1963.*

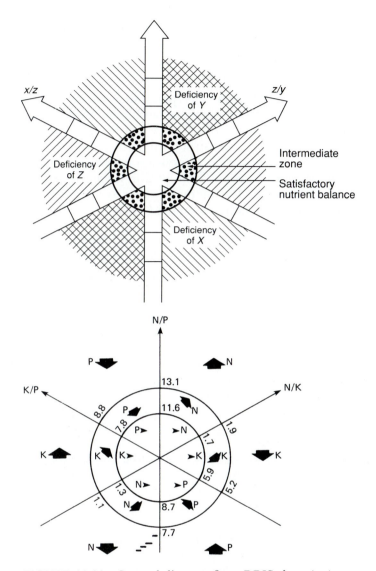

FIGURE 11.11   General diagram for a DRIS chart *(top)*
and a specific example *(bottom)* for obtaining the
qualitative order of requirements for N-P-K in corn.
Means of significant expressions (value at origin in chart)
are: N/P = 10.04, N/K = 1.49, and K/P = 6.74.
Sumner, Solutions, **22:68, 1978.**

It is not possible to determine from the ratio alone which of the three situations
is occurring in the plant. All that can be said is that the two nutrients are in
relative balance.

Where the N/S ratio is either above or below the optimal range, two possibilities exist in each case:

$$\frac{N}{S} = \ \uparrow \qquad \frac{N \rightarrow}{S \downarrow} \qquad \qquad or \qquad \frac{N \uparrow}{S \rightarrow}$$

$$\text{S insufficiency} \qquad\qquad \text{N excess}$$

$$\frac{N}{S} = \downarrow \quad \frac{N \rightarrow}{S \uparrow} \qquad \text{or} \qquad \frac{N \downarrow}{S \rightarrow}$$

$$\text{S excess} \hspace{5.5cm} \text{N insufficiency}$$

With N/S above the optimal range, a response to S will be obtained only if S is lacking. If N is excessive and S is normal, additional S may not improve the yield. The same is true with respect to N when the N/S ratio is below the optimal range. This analysis demonstrates why, when a ratio has a given value outside the optimum range, a yield response is not always obtained.

Consideration of more than one ratio at a time improves the chances of making a correct diagnosis. Use of many ratios is an integral part of the Diagnosis and Recommendation Integrated System.

DIAGNOSIS AND RECOMMENDATION INTEGRATED SYSTEM (DRIS)   DRIS is a system that identifies all the nutritional factors limiting crop production and, thus, increases the chance of obtaining high crop yields by improving fertilizer recommendations. Index values that measure how far particular nutrients in the leaf or plant are from the optimum are used in the calibration to classify yield factors in order of limiting importance. To develop a DRIS for a given crop, the following requirements must be met whenever possible:

1. All factors suspected of having an effect on crop yield must be defined.
2. The relationship between these factors and yield must be described.
3. Calibrated norms must be established.
4. Recommendations suited to particular sets of conditions and based on correct and judicious use of these norms must be continually refined.

Establishment of DRIS Norms. A survey is first employed in obtaining the data required to establish DRIS norms. A large number of sites where a crop is growing are selected at random in order to represent the whole production area of a county, state, or district. At each site, plant and soil analyses for all essential nutrients are conducted. Other parameters likely to be related directly or indirectly to yield are also recorded. In addition, details of soil treatments (fertilizers, herbicides, etc.), climatic conditions (rainfall, etc.), cultural practices, and any other relevant types of information are recorded and stored in a computer for ready access.

Second, the entire population of observations is divided into two populations (high and low yielders) based on vigor, quality and yield. Each nutrient in the plant is expressed in as many ways as possible. For example, the percentage of N in the dry matter, or ratios N/P, N/K, or products N-P, N-K, and so on, may be used. The mean of each ratio for each population is calculated. Each ratio that significantly discriminates between the high- and low-yielding populations is retained as a useful diagnostic parameter. The mean values for each of these ratios then constitute the diagnostic norms. Using N-P-K in corn leaves as an example, the significant ratios have been found to be N/P, N/K, and K/P.

Determination of Relative N-P-K Requirements. An example follows on how to diagnose the relative N-P-K requirements of corn using the DRIS chart illustrated in Figure 11.11. The chart is constructed of three axes for N/P, N/K, and K/P, respectively, with the mean value for the population of high yielders (>160 bu/a) located at the point of intersection for each ratio (center of the circle). These values are N/P = 10.04, N/K = 1.49, and K/P = 6.74. The point of intersection of the three axes represents the optimum nutrient composition and where the highest yield is obtained, depending on limiting factors other than N, P, and K. The concentric circles are confidence limits, the inner one being set at the mean 15% and the outer one at the mean 30% for each ratio.

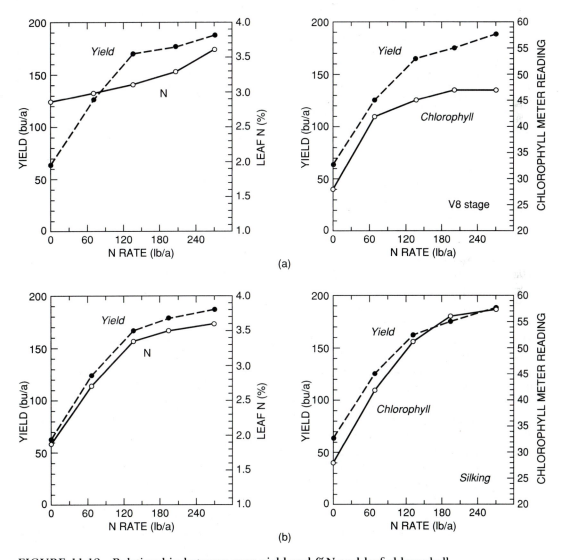

FIGURE 11.12   Relationship between corn yield and %N and leaf chlorophyll readings at V8 growth stage (a) and at silking (b) for irrigated corn.   *Schepers et al., Proc. Great Plains Soil Fert. Conf., p. 42, 1992.*

A qualitative interpretation can be made by using arrows in the following manner: horizontally for values within the inner circle of the chart (between 1.30 and 1.71 for N/K, between 8.73 and 11.55 for N/P, and between 5.86 and 7.75 for K/P) corresponding to a nutritionally balanced situation; diagonally [↘ ↗] for values between the two circles representing a tendency to imbalance (e.g., between 1.15 and 1.30 or between 1.71 and 1.94 for N/K); and vertically [↑ ↓] for values found beyond the outer circle (e.g., beyond 1.15 or 1.94 for N/K) representing nutrient imbalance. Practical use of this chart will be illustrated by the example of N, P, and K concentrations for corn leaves presented in Table 11.4.

Because an excess of one nutrient corresponds to a shortage of another, by convention only insufficiencies are recorded for the purpose of diagnosis. Using the data in the first line of Table 11.4, one finds that the value of the function N/P (13.33) lies in the zone of P insufficiency, giving: (1) N P ↓ K; while the value of N/K (1.27) lies between the two circles, indicating a tendency for N stress: (2) N ↘ P ↓ K; while that of K/P (10.48) lies in the zone of P insufficiency, giving: (3) N ↘ P ↓ ↓ K.

Once the three forms of expression have been read, the remaining element is assigned a horizontal arrow. The final reading then becomes (4) N ↘ P ↓ ↓ K→, which gives the order of N-P-K requirements of the crop in terms of limiting importance on yield: P > N > K.

Using DRIS Indices to Diagnose N-P-K Requirements. The arrow notation just illustrated can be quantified by calculating DRIS indices, although the equations used for calculating these indices are not shown here. When using these indices, the most negative DRIS index represents the most limiting nutrient, while the most positive indicates the one least needed. For example, P is the most limiting nutrient in the unfertilized crop (0-0-0) in Table 11.4. In order to see the response to P, consider the treatment in which P was applied (0-50-0). P application increased the yield and raised the DRIS index for P, but K is now the most deficient nutrient.

TABLE 11.4   Use of DRIS Norms in Diagnosing N-P-K Requirements of Corn Based on the Selection of a Plot in an N-P-K Factorial Experiment, Diagnosing the Requirement, and Satisfying It by Selecting the Plot in Which the Required Element Is Applied (Leaf Sample Taken at Tasseling)

Treatment with N-P_2O_5-K_2O (lb/a)	Leaf Composition (%)			Forms of Expression			Chart Reading			DRIS Indices			Corn Yield* (%)
	N	P	K	N/P	N/K	K/P	N	P	K	N	P	K	
0-0-0	2.80	0.21	2.20	13.33	1.27	10.48	↘	↓↓	→	7	− 22	15	28
0-50-0	3.20	0.28	1.00	11.43	3.20	3.57	→	→	↓↓	31	13	− 44	49
0-50-60	2.93	0.28	2.60	10.46	1.13	9.29	↘	↓	→	− 6	− 9	15	55
0-100-60	2.60	0.26	2.44	10.00	1.07	9.38	↓	↓	→	− 9	− 8	17	60
100-100-60	3.16	0.33	2.45	9.58	1.29	7.42	↘	→	→	− 5	− 1	6	75
200-100-60	3.40	0.34	2.40	10.00	1.42	7.06	→	→	→	− 1	− 1	2	100

*Percentage of the highest value.

SOURCE: Sumner, *Solutions*, **22**(5):68 (1978).

When K is added (0-50-60), the yield increased but P is again the most limiting nutrient. When more P is applied (0-100-60), the yield increases further and N becomes the most limiting nutrient. Addition of N (100-100-60 and 200-100-60) results in the maximum yield and a relatively well-balanced N-P-K nutritional status, which means that factors other than N-P-K are limiting the yield.

The same diagnosis is made by both the qualitative (arrow) and quantitative (index) procedures. One can see that requirements for P (treatment 0-50-60) and for N (treatment 100-100-60) are correctly predicted by the DRIS system when the %N and %P in the leaf are above the critical values discussed earlier. In many cases, the DRIS system is capable of diagnosing requirements that would not be obvious when using the critical or sufficiency level approach.

Nutrient diagnosis using DRIS norms can be made over a range of plant age (Table 11.5). Irrespective of crop age, the DRIS approach indicates that P is required more than K, which is required more than N (e.g., P > K > N). Using the critical value norms given earlier, the diagnosis would have been possible only after the 80-day stage. Thus the DRIS approach results in greater flexibility in diagnosis. Another advantage is that the DRIS method of determining N-P-K requirements is less sensitive to leaf position on the plant or cultivar. The DRIS approach has been successfully applied to other crops and nutrients.

In summary, the DRIS system has several advantages over the the critical level approach in making diagnoses for fertilizer recommendation purposes:

1. The importance of nutritional balance is taken into account in deriving the norms and making diagnoses.
2. The norms for the nutrient content in leaves can be universally applied to the particular crop, regardless of where it is grown.
3. Diagnoses can be made over a wide range of stages of crop development, irrespective of the cultivar.
4. The nutrients limiting yield, through either excess or insufficiency, can be readily identified and arranged in order of their limiting importance for yield.

TABLE 11.5   Effect of Age of Crop Sampled on the DRIS Diagnosis of N-P-K Requirements of Corn

Age of Crop at Sampling (days)	Leaf Composition (%)			Forms of Expression			Chart Reading			DRIS Indices		
	N	P	K	N/P	N/K	K/P	N	P	K	N	P	K
30	4.6	0.30	3.4	15.33	1.35	11.33	→	↓↓	→	16	− 32	16
60	3.9	0.26	2.4	15.00	1.63	9.23	→	↓↓	→	19	− 25	6
80	3.4	0.24	1.9	14.17	1.79	7.92	→	↓↘	↘	19	− 18	− 1
110	3.0	0.20	1.8	15.00	1.67	9.00	→	↓↓	↘	20	− 24	4

SOURCE: Sumner, *Solutions*, **22**(5):68 (1978).

Chlorophyll Meters for Plant N Status

Evaluation of the nutrient status of the plant may not always involve destructive sampling of the plant or leaf. Use of a hand-held chlorophyll meter may provide an indication of the leaf N status of the crop. Leaf chlorophyll content is related to N nutrition (in addition to other nutrients); therefore measuring the relative chlorophyll content can provide an indication or index of the N status in the plant. Chlorophyll meter readings do not directly indicate the chlorophyll content, but the value recorded can be related to %N in leaf and grain yield as influenced by N rate (Fig. 11.12). Increasing the rate of fertilizer N increases grain yield and both %N in the leaf and leaf chlorophyll meter reading at V8 and silking growth stages in irrigated corn. For N management purposes, chlorophyll readings have greater value at the V8 stage, because a greater corn yield response to additional fertilizer N at this growth stage would occur than with N applied later at silking.

Crop Logging

An excellent example of the use of plant analyses in crop production is the crop logging carried out for sugarcane in Hawaii. The crop log, which is a graphic record of the progress of the crop, contains a series of chemical and physical measurements. These measurements indicate the general condition of the plants and suggest changes in management that are necessary to produce maximum yields. A critical nutrient concentration approach is used in the crop log system, and nutrient concentrations in leaf sheaths are utilized for diagnosis of macro- and micronutrient deficiencies.

During the growing season, plant tissue is sampled every 35 days and analyzed for N, sugar, moisture, and weight of the young sheath tissue. Analyses are made for P and K at critical times, and adjustments in management practices are introduced as needed (Fig. 11.13). Knowledge of the percentage of moisture makes it possible to regulate irrigation, particularly during the ripening period. For example, when the moisture content drops to around 73%, the crop is ripe and irrigation is regulated accordingly. Many plantations in Hawaii are using crop logging, because productivity increases under this system of record keeping.

Grain Analysis for N Sufficiency

Grain samples are often collected to provide nutritional information to the grower. For example, wheat grain protein can be used to indicate whether sufficient N was available for optimum yield or if insufficient N was applied to the crop (Fig. 11.14). These data show that when grain protein was <11.5%, the wheat crop most likely would have responded in grain yield to additional fertilizer N. Alternatively, >11.5% grain protein indicated sufficient N availability for maximum grain yield. Although grain analysis can be very helpful in N management, it is a postmortem analysis. However, monitoring grain protein for several consecutive years will help growers to identify more accurately the appropriate N rate for a specific crop.

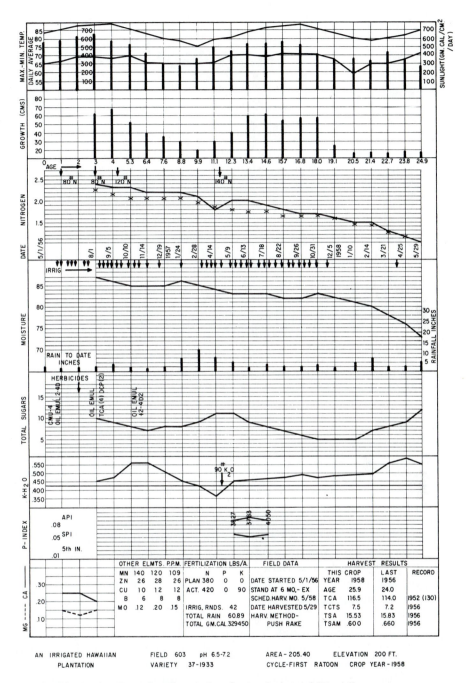

FIGURE 11.13 Completed crop log for an irrigated Hawaiian sugarcane plantation. This approach has been valuable in a complete diagnostic approach. *Clements, in Walter Reuther,* Plant Analysis and Fertilizer Problems, *p. 132. Copyright 1960 by the American Institute of Biological Sciences.*

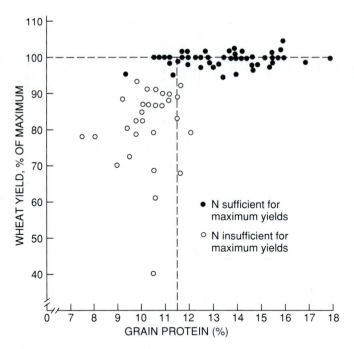

FIGURE 11.14   Relationship between grain protein and
N sufficiency in winter wheat.   *Goos et al., Agron J.,*
74:130–133, 1982.

Biological Tests

Field Tests

The field-plot method is essential for measuring the crop response to nutrients. After the specific treatments are selected, they are randomly assigned to
an area of land known as a *replication*, which is representative of the conditions.
Several replications are used to obtain more reliable results and to account
for variations in soil.

For example, when various rates of N are applied, the yield results are
helpful in determining N recommendations. When large numbers of tests
are conducted on soils that are well characterized, recommendations can be
extrapolated to other soils with similar characteristics. Field tests are expensive
and time-consuming; however, they are valuable tools and are widely used by
scientists. They are used in conjunction with laboratory and greenhouse studies in the calibration of soil and plant tests. Field experiments are essential in
establishing the equation used to provide fertilizer recommendations that will
optimize crop yield, maximize profitability, and minimize the environmental
impact of nutrient use. Plant analysis of samples collected from the various
treatments also can help establish CNR and DRIS indices.

Strip Tests on Farmers' Fields

Narrow field strips treated with selected fertilizer treatments can help verify the accuracy of recommendations based on soil or plant tests. The results of these tests must be interpreted with caution if they are unreplicated. Repetition of strip tests on several farms is also helpful.

Laboratory and Greenhouse Tests

Simpler and more rapid laboratory/greenhouse techniques utilize small quantities of soil to quantify the nutrient-supplying power of a soil. Generally, soils are collected to represent a wide range of soil chemical and physical properties that contribute to the variation in availability for a specific nutrient. Selected treatments are applied to the soils and a crop is planted that is sensitive to the specific nutrient being evaluated. Crop response to the treatments can than be determined by measuring total plant yield and nutrient content. Figure 11.15 illustrates the use of a greenhouse test to separate Fe-deficient and -sufficient soils. Soils were selected to represent a range in DTPA-extractable Fe. Sorghum plants show decreasing Fe deficiency as DTPA-extractable Fe increases.

NEUBAUER SEEDLING METHOD   The Neubauer technique is based on the uptake of nutrients by a large number of plants grown on a small amount of soil. The roots thoroughly penetrate the soil, exhausting the available nutrient

FIGURE 11.15   Greenhouse test used to evaluate the ability of DTPA to separate Fe deficient and sufficient soils. Sorghum was used as an indicator crop. Fe stress in sorghum decreased with increasing DTPA extractable Fe.

supply within a short time. The total nutrients removed are quantified, and tables are established to give the minimum values of macro- and micronutrients available for satisfactory yields of various crops.

MICROBIOLOGICAL METHODS   In the absence of nutrients, certain microorganisms exhibit behavior similar to that of higher plants. For example, growth of *Azotobacter* or *Aspergillus niger* reflects nutrient deficiency in the soil. The soil is rated from very deficient to not deficient in the respective elements, depending on the amount of colony growth. In comparison with methods that utilize higher plants, microbiological methods are rapid, simple, and require little space.

Soil Testing

Although plant analyses are extremely valuable in diagnosing nutrient stress, analysis of the soil is essential in determining the supplemental nutrient requirement of a crop. A soil test is a chemical method for estimating the nutrient-supplying power of a soil. Compared to plant analysis, the primary advantage of soil testing is its ability to determine the nutrient status of the soil *before* the crop is planted.

A soil test measures part of the total nutrient supply in the soil and represents only an index of nutrient availability. Soil tests do not measure the exact quantity of a nutrient potentially taken up by a crop. To predict the nutrient needs of crops, the soil test must be calibrated against nutrient rate experiments in the field and in the greenhouse.

Objectives of Soil Tests

Information gained from soil testing is used in many ways:

1. *To provide an index of nutrient availability or supply in a given soil.* The soil test or extractant is designed to extract a portion of the nutrient from the same "pool" (i.e., solution, exchange, organic, or mineral) used by the plant.
2. *To predict the probability of obtaining a profitable response to lime and fertilizer.* Although a response to applied nutrients will not always be obtained on low-testing soils because of other limiting factors, the probability of a response is greater than on high-testing soils.
3. *To provide a basis for recommendations on the amount of lime and fertilizer to apply.* These basic relations are obtained by careful laboratory, greenhouse, and field studies.
4. *To evaluate the fertility status of soils on a county, soil area, or statewide basis by the use of soil-test summaries.* Such summaries are helpful in developing both farm-level and regional nutrient management programs.

Expressed simply, the objective of soil testing is to obtain a value that will help to predict the amount of nutrients needed to supplement the supply in the soil. For example, a high test value soil will require little or no addition of nutrients, in contrast to soil with a low test value (Fig. 11.16).

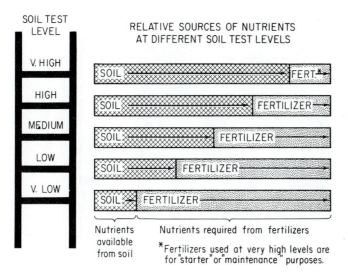

FIGURE 11.16   As the soil tests higher in a plant nutrient, the amount needed from fertilizers becomes less. The purpose of soil testing is to determine the levels of nutrients.   *Courtesy of the Potash & Phosphate Institute, Atlanta, Ga.*

Sampling the Soil

The most critical aspect of soil testing is obtaining a soil sample that is representative of the field. Usually, a composite sample of only 1 pint of soil (about 1 lb) is taken from a field, which represents, for example, a 10-a field or about 20 million lb of surface soil. There is considerable opportunity for sampling error; thus, it is essential that the field be sampled correctly. If the sample does not represent the field, it is impossible to provide a reliable fertilizer recommendation. The sampling error in a field is generally greater than the error in laboratory analyses.

Soils are normally heterogeneous, and wide variability can occur even in "uniform" fields (Fig. 11.17). Intensive soil sampling is the most efficient way to evaluate variability. Despite the sampling problems associated with soil variability, the quantity of nutrient extracted by the soil test should be closely related, but not equal, to the quantity of nutrient absorbed by the crop. For example, the Bray-1 extractable P in a 6-a (2.5-ha) field varies from about 20 to over 80 ppm (Fig. 11.18). If the P soil test represents the P-supplying power of the soil, then the variability in %P in the crop should reflect the variability in Bray-1 extractable P. Comparison of the two graphs in Figure 11.18 shows that the spatial distribution of P concentration in wheat grain reflects the distribution of plant-available P as measured by soil test P. Specifically, high and low soil test P results in high and low grain P, respectively. These data demonstrate the ability of a soil test to provide a reliable index of plant-available nutrient, in this case P. They also illustrate the spatial variability in available plant nutrient levels in a field.

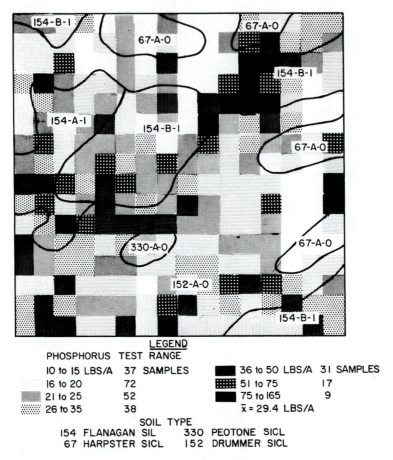

FIGURE 11.17   Variability of soil P in this 40-a Illinois
field helps to illustrate the problem of accurately
sampling a soil.   *Peck and Melsted, in Walsh and Beaton,*
Eds., Soil Testing and Plant Analysis, *p. 67. Madison,*
Wisc.: Soil Science Society of America, 1971.

TOOLS   There are two important requirements of a sampling tool: first, that
a uniform portion be taken from the surface to the depth desired and, second,
that the same volume of soil be obtained from each area. Soil tubes in general
meet these two requirements very well. Soil sampling tubes have 5 to 15 in.
cut away on one side, except at the cutting head, which has 1 in. of solid tube
of a smaller bore. These tubes work well under most conditions, except in dry
or gravelly ground. Other tools used are trowels, augers, spades, or power-
driven samplers. Hydraulically operated sampling equipment is especially
helpful in obtaining deep samples and taking large numbers of samples.

AREAS TO SAMPLE   The size of the area from which one sample may be
taken varies greatly but usually ranges from 5 to 20 a or more. Areas that vary
in appearance, slope, drainage, soil types, or past treatment should be sampled
separately (Figure 11.19), and small areas that cannot be treated separately

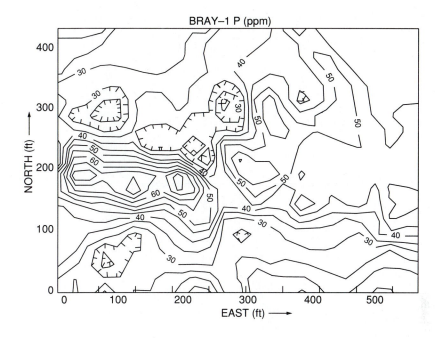

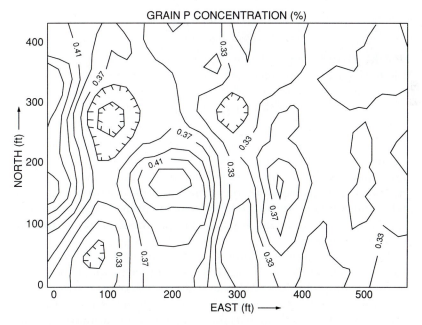

FIGURE 11.18   Spatial distribution of Bray-1 P and
wheat grain P concentration over a 6-a (2.5-ha) field.
Areas of high soil test P correspond to areas of higher
%P in the grain.   *Havlin and Sisson,* Proc. Dryland
Farming Conf., *p. 406–408, 1990.*

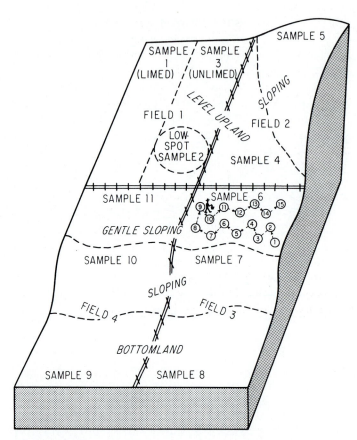

FIGURE 11.19 Samples that are representative of the field to be fertilized are important. The sampling pattern recommended by the various agricultural agencies should be followed. *Courtesy of the Nebraska Agricultural Extension Service.*

by lime and fertilizer applications might be omitted from the sample. On the other hand, with the trend toward higher production and more uniform crop growth, many growers are directing attention to these small spots in their fields and treating them as needed. The idea is to have every acre yield as well as every other acre. Hence separate samples from localized areas of poor crop growth are helpful.

Use of application equipment capable of variably applying fertilizer according to variability in soil test levels for a specific nutrient requires intensive field sampling. To accurately quantify the variability in soil test level, fields are sampled on a 1- to 5-a grid, depending on the field and the nutrient being evaluated. The resultant variability is generally displayed similarly to that shown in Figure 11.18 and subsequently used to provide variable fertilizer recommendations used with variable fertilizer applicators.

NUMBER OF SITES TO SAMPLE FOR COMPOSITES   Each soil sample is a composite consisting of the soil from cores taken at several places in the field.

The purpose of this procedure is to minimize the influence of any local nonuniformity in the soil. For example, in fields in which lime or fertilizer applications have been made in the last 2 or 3 years, the plant nutrients may be incompletely mixed with the soil. In addition, there may be spots where the fertilizer or lime was spilled or dumped, or where plant refuse was burned, or where crop residue from harvesting was concentrated. A sample taken entirely from such an area would be completely misleading. Consequently, most recommendations call for sampling 15 to 20 locations over the field for each composite sample (Fig. 11.19). The soil samples from each area are mixed well, and a subsample is sent to the laboratory for analysis.

Band application of immobile nutrients in the soil (i.e., P and K) often results in residual available nutrient in the old fertilizer bands for several years after application. Residual availability depends on the rate of application, soil chemical and physical properties, quantity of nutrient removed by the crop, crop rotation and intensity, and time after application. For example, the variation in soil test P level with P rate and method of placement is shown in Figure 11.20. Increasing the broadcast P rate increased soil test P. Similarly, increasing band-applied P increased soil test P in the band. Band-applied P increased soil test P more than the same rate of broadcast P. Thus, if only the

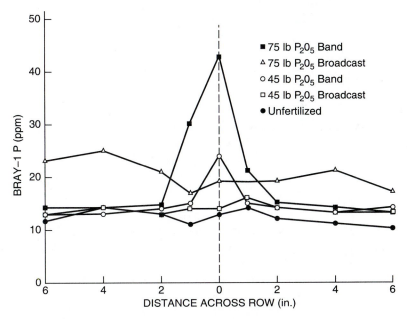

FIGURE 11.20   Effect of P rate and placement on Bray-1 P levels measured in 1-in increments across the wheat row. Fertilizer was broadcast and band (2in. below the seed)applied at 15, 45, and 75 lb P_2O_5/a at planting in September 1986. Soils were sampled in August 1988. The soil test levels for the 15 lb P_2O_5/a rate are not shown because they were similar to the unfertilized "check" treatment.   *Havlin et al.,* Proc. Fluid Fert. Foundation, *p. 193, 1989.*

bands are sampled; the soil test P is much higher than if none of the bands are sampled (i.e., if only the between-band areas are sampled). Few guidelines have been established for soil sampling fields where immobile nutrients have been band applied. In wheat-fallow-wheat systems the following recommendations have been developed:

$$S = \frac{8 \text{ (row spacing)}}{30}$$

where S = ratio of off-band to on-band samples Thus, for 30-cm (12-in.) band spacing, eight samples between the bands are required for every sample taken on the band. If similar recommendations do not exist in other regions, then increasing the sampling intensity should provide an adequate estimate of the average soil test level in the field.

To obtain good representative samples, about 25 sampling sites are required for each average-sized field. For small fields, a minimum of eight places should be sampled and more intensive sampling is encouraged. In very large fields, at least one site should be sampled for each 5 to 7.5 a. These guidelines are for situations in which there is no extreme variability within the field.

DEPTH OF SAMPLING   For cultivated crops samples are ordinarily taken to the depth of tillage, which can vary from 6 to 12 in. (Fig. 11.21). Tillage will

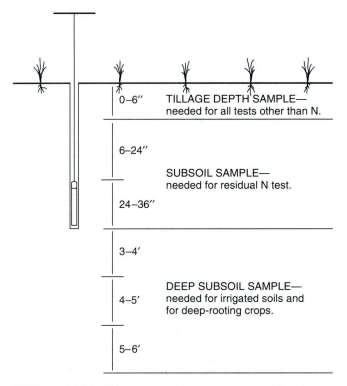

FIGURE 11.21   Diagrammatic representation of various soil sampling depths used to collect samples for soil nutrient analyses.

TABLE 11.6   Influence of Tillage on Stratification of Soil Test P and K

Depth	Plow	Chisel	No-till	Plow	Chisel	No-till
Inches	-------- Bray P$_1$, ppm --------			-------- Exch. K, ppm --------		
0–3	37	85	90	150	230	285
3–6	47	35	27	165	105	100
6–9	30	15	18	140	100	100
9–12	8	8	8	100	100	100

SOURCE: Mengel, *Agron. Guide AY-268*. Purdue University (1990).

generally mix previous lime and fertilizer applications within the tillage layer. When lime and fertilizer are broadcast on the surface for established pastures and lawns, a sample from the upper 2 in. is satisfactory. With no-till or minimum till, it is best to take a sample from the surface 2 in. and another sample from the 2- to 8-in. layer. Considerable nutrient stratification occurs with reduced and no-till systems, as shown in Tables 11.6 and 11.7. Although plants obtain nutrients from below the tillage layer, little information is available for interpreting the analyses of subsoil samples.

Soil sampling 2 to 6 ft in lower-rainfall areas is done to measure NO_3^- and moisture in the profile. The soil profile NO_3^- content is generally correlated with the crop response to N fertilization; thus, profile NO_3^- measurements are commonly used to predict the fertilizer N requirements of crops (Fig. 11.22).

TIME OF SAMPLING   Ideally, samples should be taken just prior to seeding or when the crop is growing. However, these times are largely impractical because of constraints in taking samples, obtaining test results, and supplying the needed lime and fertilizer. Consequently, samples are customarily taken any time soil conditions permit. For spring-planted crops, sampling in the fall after harvesting is often practiced. In drier regions where NO_3^- levels are used to assess the N status of soil, sampling in the fall to diagnose the needs of annual spring-seeded crops is often postponed until the surface soil temperature drops to 5°C.

Most recommendations call for testing each field about every 3 years, with more frequent testing on coarse soils. In most instances, this is often enough to check soil pH and to determine whether the fertilization program is adequate for the crop rotation. For instance, if the P level is decreasing, the rate of application can be increased. If it has risen to a satisfactory level, application may be reduced to maintenance rates.

TABLE 11.7   Influence of Tillage and N Rate on Soil pH

Yearly N Rate lb/a	No-tillage		Plow Tillage	
	0–2 in.	2–6 in.	0–2 in.	2–6 in.
0	5.75	6.05	6.45	6.45
75	5.20	5.90	6.40	6.35
150	4.82	5.63	5.85	5.83
300	4.45	4.88	5.58	5.43

SOURCE: Blevins et al. *Soil Tillage Res.* **3**:136–146 (1983).

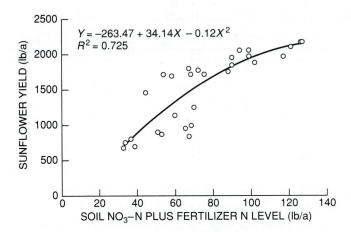

FIGURE 11.22   Influence of soil profile NO_3-N and fertilizer N on sunflower seed yield.   *Black and Bauer*, Proc. Great Plains Soil Fert. Conf., *1992*.

Analyzing the Soil

Many chemical extractants have been developed for use in soil testing, and several of them have been discussed in the chapters on specific nutrients. The ability of an extractant to extract a plant nutrient in quantities related to plant requirements depends on the reactions that control nutrient supply and availability (Fig. 11.23). For example, exchangeable K, Ca, and Mg are predominant forms of plant-available K, Ca, and Mg; thus, the soil extractant would be one that displaces these nutrients from exchange sites during the extraction. Therefore, for K^+, 1 M NH_4OAc is the most common extractant, where the NH_4^+ exchanges for K^+ on the CEC. In contrast, the DTPA soil

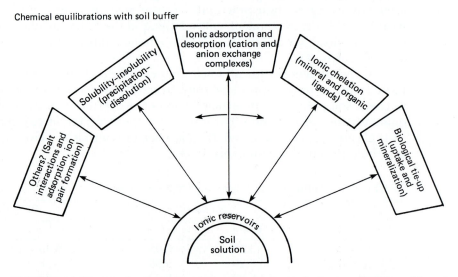

FIGURE 11.23   Mechanisms regulating reactivity and plant availability of ions in soil.   *McLean*, Commun. Soil Sci. Plant Anal., *13(6):411, 1982*.

test for micronutrients involves complexing Fe^{3+} in solution, for example, similar to the effect of natural chelation on Fe transport to roots (Fig. 9.9). See Chapter 6 for a description of P soil tests. Therefore, an effective soil test will, to some extent, simulate plant removal of the specific nutrient, with subsequent resupply to solution from the nutrient pools (i.e., exchange, OM, mineral, etc.) that control availability. Table 11.8 lists some of the common soil tests for specific macro- and micronutrients in soils. See the chapters on specific nutrients for additional description of the soil test extractants and chemistry.

Calibrating Soil Tests

Although the chemical analysis of soils presents some difficulties, perhaps the greatest problem in a testing program is the calibration of the tests. It is essential that the results of soil tests be calibrated against crop responses from applications of the nutrients in question. This information is obtained from field and greenhouse fertility experiments conducted over a wide range of soils. Yield responses from various rates of applied nutrients can then be related to the quantity of available nutrients in the soil indicated by the soil test.

Many of the testing laboratories classify the fertility level of soils as very low, low, medium, high, or very high, based on the quantity of nutrient extracted, although the exact quantities also are reported. In general, the very low to very high classification is easily understood by the grower. However, crops differ in their requirements; and what is low for potatoes may be high for small grain; and what is low for a clay loam may be high for a sandy loam. In any case, it is important that the grower know the meaning of the results reported.

Some laboratories use a fertility index that expresses the relative sufficiency as a percentage of the amount adequate for optimum yields (Table 11.9). The probability of a response to fertilization increases with decreasing soil test level (Fig. 11.24). Soil fertility is only one of the factors influencing plant growth, but in general there is a greater chance of obtaining a response from a given nutrient with a low soil test result. For example, more than 85% of the fields

TABLE 11.8   Common Soil Test Extractants and the Nutrient Source or "Pool" Extracted in the Soil

Plant Nutrient	Common Extractants	Nutrient Source
NO_3^-	KCl, $CaCl_2$	Solution
NH_4^+	KCl	Solution/CEC
$H_2PO_4^-/HPO_4^{2-}$	NH_4F/HCl (Bray-P)	Fe/Al mineral solubility
	$NaHCO_3$ (Olsen-P)	Ca mineral solubility
	HCl/H_2SO_4 (Mehlich-P)	Fe/Al and Ca mineral solubility
K^+	NH_4OAc	CEC
SO_4^{2-}	$Ca(H_2PO_4)_2$, $CaCl_2$	Solution/AEC
Zn^{2+}, Fe^{3+}, Mn^{2+}, Cu^{2+}	DTPA	Chelation
$H_3BO_3^0$	Hot Water	Solution
Cl^-	Water	Solution

TABLE 11.9   Fertility Index as Related to Soil Tests, Responses, and
Recommendations

Soil Test Rating	Crop Response	Fertility Index	Recommendations
Very low	Highly probable	0–10	Crop requirement + substantial buildup
Low	Probable	10–25	Crop requirement + buildup
Medium	Possible	25–50	Crop requirement + modest buildup
High	Unlikely	50–100	Maintenance
Very high	Highly unlikely	100 +	Vegetable crops—some maintenance

SOURCE: D. W. Eaddy, North Carolina Dept. of Agriculture, personal communication.

testing very low may give a profitable increase; in the low range 60 to 85%
may give increases, whereas in the very high range less than 15% will respond.
These values are arbitrary, but they illustrate the concept of the probability
of a response.

For example, the yield response of barley to applied N and P is greatest at
low soil-test levels, although some minor yield benefits also occur at high soil-
test levels (Fig. 11.25).

Response probability also can vary with fertilizer placement. The data in
the tabulation on page 439 indicate a greater probability of response at a
given soil test level when the fertilizer is band applied than when broadcast.
Similarly, the barley response to K is greater at low levels of exchangeable K
than at high soil test K levels (Table 11.10).

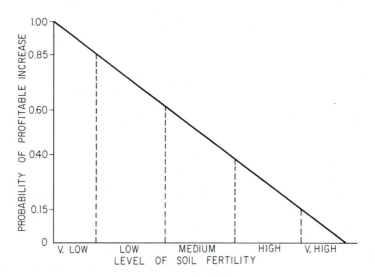

FIGURE 11.24   There is a greater probability of
obtaining a profitable response from fertilization on soils
testing low in an element than from soils testing high in
that element.   *Fitts*, Better Crops Plant Food, *39(3):17,
1955.*

	Probability of Response (%) to P and K	
	Broadcast*	Banded†
Very low	95–100	—
Low	70–95	95–100
Medium	40–70	65–95
High	10–40	30–65
Very high	0–10	10–30

*Purdue University.

†University of Minnesota.

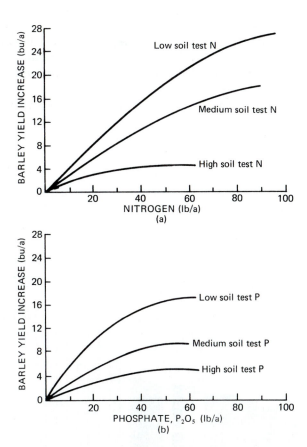

FIGURE 11.25   Expected barley yield increases from (a) N and (b) P at different soil test levels in central Alberta.   *McLelland, 114/20-1.*, Alberta Agriculture, *1982.*

TABLE 11.10   Yield Response of Barley to Added K
Fertilizer at Different Levels of Exchangeable K

Exchangeable K (lb/a) in Soil 0–6 in. Depth	Percent of Test Sites Giving Yield Response	Percent Increase in Yield from K Fertilizer*
<50	100	>1000
51–100	75	242
101–150	66	47
151–200	24	30
201–250	18	34
>250	3	11

*Percent yield increase calculated for those field sites that gave yield increases from K fertilizer.
SOURCE: Walker, *Better Crops Plant Food,* **62:**13 (Summer 1978).

An example of calibration of the soil test for Zn with crop response is shown in Figure 11.26. Ninety percent of the soils testing below 0.65 ppm Zn responded to Zn, whereas 100% above this level did not respond.

Soil test calibration is complicated by other factors that influence the response to fertility level. Temperature, water, soil properties, stand, cultural practices, crop and/or variety, and pests are more readily controlled in the greenhouse than in the field.

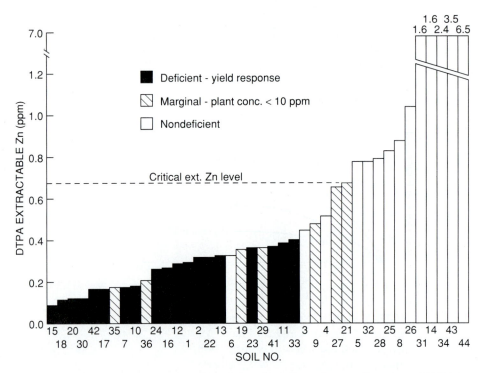

FIGURE 11.26   Corn response of 40 Colorado soils to Zn as a function of DTPA soil test levels.   *Havlin and Soltanpour, SSSAJ, 45:70–75, 1981.*

Therefore, controlled experiments are initially conducted in the greenhouse to provide information about (1) the ability of a soil test extractant to extract a nutrient in quantities related to the amount removed by the plant (i.e., to identify the best extractants), (2) the relationship between soil test level and relative yield and the CNR for various crops, and (3) the range in soil test levels and crop responsiveness for the major soil types in the region.

After greenhouse studies have been completed, field calibration experiments are conducted on the major soil series and crops in the region. For example, if a P soil test is being calibrated, four to six rates of P will be applied and the crop response quantified by measuring yield (forage, grain, fruit, etc.) and P content in the whole plant or plant part. The yield response data can be plotted in terms of percent yield or yield increase (Fig. 11.27). In Figure 11.27a, percent yield represents the ratio of the yield in the unfertilized soil to the yield obtained where P is nonlimiting (fertilized soil). For example, 70% yield means that the crop yield with the unfertilized soil is 70% of the yield obtained at the optimum level of P. Similarly, yield increase represents the

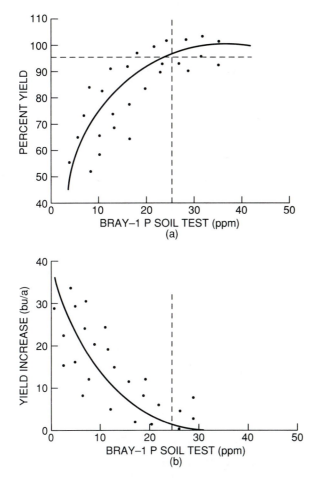

FIGURE 11.27   Relationship between percent yield (a) and yield increase (b) and Bray-1 P soil test level.

increase in yield obtained with optimum fertilization (Fig. 11.27b). Thus, as soil test P increases, percent yield increases to 100%, which represents the soil test level where there is no difference in yield between fertilized and unfertilized soil. Alternatively, as soil test P level increases, the yield increase to P fertilization decreases to 0 (Fig. 11.27b).

Generally, when percent yield reaches 95 to 100% or when yield increase reaches 0 to 5%, the critical soil test level (CL) is obtained. The CL represents the soil test level above which no yield response to fertilization will be obtained. Soil test CLs can vary among crops, climatic regions, and extractants. For example, the CL for the Bray-1 P, Olsen-P, and Mehlich-P tests is approximately 25, 13, and 28 ppm, respectively.

Soil test calibration studies also provide the data to establish fertilizer recommendations. For example, at each location, the P rate required for optimum yield can be determined and the results displayed similarly to Figure 11.28. Increasing soil test level corresponds to decreasing P rate required for optimum yield. These diagrams can be used to establish the fertilizer rates associated with very low, low, medium, and high soil test levels; however, most laboratories use an equation that describes the relationship in Figure 11.28.

Interpretation of Soil Tests

Soil test interpretation involves an economic evaluation of the relation between the soil test value and the fertilizer response. However, the potential response may vary due to several factors, including soil, crop, expected yield, level of management, and weather (Fig. 11.29). Factors *A–D* in Figure 11.29 represent

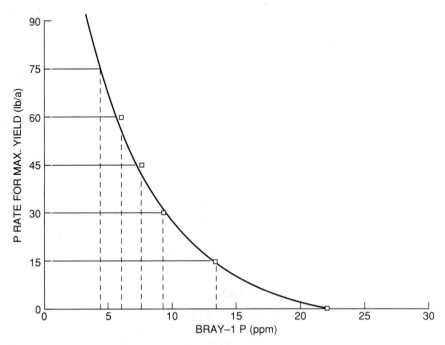

FIGURE 11.28   Influence of soil test P level on the fertilizer P rate required for maximum yield.

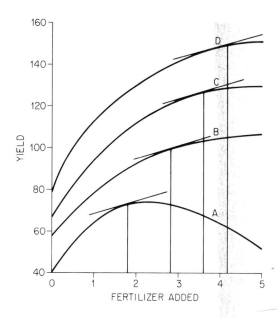

FIGURE 11.29   The yield response to fertilizer depends on the potential yield level, with *A* being the poorest and *D* the greatest potential.   *Barber, in Walsh and Beaton, Eds.,* Soil Testing and Plant Analysis, *p. 203, Madison, Wisc.: Soil Science Society of America, 1971.*

increasing yield potential or yield goal where *A* and *D* are low and high yield levels, respectively. However, *A–D* also represent other factors, like climate and crop/variety. For example, increasing rainfall from drought (*A*) to optimum rainfall (*D*) increases the yield potential and, therefore, the nutrient requirement. Different crops often respond differently to applied nutrients, where *A* might be a relatively unresponsive crop (e.g., soybeans) while crop *D* is extremely responsive (e.g., alfalfa or wheat). In this case, however, the *y*-axis would represent yield increase instead of yield. An example of the difference in crop responsiveness to soil P levels is shown in Figure 11.30. These data show that at 15 ppm Bray-1 P soybeans do not respond to P fertilization, whereas wheat and alfalfa are very responsive. Some laboratories may vary recommendations with expected yield level (Table 11.11).

Many laboratories make one recommendation, assuming best production practices for the region, and the grower may make adjustments as necessary. As technology and management practices improve or as economic incentives increase, yield potential and recommendations increase. *For the commercial grower, the goal is to maintain plant nutrients at a level for sustained productivity and profitability, which means that nutrients should not be a limiting factor at any stage, from plant emergence to maturity.*

Several fertilizer management programs can be adopted, depending on the nutrient. Nutrients that are immobile in the soil (i.e., P, K, Zn) require different management than mobile nutrients (e.g., N, S, Cl).

IMMOBILE NUTRIENTS   For immobile nutrients, Figure 11.31 represents several management options.

1. Buildup.   When the soil test is below the CL, it may be desirable to apply rates of a nutrient to increase the soil test to the CL or above (*A* in Fig. 11.31). Generally, applications of 10 to 30 lb/a of P_2O_5 are required to increase the

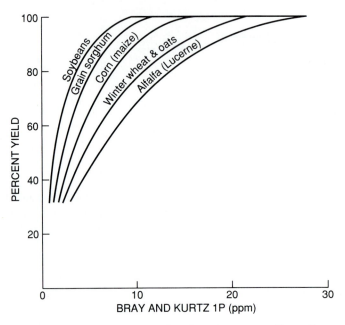

FIGURE 11.30   Response of various crops to soil test P level. *Olsen et al.*, Nat. Corn Handbook., *NCH-2, 1984*.

soil test P level 1 ppm, depending on the soil (Fig. 11.32). Similarly, to increase the soil test K level 1 ppm, 5 to 10 lb/a of K_2O is needed.

When the soil test is above the CL (*D* in Figure 11.31), continued applications to increase it further are advisable only if the grower or consultant believes that an economic yield response is probable. In contrast, if the soil test is above the CL, it may be possible, depending on the crop, to "drawdown" (*B*) the soil test level by not fertilizing the crop. Although this can be an viable option, soil test levels can decline rapidly; thus, annual soil testing would be required.

TABLE 11.11   P and N Recommended for Corn

Soil Test-Bray P_1 (lb P/a)	Yield Goals (bu/a)		
	80	120	160
	Annual Application of P_2O_5 (lb/a)		
5	55	70	85
15	45	60	75
25	35	50	65
30–60	30	45	60
75	20	30	45
90	20	20	30
	Annual Application of N (lb/a)		
Continuous corn	40	115	200

SOURCE: *Ohio Agronomy Guide 1983–1984*. Columbus, Ohio: Cooperative Extension Service, Ohio State Univ., 1983.

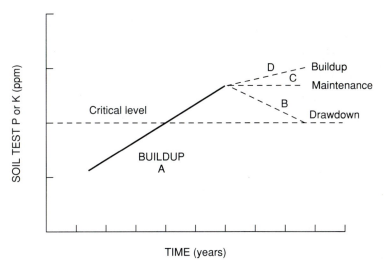

FIGURE 11.31 Hypothetical diagram representing several management options for an immobile nutrient. When the soil test level is increased above the CL, a grower could utilize residual fertilizer P reserves, which would cause the soil test level to decline (drawdown—B). Sufficient fertilizer P could be applied to maintain the soil test level (maintenance—C) or the soil test could be further increased with higher nutrient rates (buildup— D). The later program would be advisable only if the grower knew an economic crop response to the higher rate was probable or if a crop with a higher P requirement followed in the cropping sequence.

2. Maintenance. When the soil test is at or above the CL, it can be maintained by fertilizer rates that replace losses by crop removal, erosion, and fixation. This approach demands well-calibrated tests, and there is a considerably greater chance of error. When capital is limited, the area is new, or the land is being rented, the maintenance program is probably the preferred method.

In some rotations, such as corn-soybeans, fertilizing the corn at sufficient rates for both crops is commonly practiced. The favorable effect of applying high rates of K to corn on the following soybean crop is readily apparent in Figure 11.33.

Double cropping, or two crops in one year, is practiced in some areas. This may be small grain-soybean or small grain silage-soybean. In such instances, sufficient P and K is recommended for both crops before the small grain is planted. Again, the removals, losses, and any buildup must be considered in making the recommendations.

With any fertilizer management program for an immobile nutrient, the soil must be monitored periodically to determine if the soil fertility level is decreasing or increasing. The data in Figure 11.34 show that soil test P declined below the critical level when the irrigated alfalfa was not "fertilized" or

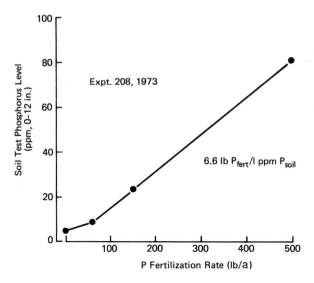

FIGURE 11.32   Effect of P fertilization on sodium bicarbonate-extractable P levels in a calcareous Portneuf silt loam soil, Kimberly, Idaho. *Westermann,* Proc. 28th Annu. Northwest Fert. Conf., *pp. 141–146, 1977.*

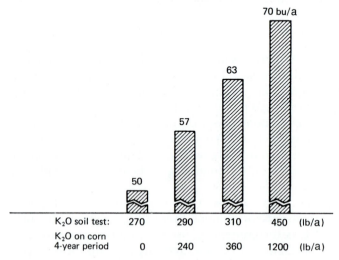

FIGURE 11.33   Carryover K applied the previous 4 years on corn increases soil test and soybean yield.   *Welch,* Better Crops Plant Food, *58(4):26, 1974–1975.*

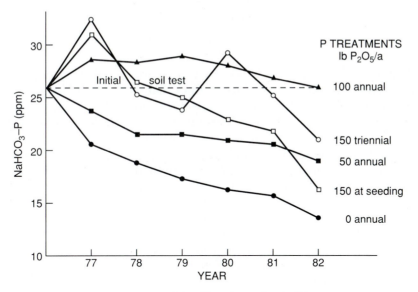

FIGURE 11.34   Influence of fertilizer P on the buildup, maintenance, and decline of soil test P in irrigated alfalfa.   *Havlin et al., SSSAJ* **46:331–336, 1984.**

was fertilized at a low P rate (50 lb/a of P_2O_5). In contrast, the P soil test was maintained at or slightly above the initial level with annual applications of 100 lb/a of P_2O_5. Notice how soil test level increased with a buildup rate of P (150 lb/a initially and every 3 years) and subsequently decreased with crop removal. These data clearly illustrate the value of annual soil testing.

MOBILE NUTRIENTS With mobile nutrients like NO_3^-, SO_4^{2-}, and Cl^-, buildup and/or maintenance programs are not practical, because these nutrients can readily leach below the root zone in many soils. Preventing potential NO_3^- contamination of groundwater by leaching while providing sufficient N for profitable crop production requires accurate N recommendations. N recommendations require knowledge of the quantity of N needed by the crop and supplied by the soil. In general, N recommendations are based on the following formula:

$$N_{fert} = N_{crop} - N_{soil} - (N_{org\,mat} + N_{prev\,crop} + N_{manure}) \qquad (1)$$

where

$$N_{fert} = \text{fertilizer N recommendation}$$

$$N_{crop} = \text{yield goal} \times \text{N yield}$$

$$N_{soil} = \text{preplant soil profile } NO_3^- \text{ content}$$

$$N_{org\,mat} = \text{organic N mineralization}$$

$$N_{\text{prev crop}} = \text{legume N availability}$$

$$N_{\text{manure}} = \text{manure N availability}$$

The N_{crop} represents the N required by the crop and involves predicting the crop yield and the N needed to produce that yield. Underestimating the yield goal can cause considerable yield loss due to underfertilization. Alternatively, overestimating the yield goal results in overfertilization, which can greatly increase the profile N content after the harvest and increase the potential for groundwater contamination if the N leaches below the root zone.

Many growers commonly overestimate their yield goals and thus apply N in excess of the crop requirement (Fig. 11.35). These data illustrate that only 10% of growers attained their desired yield goals, and only 50% of growers reached 80% of their yield goal. More important, overestimating the yield goal by these producers resulted in N recommendations that were 40 lb N/a greater than those required for a more realistic yield goal.

The quantity of N needed to produce the yield goal also varies among crops, regions or climates, and laboratories making the N recommendation. For example, 1 bu of corn contains 0.7 lb N (N yield in Eq. 1), assuming that all the N applied enters the corn grain. Unfortunately, N recovered by the grain can vary between 40 and 75%; thus, usually 0.9 to 1.7 lb N/bu grain is used for "N yield" to determine N_{crop}. With winter wheat, for example, a 1.8 to 2.4 lb N/bu yield goal is used to estimate the N required. Thus, if the yield goal is 60 bu/a, the total N required by the crop would be 120 lb/a (using 2.0 lb/bu).

Once N_{crop} is estimated, this value is reduced by the N credits or potential N available in the soil. Generally, in regions where evapotranspiration demand by the crop exceeds annual precipitation, measuring the profile NO_3^- content is valuable in predicting fertilizer N (Fig. 11.22). However, in humid regions where annual precipitation normally exceeds evapotranspiration, leaching can

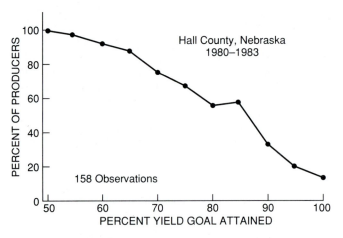

FIGURE 11.35   Success of producers in attaining the yield goal for irrigated corn. *Schepers and Martin,* Proc. Agric. Impacts on Groundwater. Well Water Journal Publ. Co., *Dublin, Ohio, 1986.*

occur; thus, profile N measurements have not been reliable in predicting fertilizer N requirements. Laboratories that utilize residual profile NO_3^- in making N recommendations are generally in regions where little or no water percolates below the root zone (Fig. 11.36).

After adjustments for soil profile NO_3^- content, N_{crop} is adjusted for

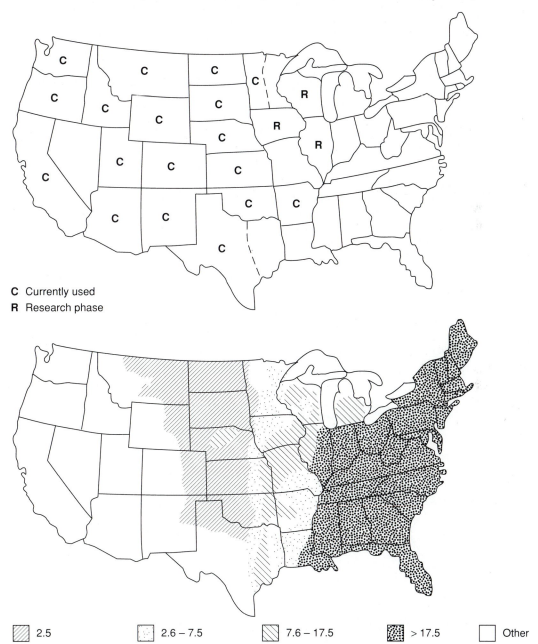

C Currently used
R Research phase

| | 2.5 | | 2.6 – 7.5 | | 7.6 – 17.5 | | > 17.5 | | Other |

FIGURE 11.36   Use of residual profile NO_3 analysis in the U.S. relative to average annual potential percolation (shading indicates values in centimeters) below the root zone.   *Hergert*, SSSA Spec. Publ. No. 21., *1987*.

potential N mineralization from soil OM, legume N, and manure N. The N credit from manure varies with the rate applied and is approximately 5 lb N/ton in the first year following manure application. Generally, 50% of the manure N is available the first year, 25% the second year, and none in the third year.

The N credit for previous legume crops depends on the legume crop, the productivity of the legume (i.e., yield), and the length of time after the legume crop was rotated to the nonlegume crop (see Chapters 5 and 15). N credits from forage legumes are generally much greater than those from grain legumes, although low-yielding forage legumes can fix less N_2 than high-yielding grain legumes (Table 5.2).

The generalized relationship between N required by corn grown continuously or in rotation with alfalfa and mixed hay is shown in Figure 11.37. When nonlegume crops are grown on soils previously cropped to a forage legume, potential legume N mineralization decreases with time (Fig. 5.2). Thus, very little fertilizer N is required in the first year following the legume compared to subsequent years. The number of years a legume N credit is used in adjusting the fertilizer N rate also depends on the quantities of N fixed, which depends on the specific legume and its productivity. The N credit for soybeans, a grain legume, is usually 1 lb N/bu soybean yield.

Few laboratories use a term for $N_{org\ mat}$ in their N recommendation models because of the difficulty in accurately estimating the quantity of N mineralization under variable climate (moisture and temperature) conditions from year to year. Although many tests have been evaluated, estimates of potential N

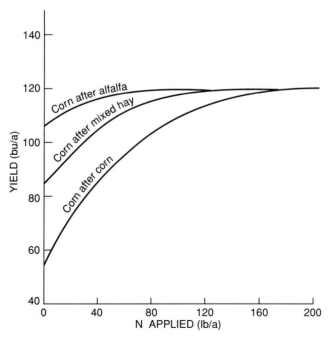

FIGURE 11.37 Influence of previous crop on corn grain yield response to N fertilization. *Barber*, SSSA Spec. Publ. No. 2., *1967.*

mineralization from soil OM have not always been highly correlated with N availability. Some laboratories reduce the coefficient multiplied by yield goal (1.8 lb N/bu instead of a higher value, for example) to account for some N mineralization. Alternatively, some laboratories use a measure of soil OM as an indicator of potential mineralizable N. Credits for $N_{org\ mat}$ generally range from 20 to 80 lb N/a and are primarily based on the %OM content in the soil. For example, as %OM increases between soils, $N_{org\ mat}$ increases; thus, the N recommendation will decrease (Table 11.12).

In general, when the fertilizer N rate exceeds the N requirement for optimum crop yield, considerable quantities of residual fertilizer N exist after harvesting, which represents potentially leachable NO_3^- (Fig. 11.38). However, when the fertilizer N recommendation is estimated by accurately quantifying N_{crop} and the N credits, the N rate should maximize the recovery of fertilizer N by the crop. The data in Figure 11.38 illustrate that when fertilizer N exceeded the N rate needed for maximum yield (200 lb N/a), residual profile NO_3^- after harvesting increased greatly. Since rooting depth is only about 5 to 6 ft for corn, a portion of the profile NO_3^- moves below the root zone and, thus, is unavailable to the next crop. Eventual contamination of the groundwater could occur as this N continues to leach down through the profile. These data demonstrate the importance of accurately quantifying fertilizer N requirements for optimum production and minimum environmental risk.

PRE-SIDE-DRESS SOIL TEST   Recent development of an N soil test for corn may improve the accuracy of fertilizer N recommendations in humid regions. In this test, surface 12-in. soil samples are collected between rows when corn is 12 in. high and analyzed for NO_3^-. The data in Figure 11.39 suggest that when NO_3^- exceeds 20–24 ppm, fertilizer N is not required. The theoretical basis of this test is that the soil sample is collected at or near maximum N mineralization; thus, the contribution of $N_{org\ mat}$ (Eq. 1) is more accurately

TABLE 11.12   Influence of Soil Profile NO_3^-
and OM Content on Fertilizer N Recommendations for
Dryland Winter Wheat in Colorado

NO_3^-* Soil Test (ppm)	Soil OM % (Surface Sample)		
	0–1.0	*1.1–2.0*	*>2.0*
	Fertilizer N (lb/a)		
0–4	75	75	65
5–8	75	75	50
9–12	75	60	35
13–16	70	45	20
17–20	60	30	5
21–24	45	15	0
25–28	30	0	0
29–32	15	0	0
>32	0	0	0

*0- to 2-ft sample. *Guide to Fert. Recommendations*, XCM-37. Colo. State Univ. (1985).

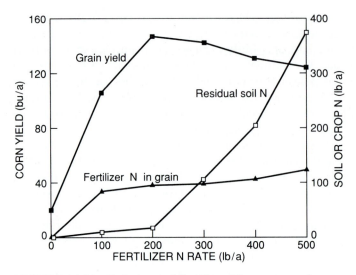

FIGURE 11.38   Influence of fertilizer N on corn grain
yield, fertilizer N recovered in the grain, and residual
profile NO$_3$ after harvest.   *Broadbent and Rauschkolb,*
Calif. Agric., *31:124–125, 1977.*

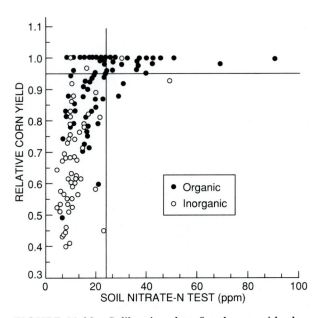

FIGURE 11.39   Calibration data for the pre-side-dress
soil NO$_3$ test for Pennsylvania. "Organic" represents soils
with a history of manure or legumes, whereas
"Inorganic" indicates soils without this history.   *Beegle,*
Proc. Indiana Ag. Chem. Conf., *1982.*

quantified (Fig. 11.40). Applying a portion of the fertilizer N requirement at an early crop growth stage could increase fertilizer N efficiency through synchrony of fertilization with the period of maximum crop N uptake.

RESPONSES ON HIGH TESTING SOILS   Many factors affect crop response to a nutrient—temperature, moisture, time, placement, tillage, yield goal, crop, and so on. Hence, even on a soil testing very high, a starter fertilizer may be recommended for some crops, especially if they are planted early under cool, wet conditions.

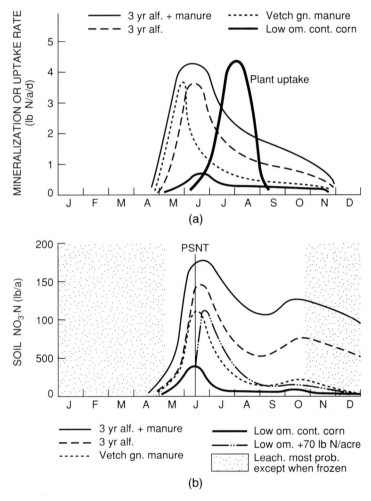

FIGURE 11.40   Hypothetical synchronization of N mineralization and crop N uptake (*a*) and subsequent accumulation of NO_3 (*b*) as influenced by previous crop. Theoretically, the pre-side-dress NO_3 test (PSNT) is determined from a soil sample collected at maximum N mineralization/NO_3 production, which corresponds to 12-in.-high corn.   *Magdoff*, J. Prod. Ag. *4:297–305, 1991.*

Importance of Soil Tests to the Farmer and to the Lime and Fertilizer Industries

It should be recognized that soil testing is not an infallible guide to crop production. The problems of representative samples, accurate analyses, correct interpretation, and environmental factors that influence crop responses strongly influence nutrient recommendations and crop recovery of applied nutrients. However, the soil test helps to reduce the guesswork in fertilizer practices. The test may be used to monitor the soil periodically to determine general soil fertility levels. Rates of lime and fertilizer are then applied to supply adequate quantities to the current crop or crops. Consultants and fertilizer dealers recognize the importance of soil tests, as well as other diagnostic techniques, in helping to predict the plant nutrients needed for profitable crop production.

Fertilizer recommendations based on soil analysis results will differ, depending on the interpretation or philosophy (buildup, maintenance, etc.) and therefore are only a good guide in estimating the fertilizer rate required. However, a regular soil testing program is essential for maximizing profitability, sustaining long-term soil/crop productivity, and minimizing the environmental impact from fertilizer and manure use. It is no coincidence that the most skilled, efficient, and profitable growers include a regular soil testing program in their soil/crop management system.

Soil Test Summaries

The purpose of soil testing is to give the individual growers dependable information regarding the fertilizer and lime needs of their fields. Soil test laboratories often summarize the soil test results obtained over time for various purposes. These include the following:

1. Identifying specific areas of relative nutrient deficiency/sufficiency (Table 11.13).
2. Quantifying the distribution of crops and yield levels produced at various soil test levels (Table 11.14).

If these summaries are prepared periodically, changes in soil test levels over time can be identified (Fig. 11.41). These data show that P soil test levels have slowly increased over time, which may represent increasing use of P fertilizers.

TABLE 11.13   Summary of K Tests (Percent) for Field Crops by Soil Management Group, Michigan

Soil Management Group	Soil Test K (lb K/a)					
	<110	110–159	160–209	210–249	250–299	>300
Organic	12.8	13.2	16.7	10.5	7.5	39.4
Clay loam or loam	11.1	19.7	26.0	14.8	11.1	17.2
Sandy loam	22.0	23.6	24.1	11.1	8.3	11.0
Loamy sand	43.4	23.8	16.9	6.3	3.9	5.8

SOURCE: V. W. Meints, Michigan State Univ., personal communication.

TABLE 11.14 P Level Related to the Crop to Be Grown in North Carolina (1991)

Crop	Number Tested	VL	L	M	H	VH
		Percent of Samples Testing				
Legume–grass pasture	2,773	15.9	19.6	23.1	23.1	18.3
Soybeans	24,394	1.9	9.1	23.6	30.7	34.1
Small grains	14,101	1.5	7.2	24.3	33.2	33.8
Corn	42,976	1.8	9.3	26.8	32.5	29.6
Tobacco (FC and Bur.)	14,746	1.2	3.4	9.2	24.3	61.9
Potatoes, Irish	2,276	0.6	2.9	14.3	35.1	47.1

SOURCE: D. W. Eaddy, North Carolina Dept. of Agriculture, personal communication.

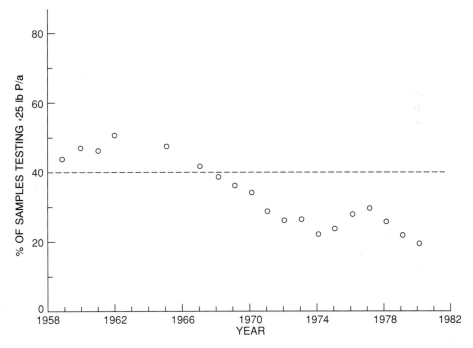

FIGURE 11.41 Changes in soil test P from 1958 to 1982 in central Kansas. NCR Res. Publ. No. 310, *1986.*

Therefore, soil test summaries can be useful in alerting growers, consultants, and others about potential nutrient problem areas.

Remote Sensing

Remote sensing is the collection and analysis of data on features of the earth's surface. The information is acquired through cameras and other imaging systems mounted in or on aircraft or space vehicles. Although remote sensing has progressed beyond the acquisition and analysis of aerial photographs, the field of aerial photography is still a basic component.

The scope of aerial photography is limited by the spectral sensitivities of photographic film emulsions, the spectral resolution of photographic filters, and the logistical problems associated with obtaining photographs via unmanned space vehicles. These limitations have been overcome through the development of various nonphotographic imaging systems installed on earth-orbiting satellites. Data acquired by means of satellite imaging systems provide scientists with views of the earth not achievable with conventional aerial photography. Satellite systems provide repetitive coverage of very large areas with a spatial resolution that allows each half hectare to be studied.

Through satellite imaging it is feasible to monitor land cover changes, both natural and those resulting from human activities, that would otherwise require thousands of aerial photographs if conventional methods were used. Interpretation of satellite images can assess crop conditions and indicate the location and size of weather systems. Figure 11.42 is a satellite image of the Prairie Provinces of Canada and part of the northern United States, showing the areas affected by a late summer frost.

Aerial photography and field surveys are still required for detailed studies of relatively small areas. In recent years, aerial photography has been used successfully as a research tool for studies of population densities in livestock and wildlife, forest inventories, movement of ocean currents, land-use patterns over large areas, physical and chemical characteristics of soils, plant diseases, insect infestations, crop vigor, and winter injury to perennial crops such as alfalfa and to fall-seeded small grains.

Infrared aerial crop photography is a reliable means of monitoring crop production systems. Near-infrared film is used to take vertical photographs at approximately 8,000 ft above sea level. The photographs will often cover as much as 1 sq mi of land.

Healthy green plants reflect a large amount of infrared light, whereas plants under stress caused by such factors as drought, nutrient deficiency, disease, weeds, and chemical spray damage do not reflect infrared light. These differences can be clearly recorded on infrared film. Other crop management problems that can be detected by this technique include inadequate drainage, poor stand establishment, and uneven application of fertilizers and herbicides. A number of such problems are identified in Figure 11.43, which is a black-and-white reproduction of an aerial infrared photograph.

Summary

1. The selection of adequate lime and fertilization practices depends on crop requirement, yield goal, weather, and soil characteristics, and necessitates finding a method that will reveal the deficiencies in the soil.

2. Although diagnostic approaches are used in troubleshooting, they are more important as preventive measures.

3. Deficiency symptoms are helpful guides in new fertilizer-using areas. In high-use areas they are a sign of mediocre farming practices and are often difficult to interpret because of their complication with many other problems.

4. Hidden hunger is insidious, but careful plant and soil tests will help to avoid it.

5. The plant integrates all factors in the environment into itself, and plant

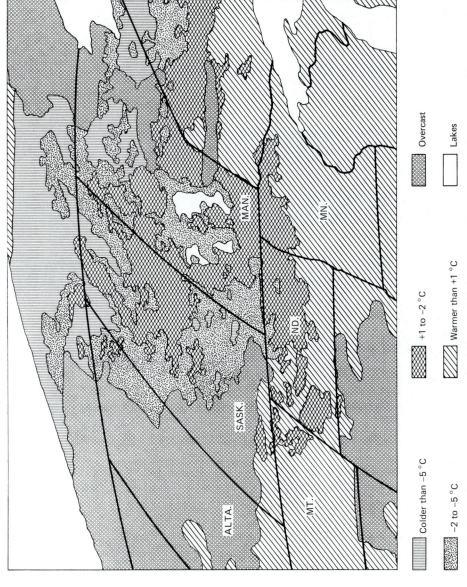

Colder than −5 °C

−2 to −5 °C

+1 to −2 °C

Warmer than +1 °C

Overcast

Lakes

FIGURE 11.42 Geostational Operational Environmental Satellite (GOES) image of the Prairie Provinces of Canada, showing crop production areas affected by severe freezing on August 27, 1982. The impact of this unseasonable weather was greatest in the southwestern corner of Manitoba and much of the grain-growing area of Saskatchewan. The rest of the area either had milder temperatures or was protected from the cold front by cloud cover.   *Canadian Wheat Board, Grain Matters (September 1982)*.

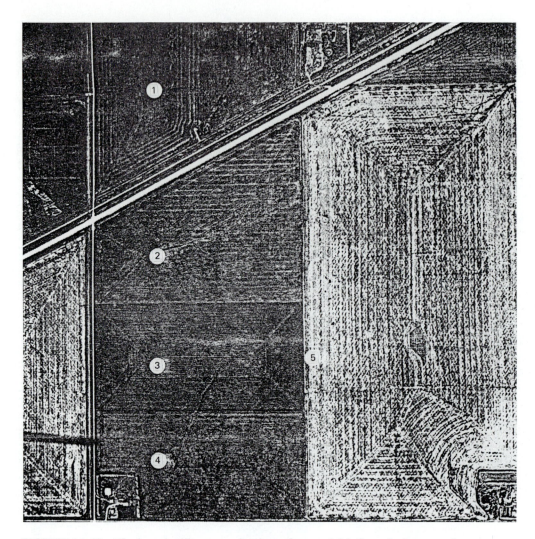

FIGURE 11.43   Black-and-white reproduction of an aerial infrared photograph
showing the effects of improper fertilization and poor drainage on crop growth.
The dark black areas correspond to a deep red color on the infrared photograph
and they represent healthy plant growth. Gray portions of the reproduction
equivalent to a pinkish color on the original are indicative of stressed plants. White
areas in this figure are the counterpart of a greenish tint on the infrared
photograph and they signify exposed soil showing through a seedling crop or
sparse plant growth. 1, caused by uneven spreading of granular fertilizer; 2, liquid
fertilizer misses; 3, N fertilizer not applied to center portion of field; 4, drainage
problem in a flax field; 5, P fertilizer not applied to outer edges of field.   The
Furrow *(John Deere), p. 27 (January 1978)*.

tests can be highly revealing. Quick tissue tests in the field on growing plants are useful, but careful interpretation is essential.

6. As a nutrient is added, the percentage of that nutrient in the plant usually increases. It is important to identify the point at which there is no further economic yield increase.

7. Critical nutrient range rather than a specific critical nutrient concentration seems more meaningful in the evaluation of plant nutrient status.

8. Caution must be exercised in the interpretation of plant nutrient ratios.

9. The Diagnosis and Recommendation Integrated System (DRIS) is a greatly improved approach to plant analysis.

10. Plant analyses are of great value in making surveys of incipient micronutrient problems in a given area.

11. Balance among nutrients in the plant may be just as important as actual amounts. The relationships among Ca, Mg, K, and NH_4, Mn and Fe, and Zn and P are examples.

12. There are many biological short-term methods in which both higher and lower plants are used for determining nutrient needs. Eventually, all of these methods must be related to field responses for calibration.

13. The principle of soil testing is to obtain a value that will help to predict the amount of nutrients needed to supplement the supply in the soil. Soil tests are of little use in themselves. They must be calibrated against nutrient rate experiments in the field and in the greenhouse. Also, soil fertility is only one of many factors influencing crop production.

14. Soil is very heterogeneous or nonuniform. This characteristic must be adequately taken into account in soil sampling and application of plant nutrients.

15. The physical properties of soils become more and more important as the most profitable yields are approached, but much more work is needed to identify favorable and unfavorable physical conditions.

16. Soil tests can be classified on the basis of probability of response. For example, in a low test for P there would be a high probability of response to this element, although in some cases it might not be obtained because of other limiting factors.

17. Most calibration experiments need to be rerun using the new technology in crop production, to obtain maximum yields for the soil and environment, and to provide greater opportunity for response. These results will have more meaning for the commercial grower.

18. Two or more levels of recommendation or provision for higher yields, made by some laboratories, help to provide benefits from soil testing for the leading growers. The goal is to *maintain soil fertility at a level for optimum profit yields.*

19. In general, there are four approaches in recommendations: (a) *buildup* with heavy broadcast rates plus maintenance; (b) *annual application* with modest additions to each crop in the rotation; (c) *rotational fertilization*, a combination of the first two; (d) *replacement* of nutrients removed in crops.

20. Prescription and debit and credit methods, in which the contributions of the soil, crop residues, fertilizer, and manure are related to crop needs, are useful. Although the values obviously are influenced by many factors, the methods serve to point out broad needs and possible adjustments.

21. The lime and fertilizer industries can utilize soil and plant tests in a complete service program to provide better identification of the nutrient needs of farmers' fields.

22. Summaries of soil tests help to determine the most needed fertilizer ratios and corrective applications. It is possible to develop more realistic general fertilizer recommendations for use by the growers, the vast majority of whom still do not have their soils tested.

23. P and K levels are being built up in some soils. Excessive buildup must be studied carefully. However, increased fertility levels encourage higher yields and aid the plant in periods of stress. Nutrient balance must be continually evaluated.

Questions

1. For what reason may a plant develop an unusual red or purple color? What factors encourage this change in color? Distinguish between N- and K-deficiency symptoms in corn.
2. What factors must be taken into consideration in interpreting tissue tests?
3. What are critical nutrient concentrations (CNC) and critical nutrient ranges (CNR)?
4. What is the Diagnosis and Recommendation Integrated System (DRIS)? What are its advantages?
5. Is it possible to misinterpret plant nutrient ratios in diagnosing nutritional problems?
6. What is the difference between tissue analysis and plant analysis?
7. Why cannot just any part or growth stage of crops be used in most systems of plant analysis? Are there approaches that eliminate such restrictions?
8. How does cation balance influence the interpretation of plant analysis for a given cation?
9. Why must soil tests be calibrated with crop response? How would you set up a series of experiments to determine the calibration of the P test for corn soils in your area?
10. Explain why a response to P would be more generally expected in the northern United States than in the southern states.
11. From the standpoints of the grower and the agencies making the tests, what are the greatest problems in the soil-testing program in the area in which you live? What can be done to remedy the situation?
12. Would you apply a given nutrient if there were a 50% chance of obtaining a response? A 25% chance? Why or why not?
13. Of what value to a county agent or district agriculturist is a summary by crops of the soil test results? How can a summary be prepared?
14. Name some advantages and disadvantages of deficiency symptoms, tissue tests, and soil tests for detecting plant-nutrient needs.
15. What is crop logging?
16. Ten percent of a grower's field is black lowland soil and the remainder is light-colored upland soil. How should the field be sampled? How frequently should it be resampled?
17. What complicates the securing of a good correlation of soil tests for N with the response to N in the field?
18. Why does the probability of a response to a nutrient vary at a given soil test level?
19. Can soil variability be a problem in soil fertility research?

20. Explain the percentage maximum yield concept. Under what conditions might it not be accurate?

21. What are four general types of recommendations? List advantages and disadvantages of each in your area.

22. If you were a fertilizer dealer, what kind of a crop production service would you have for your customer? (Remember that it costs money.)

23. Are soil fertility levels increasing in your area? What are the advantages? Disadvantages?

24. Is deep soil sampling a useful practice? State reasons for your answer.

25. Does band application of nutrients such as P and K complicate soil sampling procedures? Explain your answer.

26. Under what conditions might there be a response to added nutrients on a soil high in available nutrients?

27. Is there a reliable N soil test in high-rainfall areas?

28. How often should farm fields be soil sampled?

Selected References

BROWN, J. R. (Ed.). 1987. *Soil Testing: Sampling, Correlation, Calibration, and Interpretation*. Special Publication No. 21. Soil Science Society of America, Madison, Wisc.

WALSH, L. M., and J. D. BEATON. 1973. *Soil Testing and Plant Analysis*. Soil Science Society of America, Madison, Wisc.

WESTERMANN, R. L. (Ed.). 1990. *Soil Testing and Plant Analysis*. No. 3. Soil Science Society of America, Madison, Wisc.

WHITNEY, D. A., J. T. COPE, and L. F. WELCH. 1985. Prescribing soil and crop nutrient needs. *In* O.P. ENGELSTAD (Ed.), *Fertilizer Technology and Use*. Soil Science Society of America, Madison, Wisc.

CHAPTER 12

Fundamentals of Fertilizer Application

Adequate fertilization programs supply plant nutrients needed to sustain maximum crop productivity and profitability while minimizing environmental impact from nutrient use. In essence, fertilizers are used so that soil fertility is not a limiting factor in crop production. The major factors influencing the quantity of nutrients to apply are crop characteristics; soil characteristics; fertilizer placement; climate, especially moisture and temperature; yield goal; and economics.

Crop Characteristics

Nutrient Utilization

The approximate amounts of nutrients in crops (Table 12.1) vary considerably, depending on (1) variety or hybrid, (2) moisture availability, (3) temperature, (4) soil type, (5) nutrient levels and their balance in the soil, (6) plant population, (7) tillage practices, and (8) pest control. Harvesting of the complete crop will result in greater removal of nutrients; however, 65 to 75% of total N and P uptake can be accounted for in the grain, with much smaller proportions of K, Mg and other nutrients.

The percentage of total uptake during various growth stages for corn, soybeans, and sorghum is shown in Table 12.2. During the first two growth stages, all three crops take up a higher percentage of their total K than of N or P. This illustrates the importance of an adequate supply of K early in the life of the plant. In the last two growth periods, the %N and %P uptake is greater than for K. It is important to have a sufficient nutrient supply in the soil to provide crop needs during the entire growing season, especially during periods of peak nutrient demand.

Root Characteristics

The growth and appearance of the above- and belowground portion of plants vary considerably among and within species, depending on the environment.

TABLE 12.1 Approximate Utilization of Nutrients by Selected Crops

Plant	Yield per Acre	N	P	K	Mg	S
				lb		
Alfalfa	10 tons	600	52	500	53	51
Orchard grass	6 tons	300	44	313	25	35
Coastal Bermuda	10 tons	500	61	350	50	50
Clover-grass	6 tons	300	39	300	30	30
Corn						
Grain	200 bu	150	40	48	18	15
Stover	8,000 lb	116	12	174	47	18
Sorghum						
Grain	8,000 lb	120	26	25	14	22
Stover	8,000 lb	130	13	142	30	16
Corn silage	32 tons	266	50	222	65	33
Cotton						
1500 lb lint, 2,250 lb seed		94	17	37	11	7
Stalks, leaves, and burrs		86	11	68	24	23
Oats						
Grain		80	11	17	5	8
Straw	100 bu	35	7	104	15	11
Peanuts						
Nuts	4,000 lb	140	10	29	5	10
Vines	5,000 lb	100	7	125	20	11
Potatoes, Irish						
Tubers	500 cwt	173	32	233	14	15
Vines	300 cwt	96	7	221	36	7
Potatoes, sweet						
Roots		73	15	140	8	—
Vines	300 cwt	83	13	120	10	—
Rice						
Grain	7,000lb	77	20	23	8	5
Straw	7,000lb	35	6	100	6	7
Soybeans						
Grain	60 bu	240	20	70	17	12
Straw	7,000 lb	84	7	48	10	13
Tomatoes						
Fruit	30 tons	100	10	180	8	21
Vines	4,400 lb	80	11	100	20	20
Wheat						
Grain	80 bu	92	19	23	12	5
Straw	6,000 lb	42	4	113	12	15
Barley						
Grain		110	17	29	8	10
Straw	100 bu	40	7	96	9	10
Sugar beets						
Roots	30 tons	125	7	208	27	10
Tops	16 tons	130	11	250	53	35
Sugarcane						
Stalks		160	39	279	40	54
Tops and trash	100 tons	200	29	229	60	32

SOURCE: *Better Crops Plant Food*, **63**:4 (Spring 1979).

TABLE 12.2   Percentage of the Total Nutrient
Requirement Taken Up at Different Growth Stages

| | Corn Growth Periods (days) | | | | |
	0–25	26–50	51–75	76–100	101–115
N	8	35	31	20	6
P	4	27	36	25	8
K	9	44	31	14	2
	Soybean Growth Periods (days)				
	0–40	41–80	81–100	101–120	121–140
N	3	46	3	24	24
P	2	41	7	25	24
K	3	53	3	21	20
	Sorghum Growth Periods (days)				
	0–20	21–40	41–60	61–85	86–95
N	5	33	32	15	15
P	3	23	34	26	14
K	7	40	33	15	5

SOURCE: Basic data on soybeans and sorghum from North Carolina and Kansas, respectively. Corn, soybean, and sorghum data
appeared in *Better Crops Plant Food,* **56:**2 (1971) and **57:**4 (1973).

Since the majority of plant nutrients are absorbed by roots, understanding
rooting characteristics is important in developing efficient fertilization practices. Root systems are usually either fibrous or tap, and both occur with
annuals, biennials, or perennials. The ability of roots to exploit the soil for
nutrients and water is dependent on the morphological and physiological
characteristics of roots. Root radius, root length, root surface/shoot weight
ratio, and root hair density are the main morphological features. The presence
of mycorrhizae also may be important.

SPECIES AND VARIETY DIFFERENCES   Knowledge of early rooting characteristics is helpful in determining the most effective method of fertilizer placement. If a vigorous taproot is produced early, applications may best be placed
directly under the seed. If many lateral roots are formed early, side placement
may be best.

A diagram of the extent of root development of several crops 2 and 3 weeks
after planting shows that at 2 weeks the corn root system is more extensive
than that of cotton (Fig. 12.1). The root development of cotton suggests that,
at least for early absorption of nutrients, the presence of nutrients under the
plant is important. At 3 weeks, corn has the most extensive root system and
cotton the most restricted one. These differences tend to persist as long as 3
months after planting.

Carrots show considerable root activity at 33 in. soil depth (Fig. 12.2), much
more than that of onions, peppers, and snap beans. The reduction in activity
at 13 in. marks the beginning of the mottled compact subsoil in the Blount
soil. There is a marked effect of soil, and activity is much less at 33 in. in peat.

Corn depends heavily on P near the roots early in the season. However,

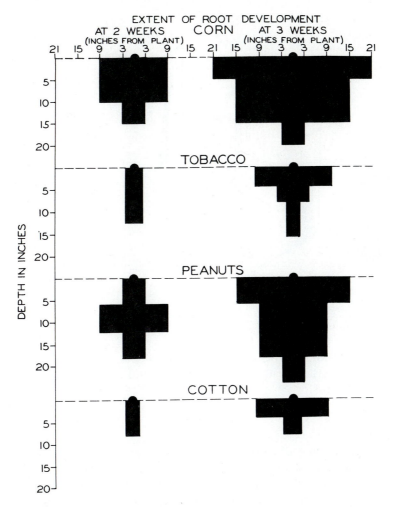

FIGURE 12.1   Root development at 2 and 3 weeks of
corn, tobacco, peanuts, and cotton.   North Carolina Agr.
Expt. Sta. Tech. Bull. 101, *1953*.

corn develops an extensive root system and has a great capacity for utilizing
the nutrients distributed throughout a large soil zone. The root systems of
corn and soybean exploit the soil thoroughly, in contrast to those of carrots,
cotton, potatoes, and other shallow-rooted crops (Fig. 12.3). Potatoes have a
limited root system, often being confined by the hilled row, where the roots
may penetrate only 10 to 20 in. below level ground.

Early root growth of most crops occurs mainly in the topsoil (Fig. 12.4).
Both root surface per plant and root density in the subsoil, however, increase
considerably with plant age, and by harvest they are equal to or greater than
those in the topsoil. On sandy soils, corn roots can reach a depth 8 ft or more
and completely extract available soil moisture down to 6 ft. Corn root weights
of 3,000 lb/a or more have been found.

Small grains have an extensive root system (Figs. 12.5, 12.6, and 12.7).
The development of wheat roots from tillering to grain filling shows that, at

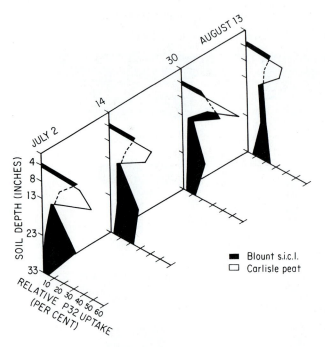

FIGURE 12.2   Root activity of carrots on Blount silty
clay loam and Carlisle peat.   *Hammes et al., Agron. J.,*
55:329, 1963.

maturity, most of the wheat roots are concentrated in the surface soil. An
early response of small grains to P placed near the seed even on medium- to
high-P soils is commonly observed. Alfalfa roots may penetrate 25 ft if soil
conditions are favorable (Fig. 12.5). Depths of 8 to 10 ft are common even on
compact soils. One of the advantages of deep-taprooted crops such as alfalfa
and sweet clover is that they loosen compact subsoils by root penetration and
subsequent decomposition. Also, legumes in pastures provide more animal
feed during drought periods than do shallow-rooted grasses.

There is a tendency for root systems of the same species not to interpene-
trate, which suggests an antagonistic or toxic effect (Fig. 12.8). Thus, with
narrow row spacing and high populations, the characteristic root pattern is
altered and there may be deeper rooting if soil conditions permit.

The tillage system also affects root distribution with depth (Table 12.3).
When soil is cultivated annually, corn roots develop more extensively below
10 cm than with no-till, while intermediate root distribution occurs with rototill
and chisel. When residues are removed, there is greater root growth in the
surface 15 cm. Thus, residue decomposition products may inhibit root growth.

Differences in the extent of root development among varieties of corn and
wheat are evident in Figures 12.4 and 12.7, respectively. Numerous other
crops, including soybeans and sweet potatoes, exhibit differences in root development related primarily to deeper rooting.

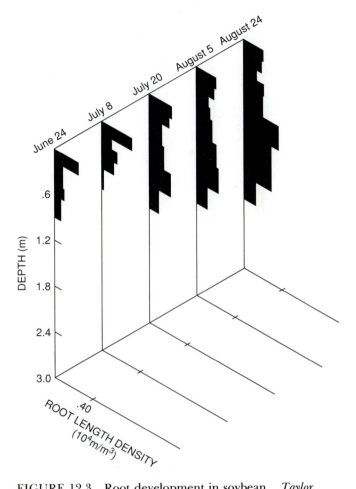

FIGURE 12.3   Root development in soybean.   *Taylor,*
Agron. J., *72:543–547, 1980.*

NUTRIENT-EXTRACTING POWER   Since roots occupy about 1% of the topsoil
volume and much less in the subsoil, nutrient absorption characteristics and
root–soil interactions can influence nutrient uptake. The exchange capacity of
roots of dicotyledonous plants is much higher than that of monocotyledonous
plants. Nonlegumes have a lower requirement for divalent cations and take
up more monovalent cations. The relative absorption of cations and anions
by the root is related to the release of H^+ or HCO_3^- by the root. Acidity
develops from the release of H^+ from the root in response to absorption of
NH_4^+, whereas pH increases with the release of HCO_3^- and/or OH^- following
uptake of NO_3^-. Changes in rhizosphere pH affect the solubility and availability of many plant nutrients.

   Mycorrhizal fungi are often associated with plant roots grown under low
soil fertility conditions and can increase the ability of plants to absorb nutrients
(Fig. 12.9). Addition of N and/or P fertilizers and soil tillage can reduce the
contribution of mycorrhiza-related nutrient uptake.

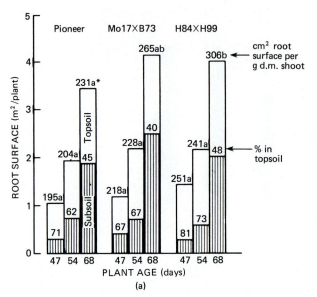

*Different letters indicate significant differences at
particular plant age.

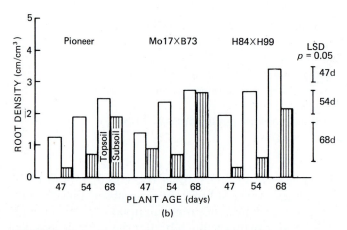

FIGURE 12.4   Root surface per plant (a) and root
density (b) in topsoil and subsoil for three corn genotypes
at three harvest dates.   *Schenk and Barber,* Plant Soil,
54:65, 1980.

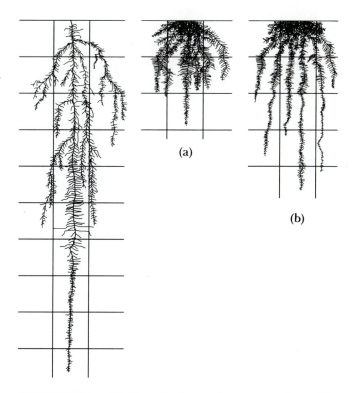

FIGURE 12.5   Alfalfa and winter wheat roots, where (a)
and (b) are wheat roots under dryland and irrigated
conditions, respectively. Grid lines are 30 cm
apart.   *Russel,* Plant Root Systems, *McGraw-Hill, New
York, 1977; taken from Weaver,* Root Development of Field
Crops, *McGraw Hill, New York, 1926.*

Soil Characteristics

Soil characteristics have a pronounced influence on the depth of root penetration, as soils with compact B horizons are highly restrictive. Effects of soil
series in Illinois on corn root development are shown in the accompanying
table. A number of factors, including bulk density, OM content, acidity and
plant nutrient content, old root channels, and O_2 supply caused the rooting
differences.

	Soil Series				
	Flanagan	*Muscatine*	*Clarence*	*Wartrace*	*Cisne*
Corn root weights (lb/a)	1,846	2,008	1,758	3,136	2,647
Depth of penetration (in.)	60	66	38	48	60
Water-holding capacity to rooting depth (in.)	12.8	17.4	6.4	17.1	17.3

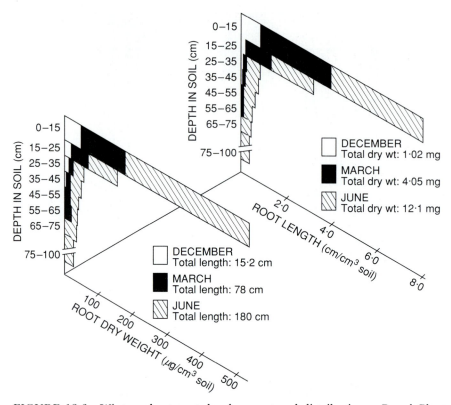

FIGURE 12.6   Winter wheat root development and distribution.   *Russel*, Plant
Root Systems, *McGraw-Hill, New York, 1977; based on Welbank et al.,* Rep.
Rothamsted Exp. Stn., *1974.*

The yield of a crop is often directly related to the availability of stored water
in the soil. This amount is related to soil characteristics. The following yields
of corn were produced at different levels of plant available H_2O:

Available H_2O in Profile (in.)	Corn Yield (bu/a)		
	Lafayette, Indiana	Urbana, Illinois	Ames, Iowa
4	79	85	88
8	121	128	129
12	130	136	135

Basically, the soil is a rooting medium and a storehouse for nutrients and
water. Hence it is essential that the roots fully exploit the soil in order that
the plants not only obtain nutrients but also root deeply enough to reduce
water stress.

In drought-prone areas with root restrictive soil layers, subsoiling can increase rooting depth and plant-available water, especially in crops like soy-

WIEGHT OF ROOTS PER PLANT

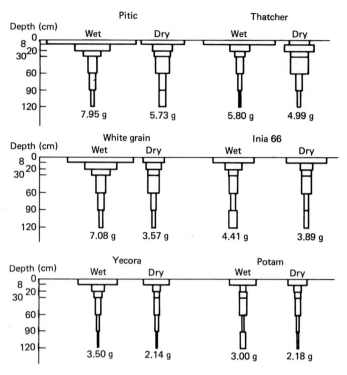

FIGURE 12.7   Weight of roots of wheat cultivars grown at high and low moisture levels.   *Hurd, Agron. J., 60:201, 1968.*

beans that have limited ability to penetrate even moderately compacted soil layers. Both soybeans and corn have responded to in-row subsoiling.

In some soils, no-till management can cause some restricted root growth because of increased bulk density compared to full tillage management (Fig. 12.10). Attempts to loosen plowpans or heavy subsoils have not been entirely successful. Subsoiling is most effective when the subsoil is dry so that shattering of the soil occurs; however, in most cases, there is a rapid resealing of the subsoil. One cultivation with a disk or another such implement may almost

TABLE 12.3   Effect of Tillage Treatment on the Corn Root-Weight Distribution (mg/100 cm³ of Soil) with Depth

Depth (cm)	Tillage Treatment				
	Conventional	Conventional, No Residues	Chisel	Rototill	No-till
0–5	29	49	26	69	137
5–10	38	52	104	136	100
10–15	96	218	110	137	74
15–30	85	111	68	88	73
30–45	73	61	47	52	28
45–60	61	59	35	52	37

SOURCE: Barber, *Agron. J.,* **63:**724 (1971).

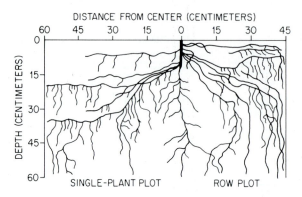

FIGURE 12.8   Contrasting rooting patterns of soybeans
in single-plant plots and row plots.   *Raper and Barber,*
Agron. J., *62:581, 1970.*

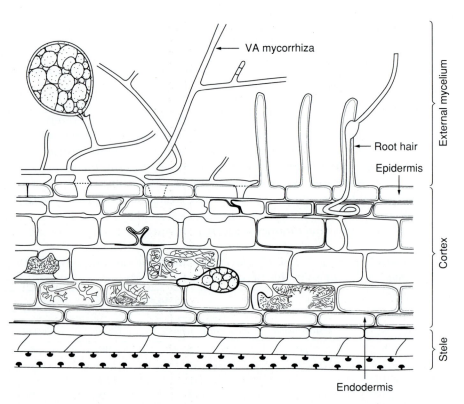

FIGURE 12.9   Schematic diagram of a root infected by vesicular arbuscular
mycorrhizal fungi.   *Russel,* Plant Root Systems, *McGraw-Hill, New York, 1977.*

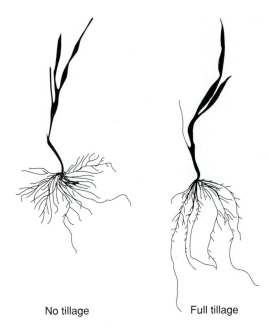

No tillage   Full tillage

FIGURE 12.10   Barley root growth under no-tillage and full-tillage systems. *Russel,* Plant Root Systems, *McGraw-Hill, New York, 1977; after Ellis,* J. Agric. Sci., *1977.*

eliminate any effect of subsoiling. Vertical mulching, where chopped plant residues are blown into the slit behind the subsoiler, serves to keep the channel open and improve water uptake.

Crop production on sodic claypan or Solonetzic soils may be only about one-third of that possible on Na-free soils in the same area. Deep plowing to depths of 16 to 24 in. and subsequent mixing of surface soil, the hardpan layer, and the lime-salt layer in approximately equal proportions can increase root penetration and crop growth.

Plant Nutrient Effects

Adequate fertilization of surface soil encourages not only greater top growth but also a more vigorous and extensive root system. The proliferation of roots resulting from contact with localized zones of high nutrient concentration in infertile soil was illustrated in Figure 3.3. This stimulation of root development is related to the buildup of N and P in the cells that hastens division and elongation (Fig. 12.11).

Plants absorb nutrients only from those areas in the soil in which roots are active. Plants cannot absorb nutrients from a dry soil; thus, root systems modified by shallow applications of fertilizer may be less effective in time of drought (Fig. 12.11). In general, fertilizer should be placed in that portion of the root zone where stimulation of root growth is wanted; therefore, deep placement may be necessary in frequently droughty soils.

An example of the effect of proper management on corn root growth is shown in Figure 12.12. This soil has a claypan and is low in native fertility. The use of adequate plant nutrients with proper cropping systems, including legumes, was effective in the development of a much deeper root system.

FIGURE 12.11   Influence of P placement on root
growth of dryland wheat grown on a low P soil. Plant on
the left is unfertilized. Plant in the middle received 30 lb
P_2O_5/a broadcast, while the plant on the right received 30
lb P_2O_5/a banded 2 in. below the seed.   *Havlin, 1988.*

The influence of plant nutrition on resistance to winter killing is important.
With the addition of adequate plant nutrients to the soil, alfalfa is now being
grown on many soils in which growth was once thought to be impossible.
Similar effects of plant nutrients on extending the root growth and winter
survival of wheat have been observed (Fig. 12.13).

The previous discussion indicates that numerous soil physical and chemical
factors influence the growth of roots and their ability to absorb water and
nutrients in quantities sufficient for optimum productivity. Any management
factor that improves the soil environment for healthy root growth will help
ensure maximum yields. The data in Figure 12.14 illustrate the relationship
between root length and soybean grain yield.

Fertilizer Placement

Determining the proper zone in the soil in which to apply the fertilizer is just
as important as choosing the correct amount of plant nutrients. Placement
decisions involve intimate knowledge of crop growth and soil characteristics,
whose interactions determine nutrient availability. Numerous placement

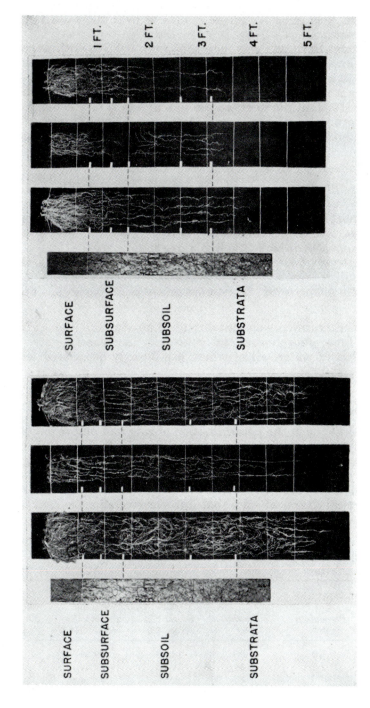

FIGURE 12.12 Soil treatment affects root growth. A rotation including corn, small grains, and legumes was followed. The corn roots on the left, however, were grown in soil receiving adequate lime, P, and K. Those on the right were grown in soil receiving no fertilizer or lime. *Fehrenbacher et al.*, Soil Sci., *17:281. Copyright 1954 by The Williams & Wilkins Co., Baltimore.*

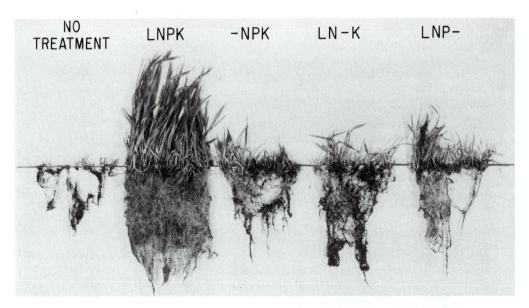

FIGURE 12.13 Balanced fertility aids winter survival of wheat. Early spring vigor means more stooling and more yield. *Courtesy of the Brownstown Agronomy Research Center, Ill.*

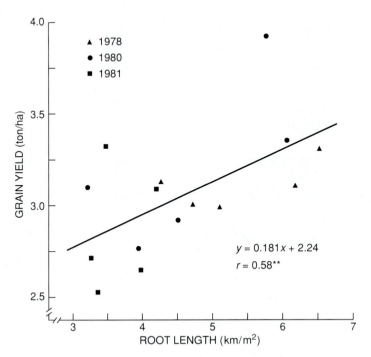

FIGURE 12.14 Relationship between soybean grain yield and root length. *Barber and Silberbush, ASA Spec. Publ. No. 49, p. 86, 1984.*

methods have been developed, and the following factors should be considered with fertilizer placement decisions:

1. *Efficient use of nutrients from plant emergence to maturity.* Vigorous seedling growth (i.e., no early growth stress) is essential for obtaining the desired yield potential and maximizing profitability. Merely applying fertilizer does not ensure that it will be taken up by the plant.

2. *Prevention of salt injury to the seedling.* Soluble N, P, K, or other salts close to the seed may be harmful, although the potential for salt injury depends on the fertilizer source and the crop sensitivity to salts. In general, there should be some fertilizer-free soil between the seed and the fertilizer band, especially for sensitive crops.

3. *Convenience to the grower.* Timeliness of all crop management factors is essential for obtaining the desired yield potential and maximum profit. In many areas, delay in planting after the optimum date often reduces yield potential. Consequently, growers often reject fertilizer placement options, even when they may increase yield, to avoid delays in planting. However, fertilizer placement decisions also influence yield potential; thus, planting date and fertilizer placement effects on yield must be carefully evaluated.

Methods of Placement

Fertilizer placement options generally involve surface or subsurface applications before, at, or after planting (Fig. 12.15). Placement practices depend on the crop and crop rotation, degree of deficiency or soil test level, mobility of the nutrient in the soil, and equipment availability.

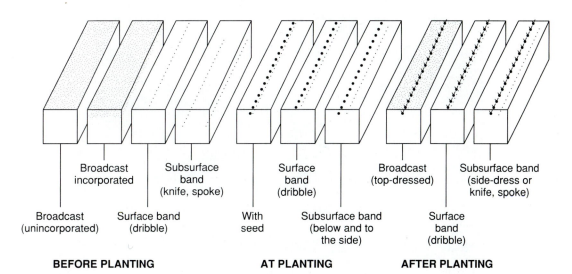

| Broadcast incorporated | Subsurface band (knife, spoke) | | Surface band (dribble) | | Broadcast (top-dressed) | | Subsurface band (side-dress or knife, spoke) |

| Broadcast (unincorporated) | Surface band (dribble) | With seed | Subsurface band (below and to the side) | Surface band (dribble) |

BEFORE PLANTING   **AT PLANTING**   **AFTER PLANTING**

FIGURE 12.15 Cross section of soil profile showing fertilizer placement. *Adapted from Robertson,* Agdex 542-5, *Alberta Agriculture, August 1982.*

Preplant

1. *Broadcast.* The fertilizer is applied uniformly over the field before planting the crop, and it is incorporated by tilling or cultivating. Where there is no opportunity for incorporation, such as on perennial forage crops and in no-till cropping systems, fertilizer materials may be broadcast on the surface. However, broadcast applications of N in no-till systems can greatly reduce fertilizer N recovery by the crop due to immobilization, denitrification, and volatilization losses (Table 5.11). Crop recovery of N, P, K, and other nutrients can be increased with subsurface band applications to no-till crops.

2. *Subsurface Band.* Increasing the fertilizer efficiency or crop recovery of nutrients can be increased with subsurface banding. The depth of placement varies between 2 and 8 in., depending on the crop and the fertilizer source. Subsurface "point or spoke injection" can be effective, especially with application of immobile nutrients (Fig. 12.16). Point injection of N in no-till systems is also more efficient than broadcast N.

3. *Surface Band.* Surface band- or "dribble"-applied fertilizers can be an effective method of preplant application. However, if not incorporated, dry surface soil conditions can reduce nutrient uptake, especially with immobile nutrients. Surface band applications of N also can improve N availability compared to broadcast application in some soils and cropping systems.

At Planting

1. *Subsurface Band.* Fertilizer placement can occur at numerous locations near the seed, depending on the equipment and crop. Solid and fluid fertilizer sources can be used, although subsurface application of fluid sources is more

FIGURE 12.16   "Point" or "spoke" injection application for fluid fertilizers.

common. At planting, the fertilizer is applied 1 to 2 in. directly below the seed or 1 to 3 in. to the side and below the seed, depending on the equipment (Fig. 12.17).

2. *Seed Band.* Fertilizer application "with the seed" also is a subsurface band but is commonly used as "starter" or "pop-up" applications. These are generally used to enhance early seedling vigor, especially in cold, wet soils. Starter fertilization can also be placed near the seed instead of with the seed. Usually low rates of fertilizer are applied to avoid germination or seedling damage. Fluid or solid sources can be used.

3. *Surface Band.* Fertilizers can be surface applied or dribbled at planting in bands directly over the row or several inches to the side of the row (Fig. 12.18). Application over the row can be an effective method of placement of immobile nutrients with a "hoe" opener because soil can "slough off" over time and cover up the fertilizer band. Thus, the surface-applied band becomes a subsurface band placed slightly above the seed (Fig. 12.19).

AFTER PLANTING

1. *Top-dressing.* Top-dress applications of N are common on small grains and pastures; however, N immobilization in high surface residue systems can

FIGURE 12.17   Band application of fertilizer to the side and somewhat below the seed helps to avoid injury to the plant and permits more efficient use of fertilizer. *Courtesy of the Potash & Phosphate Institute, Atlanta, Ga.*

FIGURE 12.18   Surface band or 'dribble' applicator for
fluid fertilizer. In this case fertilizer is applied over the
row after the press wheel.

reduce the efficiency or recovery of top-dress N. Top-dressed P and K is not
as effective as preplant applications. Both solid and liquid sources can be used.

2. *Side-dressing.* Side-dress application of N is very common with corn, sorghum, cotton, and other crops and is done with a standard knife or point
injector applicator. Anhydrous NH_3 and fluid N sources are most common.
Fluid fertilizer also can be surface band applied or dribbled beside the row
after planting. Side-dressing allows a grower more flexibility in application
time since side-dress applications can be made almost anytime the equipment

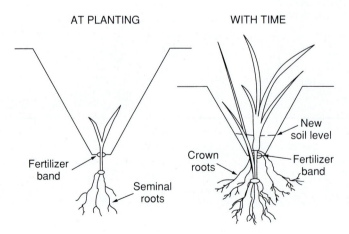

FIGURE 12.19   With surface band application of
nutrients, soil will slough into the furrow and bury the
fertilizer.   *Westfall, et al.,* J. Fert. Issues, *4:114–121, 1987.*

can be operated without damage to the crop. Subsurface side-dress applications with a knife too close to the plant can cause damage by either root-pruning or fertilizer toxicity (i.e., anhydrous NH_3). Side-dress application of immobile nutrients (i.e., P and K) is not recommended because most crops need P and K early in the season and during the reproductive growth stage.

Movement of Fertilizer

Soluble fertilizer salts concentrate in the soil solution surrounding the zone of fertilizer application. The rate and distance of movement of the salts from point of application depend on the fertilizer salt, application rate, soil properties, and climatic conditions.

The movement of P from the point of placement is generally limited, for the $H_2PO_4^-$ ion is only slightly mobile in the soil. N salts move in the soil solution, depending on the direction of water movement. Nitrate moves more readily than NH_4^+-N, which is adsorbed to the CEC. Like NH_4^+, K^+ is relatively immobile, because it is also adsorbed on the CEC.

As the soil dries out, the salt concentration of the soil solution increases and soil water moves upward by capillary movement, carrying the salts with it. In some instances, the salts may be deposited on the surface just above the fertilizer band as a white or light brown deposit coming from dispersed OM. Rain immediately after planting followed by a long dry period is conducive to upward salt movement. With sufficient rain the soluble salts move down. On soils having a relatively low water-holding capacity, the increase in soil solution concentration is greater than on fine-textured soils with a large water-holding capacity.

Excessive concentration of soluble salts in contact with roots or germinating seeds causes injurious effects through plasmolysis, restriction of moisture availability, or actual toxicity. The term *fertilizer burn* is often used. The plant dessicates and exhibits symptoms similar to those of drought.

Some N fertilizers contribute more to germination and seedling damage than is explained by the osmotic effects. Free NH_3 is toxic and can move freely through the cell wall, whereas NH_4^+ cannot. Urea, DAP, $(NH_4)_2CO_3$, and NH_4OH may cause more damage than MAP, $(NH_4)_2SO_4$, and NH_4NO_3. Broadcast application or placement to the side and below the seed is an effective method of avoiding salt injury.

The varying effects of large quantities of fertilizer applied near the seed row on increasing the salt concentration or electrical conductivity in a calcareous, alluvial soil are obvious in Figure 12.20. These elevated conductivities are inversely related to the clay content of soil. Consequently, potential problems related to fertilizer salts are likely to be greatest in coarse-textured soils. High initial NH_4^+ concentrations resulting from ammoniacal sources will also increase osmotic suction and favor the temporary accumulation of NO_2^--N, which is toxic to plants. The effect of seed placed and 1.5×1.5 in. placement of N-P-K fertilizer is shown in Figure 12.21.

Salt Index

The salt index of a fertilizer is determined by placing the material in the soil and measuring the osmotic pressure of the soil solution in atmospheres. The

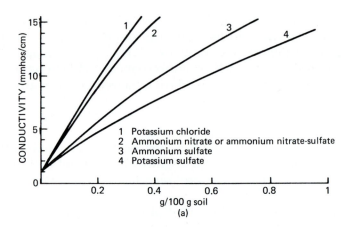

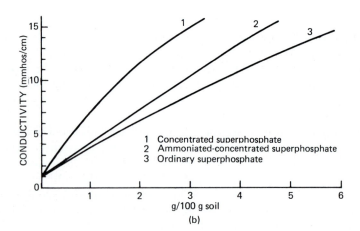

FIGURE 12.20   Effect of (a) K and N, and (b) P
fertilizers on the conductivity of saturation extracts of
Norwood silt loam.   *Chapin et al.,* Soil Sci. Soc. Am. J.,
28:90, 1964.

salt index is the ratio of the increase in osmotic pressure produced by the
fertilizer to that produced by the same weight of $NaNO_3$, based on a relative
value of 100.

   Fertilizer salts differ greatly in their effect on the concentration of the soil
solution. Mixed fertilizers of the same grade may also vary widely in salt index,
depending on the carriers from which they are formulated. Higher-analysis
fertilizers will generally have a lower salt index per unit of plant nutrients
than lower-analysis fertilizers because they are usually made up of higher-
analysis materials. For example, to furnish 50 lb of N, 250 lb of $(NH_4)_2SO_4$
would be required, whereas with urea, 110 lb would be required. Hence the
higher-analysis fertilizers are less likely to produce salt injury than equal

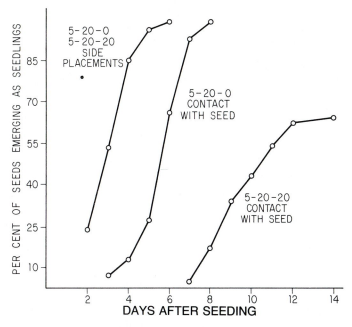

FIGURE 12.21   Placement 1.5 in. below and 1.5 in. to
the side resulted in faster emergence of wheat seedlings
under greenhouse conditions in contrast to placement in
contact with the seed.   *Lawton et al.,* Agron. J., *52:326,
1960.*

amounts of lower-analysis fertilizers. In addition, increasing the row width
increases the quantity of fertilizer applied in a row, assuming that equal rates
are applied. For example, with the same fertilizer rate, fertilizer placed per
unit length of row is twice as great in 30-in. rows than in 15-in. rows.

N and K salts have much higher salt indices and are much more detrimental
to germination than P salts when placed close to or in contact with the seed.
The salt index of representative fertilizers is shown in Table 12.4. In considering the placement of fertilizers at planting, differences in salt effects among
fertilizers are important. Obviously, if high rates of high-analysis fertilizers
are placed near the seed, salt damage can occur.

General Considerations

Band Applications

Early stimulation of the seedlings is usually advantageous, and it is desirable
to have N-P-K near the plant roots. The early aboveground growth of the
plant is essentially all leaves. Since photosynthesis is carried on in the leaves,
the number of leaves produced in this period will influence the yield. It is
important to have a small amount of nutrients near young plants to promote
early growth and the formation of large healthy leaves (Fig. 12.22).

TABLE 12.4   Salt Index per Unit of Plant Nutrients
Supplied for Representative Materials

Material	Analysis*	Salt Index per Unit of Plant Nutrients
Nitrogen carriers		
Anhydrous ammonia	82.2	0.572
Ammonium nitrate	35.0	2.990
Ammonium sulfate	21.2	3.253
Monammonium phosphate	12.2	2.453
Diammonium phosphate	21.2	1.614
Nitrogen solution	40.6	1.930
Potassium nitrate	13.8	5.336
Sodium nitrate	16.5	6.060
Urea	46.6	1.618
Phosphorus carriers		
Superphosphate (single)	20.0	0.390
Superphosphate (triple)	48.0	0.210
Monoammonium phosphate	51.7	0.485
Diammonium phosphate	53.8	0.637
Potassium carriers		
Manure salts	20.0	5.636
Potassium chloride	60.0	1.936
Potassium nitrate	46.6	1.580
Potassium sulfate	54.0	0.853
Potassium magnesium sulfate	21.9	1.971

*By analysis is meant the percentage of N in N carriers, of P_2O_5 in
P carriers, and of K_2O in K carriers.
SOURCE: Rader et al., *Soil Sci.,* **55:**201. Copyright 1943 by The
Williams & Wilkins Company.

Starter response to N-P-K can be independent of fertility level. With cool
temperatures, the early available nutrient supplies may be inadequate because
of slow mineralization of N, P, S, and micronutrients from the soil OM;
restricted release of nutrients from soil minerals; reduced diffusion of P and
K; or limited absorption of nutrients by the plant. The advantage of early
stimulation depends on the crop and seasonal conditions. Some of the factors
that might be considered are the following:

1. *Resistance to pests.*   Under adverse conditions, a fast-growing young plant is
   usually more likely to resist insect and disease attacks.
2. *Competition with weeds.*   Vigorous early growth of crops is important in reducing weed competition. Reduced weed pressure can improve herbicide
   effectiveness or reduce the number of cultivations.
3. *Early maturity.*   Particularly with vegetables, an early crop is generally very
   important. A delay of only 3 or 4 days may make the difference between
   a good price in an early market and a breakeven situation. This can be
   important in northern climates where adverse fall weather may interrupt
   and delay the harvest.

FIGURE 12.22   A readily available supply of nutrients near the young plant helps to ensure rapid early growth and the formation of large leaves essential in photosynthesis.   *Courtesy of the Potash & Phosphate Institute, Atlanta, Ga.*

Broadcast Applications

Broadcast applications usually involve large amounts of lime and/or nutrients in buildup or maintenance programs. With the trend to reduced tillage, more nutrients remain near the surface (Table 11.6). Incorporation with a tillage implement usually increases crop recovery of immobile nutrients (i.e., P and K), while rainfall or irrigation can move mobile nutrients into the root zone without incorporation. The advantages of broadcast application of nutrients include the following:

1. Application of large amounts of fertilizer is accomplished without danger of injuring the plant.
2. If tilled into the soil, distribution of nutrients throughout the plow layer encourages deeper rooting and improved exploration of the soil for water and nutrients.
3. Labor is saved during planting. The fertilizer marketing season is spread out through fall, winter, or early spring applications.
4. This method can be a practical means of applying maintenance fertilizer, especially in forage crops and in no-till cropping systems.

Uniform and accurate spreading of fertilizers and lime is essential for effective utilization by the crop. The effects of uneven application of recommended rates of N-P-K fertilizer on yield of crops are shown in Table 12.5. As might be expected, fertilizer rates well below those needed for low-fertility soils will

TABLE 12.5   Effects of Uneven Application of
Recommended Rates of N-P-K Fertilizer on Crop Yields
at Two Locations in Virginia

| | Yield | | | |
| | Capron | Orange | | |
Spread Pattern	Barley (kg/ha)	Barley, Forage (ton/ha)	Soybeans (kg/ha)	Corn (kg/ha)
No fertilizer	592	1.41	1,278	2,059
Uniform	2,809	2.38	1,345	8,060
Skewed	2,540	2.35	1,264	7,571
Single pyramid	2,454	2.29	1,318	6,993
Double pyramid	2,793	2.33	1,197	7,740

SOURCE: Lutz et al., *Agron J.*, **67**:526 (1975).

result in significant yield losses. There may also be yield reductions from overfertilization on soils medium or higher in fertility. N on small grain is an example.

Specific Nutrients

PHOSPHORUS   Since P is immobile in the soil, placement near roots is usually advantageous. Surface applications after the crop is planted will not place P near the zone of root activity and will be of little value to annual crops in the year of application.

An exception to the inefficiency of surface application is with forage-crop fertilization. Top-dressed P for maintenance purposes is an efficient method of placement. Some of the P is absorbed by the crowns of the plant, as well as by very shallow roots. In addition, such applications come in contact with less soil than applications disked in, and there is less opportunity for fixation. Injection can overcome the difficulty of placing P in the soil for a growing sod crop. Surface banding can be effective on low-P or low-K soils.

The question of band versus broadcast application is very important. When all of the P is either banded or broadcast, the relative efficiency is related to both the P status of the soil and the rate of application. As the P rate increases, broadcast applications can be equal or superior to banding (Fig. 12.23). When the application is split between band and broadcast, at no point will band application alone achieve the maximum yield, thus the advantage of building up the general soil P level. In general, differences between seed-placed and broadcast P decline with increasing levels of available soil P (Fig. 12.24).

When surface soil dries out, surface or shallow-placed P can become unavailable. Placement of nutrients at depths of 4 to 7 in. for dryland crops can improve availability because of greater soil moisture at lower depths.

Even with band placement, crops during any one season are generally able to recover only a small fraction of fertilizer P, usually less than 25%. This is in marked contrast to the recovery of N and K, which may be 50 to 75%. Bands placement of P reduces contact with the soil and should result in less fixation than broadcast application. With broadcasting and thorough mixing,

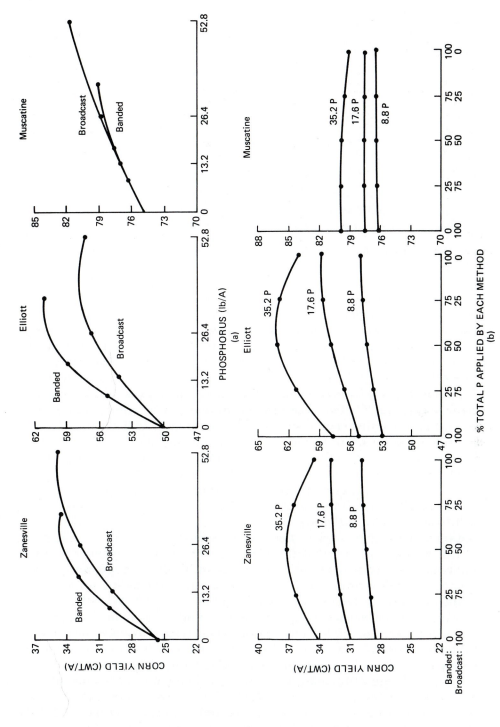

FIGURE 12.23 Corn yields at (a) various rates of P applied either banded or broadcast and (b) various percentages of the total P banded and broadcast (numbers on curves are total lb P/a). *Welch et al.*, Agron. J., **58**:283, 1966.

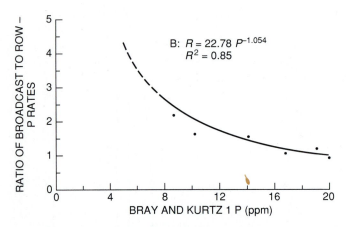

FIGURE 12.24   Influence of soil test P level on the ratio of broadcast to band P fertilizer rates required for equal grain yield.   *Peterson et al., Agron. J., 73:13–17, 1981.*

fertilizer P comes into intimate contact with a large amount of soil. Therefore, band-applied P should increase crop recovery compared to broadcast P.

P placement for small grains is often more critical than for row crops and perennials. Limited root systems, shorter growing seasons, and cooler temperature enhance the response to band over broadcast P, especially in low-P soils (Fig. 12.24). When high P rates are used in dry and/or coarse-textured soils, banding away from the seed at planting may be superior to banding with the seed (Fig. 12.25). The highest yields were obtained when P was banded below the seed rather than being placed with the seed or banded to the side and below the seed row at rates greater than 60 kg/ha of P_2O_5. The loss in efficiency of high rates of seed-placed P supplied as MAP or DAP is probably due to NH_4^+ toxicity.

In establishing forage crops, surface or subsurface band-applied P and/or K is generally superior to broadcasting (Figs. 12.26 and 12.27). This is especially true in low-P and/or low-K soils.

Although it is generally true that the most efficient use of limited quantities of P is at planting and that the highest return will be obtained by band applications, there may be some advantage in building up soil fertility in a long-term fertilizer program. The beneficial effect of building up soil test P on the value of four successive crops is shown in Table 12.6. When P is applied only once, higher net returns were achieved when high value crops such as potatoes, were grown after P fertilization. These results demonstrate that high P rates may be profitable over several crops. When high-value crops are grown on low-P soils, it is advisable to increase the soil test P with a buildup program.

In some cropping situations, a combination of broadcast P plus annual band applications will be more effective than either treatment alone (Fig. 12.23). Current-year banding plus residual effects of banding in previous years may increase yields over those obtained from residual P alone.

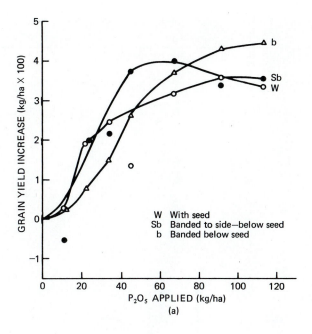

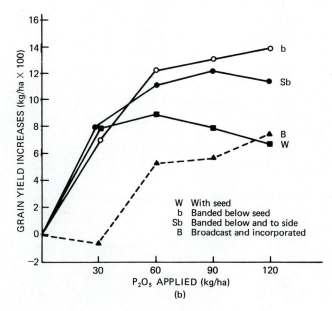

FIGURE 12.25   Effect of P placement on yields of (a) wheat (average check 2525) in Saskatchewan and (b) barley (average check 1695) in Manitoba.   *Bailey et al., Proc. Western Canada Phosphate Symp., pp. 200–229, Edmonton, Alberta: Alberta Soil and Feed Testing Laboratory, 1980.*

FIGURE 12.26   Forage plants on the left resulted from broadcasting both seed and fertilizer. Plants on the right resulted from drilling seed and banding fertilizer 1 in. below the seed. Both plots were planted September 17 and photographed March 31.   *Courtesy of Wagner et al., Natl. Fert. Rev., 29:13, 1954.*

(a)

(b)

FIGURE 12.27   Alfalfa response to P and K placement
in the year of establishment. 80 lb/a of P_2O_5 and K_2O
were applied by (a) surface broadcast, (b) surface band or
dribble, and (c) subsurface band or knife (p. 492) prior to
planting. Early alfalfa growth is greater with band-
applied P and K compared to broadcast.   *Sweeney, 1989.*

(c)

FIGURE 12.27c

NITROGEN   Small amounts of N are important in early seedling vigor, but
because of its mobility and potential salt effects, high rates of N fertilizers
should be applied before planting and at some distance from the seed/seed-
ling, especially on sandy soils. The total amount of N fertilizer needed could
be reduced with surface band applications rather than broadcasting over the
entire area. Both downward and lateral movement of N from the fertilized
strips, combined with root extension into the areas of high N concentration,
compensate for lower N rates applied in a band. Similar results from strip or
dribble application of fertilizer on forages have occurred.

The addition of NH_4^+-N to the fertilizer at planting has beneficial effects
on absorption of P by the plant (see Chapter 6). Controversy exists over
the exact mechanisms responsible for this interaction and over its relative
importance under field conditions. Although some have credited this interaction with being a major cause of the effectiveness of dual placement of N and

TABLE 12.6   Crop Values from P Fertilization over a Four-Year Cropping Period Under
Irrigation in Idaho

P *Applied (lb/a),* *Fall, 1972**	*Gross Crop Value ($/a)*					*P* *Fertilizer* *Costs ($/a)*	*4-Year* *Net* *Returns ($/a)*
	1973 *Sugar-* *beets*	*1974* *Spring* *Wheat*	*1975* *Potatoes*	*1976* *Silage* *Corn*	*Total*		
0	577	240	1,214	232	2,263	0	—
60	661	267	1,204	250	2,382	24	95
150	655	291	1,277	239	2,462	60	139
500	660	306	1,512	252	2,730	200	267

*Initial soil test P level of 5.6 ppm P (0–12 in.).

SOURCE: Westermann, *Proc. 28th Annu. Northwest Fert. Conf.*, pp. 141–146 (1977).

P, others believe that more favorable positional availability is the principal factor.

Although dual application of N and P may not increase the yield in all soils, positive responses have frequently been observed, especially with winter wheat in the Great Plains. The data in Figure 12.28 show that dual applied N, as either UAN or anhydrous NH_3, increased dryland winter wheat grain yield compared to the yield achieved with P placed with the seed (SD) or broadcast, but separately from the N as UAN.

It is usually undesirable to apply all N in the row at planting because of possible injury to the crop. Thus, most N is applied either before planting or as a top- or side-dressing after the crop is growing. Water movement carries the N down to the plant roots. If NH_4-N is used, it must be nitrified before it moves down in appreciable quantities, except on low-CEC soil. More efficient use of N for row crops can be obtained by side-dressing part of the N. This is particularly applicable on coarse-textured soils but also can be important on medium and fine-textured soils. Prediction of the quantity of side-dress N rate can be improved by use of the pre-side-dress NO_3^- soil test, as discussed in Chapter 11 (Fig. 11.40).

Production from permanent grassland and native range in semiarid regions is limited primarily by moisture and N availability. Low rates of N (i.e., less than 150 lb/a) are generally ineffective, because considerable N is immobilized by the high C/N grass residue (Fig. 12.29).

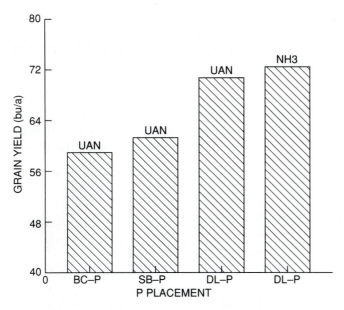

FIGURE 12.28   Winter wheat response to dual N-P application compared to N and P applied separately. BC and SB represent broadcast and seed placed P, respectively, while the UAN was knife applied separately. DL represents dual N-P. *Leikam, SSSAJ, 47:530–536, 1983.*

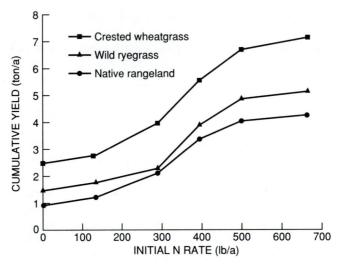

FIGURE 12.29 Effect of fertilization with high rates of N on production from permanent tame grassland and native range. *Leyshon and Kilcher,* Proc. 1976 Soil Fert. Workshop, *Soil Manag. 510, Publ. 244. Univ. of Saskatchewan, Saskatoon, 1976.*

POTASSIUM K salts are much less mobile than the NO_3^- but more mobile than $H_2PO_4^-$. Although some leaching on sandy soils may occur, losses from most soils are negligible. Because fertilizer K salts cannot be placed in contact with the seed in great quantity, they should be placed in a band to the side and below the seed (Fig. 12.17). In contrast, salt-tolerant crops such as barley and other small grains will respond to rates of 15 to 30 lb/a of K_2O placed with the seed.

Starter responses from K, similar to those from N and P, occur with many crops planted under cool, wet conditions even on high-K soils (Table 12.7). This response is unlikely to occur every year, and it should be noted that the results here are only for 1 year. Nevertheless, this example shows the value of using K on a high-K soil.

TABLE 12.7 Starter K Overcomes Cold, Wet Soil in a Normal Spring to Produce More Corn on a Kansas Soil Testing Very High in Available K (>700 ppm)

Nitrogen (lb/a)	Yield (bu/a) with:		
	No K_2O	20 lb/a Banded K_2O	Increase from Starter
0	72	80	8
75	128	137	9
150	167	182	15
225	166	182	16
300	167	185	18

SOURCE: H. Sunderman, Report of Progress 382. Colby Branch Exp. Sta. Kansas State Univ., 1980.

Broadcast K is usually less efficient than banded K; however, as soil-test K increases, there is generally less difference between placement methods. The importance of placement also decreases as higher rates of fertilizer are used.

Conservation Tillage

With reduced tillage systems, nutrients concentrate in the upper 2 to 4 in. of soil (Table 11.6). Periodic tillage (i.e., every 5 or 6 years) will distribute these nutrients more uniformly throughout the root zone. Wherever feasible, soils low in fertility should be brought up to medium or higher fertility before initiating no-till.

Broadcast-applied P and K are effective under many conditions, particularly in the more humid areas. With surface residues there is more moisture near the surface and increased root growth; however, under low-fertility conditions and/or in cooler and drier areas, surface-applied P and K may not be sufficiently available (Table 12.8). Lower K in corn leaves at silking occurs with no-tillage, while K application increases K in the leaves under both tillage systems and reduces the yield loss from no-till. The same principle holds for P.

Yield increases from band-applied fertilizer are generally greater under no-till systems than under plowed systems (Fig. 12.30). With conservation tillage there are greater amounts of surface residues, which leads to cooler and wetter conditions at planting and lower availability of nutrients in the soil.

As indicated earlier, a large portion of broadcast-applied N in reduced tillage systems can be immobilized by the surface crop residues. Therefore, maximizing crop recovery of fertilizer N requires placement below the residue. The data in Figure 12.31 illustrate increased N response and recovery of applied N with subsurface N compared to broadcast and dribble N.

Time of Application

Application timing depends on the soil, climate, nutrients, and crop. Fertilizers are applied, however, at times during the year that may not be the most efficient agronomically but that are better suited to workload or distribution constraints of both the grower and the dealer. Despite these considerations, growers should apply nutrients at a time that will maximize recovery by the crop and reduce the potential for environmental problems.

TABLE 12.8   K Fertilization Helps Compensate for Losses in Corn Yields Due to Reduced Tillage of a Wisconsin Soil Medium in Available Soil K

K_2O Applied Annually 1973–1976 (lb/a)	Yield Loss (bu/a) from Not Plowing (Plowed-Unplowed)	% K in Ear Leaf Tissue	
		Plowed	No-Till
0	37	0.73	0.59
80	26	1.40	1.04
160	13	1.71	1.42

SOURCE: Schulte et al., in Soils, Fertilizer and Agricultural Pesticides Short Course, Minneapolis, Minn. (December 12–13, 1978).

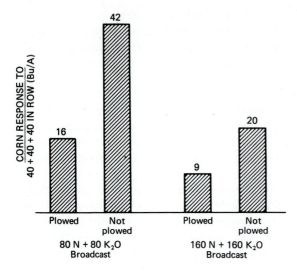

FIGURE 12.30   Use of row fertilizer greatly reduced the yield difference between plowed and unplowed treatments on the Plano silt loam at Arlington, Wisconsin.   *Schulte*, Better Crops Plant Food, *63:25, Fall 1979.*

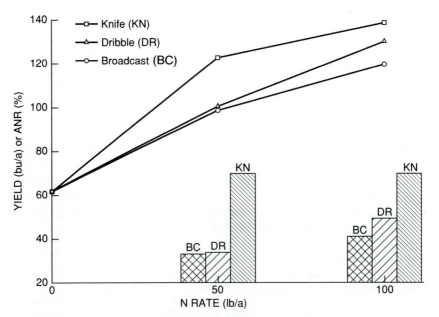

FIGURE 12.31   Influence of N rate and placement on no-till grain sorghum yield and apparent N recovered (ANR) by the grain. Placing the N below the crop residue increased crop yield.and fertilizer N recovery compared to broadcast or dribble N. Lines represent yield and bars represent ANR.   *Lamond et al., KSU Rep. of Prog., 1989.*

PHOSPHORUS   In general, P should be applied just before or at planting because of the conversion of soluble P to less available forms. The magnitude varies greatly with the fixing capacity of the soil (see Chapter 6). On soils of low to moderate fixing capacity, broadcasting in the fall for a spring-planted crop is one of the most effective methods. On low-P and/or high-P fixing soils, band-applied P as close to planting as possible will be the most efficient and should maximize recovery of P by the crop. Low rates can be placed directly with the seed of many crops, but rates for fertilizer-salt-sensitive crops should normally not exceed about 20 lb/a of P_2O_5. On medium- to high-P soils, the time and method are less important (Figs. 12.23 and 12.25), and proportionately larger applications every 2 to 4 years may be recommended.

NITROGEN   In contrast to P, the numerous N loss mechanisms (see Chapter 5) must be considered in selecting the time of application. Theoretically, it would be desirable to apply N as close as possible to the time of peak N demand of the crop; however, this is seldom feasible except with side-dress N.

Because of N mobility in soils, the amount and distribution of rainfall are important considerations. Figure 12.32 shows the relation of annual water surplus to geographical area. To obtain the annual surplus, the potential evapotranspiration is subtracted from the annual precipitation. The greater the surplus, the greater the possibility of loss of N through leaching if the crop is not growing vigorously or if the land is not protected by a plant cover. In addition, conditions conducive to denitrification are likely to occur when soils become waterlogged due to excessive amounts of water.

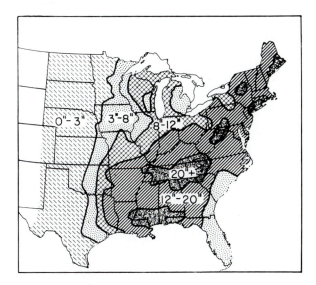

FIGURE 12.32   Average annual water surplus in inches. The surplus is the amount precipitation exceeds evapotranspiration.   *Nelson et al., SSSA Proc., **19**:492, 1955.*

Temperature should also be considered because, farther south, the temperatures are more nearly optimum for nitrification during a greater portion of the year. Ammoniacal N applied before planting would thus be more subject to nitrification and leaching (see Chapter 5). In practice, NH_4^+-N is generally recommended in the fall in the north central United States when the temperature drops to 45 to 50°F at 4 in. except on sandy or organic soil. However, compared to fall-applied NH_4^+-N, spring applications are 5 to 10% more efficient on fine- and medium-textured soils and 10 to 30% more efficient on coarse-textured soils.

Many growers apply N in the fall; however, side-dressing after crop emergence can often be superior to the fall application (Fig. 12.33). Not shown in this figure are the results of spring preplant incorporation, which are similar to those for side-dressing. Differences between the two N sources are small.

With fall-planted small grain on heavy soils in cool climates or in drier areas, application of all or most of the N occurs in late summer or fall. In warmer, humid regions, yields will be somewhat below those obtained by top-dressing N in late winter because of leaching or gaseous losses. However, there are several important advantages to fall applications on small grains. In late winter the ground may be too wet for machinery to be operated, and spring N application after jointing is usually too late for small grain to respond in yield to the applied N.

Fertilizer N recommendations for many crops in dry regions are determined in part by the amount of NO_3-N in the soil profile (see Chapter 11). Both soil and fertilizer NO_3-N are equally effective when present in the zones of maximum root activity. These zones are normally in the upper parts of the profile, although profile NO_3-N can be less effective than fertilizer N. Nitrate N from spring-applied fertilizer will initially be close to the soil surface when plant demands are high, whereas soil NO_3-N at deeper depths will not be within early reach of plants.

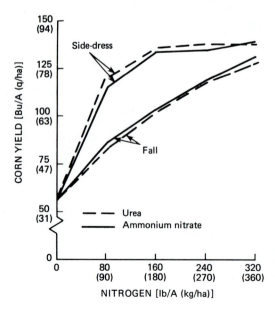

FIGURE 12.33   Corn yields related to source and rate of N and time of application, DeKalb, Illinois, 1969–1972.   *Welch,* Bull. 761., *Univ. of Illinois at Urbana-Champaign, 1979.*

POTASSIUM   K is commonly applied and incorporated before or at planting, which is usually more efficient than side-dressing. Because K is relatively immobile, side-dressed K is less likely to move to the root zone to benefit the current crop. Fall-applied K is even more dependable than for either P or N because fewer loss mechanisms exist with K.

Under some cropping practices, K fertilizers may be broadcast once or twice in the rotation. Fall incorporation of K is generally made before planting K responsive crops, such as corn and legumes. In a corn-soybean rotation, K may be applied after corn because it will be incorporated before the soybeans are planted. Less tillage is normally practiced following soybeans because of potential soil erosion. K may be included in the starter but, because of potential salt injury, N + K rate should be <20 lb/a under row crops. When rates of 30 to 50 lb/a of K_2O are used, they should be placed at a safe distance from the seed row.

Maintenance application on forage crops can be made at almost any time. Fall applications are generally desirable, because the K will have had time to move down into the root zone. On hay crops, an application made after the first cutting and/or before the last cutting is desirable.

Fertilization of the Rotation

Fertilizer management decisions in a rotation involve splitting the application of fertilizer among all the crops in the rotation, applying all at one time, or treating only a few specific crops. Actually, there are many times at which to fertilize in the rotation (Fig. 12.34). Any added effectiveness gained from small, frequent applications must be balanced against the extra time and cost of making them.

MANY POSSIBLE TIMES TO APPLY FERTILIZERS IN ROTATION

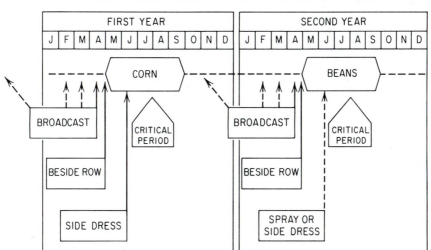

CONSIDER SAFETY, EFFICIENCY, AND LABOR

FIGURE 12.34   There are many times when fertilizer can be applied in a rotation. An example of a corn—soybean rotation is shown.

Residual Fertilizer Availability

Applications of nutrients will result in a certain portion of these nutrients being left in the soil after harvesting. The amounts remaining will depend on the amounts added, the yield, the portion of the crop harvested, and the soil.

Long-term residual benefits of N do not usually receive as much recognition as do those of P and K. However, residual effects of fertilizer N applications are observed. The data in Figure 12.35 show that 50 lb/a of N with no residual N was equivalent to the residual value of 150 lb/a. Carryover of N equal to one-fourth of the preceding year's application is common.

The residual effect of P is well known. The availability of large initial applications of P can be observed for many years, depending on P rate and the P fixation potential of the soil (Fig. 12.36). In Figure 11.34, the residual availability of 150 lb/a of P_2O_5 was observed for 5 years.

An example of the residual effect of K on soybeans is shown in Figure 12.37. The response by soybeans to residual fertilizer is generally quite good, provided that enough K is initially applied to the previous crop.

Beneficial carryover effects can also occur with other nutrients. S applied as gypsum at seeding of winter wheat in the first year of a wheat-pea rotation lasted for 8 years (Fig. 12.38).

As the fertilizer application rate increases, the residual value also increases. In many cases, the cost of fertilization is charged to the crop treated. However, residual fertilizer availability should be included in evaluation of fertilizer economics.

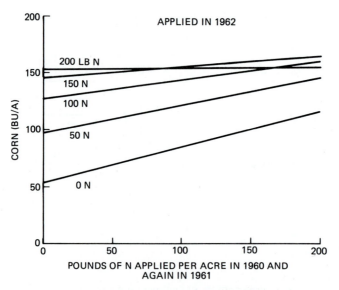

FIGURE 12.35 Residual effects of 1960–1961 applied nitrogen on corn were quite marked in 1962 on this black prairie soil in Indiana. *Courtesy of S. A. Barber, Purdue Univ.*

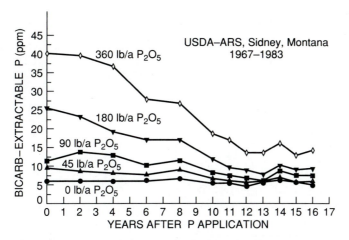

FIGURE 12.36   Effect of high initial broadcast P
applications on soil test P levels.   *Halvorson and Black,*
SSSAJ, 49:928–933, 1985.

Micronutrients

Micronutrient deficiencies are increasing and can be expected to continue.
Higher yields are being obtained and are putting a greater demand on all
nutrients. Interaction among macro- and micronutrients will assume greater
importance. Specific micronutrients are applied in areas known to be severely
deficient or to crops known to have especially high micronutrient require-

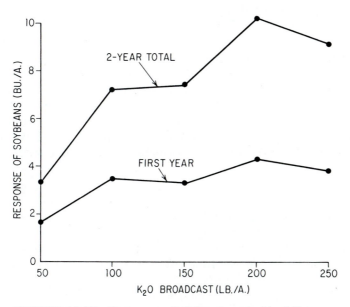

FIGURE 12.37   Response of soybeans to residual K on a
soil low in that element. Note the marked response in the
second year.   *Miller et al.,* Soybean Dig., *21(3):6, 1961.*

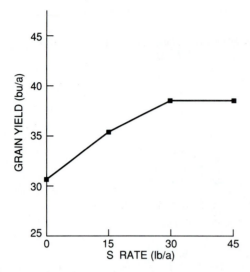

FIGURE 12.38   Yield response (bu/a) of the fourth crop
of soft white winter wheat to S applied at the seeding of
the first wheat crop in a wheat–pea rotation. The fourth
wheat crop was the seventh crop grown in a wheat–pea
rotation following S application. Yields are derived from
13 field experiments conducted over an 8-yr period.
Ramig and Rasmussen, Proc. 23rd Annu. Northwest Fert.
Conf., *pp. 125–137, Boise, Idaho (July 17–20, 1972.*

ments. The micronutrient may be added to a mixed fertilizer or may be
applied separately as a broadcast application, or as a foliar spray.

Micronutrients added to N-P-K fertilizer should be placed in bands about
2 in. away from the seed to prevent fertilizer injury. B should not be band
applied to crops such as beans or small grain, which are easily injured.

As pointed out in Chapters 8 and 9, the addition of certain acid-forming,
S-containing fertilizers in a band near the seed or transplant may also serve
to correct micronutrient deficiencies that are induced by high pH values.
Elemental S, ammonium polysulfide, ammonium thiosulfate, and H_2SO_4 are
effective under such conditions. Banding of N sources that have a strong
acidifying action will markedly increase available levels of some micronutrients
in soil.

Utilization of Nutrients from the Subsoil

The utilization of nutrients from the subsoil depends on numerous factors,
which include soil structure, aeration, pH, drainage, and root distribution.
Two general aspects are discussed:

1. Utilization of the native nutrients from the subsoil.
2. Addition of nutrients to the subsoil.

Native Nutrients

In the most humid regions subsoils are generally acid and low in fertility. In semiarid or arid areas, however, the subsoil may be calcareous and low in fertility. Low-fertility subsoil will contribute very little to total nutrient supplies. However, on certain loess soils, the B and C horizons may be fairly high in P. Deep-rooted crops like alfalfa or sweet clover increase available P in the surface by upward transfer from the subsoil as the organic residues are returned to the surface soil and decomposed. The surface horizons of forest soils are commonly higher in nutrients than the subsoil horizons because of this upward transfer and accumulation.

Loess or alluvial soils can be high in K throughout the profile, and subsoil K can be utilized by deep-rooted plants. In some areas, difficulty is experienced in correlating soil test results for P or K in the surface soil with the P or K response. When the content of P or K in the subsoil is considered, the relation between extractable P or K and crop response can be improved. Some states have made a systematic analysis of the subsoils of major series (Fig. 12.39). These data could be helpful in making more accurate fertilizer recommendations.

In calcareous soil, soil test K is usually high in both surface soil and subsoil, however, most subsoils are low in plant-available P. In addition, subsoil pH can be very high in calcareous soil, where micronutrients like Zn and Fe can be low. P and micronutrient fertilization of the surface soil is generally adequate to correct inadequate P and micronutrient availability.

Added Nutrients

Deeper application of nutrients may sometimes be desirable. Application of nutrients to a deficient soil zone will greatly enhance root development in the treated zone. Lime added to an acid subsoil will not only supply Ca and Mg but will also reduce the quantities of Al, Fe, and Mn in solution.

Much of the early work on subsoil fertilization involved the placement of fertilizer at a depth of 18 to 24 in. behind a subsoiler. Results were inconsistent because of the wide spacing between fertilizer bands. In some cases, subsoiling alone can increase crop yields, although subsoil incorporation of fertilizers can further increase yields. The data in Table 12.9 show that subsoiling to a depth of 11 to 22 in. increased the 4-yr mean yield of barley by 24%, and subsoil incorporation of P and K increased barley yield an additional 20%.

Root penetration may be impeded for a number of reasons, including cementation of soil particles, compaction, poor aeration, soil acidity, nutrient toxicity, or nutrient deficiency. In some soils, deep tillage (24 to 36 in.) can improve root growth and crop yield without subsoil fertilization. In this situation, the principal benefit may be related to more efficient use of annual precipitation. More water should enter the soil, and the roots should penetrate more deeply. If the plant is utilizing water from a layer 24 to 36 in. thick, in contrast to one only 12 in. thick, the probability of drought stress will be reduced. Under some conditions, turning up heavy clay subsoil material may cause the surface soil to seal off more rapidly and decrease water intake.

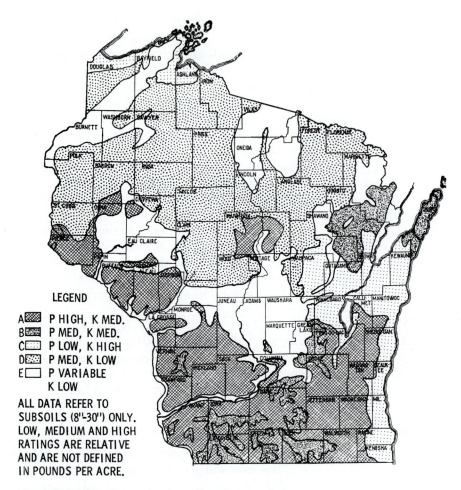

LEGEND

A ▨ P HIGH, K MED.
B ▨ P MED, K MED.
C ▨ P LOW, K HIGH
D ▨ P MED, K LOW
E ☐ P VARIABLE
        K LOW

ALL DATA REFER TO
SUBSOILS (8"-30") ONLY.
LOW, MEDIUM AND HIGH
RATINGS ARE RELATIVE
AND ARE NOT DEFINED
IN POUNDS PER ACRE.

FIGURE 12.39   General subsoil fertility groups in Wisconsin have been established.   *Courtesy of M. T. Beatty and R. B. Corey, Univ. of Wisconsin.*

TABLE 12.9   Effect of Subsoiling and Deep Incorporation of P and K on the Yield of Barley Grain

	Grain Yield (tons/ha at 85% dry matter)				
Treatment	*1974*	*1975*	*1976*	*1977*	*Mean*
None	4.89	2.30	3.43	2.90	3.38
Subsoiled alone	5.23	3.79	4.46	3.53	4.25
Subsoiled + P and K	6.21	4.69	4.51	4.34	4.94
P and K to topsoil	4.53	1.90	3.77	3.20	3.35
SEM of a difference ⇒	1.075	0.240	0.261	0.338	0.302

SOURCE: McEwan and Johnston, *J. Agr. Sci. (Camb.),* **92:**695 (1979).

From this discussion, it is apparent that the effects of physical manipulation of the soil profile vary according to the soil. Deep plowing or chiseling to break up plowpans and improved management practices to encourage deeper rooting are important for improved productivity.

Fluid Fertilizers

The crop response to surface or subsurface application of fluid fertilizers does not differ from that of solid fertilizers if placement is the same. Certain other aspects will be considered here.

Fertigation

Fertigation is the application of fertilizer in irrigation water in either open or closed systems. The open systems include lined and unlined open ditches and gated pipes that are used for furrow and flood irrigation methods. Sprinkler, spitter, trickle, drop, and dual-wall tubing systems are the main types of open systems.

N and S are the principal nutrients applied by fertigation. P fertigation has been less common because of concerns over the precipitation of P in high-Ca and high-Mg waters. Application of soluble S through sprinkler irrigation systems is effective.

Application of anhydrous NH_3 or other fertilizer materials such as UAN solutions containing free NH_3 to irrigation waters high in Ca^{2+}, Mg^{2+}, and HCO_3^- may result in precipitation of $CaCO_3$ and/or $MgCO_3$, causing scaling and plugging problems in equipment. Their formation can be prevented or corrected by the addition of H_2SO_4. Anhydrous and aqua NH_3 are not usually applied in closed irrigation systems because they tend to volatilize at the discharge point, resulting in high N losses.

The advantages of fertigation are as follows: (1) nutrients, especially N, can be applied close to the time of greatest plant need and (2) one or more field operations are eliminated. Corn, for example, has two intense periods of N uptake, during vegetative growth stages V12 to V18 and during reproductive growth or grain-fill. Providing adequate N at these stages is important in sandy soils and is just as important on heavier soils. Labor is saved, for it takes little effort to meter fluid fertilizer from a tank or nurse wagon into the water, allowing it to do the work. With rank-growing, long-season crops such as sugarcane, it is difficult to get through the field to distribute fertilizer. Finally, midseason deficiencies in crops can be corrected by fertigation.

The question of uniformity of application of nutrients in irrigation water is sometimes raised. This should not be a problem with skilled irrigation management and properly designed irrigation systems, since the dissolved nutrients accompany the water wherever it goes. However, unsatisfactory distribution of nutrients can occur under some conditions and with low rates of fertilization. Under row irrigation a large proportion of the nutrients may be deposited near the inlet. With sprinkler irrigation the nutrients, of course, fall with the water.

To prevent nutrients from being leached beyond the root zone or from accumulating near the surface, inaccessible to the crop, they should not be introduced at the initiation of irrigation. Best results are obtained when the fertilizer materials are supplied toward the middle of the irrigation period and their application terminated shortly before completion of the irrigation.

Foliar Applications

Certain fertilizer nutrients that are soluble in water may be applied directly to the aerial portion of plants. The nutrients must penetrate the cuticle of the leaf or the stomata and then enter the cells. This method provides for more rapid utilization of nutrients and permits the correction of observed deficiencies in less time than would be required by soil treatments. However, the response is often only temporary. When problems of soil fixation of nutrients exist, foliar fertilization constitutes the most effective means of fertilizer application.

So far, the most important use of foliar sprays has been in the application of micronutrients. The greatest difficulty in supplying N, P, and K in foliar sprays is in the application of adequate amounts without severely burning the leaves and without an unduly large volume of solution or number of spraying operations. Nutrient concentrations of generally less than 1 to 2% are employed to avoid injury to foliage. Nevertheless, foliar sprays may be excellent supplements to soil applications.

Foliar fertilization can be accomplished by means of overhead sprinkler systems and by application through equipment customarily used for spraying pesticides. Ground-spray equipment used for foliar feeding is usually of the high-pressure, low-volume type, designed for uniform spraying of foliage and for keeping water volume to a minimum. The nutrient spray may be applied through single- or multiple-nozzle hand guns; multiple-nozzle booms; or multiple-nozzle oscillating or stationary cyclone-type orchard sprayers. Droplet size must be carefully controlled, since it will affect the crop response.

Micronutrients readily lend themselves to spray applications because of the small amounts required. Foliar applications have been found to be many times more efficient than soil applications for fruit trees and other crops. Soil-applied Fe is often not effective on high-pH soil because of precipitation of $Fe(OH)_3$ (see Chapter 9).

Efforts to correct Fe chlorosis have not always been successful, and more than one application may be needed on some crops. Chlorosis is a common problem on soybeans grown on high-pH soils under low-rainfall conditions.

Foliar application of urea has been successful in apples, citrus, pineapple, and other similar crops, because N is absorbed more rapidly than with soil applications.

Foliar applications of P are used less than with N, largely because most P compounds are damaging to leaves when sprayed on in quantities large enough to make the application beneficial. The maximum concentration of P is 0.5% for corn and 0.4% for soybeans.

Various environmental factors, including temperature, humidity, and light intensity, also affect the rate of absorption and translocation of nutrients applied to the foliage. To be most effective, two or three spray applications repeated at short intervals may be needed, particularly if the deficiency has caused severe stunting. Care must be taken to identify the nutrient needed, or additional problems may develop. The microelements are usually required in only very small amounts, and too much of one element may be detrimental.

Summary

1. Although nutrient uptake by crops cannot be used as an accurate guide for fertilizer recommendations, it does indicate differences among crops and provides insight into the rate at which the nutrient reserves in the soil are being depleted.

2. Root development helps to indicate the most effective placement of fertilizer. For example, potato roots are much less extensive than corn roots; hence, potatoes can utilize nutrients closer to the plant more effectively.

3. Varietal differences in root characteristics may also result in large variances in extracting power of plants.

4. Soil characteristics influence depth of rooting, and yields may be directly proportional to the water available to the roots.

5. Proper fertilizer placement is important in the efficient use of nutrients from emergence to maturity, prevention of salt injury, enhancement of deeper rooting to compensate for dry conditions at the soil surface, and convenience to the growers.

6. Movement of some fertilizer salts in soils is appreciable. Nitrates move most freely, but NH_4^+ is adsorbed by the soil colloids and moves very little until converted to NO_3^-. Potassium is also adsorbed and moves little except in sandy soils. P movement is generally limited but can be appreciable on sandy irrigated soils, in the presence of large quantities of organic residues, and when heavy batch applications are made.

7. The more concentrated materials have a lower salt index per unit of plant nutrient and, when placed close to the roots, have less salt effect on young plants.

8. Band application at planting is important in providing a rapid start in physiological processes and large, healthy leaves. Under cool conditions, N, P, K, and Zn are generally less available to the young plants, and band placement will enhance their absorption.

9. Broadcast application is a means of applying large quantities of nutrients that cannot be conveniently added at planting. The nutrients are often mixed through the plow layer, which helps to provide a continuing source of nutrients later in the growing season. They are more likely to be in a moist zone for a greater part of the year. Because elements such as P and K move to the roots by diffusion through water films, this point is important.

10. With reduced tillage, nutrients tend to accumulate in the top 1 to 2 in. of soil. This may necessitate higher application rates, more banding, and plowing every 4 or 5 years if conditions permit.

11. Because of the limited mobility of P, it should, when used at low rates, be placed in the zone of root development. Band applications are generally most efficient when soil test P is low and when low rates are applied. For high yields of many crops, however, it is essential to build up the soil level. Forages

are an exception in that they can use top-dressed P by absorption through the crowns.

12. In no-till crop production where planting is done in killed sod or plant residues, surface applications of P and K are generally available to the crop in humid regions. The soil is moist under the residues, and the roots develop near the surface. If the soil is low in these elements, building the soil fertility of the plow layer is desirable before beginning no-till. In drier regions, surface applications may not be as effective because of less moisture under the residues.

13. Small amounts of N in the fertilizer at planting encourage absorption of P. Because N is mobile or becomes mobile after nitrification, application of the major portion before planting or side-dress applications after planting are both effective. In general, the nearer the time of application to peak N demand, the more efficient the utilization. The amount and distribution of rainfall must be considered in connection with soil texture.

14. Leaching losses of K are insignificant except on sandy or organic soils under heavy rainfall. Hence band applications at planting and broadcast applications before planting or at some point in the rotation are effective. Starter responses from K similar to those from N and P are obtained on low-K soils.

15. In determining the method and time of application, convenience to the grower must be considered, along with efficiency and safety.

16. Care must be taken to spread fertilizers uniformly, since yield losses can result from uneven application.

17. As higher rates of fertilizer are used in conjunction with excellent management, more attention is given to fertilization for the entire rotation. Bulk applications of P and K may be used once or twice in a 4-year rotation, in addition to starter applications at planting for the more responsive crops. N is applied annually to nonlegume crops.

18. Year-round fertilization implies fertilization any time the soil, crop, or weather permits and is becoming a must with increased volumes of fertilizer, declining transportation facilities, and the need to save labor. As soil fertility levels increase, the place and time of application of P and K decline in importance. The point is to apply adequate quantities for maximum profit.

19. Application of nutrients for the most profitable yield will result in a portion remaining in the soil. In many cases, the cost of fertilization is charged to the crop treated. However, if a critical evaluation of fertilizer use is made, the carryover value must be considered.

20. Two approaches are followed in the application of micronutrients: (a) addition to take care of specific needs and (b) addition to the fertilizer of a small amount of a mixture of micronutrients for general use. The latter is a method of insurance. Development of suitable diagnostic tools for the recognition of needs and to help predict responses is a major problem. Plant analysis is a good guide.

21. The theory of subsoil fertilization is the promotion of deeper rooting, greater water penetration, and more efficient use of water. Field results from fertilization with a subsoiler have been variable.

22. With problems of soil fixation of nutrients, foliar fertilization may constitute the most effective application method, particularly for certain micronutrients.

23. Fertigation, application of plant nutrients in irrigation water, is an effective method of applying N. Other nutrients applied less frequently by this means include P, S, K, Zn, and Fe

Questions

1. Why can P materials be placed close to the seed or plant? Why is it usually important that P be close to the seed or young plant? How do you account for the marked response of legumes to band seeding?
2. Why is root growth stimulated in response to plant nutrients on an infertile soil?
3. What root characteristics influence the ability of crops to exploit soil for moisture and nutrients?
4. In what part of the root zone does most of the early root growth take place?
5. Are there differences in extent of root development among crop varieties?
6. What soil conditions might affect depth of crop rooting?
7. Under what soil texture conditions would NH_4^+-N be more likely to move? Why? What soil environmental conditions would favor rapid transformation to NO_3^--N?
8. What crops in your area are being underfertilized with P or K? Overfertilized?
9. What materials may be used in a low-analysis fertilizer such as 5-4.4-8.3 (5-10-10) to give it a lower salt index per pound of nutrients than a 10-8.8-16.6 (10-20-20)?
10. An experiment is being conducted on a sandy soil to determine the effects of N and K on snap-bean production. The fertilizer is placed in bands 2 in. to each side and 2 in. below the seed. In addition to P, N, and K are applied in quantities to furnish 50 lb/a of N as NH_4^+ and 60 lb/a of K. On all plots in which the complete fertilizer was applied the stand was poor; when N was omitted, the stand was poor, but when K was omitted, the stand was good. Explain what happened. What would you do to avoid this trouble?
11. Explain specifically why crops are more likely to experience salt injury on a sandy soil than on a silt loam. Why does K not move appreciably in a silt loam?
12. Why might the nature of the root system of the crop being grown affect the decision to build up the fertility level of the soil versus applying fertilizer in the row? How would the economic status of the farmer affect the decision?
13. Explain how band and broadcast applications complement each other in encouraging efficient crop production.
14. You are planning to apply P broadcast. You have the choice of broadcasting and plowing down, broadcasting and disking in after plowing, or subsurface application in a broad band. Which procedure would be most desirable? Explain fully.
15. What is meant by *soil building* and *maintenance applications of fertilizer*?
16. Under what relative levels of available soil-test P and K would you approve of making broadcast maintenance applications of these nutrients?
17. Under what conditions is surface broadcast P and K taken up by the plant? Explain.
18. What cropping systems exist in your

area in which it might be desirable to apply all the P and K to one crop in the rotation?

19. Under what specific conditions in your area do you believe that all the N could be applied before planting? Under what conditions should none be applied before planting?

20. Why does NH_4^+-N applied with P cause more P to be absorbed by the plant?

21. Under what conditions would you advocate fall fertilization in your area?

22. What are the possibilities for summer, fall, winter, and spring application of fertilizer in your area? Why is there a need to spread the fertilizer season?

23. Calculate the removal of P and K in a corn-corn-soybean-wheat-alfalfa rotation and in a corn-soybean-wheat-alfalfa-alfalfa rotation. Assume the yields given in Table 13.1 and

assume that corn, soybeans, and wheat are harvested for grain only.

24. What is meant by *carryover fertilizer*? Why is there an appreciable amount in a properly fertilized rotation?

25. Are there residual benefits from NO_3-N in soils?

26. Do crops benefit equally from soil NO_3-N and that derived from fertilizer? Explain any differences if they exist.

27. Give the pros and cons of the two approaches used in applying micronutrients.

28. What is fertigation, and what are its advantages and drawbacks?

29. What is foliar fertilization? Discuss any limitations.

30. What is dual deep placement of fertilizers? What are its advantages and disadvantages?

31. Is the distribution of plant nutrients in the root zone modified by tillage?

Selected References

FOLLETT, R. H., L. S. MURPHY, and R. L. DONAHUE. 1981. *Fertilizers and Soil Amendments*. Prentice-Hall, Inc., Englewood Cliffs, N.J.

RANDALL, G. W., K. L. WELLS, and J. J. HANWAY. 1985. Modern techniques in fertilizer application. *In* O. P. ENGELSTAD (Ed.), *Fertilizer Technology and Use*. Soil Science Society of America, Madison, Wisc.

RENDIG, V. V., and H. M. TAYLOR. 1989. *Principles of Soil–Plant Interrelationships*. McGraw-Hill Publishing Co., New York.

CHAPTER 13

Fertilizers, Water Use, and Other Interactions

An interaction occurs when the response of one factor is modified by the effect of another factor (Fig. 2.18). A positive interaction occurs when the response to two or more inputs used together is greater than the sum of their individual responses. Negative interactions also occur. Interactions have been observed between

1. Two or more nutrients.
2. Nutrients and a cultural practice such as planting date, placement, tillage, plant population, or pest control.
3. Nutrient rate and hybrid or variety.
4. Hybrid or variety and row width or plant population.
5. Nutrients and the environment (water, temperature, etc.).

Many interactions are not significant with average yields; however, with high yields, increasing pressure or stress is being placed on the plant, and the interactions between the various factors contributing to those yields are often observed. Thus, in the future, interactions and their recognition will be the key to significant progress toward optimum yields and efficient utilization of inputs.

Plant-nutrient response is often difficult to separate from the influence of other management practices due to the numerous positive interactions between fertility and other inputs.

The interaction between the response to fertilizer and various management levels is shown schematically in Figure 13.1. Additions *B, C,* and *D* represent the introduction of improved management practices. As improved practices are implemented, yields are increased and larger quantities of fertilizer can be used more effectively. A fundamental objective of agronomic technology development and transfer must be to lead growers from one level of improved practices to the next. For example, fertilizer contributed most to the increase in yields of wheat and teff; in the case of corn, introduction of an improved variety in the recommended practice gave a marked yield increase (Table 13.1).

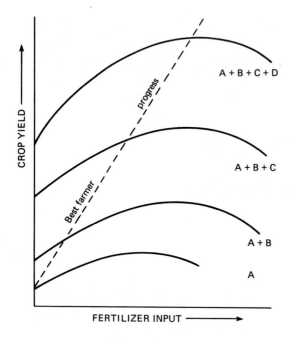

FIGURE 13.1   Relation between adoption of management practices *B*, *C*, and *D*, and response to applied fertilizer.

The potential exists for greatly increasing agricultural output through improvement and integration of the numerous components of yield. General Carlos P. Romulo, in *Strategy for the Conquest of Hunger* (1968), stated the challenge concisely:

> The test of technology is yield. Unless scientists can produce yields which are high by world standards, we can be sure that we have not yet mastered the technical problems confronting the producer. If the yields are high on the experiment stations but national average yields are low, we can be equally sure that the scientific advances are on the shelf and that for some reason the farmer is unable to make profitable use of them.

Future increases in agricultural productivity will likely be related to manipulation of the interactions between the numerous management inputs and factors. It is essential that growers and consultants recognize and take advantage of these interactions.

TABLE 13.1   Yield Improvement from Fertilization and Management in Ethiopia

	Yield Increase over Check (%)		
Crop	Farmer Practice + Fert.	Recommended Practice − Fert.	Recommended Practice + Fert.
Wheat	64	15	91
Corn	81	80	156
Teff	84	2	94

SOURCE: Mathieu, *Progress Report. Third Consultation on the FAO Fertilizer Programme.* Rome: FAO, 1978.

Nutrient—Water Interactions

Even in regions where annual precipitation exceeds growing season evapo-transpiration, water stress frequently limits crop production on the majority of agricultural lands. Stresses caused by nutrient deficiencies, pests, and other factors reduce the plants' ability to use water efficiently, which reduces productivity and profit. As pressures grow for increased industrial, recreational, and urban use of water, agriculture will have less access to water for irrigation.

Increasing water use efficiency is a major challenge to agriculture. It is estimated that overall efficiency of water in irrigated and dryland farming is 20 to 50%. In general, any growth factor that increases yield will improve the efficiency of water use.

Water Use Efficiency

Water use efficiency (WUE) is the yield of crop per unit of water—from the soil, rainfall, and irrigation. When management practices increase yields, WUE is increased. Yields of crops have increased greatly in the past 20 years on essentially the same amount of water, which is directly related to improved soil and crop management practices. For example, tillage systems that leave large amounts of surface residues conserve water by

1. Increased water infiltration.
2. Decreased evaporation from the surface.
3. Increased snow collection.
4. Reduced runoff.

In many parts of the world irrigation has stabilized production, but yields per unit of land have not increased greatly. After the lack of moisture is eliminated by irrigation, many factors may limit yields (Fig. 13.2). Because of these other factors, there can be many disappointments. If yields of 300 bu/a rather than 150 bu/a of corn or 14 tons/a rather than 7 tons/a of alfalfa are to be obtained, the nutrient removal is at least doubled. This means that the crop must obtain more nutrients from some source, whether from native soil supply, manures, or fertilizers.

How Water Is Lost from the Soil

Water in a soil is lost in three ways:

1. From the soil surface by evaporation.
2. Through the plant by transpiration.
3. By percolation beyond the rooting zone.

The sum of the water used in transpiration and evaporation from soil plus intercepted precipitation is called *evapotranspiration*. With more complete cover, less water evaporates from the soil and more goes through the plant. Adequate fertility and satisfactory stands are among those factors that help to

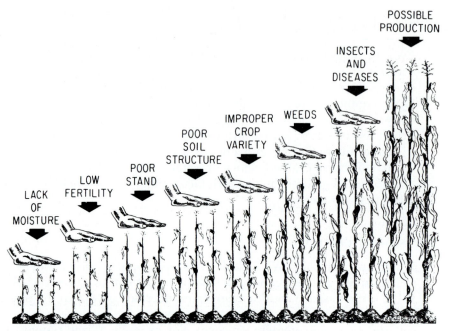

FIGURE 13.2   After the lack of moisture is eliminated
by irrigation, many factors may limit yield. Careful
attention is needed to get the greatest return from the
water, fertilizer, and other inputs.   USDA, SCS Bull.,
199.

provide more plant cover rapidly and thus realize more benefit from the
water.

With a sparse stand or growth, more sunlight will reach the soil, and a
considerable amount of water may be evaporated directly from a moist soil.
With a heavy crop canopy, the surface is shaded and less evaporative energy
reaches the soil. The soil temperature is reduced and the crop provides insula-
tion to maintain higher humidity just above the soil because of less air movement. These three effects reduce evaporation from the soil. It should be kept
in mind, however, that even with a heavy canopy a considerable amount of
energy still reaches the soil.

Nutrient availability affects plant size, total leaf area, and often the color of
the foliage. Close rows and adequate stands, along with adequate nutrition,
provide a heavy crop canopy. For example, water use would be less in 21-in.
rows than in 42-in. rows. Differences in evapotranspiration among crops may
be small once a complete cover is developed. Daily use of water with a growing
crop on the soil varies greatly from one day to another, depending on soil and
environmental conditions (temperature, moisture, and wind); however, daily
losses of 0.1 to 0.3 in. of water per acre are common.

Evaporation from the soil may account for 30 to 60% of the total water loss
in a crop year in humid areas where the soil is wet. With local droughts or in
arid regions the soil surface is dry, and very little water is lost from the soil. The
moisture films between the particles are thin, and little water is transported to
the soil surface by capillarity or diffusion of water vapor. Hence in dry soils

most of the water use is by transpiration, although most of the water received in a light shower would be evaporated quickly.

Heat advection, in which there is horizontal and vertical movement of air in a turbulent fashion, brings in more heat. In a hot, dry area with a strong wind, the heat from the air may contribute to 25 to 50% of the total evapotranspiration. In arid and semiarid areas advection is great, and thus quite variable evapotranspiration may occur.

Fertilization and Water Extraction by Roots

Most crops use water more slowly from the lower root zone than from the upper soil. The surface soil is the first to be exhausted of available water; subsequently the plant must draw water from the lower three-fourths of the root depth (Fig. 13.3).

The favorable effects of fertilization on the mass and distribution of roots when soils are nutrient deficient were illustrated in Figures 12.11 and 12.12. Under nutrient stress the plant may extract water from a depth of only 3 to 4 ft. With fertilization the plant roots may be effective to a depth of 5 to 7 ft or more. If the plant can utilize an extra 4 to 6 in. of water from the subsoil, the crop can endure droughts for a longer period of time without reducing the yield. It should be emphasized that in areas in which the subsoil is dry, increased fertilization will not help crops penetrate the soil farther to get more water.

Soil texture, soil structure, and OM content influence the water-holding capacity of soils, which may vary from less than 1 in./ft on sandy soils to more than 2 in./ft of soil on silt and clay loams. The capacities in the surface 5 ft of several soils are as follows:

Oquawka sand	5 in.
Ridgeville fine sandy loam	7 in.
Swygert silt loam	9 in.
Muscatine silt loam	12 in.

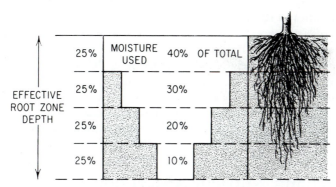

FIGURE 13.3 The top one-quarter of the root zone is the first to be exhausted of available moisture. Certain management practices, including adequate fertilization, help to develop a deeper root system to use the moisture from the lower root zone.   USDA, SCS. Bull., 199.

However, crops root differently in soils because of compact soil horizons, zones of unfavorable pH, inadequate nutrient supply, and other factors. The accompanying data illustrate the approximate rooting depth in three soils and the difference in available water.

	Depth of Roots (ft)	Water Available (in.)
Clarence silt loam	3	6.5
Saybrook silt loam	4.5	10.5
Muscatine silt loam	5+	14

The importance of adequate fertility for efficient crop water use and improvement of crop tolerance to low-rainfall conditions can be explained by the following conditions:

1. *Root exploration of the soil is increased.* Adequate fertility favors expanded root growth and proliferation. When roots explore the soil a foot deeper, another inch or two of water will be obtained.
2. *The major portion of the P and K moves to roots by diffusion through the water films around the soil particles.* Under moisture stress the films are thin and path length increases, reducing P and K diffusion to the roots. Increasing the concentration of P and K in the soil solution increases their diffusion to the roots.
3. *Increased soil moisture tension (lower moisture) exerts a physiological effect on the roots.* Elongation, turgidity, and the number of root hairs decrease with increasing tension. Mitochondria development slows, and *carrier concentration* and *phosphorylation* decrease, which reduces nutrient uptake.
4. *Adequate fertility decreases the water requirement.* K has been shown to aid in closing the stomata, thus reducing water loss by transpiration.
5. *The foliage canopy is increased and the soil is covered more quickly.* Rapid canopy development reduces water evaporation from the soil, which increases water availability to the plant.
6. *Adequate fertility advances maturity.* Advanced maturity in sorghum and corn helps ensure pollination before summer drought periods. Similarly, small grains are adversely affected when growth is delayed, so that summer drought occurs during and following heading.
7. *The amounts of plant and root residues are increased.* With any given tillage practice, a higher amount of residues will break the impact of raindrops, reduce runoff, increase water infiltration, and reduce the erosive effect of wind and water on soil.

Soil Moisture Level and Nutrient Absorption

Water is a key factor in nutrient uptake by root interception, mass flow, and diffusion. Roots intercept more nutrients, especially Ca^{2+} and Mg^{2+}, when growing in a moist soil than in a drier one because growth is more extensive. Mass flow of soil water to supply the transpiration stream transports most of the NO_3^-, SO_4^{2-}, Ca^{2+}, and Mg^{2+} to roots. Nutrients slowly diffuse from areas of higher concentration to areas of lower concentration but at distances no greater than ⅛ to ¼ in. The rate of diffusion depends partly on the soil water content; therefore, with thicker water films or with a higher nutrient content, nutrients diffuse more readily.

Nutrient absorption is affected directly by the level of soil moisture, as well as indirectly by the effect of water on the metabolic activity of the plant, soil aeration, and the salt concentration of the soil solution.

Of course, crop yield potential is greater with normal or higher moisture availability, which is clearly shown in the long-term data in Figure 13.4. Substantial responses in grain yield and WUE to fertilization occurred in the dry years, as well as in the normal rainfall years. Although the response to fertilization was less in dry years than in normal years, adequate nutrient availability greatly reduced drought-related yield losses.

Dryland Soils

Moisture is the most limiting factor in crop productivity in semiarid and arid regions. In crop-fallow systems, conserving soil water may not always increase the grain yield in some crops, but increased soil water conservation will reduce the dependence on fallowing through more intensive cropping (Fig. 13.5). These data illustrate that wheat yields in a wheat-fallow-wheat rotation are not greatly increased due to the extra water conserved in a no-till system. Although wheat yields were not increased, the additional water enabled production of wheat-corn-fallow and wheat-corn-millet-fallow rotations (two crops in 3 years and three crops in 4 years, respectively, vs. one crop in 2 years). Thus, total WUE increased over 50% in the 3-year rotation compared to the 2-year rotation.

N and P are particularly important for high yields when water is limiting.

NITROGEN   Although absorption of N is definitely reduced on dry soils, it is usually not reduced as much as that of P and K. Under drought conditions, OM decomposition and hence N mineralization are reduced. Also, when water is limiting, uptake of soluble nutrients in the water is reduced. Ammoniacal N does not move readily, but NO_3-N moves with the soil water. In heavy rains NO_3^- moves downward in the soil profile and is available for later use, unless it moves below the root zone.

The data in Figure 13.6 show that (1) fertilizer N will not increase yield without sufficient plant-available water and (2) increasing stored soil water by conservation practices will not increase production without adequate fertility.

The accompanying data illustrate that N increased wheat yield, evapotranspiration, and WUE. Without N, water extraction was largely limited to the upper 3 ft.

	0 N	60 lb N	240 lb N
Yield of wheat (bu/a)	24	46	54
Evapotranspiration (in.)	8.7	10.7	12.4
Water remaining in 7-ft profile (in.)	7.1	5.2	3.8
Bushels per inch of H_2O	5.1	6.7	6.4

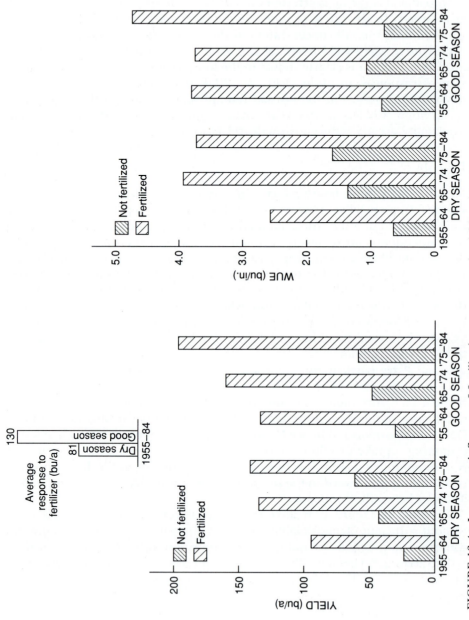

FIGURE 13.4 Long-term influence of fertilization on corn grain yield and WUE. (Morrow plots, Univ. of Illinois, 10-year averages). *Adapted from Potash & Phosphate Inst. Fert. Improves Water Use Eff., 1990.*

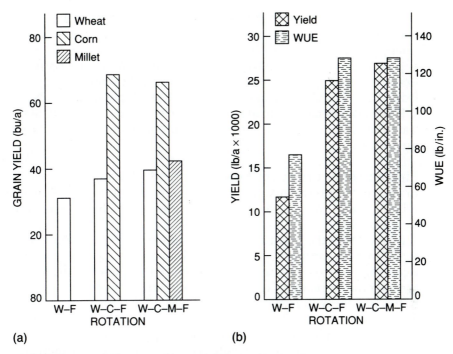

FIGURE 13.5   Influence of cropping intensity on wheat, corn, and millet grain yield (a) and on total grain production and water-use-efficiency in a 12-year cycle (b). W = wheat; C = corn; M = millet; F = fallow.   *Peterson et al., Proc. Great Plains Soil Fert. Conf., p. 47–53, 1992.*

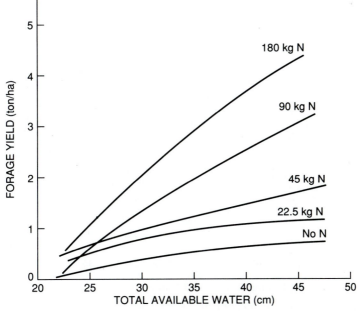

FIGURE 13.6   Interaction between soil water content and N fertilization on native grass forage production.   *Smika, Agron. J., 56:483–486, 1965.*

Some of the positive effects of fertilization on improved water use that occur under moisture stress are due to placement of the nutrients deeper in the soil where the soil is more likely to be moist. A comparison of the effects of two application techniques on the WUE of wheat is shown in Figure 13.7. Yields were increased greatly with dual injection application. WUE was improved from 3.5 bu/in. of water to 6.2 bu/in. with the highest rate of N and P.

Much additional information could be cited; however, the data shown illustrate the effect of adequate nutrients on obtaining greater efficiency in the use of water.

PHOSPHORUS   Crop yield response to P and other nutrients varies from year to year and can be related to the amount of rainfall (Fig. 13.8). The lower the rainfall, the greater the percentage response to P. The same relationship was found for K.

In low-P soils the majority of wheat response to N-P fertilization in dry years is due to P. In wet years, wheat yields dramatically increase, with both N and P contributing to the wheat response. Figure 13.9 indicates the inverse relation between the response of cereals to P and rainfall. The percent yield response to P is greater with low rainfall.

POTASSIUM   The effect of decreasing soil water on corn growth is shown in Figure 13.10. However, as %K saturation of CEC was increased, growth increased at all three moisture levels; thus, adequate K availability reduced some of the water stress. Generally, the lower the rainfall, the greater the K response, which is related to the following factors:

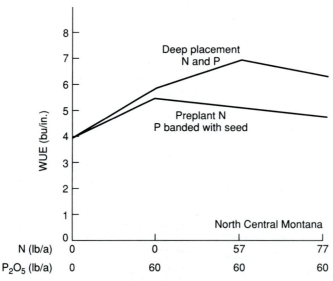

FIGURE 13.7   Deep placement of N and P improves WUE (fallow production-spring wheat).   *Houlton,* Great Plains Univ.—Ind. Soil Fert. Workshop, *Denver, Colo., March 6–7, 1980.*

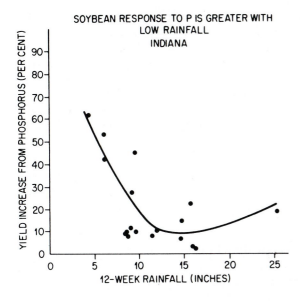

FIGURE 13.8   The less the rainfall for 12 weeks after planting, the greater the percentage yield response of soybeans to P (Indiana).   *Barber,* Better Crops Plant Food, *55(2):9, 1971.*

1. Most of the K absorbed moves to the roots by diffusion through the water films, and with low water content, K diffusion is reduced. Therefore, K fertilization will increase the K content in the water films and increase diffusion. The same is true for P.

2. In some soils the subsoil contains less K than the surface. When the surface soil is exhausted of water in dry periods, the plant roots must feed in the subsoil, where they cannot absorb as much K; this is also true for P.

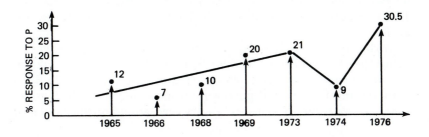

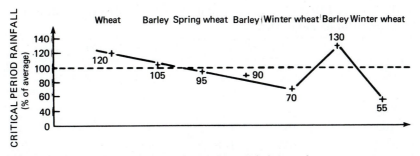

FIGURE 13.9   The response of cereals to P is inversely related to the amount of rainfall.   *Ignaze,* Phosphorus Agr., *70:85, 1977.*

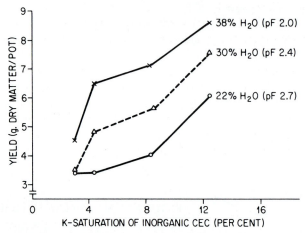

FIGURE 13.10   Dry-matter yield of corn after 3 weeks'
growth in relation to K saturation and water
supply.   *Grimme et al.,* Int. Symp. Soil Fertil. Eval. Proc.,
New Delhi: Indian Society of Soil Science, *1:33, 1971.*

In wet periods, the K response can also be large and is related to restricted aeration. Plant roots respire to obtain energy to absorb nutrients. Respiration requires O_2. Adequate K helps to meet the needs of the plant even when root respiration is restricted.

The effect of normal rainfall and dry years on the response by corn and soybeans to K is shown in Table 13.2. On a medium-K soil there was little or no yield response or profit in years of good rainfall. In stress years, K gave a 48-bu response on corn and an 18-bu response on soybeans, with excellent profits.

In dry years, the K content in the corn or soybean leaves is below the sufficiency K level even with high K rates. The inability to take up adequate amounts of nutrients probably contributes to lower yields in dry years.

Increasing fertility and peanut hulls increased corn yields in both dry and wet seasons (Table 13.3). Even in a dry year, added fertility and hulls increased

TABLE 13.2   Effect of K on Corn and Soybean Yields and Profits in Years
of Good Rainfall and in Dry Years

K_2O (lb/a)	Corn (bu/a)		Soybeans (bu/a)		K Soil Test, Initial 162 (lb/a)
	Good Year	Stress Year	Good Year	Stress Year	
0	163	81	56	30	129
50	163	113	59	42	152
100	167	121	60	48	196
200	163	129	58	48	236
Response	0	48	4	18	—
Profit ($/a)	0	87	18	104	—

SOURCE: Johnson and Wallingford, *Crops and Soils,* **36**(6):15 (1983).

yields to 157 bu/a. With good rainfall, yields were 266 bu/a. Peanut hulls greatly increased K uptake. Moisture content was increased in the 0- to 15-cm depth by reduced evaporation and runoff and increased infiltration.

MICRONUTRIENTS   Since transport of micronutrients to plant roots is by diffusion (and some mass transport), low soil moisture content will reduce micronutrient uptake in dry weather, as with P uptake. The only difference is that the quantity of micronutrients required by plants is much less than for P; thus, drought stress effects are probably not as great as for P.

Temporary B deficiency during periods of dry weather is quite common. Explanations include the following:

1. Much of the B is in the organic matter, and under dry conditions mineralization is reduced.
2. In some areas the subsoil is lower in B than the surface soil. Under dry conditions, water uptake is predominately from the subsoil; thus, plants take up less B. In contrast, in sandy soils, excessive rainfall may leach some of the available soil B.

Low soil moisture can also induce deficiencies of Mn and Mo, although Fe and Zn deficiencies are often associated with high soil moisture. Increased soil moisture results in greater amounts of Mo uptake. Mn becomes more available under moist conditions because of conversion to reduced, more soluble forms.

Placement and Nutrient Absorption

In general, deep placement of nutrients, where moist soil exists during a greater portion of the season, will increase utilization of fertilizer nutrients. Of course, if a soil is very dry, deeper placement will not be effective. As shown in Figure 13.11, deeper fertilizer placement increased N absorption under dry conditions. There was no effect under wet conditions.

Irrigated Soils

Fertility is one of the important controllable factors influencing water use in irrigated soils. As the data in Figure 13.12 show, adequate N increased the

TABLE 13.3   Effect of Adequate Fertility on Corn Yields (bu/a) in Good- and Poor-Rainfall Years

Peanut Hulls Applied Annually (tons/a)	1975 (Good Year)		1976 (Poor Year)	
	Unfertilized	100-100-100	Unfertilized	100-100-100
0	126	205	53	111
10	188	196	100	129
20	161	239	107	138
40	231	266	136	157
Average	176	226	99	134

SOURCE: Lutz and Jones, *Agron. J.*, **70**:784 (1978).

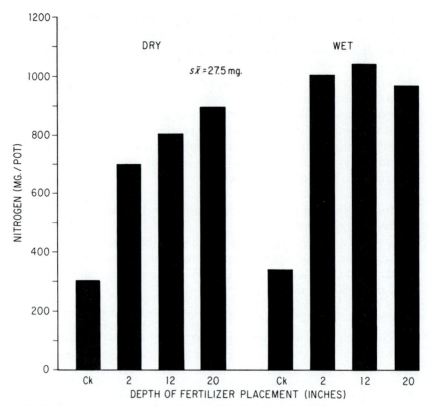

FIGURE 13.11   Effects of moisture, fertilizer, and depth
of placement on N uptake by grain sorghum.   *Eck et al.,*
Agron J., *53:335, 1961.*

forage yield, while the amount of water use decreased from 18 in./ton with
no N to 3 in./ton with 1,000 lb/a of N. Under irrigated, high-fertility, conditions, WUE in corn was 11.6 bu/in. water, compared to 8.3 bu/in. with medium
fertility (Table 13.4).

When N is deficient, increasing N fertilization will increase yield and water
use. The data in Figure 13.13 show that more irrigation water was required
with increasing N rate; however, the WUE also increased.

Stored Water and Fertilizer Recommendations

The relation of stored soil water to crop responses to fertilizer has received
much attention in low-rainfall regions. In general, a systematic survey of the
moisture in the soil profile to the rooting depth, 4 to 6 ft, is made in late fall
or early spring. This information must then be weighed against the probability
of summer rainfall. The data in Table 13.5 show that even with the lowest
moisture levels, yield responses were obtained.

Thus, N recommendations in some areas are based on stored soil moisture
and on the residual NO_3-N in the soil profile (Fig. 11.22). This approach is
based on the concept that adequate fertility will increase the use of available

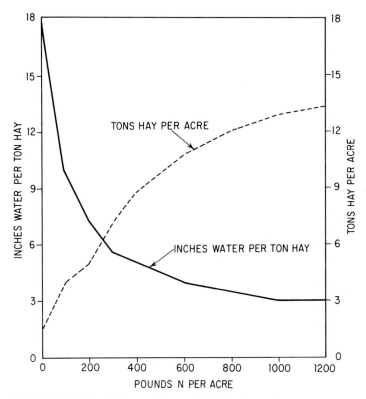

FIGURE 13.12   Effect of N on WUE by Coastal
Bermuda hay. P and K were applied in liberal
amounts.   Texas A and M College Prog. Rep. 2193,
1961, given on p. 192 of Fertilizer Salesmen's Handbook,
published and copyrighted by National Plant Food Institute,
1963, by permission of copyright owners.

water. With a low amount of water in the profile, the probability of a large
response to nutrients is reduced (Fig. 13.14).

Figure 13.15 shows that an additional water supply (stored plus rainfall)
will increase yield of wheat to a point, but if fertility is limiting, additional
water will not be beneficial. However, even with a very low supply of water,
fertilizer is beneficial.

TABLE 13.4   Effect of Fertility Level on Corn Yield, Amounts of Residues,
and Bushels per Inch of Water*

	Corn Yield (bu/a)	Residues Returned (tons/a)	Bushels per Inch of H₂O
Medium fertility, irrigated	214	5.0	8.3
High fertility, irrigated	299	6.4	11.6

*High-P and high-K soil, 2-yr average.
SOURCE: R. L. Flannery, New Jersey Agricultural Experiment Station, personal communication.

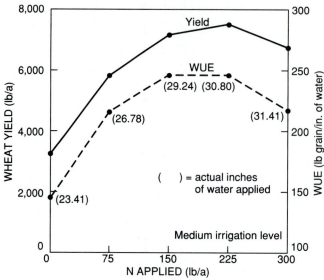

FIGURE 13.13   Adequate N fertilization increases
irrigated wheat yield and water use efficiency (Mesa,
AZ).   *Taken from PPI, 1990.*

Fertilizers have an indirect effect on the amount of stored water in the soil
profile. When there is a response to fertilizer, an increased amount of vegetative cover is produced in the growing crop. Runoff of intense rains is retarded
and infiltration increased. In the fallow season the greater amount of residues
on the soil surface helps retain the water (Fig. 13.16).

In some irrigated regions, fall irrigation after harvesting is practiced as an
insurance policy, particularly with flood irrigation. On deep soils, up to 12 in.
of water can be stored for the next year's crop. This has several advantages.
Water is usually lower in price in the fall, and sometimes water is short during
the growing season.

TABLE 13.5   Maximum Increase in Yield of Wheat (bu/a) from N-P on Nonfallow
Land as Related to Stored Moisture at Seeding and to Rainfall During the Growing
Season (66 Trials)

Rainfall from Seeding to 20 Days Before Harvest (in.)	Available Soil Moisture (in./4 ft soil)		
	0–2 (Low)	2–4 (Medium)	4–6 (High)
>8	7.1	10.0	15.0
6–8	5.0	9.5	16.4
<6	2.4	5.9	10.5

SOURCE: Norum, *Better Crops Plant Food,* **47**(1):40 (1963).

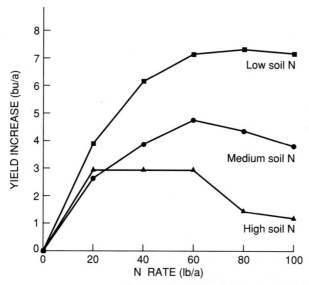

FIGURE 13.14 Interaction between wheat grain yield response to fertilizer N and soil profile NO_3. *Thompson, Kansas Agric. Exp. Sta. Bull. 590, 1976.*

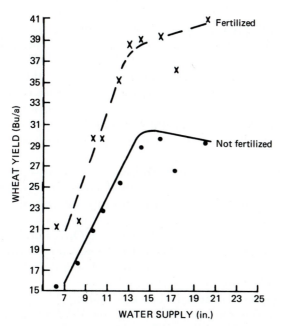

FIGURE 13.15 Wheat yields as affected by soil fertility levels and moisture supply. *Wagner and Vasey, North Dakota State Univ., Coop. Ext. Serv., Soils Fert. 1, 1971.*

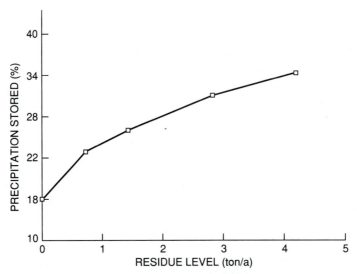

FIGURE 13.16   Influence of surface residue mass on percent water stored in the soil profile during the fallow period.   *Unger*, Dryland Agriculture, *ASA, Madison, Wisc.* p. 40, 1983.

Other Interactions

Interactions Between Nutrients

N–K and N–P interactions are commonly observed. For example, under low-yield conditions when other nutrients are limiting or management practices are inadequate, plant growth is slow, and unless K is seriously limiting, some soils will release K at a rate adequate to meet the needs of the crop. With adequate N and P and improved management practices, there is more rapid growth and the potential response to K, S, and other nutrients is greater (Fig. 13.17). With 30 kg/ha of N there was little response by rice to K; however, when 90 kg/ha of N was applied, the response to K was linear up to the highest rate applied.

Interactions with micronutrients can be dramatic. On a low-P, low-Zn soil leveled for irrigation, adding P or Zn separately decreased corn yields (Fig. 13.18). When both were applied a substantial positive interaction occurred, increasing the yield by 44 bu/a.

The effect of soil pH on the response of corn to P_2O_5 banded beside the row is shown in Figure 13.19. With a pH of 6.1 there was little response to P, but at pH 5.1 there was about a 20-bu response to 70 lb/a of P_2O_5. Also, liming alone increased yields substantially.

Crop response to N is greatly reduced when P is limiting. The data in Figure 13.20 on irrigated corn illustrate that the N rate required for an optimum yield is considerably higher with 40 lb/a of P_2O_5 (160 lb/a of N) compared with no added P (80 lb/a of N). When both N and P were adequate, crop recovery of fertilizer N was approximately 75% compared to about 40% without adequate P fertilization.

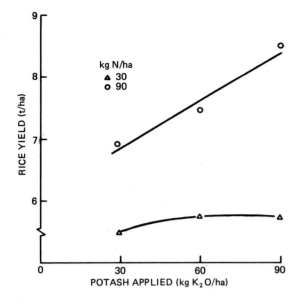

FIGURE 13.17   N level affects
response of rice to K.   *Malavolta,*
Nutrição mineral e adubação de arroz
irrigado. *São Paulo, Brazil: Ultrafertil*
S.A., 1978.

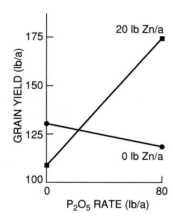

FIGURE 13.18   Interaction of P and
Zn fertilization on corn yield.   *Ellis,*
Kansas Fert. Handbook, *Kansas State*
Univ., Manhattan, Kan., 1967.

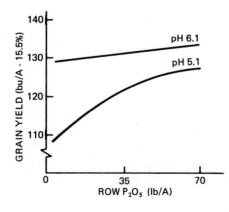

FIGURE 13.19   Soil pH and row P_2O_5
interact on corn.   *Schulte,* Better Crops
Plant Food, ***66:10, 1982.***

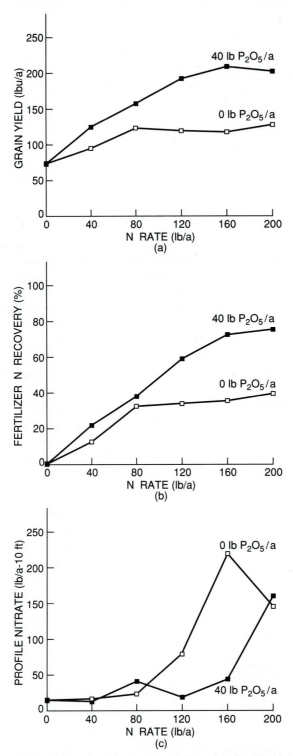

FIGURE 13.20 Interaction of N and P fertilization on irrigated corn grain yield (a), fertilizer N recovered in the grain (b), and profile NO_3 after harvest (c). *Schlegel et al.*, Proc. Great Plains Soil Fert. Conf., *p. 177–187, 1992.*

Maximizing crop recovery of fertilizer N reduces the quantity of profile NO_3^- after harvesting (Fig. 13.20). The rooting depth is about 6 ft; thus, a significant quantity of fertilizer N moved below the root zone and could potentially reach the groundwater. Thus, adequate N and P fertilization will optimize yield and profitability (see Fig. 14.7 for an economic analysis of these same data) and maximize the fertilizer N recovered while minimizing the environmental impact of fertilizer N use.

As previously discussed, the positive interaction of N and P also has been shown in wheat with N-P placement in the same band compared to separate placement (Fig. 12.28).

Many nutrient interactions occur in soils; only a few examples have been provided. The most probable nutrient interactions in a given cropping system involve those nutrients that are deficient or marginally deficient. For example, N–P or P–Zn interactions frequently occur on soils marginally deficient in P or Zn, respectively. Therefore, a good soil testing program will enable the grower or consultant to anticipate potential nutrient interactions.

Interactions Between Nutrients and Plant Population

Increasing the plant population may not optimize yield unless there is an adequate quantity of available plant nutrients. Similarly, increasing plant nutrients without a sufficient number of plants will not maximize the return. For example, increasing the plant population with 80 lb/a of N increased the corn yield 46 bu/a; however, with 240 lb/a of N, increasing the plant population increased the yield 76 bu/a (Fig. 13.21). At 12,000 plants/a, increasing N to 240 lb/a resulted in 37 bu/a increase, but with 36,000 plants the increase was 67 bu/a.

Interactions Between Plant Population and Planting Date

Plant population interacts with planting date. Generally, plants are shorter with earlier planting, and a higher population can be utilized (Fig. 13.22). With a later planting date, there was a decrease in corn yield. Plants are taller

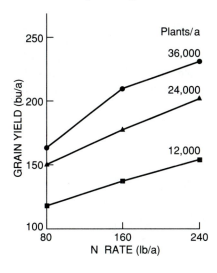

FIGURE 13.21   Interaction of N and plant population on corn yield. *Rhoades,* Quincy Res. Rep., *1978.*

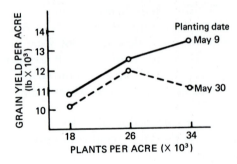

FIGURE 13.22   Planting date and population interact on corn.   *Arjal et al.,* California Agriculture, *Univ. of California, March 1978.*

with the May 30 planting date, and competition for light at this date would be expected to be higher with higher plant populations.

Interactions Between Nutrients and Planting Date

Planting date has a marked effect on the response to nutrients (Table 13.6). Earlier planting dates for spring-planted crops result in higher yields. Note the greater response of soybeans to increased K soil-test level with earlier planting. Also, the K level had a greater effect on increasing seed size and on decreasing seed disease.

Similar planting date interactions with both N and P fertilization have been observed. The increased N and P response is related to the increased yield potential associated with timely planting and the longer growth period.

Interactions Between Variety and Row Width

Varieties or hybrids may vary in their response to plant spacing. Note in Table 13.7 that soybean variety A gave a 10-bu/a response to 7-in. rows over 30-in. rows, while variety C gave a 20-bu/a response, raising the yield to 83 bu/a.

Interactions Between Nutrients and Placement

Crop response to fertilization can be greatly increased if nutrients are applied properly (see Chapter 12). Examples were provided in Figures 12.23, 12.25, and 12.28.

The response of soybeans to P placement is shown in Figure 13.23. At the highest rate, broadcast application gave only a 4-bu/a response, but drilling below the seed gave a 28-bu/a response. Under dry conditions, broadcast P

TABLE 13.6   Effect of the Planting Date on the Response of Soybeans to K

Planting Date	Yield (bu/a) at Soil K Level			Increase (bu/a)
	Low	Medium	High	
May 27	40	47	53	13
June 16	40	44	46	6
July 8	31	36	37	6

SOURCE: Peaslee, Univ. of Kentucky, personal communication.

TABLE 13.7   Effect of Row Width and Soybean
Varieties on Yield

Variety	Yield (bu/a) at Row Width of:		Increase (bu/a)
	30 in.	*7 in.*	
A	58	68	10
B	66	79	13
C	63	83	20

SOURCE: R. L. Cooper, Ohio State Univ., personal communication.

will be in the dry surface soil and positionally unavailable to the roots. When
placed below the seed, the probability is greater that the P will be in moist soil
and more available to the roots.

Interactions Between Nutrient Placement and Conservation Tillage

Soil and water conservation practices to reduce soil erosion include any system
that increases surface residues.

Surface accumulations of residues and nutrients, cooler temperatures, and
higher moisture in the spring can require a change in fertilizer use. In some
situations, higher levels of nutrients applied below the soil surface may be
needed. For example (Fig. 13.24), a 42 bu/a corn response to 40-40-40 banded
beside the row occurred where the soil was not plowed compared to 16 bu/a
where the soil was plowed. Even with higher broadcast rates, the responses
were 20 and 9 bu/a, respectively.

In general, higher rates of N and perhaps S are required under no-till

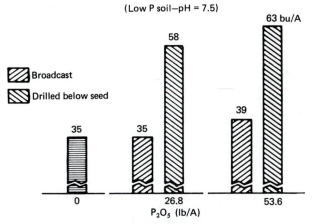

FIGURE 13.23   P rates and placement interact to
increase soybean yields.   *L. D. Bailey, 21st* Annu.
Manitoba Soil Sci. Meet., *pp. 196–200, Univ. of Manitoba,
1977.*

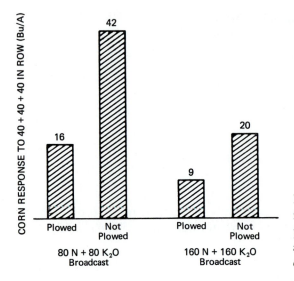

FIGURE 13.24   The response to fertilizer banded beside the row at planting was greater in the unplowed areas.   *Schulte,* Better Crops Plant Food, *63:25, 1979.*

systems than under conventional tillage (Fig. 13.25). Under no-till, the broadcast NH_4NO_3 is partially immobilized and/or denitrified. To avoid fertilizer N interactions with surface residues, N must be placed below the residue. The data in Table 13.8 show increased grain yield with N placed below the surface (knife) compared to broadcast N. Surface band-applied N (dribble) was only partially effective in minimizing N losses. These data also show that reducing immobilization/denitrification losses by subsurface N placement increased the percentage of fertilizer N recovered by the crop. Increasing fertilizer N efficiency will greatly reduce the residual profile N content after harvest and the potential for NO_3^- movement to groundwater.

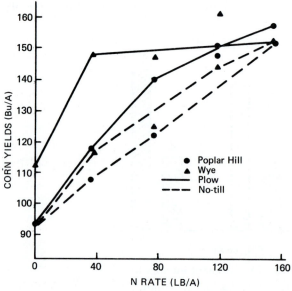

FIGURE 13.25   Tillage and N rates interact on corn yields.   *Bandel et al.,* Agron. J., *75:782, 1975.*

TABLE 13.8   N Placement Effect on Average Grain Sorghum Yield and Apparent Fertilizer N Recovery at Two Locations in Kansas

N Placement	Riley Co. (1986–88)		Greenwood Co. (1987–89)	
	Yield	AFNR*	Yield	AFNR*
	bu/a	%	bu/a	%
Broadcast	110	64	78	51
Dribble	117	70	81	56
Knife	130	87	89	65

*AFNR = apparent fertilizer N recovery.
SOURCE: Lamond et al., *J. Prod. Ag.*, **4**:531–535 (1991).

The following points summarize the available information on N management in reduced-tillage systems:

1. Subsurface placement of N can avoid N losses and increase N removal by the crop.
2. After several years in no-tillage systems, the differences in N needs between no-tillage and conventional tillage diminish.
3. Under some conditions, the yield potential is greater under no-till, thus requiring more N.
4. Soil sampling for profile NO_3^- prior to planting can help predict the fertilizer N need.

Interactions Between Nutrients and Hybrid or Variety

Within a given environment, one hybrid or variety may produce a greater response to applied nutrients than another. In Figure 13.26 the Dare soybean variety produced a higher yield and responded more to K than did the Bragg variety on this very-low-K soil.

Some corn hybrids are genetically able to produce much greater yields from higher rates of applied nutrients than others (Table 13.9). At the lower fertility level the corn hybrids differed by 19 bu/a. At the higher fertility level there was an 85-bu/a difference between the hybrids. Selection of hybrids or varieties that respond to a high-yield environment is essential for maximum productivity.

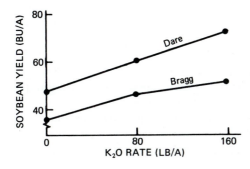

FIGURE 13.26   Interaction between soybean varieties and the response to K. *Terman*, Agron. J., *69:234, 1977.*

TABLE 13.9   Interaction of Fertility and Corn Hybrids to Increase Yields

N (lb/a)	P₂O₅ (lb/a)	K₂O (lb/a)	Yield (bu/a) of:		Increase (bu/a)
			Hybrid A	Hybrid B	
250	125	125	199	218	19
500	300	300	227	312	85

SOURCE: R. L. Flannery, New Jersey Agricultural Experiment Station, personal communication.

The importance of exploiting interactions in maximizing productivity and profitability cannot be overemphasized. When one practice or group of practices increases the yield potential, the nutrient requirement will be increased. Also, as breakthroughs occur in genetic engineering, rhizosphere technology, plant growth regulators, and related areas, they will be successful only if the technology is integrated in a manner that allows positive interactions to be expressed.

Summary

1. A good or poor season is often related to the amount and distribution of rain received during the growing period.

2. It has been known for many years that plant nutrients applied on deficient soils will increase the crop yield per inch of water.

3. Increased moisture tension, which reduces the percentage of many elements in plants, is related to thinner water films around the soil particles and, hence, less diffusion to the roots.

4. The response to K is greater in dry years and in very wet years.

5. A high ion concentration in the cell increases the osmotic pressure of the cell solute and increases the ability of the plant to withstand high water tensions in the soil.

6. Water is lost from the soil by evaporation from the soil surface and from the plant by transpiration. The sum of these two losses, plus the evaporation of intercepted precipitation held on the plant parts, is called *evapotranspiration*.

7. On moist soils, a heavy crop canopy helps to reduce direct evaporation from soils and a higher percentage of the water is used by the plant. Many factors, including higher plant stands and fertility, bring about a heavier crop canopy.

8. On dry soils, little water is evaporated directly from the surface. Low plant spacings are generally employed to reduce the amount of water used by the plants in dry areas.

9. Water use efficiency (WUE) is the yield of crop in bushels, pounds, or tons per acre-inch of water from the soil, rainfall, and irrigation. Any practice that promotes plant growth and the more efficient use of sunlight in photosynthesis to increase crop yields will increase WUE.

10. Many examples are available to show the effect of plant nutrients on increased yield of crops per inch of water.

11. Large sums of money are being spent for irrigation in many parts of

the world. Limiting factors, including lack of fertility, are often holding down yields, and disappointing results are being obtained.

12. Adequate fertility promotes a more extensive and deeper root system. The effective depth of the reservoir from which plants can draw water is thus increased. Compaction and low fertility discourage deep rooting.

13. Adequate fertility helps to avoid the drastic dips in crop yields caused by inadequate rainfall or even excessive water.

14. Growers in the past have worried about having enough moisture to get the most out of the fertilizer. In the future, they will worry about having enough fertilizer to get the most out of the moisture.

15. Conservation tillage, combined with excellent management practices over a long time period, improves water intake, creates a deeper root zone, and provides additional water for the crop to draw on during periods of moisture shortage.

16. A positive interaction occurs when the response to two or more inputs used together is greater than the sum of their individual responses.

17. Nutrients may interact with each other and with various cultural practices, such as hybrid or varieties, plant spacing, planting date, placement, tillage, and pest control.

18. Interactions are more important at higher yield levels because more stress is placed on the plant and on the various growth factors.

19. As improved practices are adopted and yields are increased, larger quantities of nutrients can be used more effectively.

20. The maximum-yield concept influences fertilizer efficiency. A large crop with a more extensive root system and greater need for nutrients takes up more of the applied nutrients.

21. It is impossible to obtain a full response to a given nutrient if other nutrients or management practices are limiting and yields are low.

22. Increasing the plant population or narrowing the row width will not give a full return if plant nutrients are lacking.

23. A proper planting date helps ensure more adequate returns from an increased plant population and increased plant nutrients. Yield potential is increased and more nutrients are required.

24. Deeper placement of plant nutrients, particularly in areas subject to drought stress, is more likely to give a greater response than placement at or near the surface. Deeper placement improves the probability that the nutrient will be in moist soil and more available to plant roots.

25. With conservation tillage and the accompanying surface accumulations of P, K, and residues, as well as cooler temperature and higher moisture in the spring, there is often need for a change in the amount and placement of fertilizer. In some situations, there may be a need for higher nutrient levels in the soil and an increased need for starter fertilizer.

26. Higher rates of N may be required in no-till, at least initially, because the surface residue helps maintain a more moist condition. This may lead to an increased microbial population, greater immobilization of surface-applied N, and faster nitrification and denitrification. Subsurface application of N could help avoid some of the losses.

27. There may be a distinct difference among hybrids or varieties in their response to applied nutrients. Because the extent of these interactions is

usually not known, several hybrids or varieties are often compared at various management levels by the researcher and the farmer.

28. The responses to nutrients may be much greater under cool temperatures because mineralization of certain nutrients from OM will be reduced and/or diffusion rates of nutrients decreased.

29. Nutrient–pesticide interactions occur. For example, inadequate plant nutrition favors certain plant or seed diseases. Added nutrients will decrease disease, as will a fungicide or a nematicide. However, both together may have an even greater effect.

30. Efficiency in the use of all inputs is vital. As yields increase, inputs such as land, labor, machinery, fuel, seed, and pesticides are used more efficiently. Proper fertilizer use is an important part of an efficient crop production system.

Questions

1. What is a positive interaction? Illustrate.
2. Why are interactions more critical at higher yield levels?
3. Why do higher yields improve fertilizer efficiency?
4. Explain why it is impossible to obtain a full response from an applied nutrient if the level of another nutrient is inadequate.
5. Describe a good or a poor season in an irrigated area and in an unirrigated area.
6. In a given soil volume, why is absorption of many nutrients by plants decreased as soil moisture tension increases?
7. Explain the greater response to K in dry years and in wet years. Why is B generally less available under dry conditions?
8. How might low soil moisture affect interpretation of the K leaf analysis?
9. Why does a higher ion concentration in the cell increase drought resistance?
10. What is evapotranspiration? How does a heavy crop cover affect the losses in the various components of evapotranspiration? Is there more total water loss with a greater yield?
11. Explain the difference in evaporation

losses from a moist soil surface and a dry soil surface.
12. Define water use efficiency (WUE). Why is it so important in agriculture? List factors that affect WUE.
13. What is the effect of adequate plant nutrients on WUE? Why does this effect occur?
14. Explain the effect of adequate nutrients on increasing the extent of the root system. Why is this important in drought periods?
15. On what soils in your area will root penetration be limited by lack of fertility and by the physical condition in the lower soil horizons?
16. How might placement of nutrients affect uptake in a dry year?
17. What soils in your area have a low water-holding capacity? Why?
18. Why is stored water important in dry regions? What advantages are there in irrigation in the fall after crops are harvested?
19. Are there irrigated farms in your region in which full returns are not being obtained from an investment in irrigation? Why?
20. Average yields of many crops are much higher than those of 20 years ago with about the same amount of rainfall. Why?

21. Explain how conservation tillage increases the water available to the crop. Why is the crop better able to withstand periods of moisture shortage?
22. How does the plant nutrient supply affect the response to an increased plant population? Why?
23. Explain why deeper placement of plant nutrients is likely to give a greater response than shallow placement in some areas.
24. Why does the planting date affect the response to the fertilizer?
25. Explain why conservation tillage often requires a change in fertilizer use.
26. Why might higher rates of N be required for no-tillage compared to conventional tillage (moldboard plow)?
27. What is the best way to determine how hybrids compare in a high-yield environment?
28. Give two reasons why responses to nutrients are usually greater under cool conditions.
29. Explain why there might be a fertilizer–pesticide interaction.
30. Why would higher yields help decrease erosion and increase water infiltration?
31. How would you go about setting up a maximum-yield-system area on a farmer's field? Whose help would you enlist?
32. Explain the various aspects of efficient fertilizer use in relation to Figure 13.1.
33. How do fertilizer and management affect the efficient use of water?

Selected References

JACKSON, T. L., A. D. HALVORSON, and B. B. TUCKER. 1983. Soil fertility in dryland agriculture. *In* H. E. DREGNE and W. O. WILLIS (Eds.), *Dryland Agriculture*. American Society of Agronomy, Madison, Wisc.

THORNE, D. W., and M. D. THORNE. 1979. *Soil, Water, and Crop Production*. AVI Publishing Co., Westport, Conn.

CHAPTER 14

Economics of Plant-Nutrient Use

Regardless of crop prices, maximum profit occurs when the last dollar spent to produce the crop returns just a dollar. Maximum profitable yield is higher than most people think, and attaining it means applying good management practices.

Higher crop yields continue to offer the greatest opportunity for reducing per-unit production costs. Modern growers have the attitude of all good business managers: that they must spend money to make money. This is certainly true of expenditures for soil amendments. The demand for fertilizer is evidence that growers recognize the returns that can be realized from added plant nutrients.

In spite of the rise in the use of plant nutrients, reliable estimates show that much land is still underfertilized and that plant-nutrient additions could be profitably increased. For the developing countries, economic development must give high priority to agriculture. As higher rates of plant nutrients are required, it becomes more important that the nutrients be applied so that they will be most efficiently utilized.

Attention must be given to water, tillage, variety, date and rate of seeding, plant spacing, fertilizer placement, cultivation, weed, insect and disease control, and harvesting practices. Proper use of fertilizers complements the effects of other management practices, and vice versa. Nutrient management practices must also be environmentally sound.

Fertilizer Use and Prices

To obtain a given level of production, farmers can vary the inputs of land, fertilizer, labor, machinery, and so on. The actual use of each depends on relative costs and returns. Production costs can vary from year to year, but costs gradually increase over time. The relative costs of many farm inputs have increased more than the costs of fertilizers and chemicals (Fig. 14.1). Although the price of fertilizers and lime will continue to rise, it may not rise as fast as the prices of some other inputs.

540

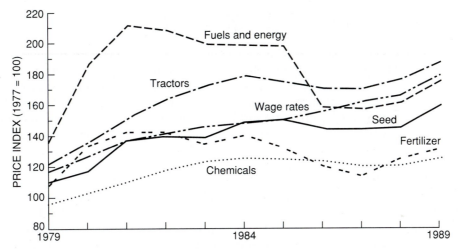

FIGURE 14.1   Index of prices paid for selected inputs.   *ERS-USDA,* 1989 Costs of Production—Major Field Crops, ECIFS 9-5.

Unfortunately, the input prices paid by farmers have increased much more than the output prices received (Fig. 14.2). Therefore, it is imperative that growers achieve optimum productivity through efficient and cost-effective use of only those inputs that will ensure a return on the investment.

Prices paid includes commodities, services, interest, taxes, and wage rates.

FIGURE 14.2   Prices paid and received by farmers in the United States.   *ERS-USDA,* 1990 Agric. Chartbook No. 689.

Maximum Economic Yield

Maximum economic yield is somewhat lower than maximum yield and is the point where the last increment of an input just pays for itself (Fig. 14.3). Maximum economic yields vary among soils, although on most farms they are much higher than those generally achieved, regardless of the soil. Growers want to maximize profits, and higher yields are the key to this goal; however, the need for increased production and yield is based on other factors.

Need for Higher Yields

The world population is expected almost to double in the next 30 to 40 years. It was pointed out in Chapter 1 that the area of cultivated land will probably increase by only about 20% during this period. So far, grain production has increased at a rate similar to that of grain utilization; however, as population increases, the number of years when utilization exceeds production will increase (Fig. 14.4). Productivity per unit area will have to increase substantially to satisfy expected food requirements.

The cost of production will continue to increase (Fig. 14.2). Crop prices are highly variable from year to year and do not always keep pace with production costs. As a result, breakeven yields will continue to increase (Table 14.1). For a production cost of $300/a with $3/bu corn, a 100-bu yield is required just to break even ($300 ÷ 100 = $3). Thus, increasing production costs demand that yield levels increase.

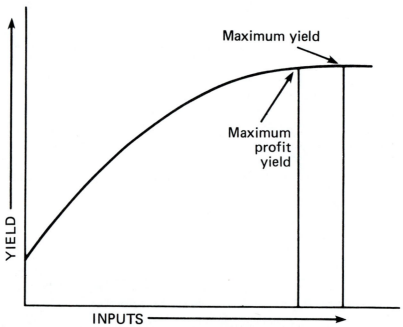

FIGURE 14.3   Maximum profit yield is slightly lower than maximum yield.

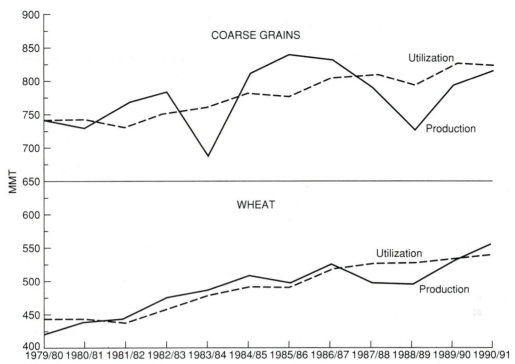

FIGURE 14.4   World coarse grain and wheat production and utilization.   *FAS-USDA*, World Grain Situation and Outlook, *1990*.

Yield Level and Unit Cost of Production

Practices that will increase the yield per unit of land usually lower the cost of producing a unit of crop, for it costs just as much to prepare the land, plant, and cultivate a low-yielding field as it does a high-yielding field. As Table 14.2 shows, increasing the yield raised the total cost of production per acre but decreased the cost per bushel and increased net profit. Land, buildings, machinery, labor, and seed will be essentially the same, whether production is high or low. These and other costs are called *fixed* and must be paid regardless of yield. *Variable costs* are those that vary with the total yield, such as the quantity of fertilizer applied, pesticides, harvesting, and handling.

TABLE 14.1   Breakeven Yields of Corn Increase with Higher Cost of Production

Cost of Production (\$/a)	Breakeven Yield of Corn* (bu/a)
300	100
400	133
500	167
600	200

*At \$3/bu.

TABLE 14.2   Effect of Increasing Yields on the Cost per
Bushel of Corn and the Net Profit per Acre ($2.75/bu)

Yield (bu/a)	Production Costs		Net Profit ($/a)
	$/a	$/bu	
100	331	3.31	−56
125	343	2.74	1
150	359	2.39	54
175	383	2.18	100

SOURCE: Adapted from Hinton, *Farm Economics Facts and Opinion*, Univ. of Illinois (January 1982).

Key factors in obtaining the most efficient use of the land and inputs are the weather and the ability of the farmer. Time of planting, proper variety selection, tillage, plant spacing, pest control, and timely harvesting are some of the factors that can be controlled by farmers. As a grower aims for increased yields, much of the initial increase will come from improved practices, not just extra fertilizer. Many of these practices cost little or nothing. Some examples follow.

1. *Timeliness.* This is important in planting, tillage, equipment adjustment, pest control, observations, and harvesting.

2. *Date of planting.* Delaying soybean planting 1 month beyond the optimum date, for example, can reduce the yield 10 to 20 bu/a in some areas, while corn yields can be reduced 1 to 2 bu/a for each day of delay in planting.

3. *Pest control.* Anticipating or identifying pest problems early will allow application of the least costly and most effective control practice.

4. *Variety selection.* Large differences in productivity, disease resistance, quality, and responsiveness to inputs exist among varieties and hybrids. Proper variety/hybrid selection can have substantial effects on yield and profitability.

5. *Plant spacing.* Planting the appropriate seeds per acre for the productive capacity of the soil and environment is a critical management decision. For example, increasing the average seeding rate in a region by 10 to 20% can have a similar effect on the yield of many crops. Many growers plant their crops below the optimum population. Row spacing decisions also can influence yield. For example, narrowing soybean rows from 30 to 7 in. has increased yields 10% or more in many areas. In humid regions, reducing wheat rows from 8 to 4 in. can increase yields 10–20%.

6. *Rotation.* Rotating crops is a valuable, no-cost input to increase crop yields and profit. Rotation may not only reduce weed, disease, and insect problems, but may also improve soil structure. Table 14.3 shows the beneficial effect of rotating peanuts with corn and how the benefit tended to increase with time (see Chapter 15).

With superior management, higher rates of nutrients can and must be used. The general relationship in Figure 14.5 shows that *A* is the most profitable rate with average management and *B* is the most profitable rate with superior management.

TABLE 14.3   Yields of Continuous Peanuts and Peanuts Following Corn

| | Peanut Yield (lb/a) | | |
Years	Continuous	Rotated	Increase
1965–1969	1,400	1,650	250
1970–1974	1,780	2,650	870
1975–1979	2,220	3,090	870

SOURCE: J. T. Cope, *Highlights of Agriculture Research,* Vol. 28, No. 1, Alabama Agr. Exp. Sta. (1981).

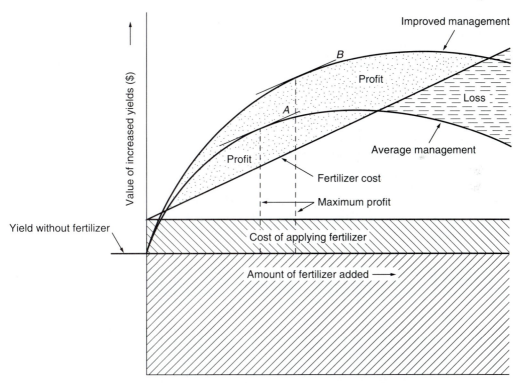

FIGURE 14.5   Diagram representing fertilizer economics associated with average management and improved management. The fertilizer rate for maximum yield occurs where the slope of the response curve is equal to 0 or is parallel with the *x*-axis. The fertilizer rate for maximum profit occurs where the slope of the response curve is parallel to the "fertilizer cost" line.   *After Miller,* Soils, *Prentice-Hall, Englewood Cliffs, N.J., 1990.*

Returns per Dollar Spent or Profit per Acre

Growers should consider the returns on each dollar spent for fertilizer, lime, and other inputs because they often have only a limited amount of cash or credit. With each input, they are interested in the maximum return that can be obtained per dollar spent. In general, as the rate of a particular nutrient is increased, the return per dollar spent decreases. This decrease is the result of a reduced response for each successive incremental input. Eventually, the point is reached at which there is no further response to increasing amounts of nutrient, a principle called the *law of diminishing returns.*

When the soil is deficient in a nutrient needed for a desired crop yield, the first added increment of the nutrient will result in a large yield increase. The next added increment may also give an increase, but not as large proportionately as the first (Fig. 14.6). Consequently responses to fertilizer increments continue diminishing to the point where the last incremental yield value just equals the cost of input. It is this application rate that gives the maximum profit.

Credit agencies are very much interested in the returns the farmer realizes for each dollar spent on fertilizer or lime. As a general rule, $1 to $3 is the expected net return for each $1 spent. Certainly, compared with current rates, the purchase of fertilizer and lime with a 100 to 300% return would be an excellent investment for the farmer and a safe one for the credit agency.

Progressive growers recognize that although returns per dollar spent are

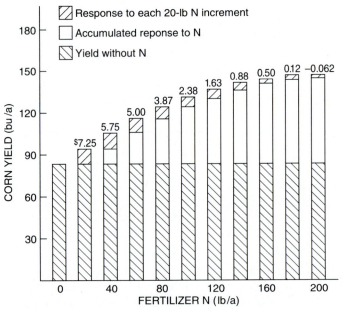

FIGURE 14.6   Diminishing returns in yield response of corn to fertilizer N. The $ values on top of each bar represent the net rate per added dollar invested.

important, the significant figure is the net return per acre. With adequate cash or credit, the farmer must select the input levels that will earn the greatest net return per acre. A low rate of fertilizer or another needed input may result in a high unit cost of production.

What Is the Most Profitable Rate of Plant Nutrients?

The most profitable rate of fertilization will change from year to year because of the variation in weather, prices of farm products, and price of fertilizer. Agronomists and economists are thus faced with a sizable calculation task if the growers' questions are to be answered each year. They must work with expected yields, prices, and costs.

Crop production has always entailed considerable risk because of such factors as floods, droughts, insects, and diseases. There will always be risk in crop production, but the grower can reduce risks through superior management and increase the probability of producing high yields and profits in the long term.

The most profitable fertilizer rate can be determined by calculating the maximum net profit or minimum cost per bushel of production. The data in Figure 14.7a for irrigated corn show that the N rate for maximum net revenue was about 10 lb/a less than the N rate for maximum yield and 10 lb/a more than the N rate for the least cost per bushel. Although these values represent a range of 20 lb/a of N, the most profitable N rate is essentially the same regardless of which parameter is used.

Although yield potential is an important parameter in determining the recommended N rate (see Chapter 11), the economic N rate usually does not vary greatly, even over a fairly wide range in actual yield potential. In the 30-year study mentioned previously, the data were grouped according to yield potential (Fig. 14.7b). Using the same costs and prices as before, the economic optimum N rate was 155 to 160 lb/a for the three yield potentials.

Other Management Factors That Influence the Profitable Response to Fertilization

For example, Figure 14.8 shows that soil levels of P and K affected the response of Coastal Bermuda grass to N. At low soil levels, the maximum profit rate was 200 lb of N and only $40 per acre. With high levels, however, the optimum rate was 310 lb, giving a profit of $79.55. This illustrates the importance of balanced soil fertility in obtaining maximum profit.

When too much fertilizer is added, the economic loss is not as great as when a crop is underfertilized by the same proportion (Table 14.4). The residual effect also must be considered and helps to compensate for the the extra fertilizer cost. Over a period of years, it appears more profitable to use the optimum amount, even if the rate would be more than optimum in unfavorable years. This, of course, applies to nutrients that do not leach from the soil.

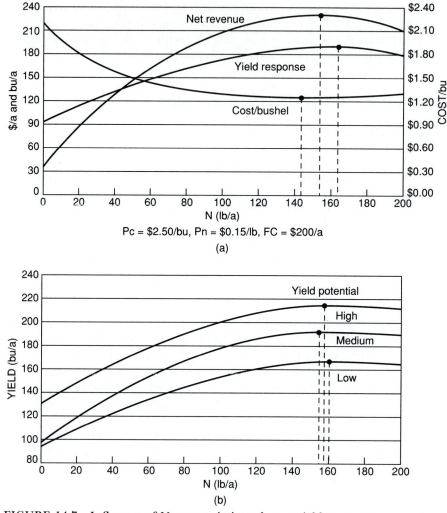

FIGURE 14.7   Influence of N rate on irrigated corn yield, net revenue, and cost
per bushel (a) and (b) the economic optimum N rate for three corn yield
potentials.   *Schlegel,* Proc. Great Plains Soil Fert. Conf., *1992.*

Price of Fertilizer in Relation to the Value of the Crop

Although the cost per pound of nutrients in a given fertilizer may fluctuate,
these variations are much less than the fluctuations in crop prices. For some
crops, however, government regulations have stabilized the floor price, and
growers may use this figure as a basis for calculations. When crop prices have
not been regulated, careful consideration of trends and outlook will be helpful
in establishing a profitable rate of fertilization.

There are several methods that take into consideration both the cost of the

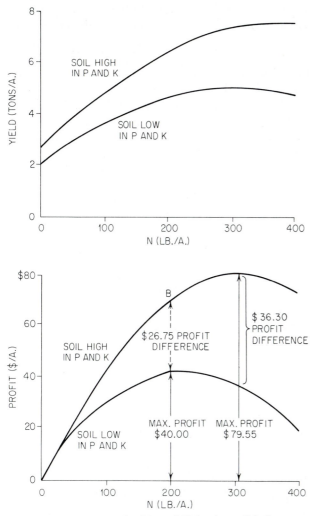

FIGURE 14.8 Level of P and K in the soil influences response of Coastal Bermuda grass to N. *Welch, cited by Engibous,* The Fertilizer Handbook. *Copyright 1972 by The Fertilizer Institute, Washington, D.C.*

TABLE 14.4 Effect of Underfertilizing versus Overfertilizing on the Net Return from Added Fertilizer

	Yield of Corn (bu/a)	*Net Return from Fertilizer (bu/a)*	*Difference from Optimum (bu/a)*
¼ less	142	50.4	−5.8
Optimum	151	56.2	0
¼ more	153	52.0	−4.2
None	79	—	—

SOURCE: S. A. Barber, Purdue Univ., personal communication.

nutrient and the price of the crop. In any case, the response of the crop to increasing nutrient rates must be known for the general soil condition. These data are obtained by research and are available from agricultural experiment stations.

One method is to use the crop/nutrient price ratio. This is shown for the corn/N price ratio in Table 14.5. If the yield potential is 160 bu and the corn/N price ratio is 15:1, the most profitable N rate is 200 lb/a.

To calculate the fertilizer rate required for maximum yield and maximum profit, the equation that describes the yield response is needed. The hypothetical yield response function for yield response Y (Fig. 14.9) is

$$Y = 70 + 1.0X - 0.0025 X^2$$

where

Y = grain yield (bu/a)
X = N rate (lb/a)

N rate for maximum yield:

1. Set the first derivative of the response function equal to zero.
2. Solve for X.

$$\frac{dY}{dX} = 1.0 - 0.005X$$

$$0 = 1.0 - 0.005X$$

Thus,

$$X = \frac{1.0}{0.005} = 200 \text{ lb/a of N}$$

3. The N rate for maximum yield is shown in Figure 14.9 and represents that point on the curve where the slope equals 0 ($dY/dX = 0$).

TABLE 14.5 Most Profitable N Rate (lb/a) for Corn Based on a Computer Model for Predicting Yield and Corn/N Price Ratios

Corn/N Price Ratio	Yield Potential (bu/a)							
	85	100	115	130	145	160	175	190
5/1	80	90	100	110	130	140	150	170
10/1	90	110	130	140	160	180	190	210
15/1	100	120	140	160	180	200	220	240
20/1	110	130	150	170	190	210	230	250
25/1	120	140	160	180	200	220	240	260

SOURCE: Vitosh et al., *Michigan State Univ. Ext. Bull. E-802* (1979).

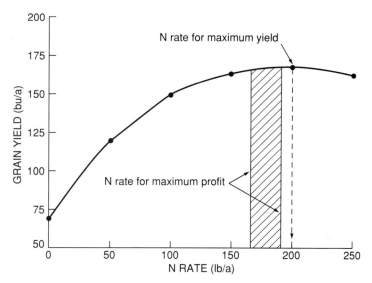

FIGURE 14.9   Hypothetical response function describing
the influence of N rate on grain yield. N rate of
maximum yield and maximum profit are shown (see the
text for the calculation). The shaded bar represents the
range in N rates for maximum profit, which varies with
crop price received and fertilizer cost (see Table 14.6).

N rate for maximum profit:

1. Set the first derivative of the response function equal to the ratio of fertilizer cost (i.e., \$0.20/lb N) to grain price (i.e., \$2.50/bu).
2. Solve for X.

$$1.0 - 0.005X = \frac{\$0.20}{\$2.50}$$

$$1.0 - 0.005X = -0.92$$

$$X = 184 \text{ lb/a}$$

3. The N rate for maximum profit is shown in Figure 14.9 and represents that point on the curve where the slope is parallel to the fertilizer cost line.

Of course, the fertilizer rate required for maximum profit will depend on the fertilizer cost and crop price (Table 14.6). As the fertilizer cost increases at constant crop price, the fertilizer rate for maximum profit decreases. Alternatively, as the crop price increases at a constant fertilizer cost, the fertilizer rate for maximum profit increases. These data also illustrate that although differences exist, changes in crop price and fertilizer cost have relatively minor effects on the fertilizer rate for maximum profit. The largest differences in optimum N rate occur when the crop price is low and the fertilizer cost is

TABLE 14.6   Effect of Crop Price
and Fertilizer Cost on the N Rate
for Maximum Profit*

Crop Price	N Fertilizer Cost ($/lb N)			
	0.15	0.20	0.25	0.30
($/bu)	--------------- lb N/a ---------------			
4.00	192	190	188	185
3.50	191	189	186	183
3.00	190	187	183	180
2.50	188	184	180	176
2.00	185	180	175	170

*Based on the response function shown in
Figure 14.9.

high, compared to a high crop price and a low fertilizer cost. Since it is difficult to predict the crop price at the time of fertilization, it is advisable to fertilize for near-maximum yields and not consider crop price and fertilizer cost factors.

Residual Effects of Fertilizers

Although discussed in Chapter 12, the residual effects of fertilizers are mentioned here because they should be included in fertilizer economics. Usually the entire cost of fertilization is charged to the current crop, whereas the lime cost is amortized over 5 to 10 years. With high rates of fertilization, however, residual effects can be substantial, especially with immobile nutrients.

At optimum fertilization rates, it has been shown that on some soils about 30% of the N may be residual for next year's crop, provided that it is not leached below the root zone. The residual value of P and K depends on the soil and how the crops are managed, but it can vary from 25 to 60%. The lower figure would apply when hay, straw, or stover is removed from the land.

The increased yield of next year's crop due to residual effects may be sufficient to pay for much of the fertilizer application on the current crop. For example, in Figure 14.10, the soil test K was increased from 272 lb/a to 448 lb/a with an increasing K rate. The average increase in corn yield for the four years was 25 bu/yr. The fifth-year soybeans were grown with no additional K, and yields were increased 20 bu/a. Residual K availability paid for most of the fertilizer. With K_2O at 13 cents/lb, the cost of the 1,200 lb/a for corn was $156. With soybeans at $6/bu, 20 bu equals $120 and there was still residual K availability after the soybean crop.

High crop yields are impossible with low levels of fertility. The fertility level of soils is a plant growth factor that is easily controlled, and up to a certain point the increasing fertility level will be profitable. However, the initial cost of building soil fertility from low to high levels may discourage growers if viewed as an annual rather than as a long-term investment.

Immobile nutrients like P and K can be built up in soils; thus, the buildup is a capital investment to be amortized over a period of years. The cost of building up soil P levels from 45 to 55 lb/a (P_1) is an example. Using a value

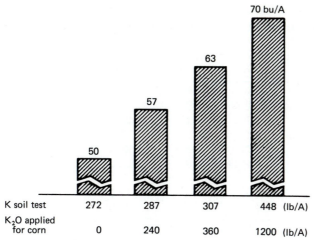

| K soil test | 272 | 287 | 307 | 448 | (lb/A) |
| K$_2$O applied for corn | 0 | 240 | 360 | 1200 | (lb/A) |

FIGURE 14.10 K applied to continuous corn for 4 years increased the K soil test and the yield of soybeans the fifth year 20 bu/a. *Welch,* Better Crops Plant Food, *63(4):3, 1974.*

of 9 lb of P$_2$O$_5$ to raise the P$_1$ test 1 lb, 90 lb/a of P$_2$O$_5$ would be required. The initial cost is $24.30/a with P$_2O_5$ at $0.27/lb. Using a payoff period of 15 years and an interest rate of 12%, the annual payment would be $3.57 (Table 14.7). A yield increase of 1.2 bu/a of $3/bu corn or 0.6 bu/a of $6/bu soybeans would pay for this cost.

Price per Pound of Nutrients

Growers are interested in the most economical source, but are accustomed to buying on the basis of cost per ton of fertilizer rather than the cost per ton of plant nutrients (Table 14.8). Wide variations in the cost per unit of nutrient

TABLE 14.7 Annual Payment Necessary to Amortize the $24.30 per Acre Initial Cost of Buildup P with Various Interest Rates and Amortization Periods

Payoff Period (yr)	Annual Payment for Payoff at an Interest Rate of:		
	8%	12%	16%
1	$26.24	$27.22	$28.19
5	6.09	6.74	7.42
10	3.62	4.30	5.03
15	2.84	3.57	4.36
20	2.47	3.25	4.10

SOURCE: Welch, *Better Crops Plant Food,* **66:**3 (Fall 1982).

TABLE 14.8   Approximate Prices of Fertilizer Materials
and Cost of Nutrients per Pound (for Illustration Only)

Material	Analysis	Price per Ton	Cost of N, P_2O_5, or K_2O (cents/lb)
Ammonium sulfate	20.5% N	$ 95	23.2
Ammonium nitrate	33.5% N	170	25.3
Urea	46% N	215	23.3
Anhydrous ammonia	82% N	220	13.4
Nitrogen solution	28% N	135	24.1
Superphosphate (triple)	44% P_2O_5	205	23.2
Muriate of potash	60% K_2O	150	12.5
Sulfate of potash	52% K_2O	250	24.0

exist, and other factors, such as the cost of application and the content of secondary nutrients, must be taken into consideration.

Growers need to decide based on the cost per unit of nutrient in mixed fertilizers, mixed and straight materials, or all straight materials. A knowledge of the cost calculation is important. For example, a farmer has a choice of 12-24-24 or 6-24-24:

6-24-24 costs $194/ton
12-24-24 costs $206/ton

Assuming that the P and K cost is the same in both mixtures, the additional N in 12-24-24 amounts to $12, or 10 cents/lb of N.

Another calculation that farmers need to make relates to the economics of high-analysis fertilizer. In a comparison of 5-10-10 and 10-20-20 fertilizer, 2 tons of 5-10-10 are required to furnish the same amount of nutrients contained in 1 ton of 10-20-20. If 5-10-10 costs $100 and 10-20-20 costs $180 per ton, the 10-20-20 will be $20 cheaper than 2 tons of 5-10-10.

Other Factors

In addition to the actual cost of the material, farmers must consider the cost of transportation, storage, and labor used in applying the fertilizer. These costs may be difficult to evaluate, but if the actual price of the nutrients from one source is the same as that from another source, growers will take the one requiring less labor. The higher-analysis goods require less labor in handling. Time is also saved because fewer stops are made in applying the material.

Liming

The returns from liming are quite high when it is applied where needed (Table 14.9). These data illustrate the high net return to liming, which, of course, will vary with lime rate, lime cost, yield response to liming, and price

TABLE 14.9 Effect of Changing Soybean Prices, Limestone Rate, and Yield Response on Net Return to Liming, $/a*

Lime Needed (ton/a)	Annual Yield Increase							
	3 bu		6 bu		9 bu		12 bu	
	$6	$8	$6	$8	$6	$8	$6	$8
1	14	20	32	44	50	68	68	92
2	10	16	28	40	46	64	64	88
3	6	12	24	36	42	60	60	84
4	2	8	20	32	38	56	56	80

*Limestone cost amortized over 5 years at 10% interest, assuming a total cost of $15/ton applied, with net return being rounded to the nearest dollar.

SOURCE: Hoeft, *Nat. Conf. Agr. Limestone*, National Fertilizer Development Center, Muscle Shoals, Ala. (1980).

received for the crop. In spite of a high return, however, lime is often neglected in the fertility program because (1) responses to lime are often not as visual as those obtained with N, P, or K unless the soil is particularly acid and (2) liming effects last for several years and the returns are not all realized in the first year.

Lime is the first step in any sound soil-management program, and broadcast applications in accordance with soil and plant requirements are essential for the greatest returns from fertilizer.

Animal Wastes

Benefits from manure, in addition to those from macro- and micronutrients, may be related to the organic components that might improve soil moisture relations and increase the downward movement of P, K, and micronutrients.

There is considerable variability in manure, depending on methods of storing and handling; however, with current fertilizer, labor, and equipment costs, it is usually profitable for the grower to use livestock manure. Because this is largely an N-K fertilizer, the best returns should be obtained by using it on nonlegumes. Hauling charges can be reduced by applying it on fields close to the source and using commercial fertilizer on the more distant fields.

In some farm management programs it is imperative that the manure be disposed of, and its value may be balanced against equipment and hauling costs. The general feeling, however, is that it is worth a little more than the cost of handling. In areas or times when fertilizers are difficult or impossible to obtain, animal manures play an important role in supplying fertilizer needs.

Plant Nutrients as Part of Increasing Land Value

When buying land, the farmer may be faced with the possibility of choosing high-priced or low-priced property. There is usually the question of which is the better buy. The higher-priced land is generally more productive, fertile, and has better improvements. The lower-priced land may actually be a good buy, however, provided that it has not been severely eroded or has no other physical limitations. Such land, however, is usually infertile and may need considerable lime and/or nutrients. Adequate liming and heavy fertilization, as indicated by soil tests and combined with other good practices, can rapidly increase productivity. Expenditures to improve fertility may be included as part of the cost of the land. If the problem is considered in this light, $100/a for liming and buildup fertility may be reasonable. Thus, with proper management, it is possible to increase land productivity and value, and the cost can be amortized over a period of years.

Additional Benefits from Maximum Economic Yields

Greater Flexibility

Although farming is a business, it differs from other businesses because yields are influenced greatly by the weather. The soil and crop management plan is designed for a specific yield goal, but weather, pests, disease, and so on may not allow the grower to achieve that yield. On the favorable side, the yield and/or price might be higher than expected; however, more often than not, lower yields and/or prices occur.

Figure 14.11 shows that a 120-bu corn yield goal allows the farmer only about 8 bu between profit and loss. A 160-bu yield goal allows him 41 bu between profit and loss. Higher yield management increases the farmer's chance for profit in the face of environmental uncertainties. Figure 14.12 shows that the farmer can take only about a 23 cents/bu price drop at the 120-bu yield before falling out of the profit zone, but about an 84 cents/bu drop at the 160-bu yield. Higher yields help make farmers much more flexible in their marketing operations and reduce risk.

Increase in Energy Efficiency

Higher yields are an effective means of improving energy efficiency in agriculture. Higher yields require more input energy per acre, but the energy cost per bushel or ton is less. The reason for this is that some costs are the same regardless of yield level. For example, it takes just as much fuel to till a field yielding 40 bu/a of soybeans as one yielding 60 bu/a.

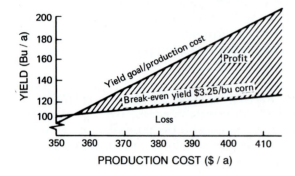

FIGURE 14.11   Higher yields increase the profit potential ($3.25 per bushel corn).   *Potash & Phosphate Institute, Maximum Economic Yield Manual, Atlanta, Ga., 1982.*

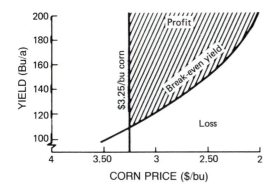

FIGURE 14.12   Higher yields mean lower breakeven prices for corn. *Potash & Phosphate Institute,* Maximum Economic Yield Manual, *Atlanta, Ga., 1982.*

Reduction in Soil Erosion

Raindrops strike the soil with surprising force, dislodging particles and increasing soil erosion. However, growing crops, crop residues, and roots absorb the impact and slow water movement, increase water intake and decrease soil erosion. The damaging effects of wind erosion are also reduced by the presence of crops and their residues. Highly productive cropping systems match perfectly with soil conservation because:

1. Crop canopy development is speeded.
2. Crop canopy density is increased.
3. More top and root residues are left.

Conservation tillage practices such as no-till and chisel plowing leave more residues on the surface than moldboard plowing. However, with any given tillage practice, higher amounts of residues will generally decrease soil losses.

Increase in Soil Productivity

Increasing soil OM is a long-term process; however, the benefits of raising OM on productivity can be substantial (see Chapter 15). Note the effects of increasing corn yields and soil OM on a loam soil:

Corn Yield (bu/a)	Soil OM (%)
50	1.7
100	3.1
150	4.0

In areas of higher temperatures and lower moisture, it is more difficult to increase OM; however, larger amounts of decomposing residues keep the soil in better physical condition. Infiltration of water is increased and the supply for the plant is improved. Thus plants are better able to withstand periods of drought. Also, less water runs off and erosion is reduced.

TABLE 14.10   Effect of Adequate K on Soybean Yield, Disease, and Dockage

K_2O (lb/a)	Yield (bu/a)	Diseased and Moldy Beans (%)	Dockage (cents/bu)	Value at $6 per Bushel ($/a)
0	38	31	54¢	$207.48
120	47	12	22¢	271.66

SOURCE: M. Kroetz, Ohio State Univ., personal communication.

Reduction in Grain Moisture

More adequate fertilizer, particularly N and P, decreases the amount of water in the grain at harvest and results in lower drying costs.

Improvement in Crop Quality

Adequate plant nutrition improves grain or forage quality. For example, increasing the grain protein of hard red spring wheat with N additions can increase the market value with protein premiums of 10 to 20 cents/bu/% protein.

As shown in Table 14.10, on a low-K soil, supplemental K not only increased soybean yields but decreased disease and mold in the seed.

Summary

1. Maximum yield is a constantly moving target because of continual technological advances. Maximum-yield research is the study of one or more variables and their interactions in a multidisciplinary system that strives for the highest yield possible for the soil and the climate of the research site.

2. Maximum profit yield is based on maximum yield. It is the point where the last unit of an input just pays for itself and gives the greatest net return per acre or unit of land area.

3. Maximum profit yields continue to offer the greatest opportunity for improving net profit per acre and for reducing production costs per bushel or per ton. Adequate fertility is a major factor in obtaining higher yields. Although fertilizer use has increased in the past few years, it is still below optimum for maximum profit yields in most areas.

4. The need for higher yields is great. The cost of production is expected to increase in the coming years. Although prices of crops will increase, they will not increase as fast as the cost of production. Hence higher and higher yields are a necessity for the farmer.

5. Fixed costs are those that remain about the same, regardless of yield. Hence practices that increase yields usually lower the cost of production per unit.

6. In a typical response curve, each successive increment of a plant nutrient gives a smaller yield increase in accordance with the law of diminishing returns. Hence the return per dollar spent decreases, an important consideration to the grower with a limited amount of capital.

7. Progressive growers recognize that profit per acre is more important than return per dollar spent. Maximum profit from fertilizer application is obtained when the added return in yield just equals the cost of the last increment of fertilizer.

8. The most profitable rate of plant nutrients is related to expected increases in yield from weather, level of management, price of fertilizer, expected price of crop, additional harvesting and marketing costs, residual effects, and soil fertility level.

9. Management level is the degree to which all the factors affecting crop production are successfully controlled. At higher rates of fertilization, with expectations of higher yields, managerial ability becomes more and more important.

10. Regardless of crop prices, it always pays to aim for the maximum profit yield. This will maximize profits or minimize losses.

11. The price of crop or fertilizer nutrients does not affect optimum fertilization appreciably.

12. Many practices cost little or nothing. Examples are timeliness, selection of variety, plant spacing, and rotations. These practices help to increase returns from the higher-cost inputs, such as fertilizer, land, labor, and machinery.

13. Residual effects of fertilizers and lime are an important part of fertilizer economics. With the increasing amounts of fertilizers being applied, the value of the residual fertilizer must be considered.

14. The cost of building soil fertility can be amortized over a period of years. The practices should be considered a long term rather than an annual investment.

15. The price per pound of nutrients varies among sources of materials. Higher-analysis mixed fertilizers are generally the most reasonable in cost and in labor required for application.

16. Personalized agronomic service to the grower by the dealer will enter more and more into the grower's choice of a fertilizer dealer.

17. Priority in the use of funds is of prime importance to the grower. It is generally more profitable to invest funds in fertilizer and lime applied in accordance with the soil-test level than in other parts of the farm business.

18. Lime usage is only about 35% adequate in the United States. Where needed, the returns from lime are much higher than those obtained from other plant nutrients.

Questions

1. What is maximum profit yield? How is it determined?
2. Why will higher crop yields be urgent in the coming years?
3. Explain how maximum profit yields minimize losses in periods of low crop prices.
4. What are fixed costs? Variable costs? How do they affect the unit cost of production?
5. When farmers have a limited amount of capital for fertilizer, will they be more interested in the rate that will give them the greatest return per dollar spent or the rate that will give them the greatest net profit? How

would soil tests help in the decision? Would the situation be changed if a farmer had unlimited capital? Explain.

6. What are the banks in your area doing in regard to credit for lime and/or fertilizer?

7. How does the use of irrigation make it easier to decide what rate of fertilizer to apply in semiarid and arid areas?

8. Discuss the factors that determine the most profitable rate of plant nutrients.

9. Why does the level of management affect the return from a given level of fertilization?

10. What are some of the yield-improving practices that cost little or nothing? How do they influence returns from high-cost inputs?

11. Explain why there is considerable leeway in the amount of nutrients that can be added after a reasonable level of application is reached.

12. Explain how the following affect the degree of carryover of fertilizer: soil, weather, application rate, yield increase, crop fertilized and harvested, and nutrient considered.

13. Why is it desirable to amortize the cost of building P and K levels or added lime over a period of years?

14. What N carrier would you choose to fertilize corn? Why? You are quoted the following prices: 3-12-12 at $120 per ton and 6-24-24 at $200 per ton. Which would you choose, and why?

15. Explain why funds invested in plant nutrients usually return more profit than investments in other phases of the farm business.

16. Obtain the average yield of corn, soybeans, or wheat in your area and the number of acres used for this production. What would a reasonable yield of corn per acre be if recommended practices were followed, and how many acres would be required to maintain the present total production? What would you estimate the profits to be under each system?

17. What is restricting lime use in your area?

18. Is farm manure a valuable source of plant nutrients in your area? Why or why not?

19. How would you evaluate the residual nutrients in your area? Considering present crop production levels, are the soil test levels suggested in the 1970s adequate now?

20. Why do maximum profit yields increase energy efficiency? Reduce soil erosion? Increase crop quality?

21. How does higher-yield management give more flexibility and increase a farmer's chance for profit?

Selected References

FOLLETT, R. H., L. S. MURPHY, and R. L. DONAHUE. 1981. *Fertilizers and Soil Amendments*. Prentice-Hall, Inc., Englewood Cliffs, N.J.

TERMAN, G. L., and O. P. ENGELSTAD. 1976. Agronomic Evaluation of Fertilizers: Principles and Practices. Bull., Y-21. Tennessee Valley Authority, National Fertilizer Development Center, Muscle Shoals, Ala.

CHAPTER 15

Cropping Systems and Soil Management

The principal objective of any soil and crop management program is sustained profitable production. The strength and longevity of any civilization depend on the ability to sustain and/or increase the productive capacity of its agriculture. Sustainable agriculture encompasses soil and crop productivity, economics, and environment and is defined by

> the integration of agricultural management technology to produce quality food and fiber while maintaining or increasing soil productivity, farm profitability, and environmental quality.

Of course, many criteria can be used to evaluate sustainable farming systems, such as:

1. Maintain short-term profitability and sustained economic viability.
2. Maintain or enhance soil productivity.
3. Provide long-term environmental quality.
4. Maximize efficient use of resources.
5. Ensure food safety, quality of life, and community viability.

Achieving agricultural sustainability depends on many agronomic, environmental, and social factors; it is more difficult to achieve in semiarid and arid climates than in humid climates (Fig. 15.1).

From a soil productivity–soil fertility standpoint, soil conservation is essential for long-term sustainability (Fig. 15.2). Soil management practices that contribute to or encourage soil degradation will reduce soil productivity and impair progress toward sustainability.

Soil erosion represents the greatest threat to sustained soil productivity. Physical removal of nutrient-rich, high-OM topsoil, and oxidation of OM with tillage, reduces the productive capacity of the soil. Exposed subsoil often is less productive because of (1) poor soil physical condition, (2) reduced water availability, (3) decreased nutrient supply, and (4) many other site-specific parameters.

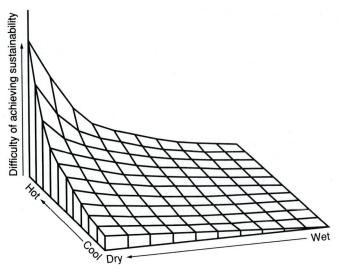

FIGURE 15.1   Agricultural sustainability is more difficult
to achieve in hot, dry climates compared to humid
climates.   *Parr et al.,* Advances in Sci., ***13:1–7, 1990.***

Although growers, consultants, and others recognize erosion on the lands
they manage, they may not be overly concerned because crop yields have
substantially increased since the 1950s. One should not confuse increasing
crop yields with increasing soil productivity, because yield increases are primarily due to technological advances in crop breeding and genetics, fertilizers
and fertilizer management, pesticides and pest management, and other agronomic technologies. Soil conservation and good soil management embrace
more than just the prevention of soil losses. Soil erosion is a *symptom* of poor soil
management, whether it be inadequate plant nutrients or improper cropping
systems.

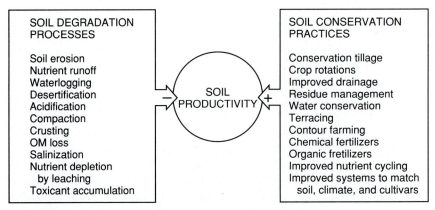

FIGURE 15.2   Soil productivity is reduced by soil
degradation processes and improved by soil conservation
practices.   *Parr et al.,* Advances in Sci., ***13:1–7, 1990.***

Soil and Crop Productivity

Crop yields in the United States have gradually increased since 1890, but most of the improvement has occurred since the 1950s (Table 1.2). In 1990 the average yields in the United States of corn, soybeans, and wheat were 116.2, 32.2, and 34.6 bu/a, respectively. Improved management, including varieties, spacing and population, pest control, tillage, and fertilization, contributes to increasing yields.

The problems of declining soil productivity are not new and have been recognized since the early 1900s.

A historical explanation for the slow increase in yields prior to the 1950s is provided in Figure 15.3. Soil productivity decreased 40% in the 60-year period from 1870 to 1930, closely related to decreased fertility levels. Soil OM and the native nutrient supply, especially N, have decreased. The nutrients removed have generally been greater than the amounts returned to the soil in manure and commercial fertilizers.

Grower adoption of many agronomic technologies developed since 1950 has increased crop productivity (Table 15.1). These data demonstrate that the technologies contributing most to the increase in average corn yields from 32 to 100 bu/a were N fertilization, breeding/genetics, weed control, and other cultural practices. In contrast, decreased used of manure and declining OM contributed to yield loss. Evaluation of the gains and losses in yield shows that, if losses had not occurred, yields would have been 158 bu/a instead of 100

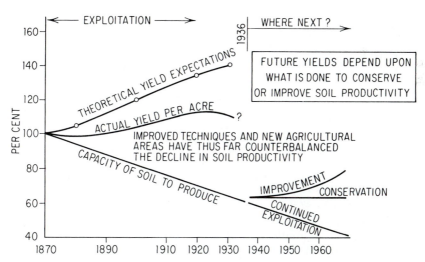

FIGURE 15.3   Improved practices in Ohio since 1870 should have resulted in yields 40 to 60% higher per acre in 1930, but the aggregate yield increased less than 15%. Improved practices only slightly more than counterbalanced the decline in the ability of the soil to produce. Yields can be increased only if proper soil-management programs are adopted.   Ohio State Agr. Ext. Serv. Bull. 175, *1936.*

TABLE 15.1   Sources of Corn Yield Increases
with Changing Production Practices in Minnesota
from 1930 to 1979

Cultural Practice or Yield Limiting Factor	bu/a	Contribution to 1979 Yield	
		kg/ha	% Net Gain/Loss
Pre-1930 yield levels	32.1	2,012	—
Productivity Gains/Losses			
Hybrids			
Double crosses	5.9	371	9
Three-way crosses	0.5	28	1
Single crosses	3.7	235	6
Genetic gain	29.1	1,825	43
Fertilizer N	31.9	2,003	47
Plant population	14.4	905	21
Herbicides	15.5	975	23
Row spacing	2.8	173	4
Planting date	5.8	364	8
Drilling vs. hill drop	5.1	322	8
Fall plowing	3.6	224	5
Rotations			
Soybeans	7.7	484	11
Alfalfa/clovers	−2.2	−136	−3
Sweet clover	−5.1	−318	−7
Interference effect	−4.6	−291	−7
Manure	−10.1	−633	−15
Organic matter	−9.1	−571	−13
Insects			
Corn borer	−3.5	−220	−5
Corn rootworm	−2.3	−145	−3
Soil erosion	−6.5	−345	−8
Unidentified negative factors	−15.5	−975	−23
Net gain	68.1	4,275	—
1977–1979 yield level	100.2	6,287	—

SOURCE: Cardwell, *Agron. J.*, **74**:984 (1982).

bu/a. Therefore, growers must take advantage of technologies that increase productivity, as well as those that minimize productivity loss.

Elimination of fertilizer would decrease yields 40 to 90%, depending on the crop, soil, and climatic region. If fertilizer were not used, about 30 to 40% more land would be needed. This would mean that less suitable land would go into production, with greater opportunity for erosion.

Although in dryland regions water is a more limiting factor than fertility, serious losses in soil fertility and productivity have often occurred as a result of summer fallow cropping systems.

Degradation of native soil fertility by not returning nutrients removed in the crops is evident in the first 40 years of dryland production (Fig. 15.4). Technological developments in varieties, water conservation, and P fertilization increased productivity during the next 50 years; however, continued soil

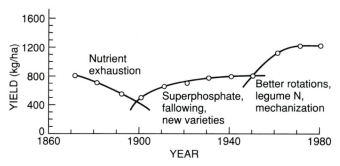

FIGURE 15.4   Changes in dryland wheat productivity in
Australia.   *Donald,* Agric. in Australian Econ., *Sydney
Univ. Press, 1981.*

erosion and OM loss again limited productivity of the wheat-fallow-wheat
cropping system. After 1950 growers adopted wheat-legume rotations to provide forage for livestock, and the increased N availability from the legume
residue dramatically increased wheat yields. Soil erosion was reduced, OM
increased, and these soils are much more productive now than in 1900.

Another example is provided in Figure 15.5, where, prior to 1940, wheat
yields were less than 10 bu/a. Average annual precipitation has remained
relatively constant, but wheat yields tripled from 1940 to 1985. The major
technological advances contributing to increased productivity were (1) improved varieties, (2) increased water conservation with stubble mulch tillage,
(3) increased N and P fertilization, and (4) improved planting and harvesting
methods.

Profitable crop production on eroded soils has been an important agricultural problem. The generally reduced crop yield of nonlegumes on subsoil is
well known. On permeable soils this decrease is largely the result of less OM
and the subsequent lower release of N.

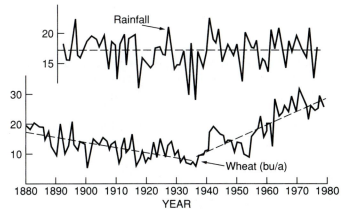

FIGURE 15.5   Changes in dryland wheat productivity in
North Dakota.   *Fanning and Reff,* NDSU Coop. Ext. Ser.
Bul. SC-710, *1981.*

Figure 15.6 shows that erosion must not be taken for granted, and that it is imperative that crop production systems minimize the destructive effects of water and wind erosion. More than one-third of the cropland in the United States is subject to erosion severe enough to shorten significantly the productive life of soils.

Water and wind erosion of topsoil can reduce productivity by exposing less productive subsoil (Fig. 15.7). The productive capacity of eroded Ulysses soil

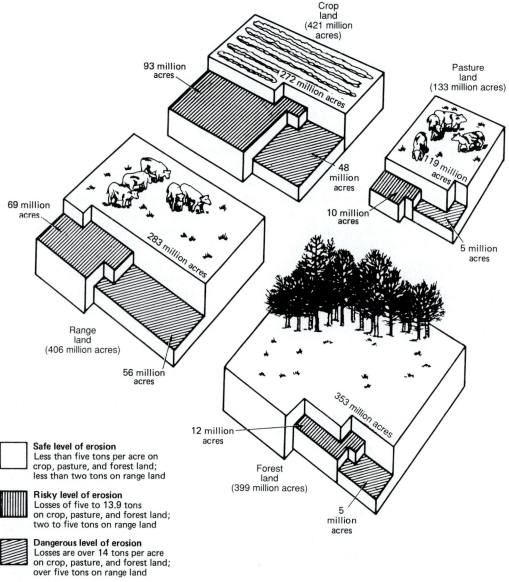

Crop land (421 million acres)

93 million acres

272 million acres

Pasture land (133 million acres)

48 million acres

119 million acres

10 million acres

5 million acres

69 million acres

283 million acres

Range land (406 million acres)

56 million acres

12 million acres

353 million acres

Forest land (399 million acres)

5 million acres

Safe level of erosion
Less than five tons per acre on crop, pasture, and forest land; less than two tons on range land

Risky level of erosion
Losses of five to 13.9 tons on crop, pasture, and forest land; two to five tons on range land

Dangerous level of erosion
Losses are over 14 tons per acre on crop, pasture, and forest land; over five tons on range land

FIGURE 15.6   How serious is erosion? Shown here are the results of the National Resource Inventory conducted by the Soil Conservation Service in 1982.

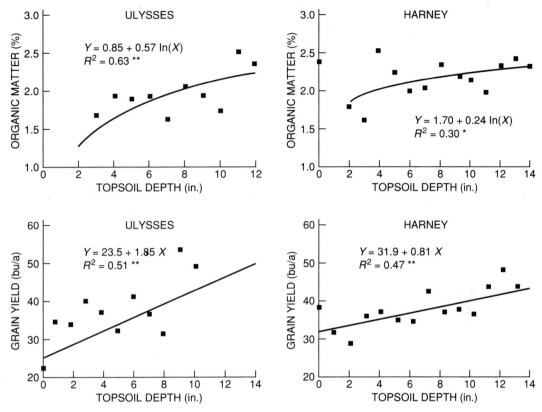

FIGURE 15.7 Loss of topsoil by wind and water erosion reduces soil OM, which contributes to wheat grain yield loss. Loss in productivity varies between soils, depending on initial topsoil depth and productivity of the subsoil. Compared to the Ulysses soil, the Harney soil has a deeper topsoil, thus productivity is not reduced as much as the Ulysses soil under equivalent topsoil loss. *Havlin et al.*, Proc. Great Plains Soil Fert. Conf., *1992*.

is less than that of eroded Harney soil because the latter is a deeper soil and has a greater OM content in the subsoil, which improves nutrient availability and water-holding capacity. Increasing topsoil loss reduces the OM content much more in the Ulysses soil than in the Harney soil. As a result, the yield loss associated with increasing soil loss also is greater in the Ulysses than in the Harney soil. Specifically, in the Ulysses soil, 1.8 bu/a yield loss per inch of topsoil loss occurs compared to 0.8 bu/a yield loss per inch of topsoil in the Harney soil.

In some years, however, moisture may limit yield because of reduced water availability related to lower water infiltration and reduced water-holding capacity. The data in Table 15.2 show the lower OM and plant-available water content on severely eroded soil compared to the noneroded/depositional areas in the field. Reduced N and water availability were primarily responsible for the 32 to 66% corn yield reduction on the eroded soil.

TABLE 15.2   Influence of Soil Erosion on Surface Soil Physical Properties of Several Sites and Their Relationship to Corn Grain Yield Loss

Site	Surface Texture	Erosion Phase	OM (%)	Dry Bulk Density (g/cm³)	Estimated Soil Profile Θva* (cm H₂O)
3	Fine	Noneroded	2.6	1.44	14.1
	sandy	Severely eroded	0.7	1.46	12.8
	loam	Depositional	5.5	1.23	18.1
6	Gravelly	Noneroded	3.7	1.37	7.4
	sandy	Severly eroded	2.2	1.68	5.1
	loam	Depositional	4.3	1.38	11.1

	1982		1983	
Site	Yield Reduction (%)	Major Cause(s)	Yield Reduction (%)	Major Cause
2	38	N, H₂O	52	H₂O
3	43	H₂O	38	H₂O
5	32	H₂O	35	H₂O
6	66	H₂O, N	59	H₂O

*Θva = estimated pre-season plant-available water in 39.4 inches of soil.
SOURCE: Battiston et al. *Erosion and Soil Prod.*, ASAE 8-85, 1985, pp. 28–320.

OM in the Soil

In addition to other factors, the OM content of a soil is intimately related to its productivity. Some of the functions of OM are as follows:

1. It acts as a storehouse for nutrients—N, P, S, and most micronutrients.
2. It increases exchange capacity.
3. It provides energy for microbial activity.
4. It increases water-holding capacity.
5. It improves soil structure.
6. It reduces crusting and increases infiltration.
7. It reduces the effects of compaction.
8. It buffers the soil against rapid changes in acidity, alkalinity, and salinity.

All these factors interact to impart *tilth* to the soil. The production of large quantities of residues, and their subsequent decay, is necessary for good soil tilth and productivity. One of the plant-nutrient problems of the Arctic is the resistance to decay of OM under low temperatures, and OM accumulates even on gravel ridges. In contrast, in the subtropics and tropics, although much OM is produced, it decays very rapidly.

Maintenance of OM for the sake of maintenance alone is not a practical approach to farming. It is more realistic to use a management system that will give sustained profitable production without degradation of OM and productivity. The greatest source of soil OM is the residue contributed by current crops. Consequently, the cropping system and the method of handling the residues are equally important. Proper management and fertilization will produce high yields, which will increase the quantity of residue and organic C returned to the soil.

Tillage of the soil produces greater aeration, thus stimulating more microbial activity, and increases the rate of disappearance of soil organic C (Fig. 15.8). When a virgin soil is cultivated, the OM decline is rapid during the first 10 years and then continues at a gradually diminishing rate for several decades. Eventually, an apparent equilibrium is reached after about 30 to 60 years, depending on the soil, climate, and cropping system.

Soil OM transformations are very dynamic. Intensive tillage systems, fallowing, and low crop productivity, combined with physical soil loss by erosion, decrease the OM content over time (Fig. 15.9). Increasing soil OM requires reducing tillage intensity and increasing the quantity of CO_2 fixed by plants and returned to the soil. Increasing the C input depends on the interaction between more productive rotations and reduced tillage.

Cropping systems differ in the amount of plant residues they contribute. Grain corn may add 3 to 7 tons/a of stover, while roots may furnish another 1 to 2 tons/a. However, silage corn results in removal of the stover and grain, drastically reducing the quantity of C returned to the soil. A grass-legume sod will produce about the same amount of residues, but most of it will be removed in hay and grazing. With a good yield of small grain, 3 tons of residue may be returned if the straw is left on the field, but only about 0.5 ton may be returned if the straw is removed. The tops, nuts, and many of the roots in peanuts are removed, encouraging much more rapid loss of OM.

The N content of the plant residues is in part related to the accumulation of soil OM. If low N residues are turned under, much of the C oxidizes to CO_2 during decomposition before the ratio approaches 10:1 or 12:1 (see Chapter 5).

Corn fertilized with N will contain more than 0.75 to 1.0% N in the stover. In contrast, corn receiving small amounts of N may have less than 0.5% N in the stover. Stover from well-fertilized corn will be more effective in main-

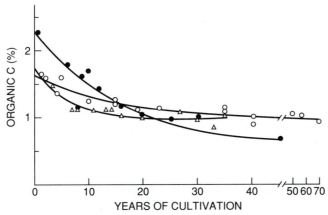

FIGURE 15.8   Loss in soil organic C over 70 years of cultivation on selected Australian soils.   *Dalal and Meyer, Aust. J. Soil Res., **45**:281–292, 1986.*

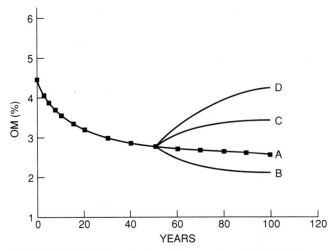

FIGURE 15.9 Hypothetical decrease in soil OM with time. At 50 years, changes in soil and crop management system can either continue (A), decrease (B), or increase (C, D) soil OM. (A) represents no change in cropping system, while (B) represents a change that would accelerate OM loss (i.e., more intensive tillage). (C) might represent adoption of reduced tillage or a crop rotation that produces more residue, whereas (D) might reflect the change in OM following adoption of a high yield no-till system or rotations that return large quantities of residue.

taining OM than that from poorly fertilized corn. Most crop residues containing about 1.5% N do not need additional N for rapid breakdown and conversion into humus (see Chapter 5).

Generally, the quantity of residue returned to the soil will have a much greater effect on increasing soil OM than the residue N content. The data in Figure 15.10 show that even though the N content of alfalfa is much greater than that of corn, the original organic C content was maintained at 1.8% C by 2 ton/a/yr of either corn or alfalfa residue. Increasing the residue produced and returned with either crop increased soil organic C. If all the residue had been left on the soil surface instead of incorporated with tillage, the increase in organic C would have been greater or the original soil C content would have been maintained with lower residue mass. The relationship between increasing residue level and tillage intensity was illustrated earlier (Fig. 5.8). These data demonstrated that the use of cropping systems that maximized residue production while maintaining the residues on the soil surface increased soil OM much greater than incorporating residue from rotations with low residue yield.

There has been some discussion on the effect of adding extra N to accelerate residue decomposition and encourage the formation of soil OM. The data in Table 15.3 show that increasing the N supply to decomposing wheat straw greatly reduced the C/N ratio by substantially increasing the %N in the residue.

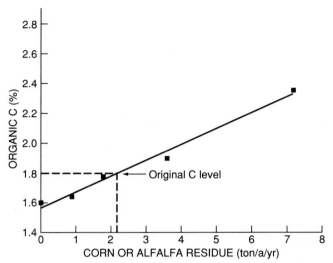

FIGURE 15.10   Influence of corn or alfalfa residue incorporated into the soil for 11 years.   *Larson et al., Agron. J., 64:204–208, 1972.*

Thus, added N will certainly encourage rapid residue degradation by microorganisms with little or no immobilization of soil inorganic N (see Chapter 5); however, as illustrated in Figure 15.10, the quantity of residue added influences OM more than the N content.

In semiarid and arid climates, where fallowing is often practiced, OM decomposition occurs over a considerable part of the year; thus, it is difficult to increase soil OM. In dry regions of the northern United States, however, the soil OM content, if depleted, can be increased more readily by certain cropping systems, particularly ones that feature less tillage and the return of more crop residues. As the data in Table 15.4 show, soil OM and total N increased in systems that reduced the frequency of fallowing and increased the production and return of organic residues.

The dryland cropping systems referred to in Figure 13.5 also showed increased soil OM content with increasing cropping intensity or reduced dependence on fallowing (Fig. 15.11). Soil OM increased as more residues were

TABLE 15.3   Effect of Adding $NH_4NO_3 \cdot (NH_4)_2SO_4$ on the C/N Ratio and N and C Concentrations in Decomposing Wheat Straw Incubated for 9 Weeks in the Absence of Soil

Treatment	Time	C/N Ratio	N (%)	C (%)
Straw only	Initially	107	0.38	40.4
Straw with added N		30	0.38	40.4
Straw only	After 9 weeks	116	0.35	40.8
Straw with added N		76	0.55	41.7

SOURCE: Cochran et al., *Soil Sci. Soc. Am. J.,* **44**:978 (1980).

TABLE 15.4   Effect of Cropping Systems and Manure on N and OM of Soils After 37 Years of Cropping

Rotation	OM (%)		Total N (%)	
	Control	Manure	Control	Manure
Fallow-wheat	3.7	4.1	0.19	0.28
Fallow-wheat-wheat	4.9	5.5	0.26	0.30
Fallow-wheat-wheat-wheat	4.7	5.5	0.25	0.28
Wheat continuous	7.2	7.6	0.36	0.38
Average	5.1	5.6	0.27	0.31
Alfalfa fallow-wheat-wheat-wheat	5.8	—	0.28	—
Grass fallow-wheat-wheat-wheat	6.3	—	0.31	—

SOURCE: Ridley and Hedlin, *Can. J. Soil Sci.,* **48:**315 (1968).

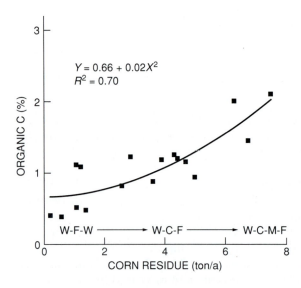

FIGURE 15.11   Increasing no-till cropping intensity increased OM, compared to wheat-fallow-wheat. W = wheat; C = corn; M = millet; F = fallow.   *Peterson and Westfall,* Proc. Great Plains Soil Fert. Conf., *1990.*

produced in the wheat-corn-fallow and wheat-corn-millet-fallow systems compared to the wheat-fallow-wheat system.

Many factors determine whether the soil OM is increased or decreased by cropping systems. *The key is to keep large amounts of crop residues (stover and roots) passing through the soil. Continued good management, including adequate fertilization, helps bring this about.*

Increasing the quantity of crop residues through adequate nutrient availability is important in OM maintenance (Fig. 5.9). In addition, higher yields mean more extensive root systems that distribute OM deeper in the soil.

OM influences many soil biological, chemical, and physical properties that favorably influence nutrient availability. Table 15.5 summarizes these properties.

Tillage

Over the years, growers have become interested in developing tillage practices to give greater protection to the soil against soil and water losses. The amount of surface residues and surface roughness both have an effect. Crop residue

TABLE 15.5   General Properties of Soil OM and Associated Effects on Soil Properties

Property	*Remarks*	*Effect on Soil*
Color	The typical dark color of many soils is caused by organic matter	May facilitate warming.
Water retention	Organic matter can hold up to 20 times its weight in water.	Helps prevent drying and shrinking. May significantly improve the moisture-retaining properties of sandy soils.
Combination with clay materials	Cements soil particles into aggregates	Permits exchange of gases stabilizes structure, increases permeability.
Chelation	Forms stable complexes with Cu^{2+}, Mn^{2+}, Zn^{2+}, and other polyvalent cations.	May enhance the availability of micronutrients to higher plants.
Solubility in water	Insolubility of organic matter is because of its association with clay.	Little organic matter is lost by leaching.
Buffer action	Organic matter exhibits pH buffering	Helps to maintain a uniform reaction in the soil.
Cation exchange	Total acidities of isolated fractions of humus range from 300 to 1400 meq/100 g	May increase the soil CEC. From 20 to 70 percent of the CEC of many soils (e.g., Mollisols) is due to organic matter.
Mineralization	Decomposition of organic matter yields CO_2, NH_4^+, NO_3^-, $H_2PO_4^-$, and SO_4^{2-}.	A source of nutrient elements for plant growth.
Combines with organic molecules	Affects bioactivity, persistence and biodegradability of pesticides	Modifies application rate of pesticides for effective control.

SOURCE: Stevenson, *Humus Chemistry*, 1982. Copyright © 1982 John Wiley & Sons, Inc.

management has been developed to leave more of the harvest residues, leaves, and roots on or near the surface.

Conservation Tillage

Conservation tillage is a term used to describe any tillage system that reduces soil and/or water loss compared to clean tillage, where all residues are incorporated into the soil (Table 15.6).

Advantages

1. Higher crop yields, except possibly in level, fine-textured, poorly drained soils.
2. Less soil erosion by water and wind.
3. Improved infiltration and more efficient use of water.
4. Increased acreage of sloping land that can safely be used for row crops.

TABLE 15.6   Conservation Tillage Methods

Rowcrop Agriculture	*Small Grain Agriculture*
Narrow strip tillage	Stubble mulch farming
No-till, zero-till, slot plant	Stirring or mixing machines
Strip rotary tillage	Disk-type implements
	• Oneway disk
Ridge planting	• Offset disk
Till plant	• tandem disk
Plant conventionally on ridge	Chisel plows
	Field cultivators
Full width—no plow tillage	Mulch treaders
Fall and/or spring disk	
Fall or spring chisel, field cultivate	Subsurface tillage
	Sweep plows
Full width—plow tillage	Rotary rodweeder
Plow plant	Rodweeder with semichisels
Spring plow–wheel–track plant	
	Ecofallow
	Direct drill

SOURCE: Mannering and Fenster, *J. Soil Water Cons.* **38**:141–143 (1983).

5. Improved timing of planting and harvesting.
6. Lower labor, machinery and fuel costs.

Disadvantages

1. More potential for rodents, insects, and diseases.
2. Cooler soil temperatures in spring, resulting in slower germination and more stand problems.
3. Greater management ability is required.

General Types of Conservation Tillage

CHISELING   A chisel implement may till 8 to 15 in. deep, with points 12 to 15 in. apart. A considerable amount of surface residues is left on the surface, and the surface is rough.

TILL PLANT AND RIDGE TILL   Till plant and ridge till is a once-over tillage–planting operation. Planter units work on ridges made the previous year during cultivation or after harvesting (Fig. 15.12). The planter pushes the old stalks, root clumps, and clods into the area between the rows. This is perhaps most useful on fine-textured, poorly drained soils. Ridges are retained in the same position year after year; hence, wheel tracks are in the same place.

STUBBLE MULCHING   Two types of tillage equipment are used—those such as disks, chisel plows, and field cultivators, which mix crop residues with the soil, and those such as sweeps or blades and rodweeders, which cut beneath

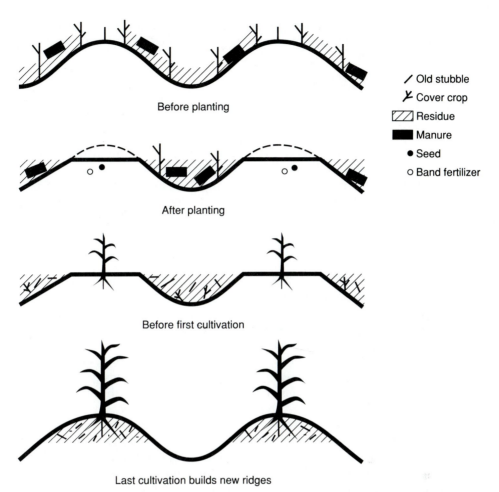

Before planting

After planting

Before first cultivation

Last cultivation builds new ridges

/ Old stubble
Ұ Cover crop
▨ Residue
▰ Manure
• Seed
○ Band fertilizer

FIGURE 15.12   Ridge tillage advantages in alternative production systems. The planter tills 2 to 4 in. of soil in a 6-in. band on top of the ridges. Seeds are planted on top of the ridges, and soil from the ridges is mixed with crop residue between the ridges. Soil on ridges is generally warmer than soil in flat fields or between ridges. Warm soil facilitates crop germination, which slows weed emergence. Crop residue between the ridges also reduces soil erosion and increases moisture retention. Mechanical cultivation during the growing season helps to control weeds, reduces the need for herbicides, and rebuilds the ridges for the next season.   *Dick Thompson, The Thompson Farm.*

the surfaces without inverting the soil. Increased moisture efficiency and reduced wind erosion are primary goals of stubble mulching.

NO-TILL   In the no-till system, all of the residues are left on the surface; therefore, it has been most successful in the better-drained soils. A seed zone 2 in. wide or less is prepared in previously untilled ground. This zone is made by a fluted coulter running ahead of a planter unit with disk or hoe openers. Seeding by this method is successful in corn or soybean residues and in small grain stubble. Narrow hoe openers or narrow-angle, double-disk openers are also used for minimum and zero-tillage seed placement.

Effects of Tillage

Effects vary greatly, depending on soil, crop, and weather conditions.

SURFACE RESIDUES   The approximate quantity of surface residue remaining after one tillage operation varies with the implement (Table 15.7). Subsurface implements that leave most of the residue on the soil surface help protect the surface against erosion. The quantity of residue incorporated or left on the surface after tillage also varies with the crop; thus, the percentage of surface residue cover also varies. Figure 15.13 illustrates the relationship between residue cover and residue mass. Crops like soybean, sunflower, and cotton provide very little surface cover compared to small and coarse grain crops.

SOIL LOSS   The quantity of residue required to prevent soil erosion depends on the following factors:

1. Soil characteristics (texture, OM, surface roughness, structure, depth, slope percentages, slope length, etc.).
2. Residue characteristics (type, quantity, orientation, etc.).
3. Rainfall characteristics (quantity, duration, intensity, etc.).
4. Wind characteristics (velocity, direction, gusts, duration, etc.).

In general, as the percentage of surface residue cover increases, the potential for soil loss decreases (Fig. 15.14). In addition, increasing the surface residue level by reducing the tillage intensity drastically reduces soil loss under a rainfall simulator (Fig. 15.15). These data also show the value of "farming" on the contour compared to up and down the slope. The same relationship can be seen in Tables 15.8 and 15.9. Soil loss was greater with all tillage systems

TABLE 15.7   Effect of Tillage Equipment on Surface Residue Remaining After Each Operation

Tillage Machine	Approximate Residue Maintained (%)
Subsurface cultivators	
Wide-blade cultivator, rodweeder	90–95
Mixing-type cultivators	
Heavy-duty cultivator, chisel, and other types of machines	50–75
Mixing and inverting disk machines	
One-way, flexible disk harrow, one-way disk, tandem disk, offset disk	25–50
Inverting machines	
Moldboard, disk plow	0–10

SOURCE: Anderson, *Great Plains Ag. Council Publ. No. 32* (1968).

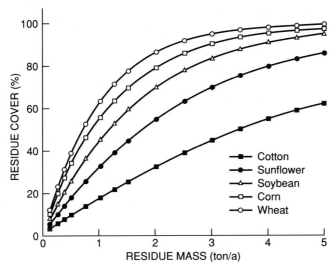

FIGURE 15.13 Relationship between residue mass and percentage of surface cover for selected crops.

when the crop was planted up and down the slope (Table 15.8) compared to across the slope (Table 15.9). In addition, surface cover was much less following soybean than corn planting, resulting in considerably higher soil loss.

SOIL TEMPERATURE, MOISTURE, AND MICROBIAL ACTIVITY   Soil temperatures early in the growing season are generally lower under conservation tillage than under conventional tillage due to the insulating effect of the unincorporated crop residues on the surface. Soil temperatures in the top 6

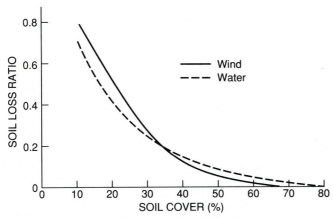

FIGURE 15.14 Relationship between soil loss with residue cover divided by soil loss from bare soil (soil loss ratio) and percentage of surface residue cover. *Adapted from Laflen et al.*, ASAE Publ. 7-81, *1981, pp. 121–133, and Fryrear,* Sci. Reviews, *Arizona Res.,* Scientific Publ., *1985, pp. 31–48.*

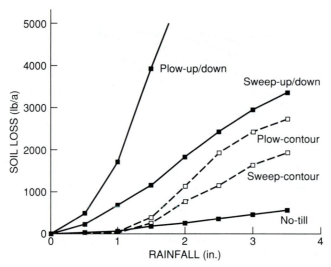

FIGURE 15.15   Influence of tillage system and direction (up/down hill or on the contour) on soil loss for a 4% slope. Initial residue level was 13,500 lb.   *Nebraska, HPAL.*

in. of the root zone during May and June can be several degrees lower under no tillage than under conventional tillage.

The influence of surface residue cover on soil temperature also depends on the crop. As Figure 15.16 shows, surface soil temperature was lower following corn than soybean planting, which was related to the higher surface cover with corn compared to that with soybeans.

Decomposition of crop residues and soil OM, with subsequent release of plant nutrients, including N, P, and S, is restricted by low soil temperatures. Thus, recycling of essential nutrients may be delayed. Further, low soil temperatures will retard root development and activity (Fig. 15.17). In tropical and semitropical regions the cooling effect of crop residues on soil temperatures may be beneficial.

Increasing residues on the surface by reducing tillage reduces runoff and soil erosion while increasing infiltration. Soils with tillage pans may require

TABLE 15.8   Surface Cover and Soil Loss from Various Tillage Systems on 4% Slope Land Tilled Up and Down Slope Following Corn and Soybeans (Harlan, Indiana)*

	Surface Cover		Soil Loss	
Tillage System	After Corn (%)	After Beans (%)	After Corn (mt/ha)	After Beans (mt/ha)
Fall moldboard plow	7	1	22.0	41.0
Fall chisel tillage	25	12	15.0	30.3
No-till	69	26	2.5	13.5

*Morley clay loam with a slope length of 10.7 m. Tests were made after overwinter weathering but prior to spring tillage. Two storms were applied at 6.25 cm of rainfall each.
SOURCE: Mannering, *Crop. Ext. Serv. Publ. AY-222,* Purdue Univ., 1979.

TABLE 15.9   Surface Cover and Soil Loss from Various Tillage Systems
on 5% Slope Land Tilled Across the Slope Following Corn and Soybeans
(Urbana, Illinois)*

| | Surface Cover | | Soil Loss | |
Tillage System	After Corn (%)	After Beans (%)	After Corn (mt/ha)	After Beans (mt/ha)
Fall moldboard plow	4	2	12.8	25.6
Fall disk-chisel tillage	50	11	1.3	7.4
No-till	85	59	1.1	3.8

*Catlin silt loam with slope length 10.7 m. Tests made after overwinter weathering but prior to
any spring tillage; 12.5 cm of simulated rainfall were applied in two storms.
SOURCE: Siemens and Oschwald, *Am. Soc. Agr. Eng. Paper No. 76-2552*, 1976.

chiseling to gradually create a deeper root zone with greater water intake and
water-holding capacity. This may provide an extra inch or two of water at
critical stages of growth (Fig. 15.18).

In semiarid regions, water conservation increases with maintenance of surface residue cover. As the data in Table 15.10 show, increasing residues with
no-till increased the total water stored, consequently improving sorghum yield
and WUE compared to residue incorporation with the disk.

The interaction of soil temperature and moisture with tillage will dramatically influence microbial activity (Fig. 15.19). When the soil is tilled, increased
aeration (along with residue C mixed in the soil) encourages microbial activity
and mineralization of OM, which eventually releases N and other nutrients.
Microbial activity is lower early in the season because of lower temperature;
however, it is slightly higher later in the season because of greater soil
moisture.

The net effect of tillage is increased mineralization of OM, which causes
gradual OM loss over time (Fig. 15.9). Reducing tillage intensity will reduce

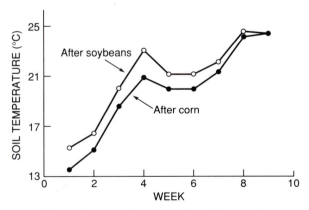

FIGURE 15.16   Weekly means for daily maximum soil
temperatures in no-till corn.   *Griffith et al.*, No Tillage
and Surface Tillage Agriculture, *p. 34. John Wiley & Sons,
1986.*

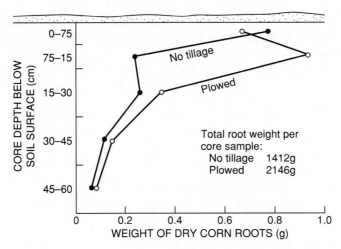

FIGURE 15.17   Effect of plowing and no-till planting on amount and distribution of corn roots.   *Griffith et al.,* No Tillage and Surface Tillage Agriculture, *p. 39. John Wiley & Sons, 1986.*

OM mineralization; thus, soil OM levels can be sustained. The data in Figure 15.20 show the percentage of total N in the surface soil relative to the total N content in an undisturbed, noncropped soil (native prairie grass), as influenced by tillage. Increasing tillage intensity from no-till to plowing increased N mineralization, which reduced soil OM over time. The influence of increasing residue cover with reduced tillage on increasing soil OM was demonstrated earlier (Fig. 5.8).

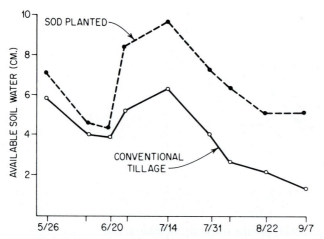

FIGURE 15.18   Available soil water in 0- to 60-cm profile as affected by tillage practice for corn with orchardgrass.   *Bennett et al.,* Agron. J., *65:488, 1973.*

TABLE 15.10   Tillage Effects on Water Storage, Sorghum Grain Yields, and WUE in an Irrigated Winter Wheat-Fallow-Dryland Grain Sorghum Cropping System in Bushland, Texas, 1973–1977*

	Water Storage		Grain Sorghum Yield			
Tillage Method	Amount (mm)	Efficiency (% of Precipitation[b])	(Mg/ha)	(bu/a)	Total water use (mm)	WUES (kg/m³)
No-till	217	35.2	3.14	47	350	0.89
Sweep	170	22.7	2.50	37	324	0.77
Disk	152	15.2	1.93	29	320	0.66

*Precipitation averaged 347 mm during the fallow period.
SOURCE: Unger and Weise, *SSSAJ* **43:**582–588 (1979).

Fertilization with Conservation Tillage

With moldboard plowing, nutrients broadcast on the surface are plowed down and mixed with the soil. Conservation tillage means little or no inversion of the soil. Hence, with broadcast applications, nutrients concentrate on or near the surface. The limited success of conservation tillage may, in some cases, be related to low fertility in the root zone.

PHOSPHORUS AND POTASSIUM   Under conservation tillage, P and K accumulate in the top few inches of soil, in contrast to moldboard plowing (Table

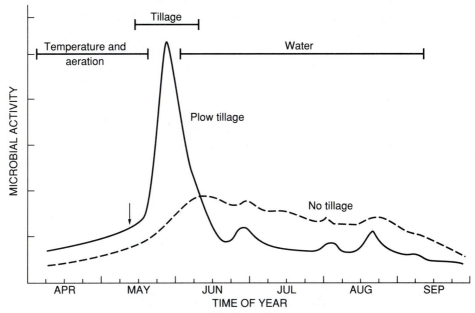

FIGURE 15.19   Hypothetical relationship between relative microbial activity and time of year in a plowed and a no-till soil; factors controlling activity are shown on top of the graph; the arrow indicates time of plowing.   *J. W. Doran, personal communication.*

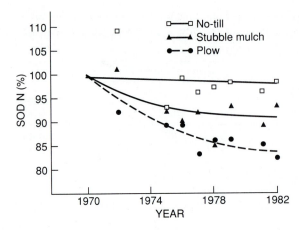

FIGURE 15.20 Influence of tillage on total soil N measured as a percentage of total N in the undisturbed, prairie soil. *Lamb et al.,* SSSAJ, *49:352–356, 1985.*

11.6). Thus, building soil-test P and K to medium or high levels before initiating conservation tillage is desirable.

With chisel plowing, more of the P and K is incorporated into the soil over the years than with till planting or no-till. With no-till, if the terrain and soil permit, periodic plowing every 4 or 5 years may be desirable for two reasons:

1. Distribution of fertility throughout the plow zone provides insurance against positional unavailability, especially in dry years, when root activity in the surface layer may be reduced.
2. Runoff and erosion of surface soils containing high amounts of plant nutrients are detrimental to the environment and an economical loss. As mentioned earlier, band applications at planting are often beneficial.

NITROGEN   In general, higher rates of surface-applied N are required with no-till than with conventional tillage (Fig. 15.21). These data illustrate that

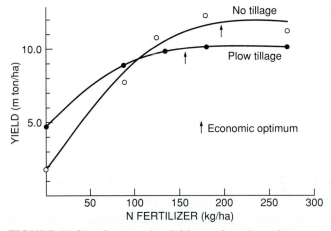

FIGURE 15.21 Corn grain yield as a function of N fertilizer rate and tillage. Poplar Hill Research Farm, Maryland, 1981. *Bandel, unpublished data.*

the economic optimum N rate for no-till corn was approximately 50 kg/ha (45 lb/a) greater than under conventional tillage. There are several possible reasons for this higher need with no-till:

1. Fertilizer N is immobilized by microorganisms in the organic layer that accumulates on surface with no-till.
2. More N is mineralized from the soil OM when plowed.
3. There is a higher yield potential with no-till in some areas.
4. There is more water movement through the soil, with greater loss of nitrates.
5. Denitrification/volatilization may reduce N efficiency.

Placement of fertilizer N below the surface residue to avoid immobilization, denitrification, and volatilization losses will increase the N recovered by the crop and crop yield (Fig. 12.31).

LIME   With no-till the surface may quickly become acid because most of the N is applied to the soil surface. Soil samples from the 0- to 2-in. layer are usually more acid than those from the lower depths (Table 11.7). Hence it is important to monitor closely surface soil pH. Certain herbicides such as the triazines are inactivated by a low pH at the surface, and weed control can suffer.

Opportunities for Adoption of No-till

No-till is destined to increase, and it is estimated that within the next 20 years, over 50% of the cropland in the United States will be farmed by no-till systems. The attractive features of no-till are as follows:

1. Row crop production on sloping lands is more feasible, with less loss of nutrients, soil, and water.
2. WUE is increased.
3. No-till corn production can readily follow soybeans.
4. Double cropping of row crops such as soybeans, sorghum, or corn planted immediately after a wheat harvest is possible in many areas.
5. Seeding legumes and/or nonlegumes in rundown pastures can be accomplished using the same principle. Fertilizer may be placed beneath the seed or broadcast. Herbicides are used to kill or retard existing grasses and weeds. Again, this technique is useful on sloping lands that should not otherwise be tilled, as well as on areas that are more level.
6. Energy, time, labor, and machinery costs are reduced.

Legumes in the Rotation

One of the reasons for including legumes in a rotation was to supply N, but with the development and availability of inexpensive fertilizer N, agriculture is no longer dependent on legumes for N (Fig. 15.22). The main purpose of

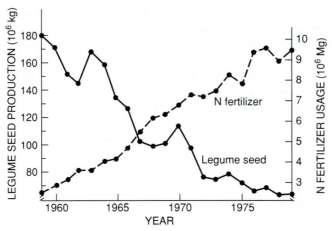

FIGURE 15.22   Inverse relationship between legume
seed production and fertilizer N use.   *Power,* Terrestrial
Nitrogen Cycles. *Swedish Natural Sci. Res. Council., p. 543,
1981.*

legumes is to supply large amounts of high-quality forage, whether hay or
pasture.

In a livestock farming system, legumes serve the dual purpose of providing
livestock feed and N for the grain crops. Legumes are generally of superior
quality, including higher protein and mineral concentrations, compared to N
fertilized grasses.

Table 15.11 shows that barley grown for a 5-year period after legumes
contained a total of 30 to 80 lb of N and yielded 60 to 70 bu more in total
than unfertilized (N) barley not following legumes. It can also be seen that
the beneficial effects of legumes were most pronounced in the first few years
after plowing down the legumes, but that some residual effects continued
even after 5 years.

TABLE 15.11   Yield and N Uptake of Barley Grown After Legumes

	Yield of Barley (bu/a)			N Uptake of Barley (lb/a)		
	No Legume	Alfalfa*	Red Clover*	No Legume	Alfalfa	Red Clover
1970	66	41	70	59.4	44	68.2
1971	27	51	51	26.4	63.8	22
1972	26	50	40	26.4	55	41.8
1973	32	52	48	26.4	46.2	33
1974	27	35	37	19.8	28.6	24.2
1975	22	31	26	—	—	—
Total	200	260	272	158.4	237.6	189.2
Mean	34	43	45	—	—	—

*Grown in 1968 and 1969.

SOURCE: Leitch, in *Alfalfa Production in the Peace River Region,* pp. C1–C5. Beaverlodge,
Alberta: Alberta Agriculture and Agriculture Canada Research Station, 1976.

In spite of these advantages of legumes in a rotation, this practice may not always be attractive to growers. Farmers in some areas may not have a ready use or market for forage crops. Thorough extraction of soil moisture reserves might also be a disadvantage in semiarid areas. Examples of various legumes and their use in the United States are shown in Table 15.12.

TABLE 15.12   Examples of Regional Use of Legumes in Cropping or Conservation Tillage Systems

Region	Legume Species	Cropping or Tillage System
Southeast	Crimson clover, hairy vetch	Winter cover crop—no-till corn
	Bigflower vetch, crownvetch, alfalfa, lupine, arrowleaf clover, red clover	Winter cover crops preceding grain sorghum and cotton
Northeast	Alfalfa, birdsfoot trefoil, red clover	Legumes grown for hay or silage in crop rotations that include conventional or no-till corn as feed grain or silage; also used as living mulches
North Central	Soybean, pea	Grown in 1-year rotation with nonlegume, possibly using conservation tillage methods; peas may precede soybeans in a double-cropping system
	Alfalfa, red clover, white clover, alsike clover	Grown for 2 years or more in 3- to 5-year rotations with small grains or corn, possibly by use of conservation tillage methods
	Birdsfoot trefoil, crownvetch, sweetclover	Used for forage, silage, or pasture
Great Plains	Native legumes (Void in genetically adapted material and economically compatible enterprises; water is the limiting factor)	Rangeland for grazing
Pacific Northwest	Dry pea, lentil, chickpea	Rotation or double cropped with grains
	Austrian winter pea	Green manure or alternated with winter wheat
	Alfalfa	Grown in rotation with winter wheat, spring barley, and winter peas
	Faba bean	Grown in rotation for silage
California	Dry bean, lima bean, blackeye pea, chickpea	Grown for grains in various rotations
	Alfalfa	Grown for seed on irrigated land and for erosion control and forage on steeply sloping soils
	Subterranean clover	Rangeland for grazing

SOURCE: Heichal, *Role of Legumes in Cons. Tillage System.* Soil Cons. Soc. Am., 1987, p. 30.

N Fixation by Legumes

Symbiotic bacteria (*Rhizobia*), fix N_2 in nodules present on the roots of legumes (see Chapter 5). This N may be utilized by the host plant, it may be excreted from the nodule into the soil and be used by other plants growing nearby, or it may be released by decomposition of the nodules or legume residues after the legume plant dies or is incorporated in the soil.

The amount of N_2 fixed by *Rhizobia* varies with the yield level, the effectiveness of inoculation, the N obtained from the soil, either from decomposition of OM or from residual N, and environmental conditions (Table 5.2). A high-yielding legume crop such as soybeans, alfalfa, or clover contains large amounts of N (Table 15.13). In general, about 25 to 80% of the total N in the plant is fixed by the nodule bacteria. The data in Table 15.13 also show that legumes grown on soils low in profile NO_3^- obtain more of their N by fixation, compared to legumes grown on high-N soils.

Soybeans remove about 1.5 lb/bu of N from the soil, and soybeans fix 40% or more of the total N in the plant. However, on lighter-colored soils, soybeans may fix 80% or more. In many environments, the quantity of N removed by soybean grain at harvest exceeds the quantity of N_2 fixed (Table 15.14). For example, when only 40% of the grain N was due to N_2 fixation, soil N exported in the grain exceeded N fixed in the grain and, thus, soil N was depleted (-74 lb/a of N). In contrast, when 90% of the N was fixed, soil N was increased ($+22$ lb/a of N).

Legumes in combination with grasses for forage generally supply N for both crops. The data in Table 15.15 show that approximately 70% of the N in the grass originated from N supplied by the legume.

The yield benefit of rotations with some legumes may not be always related to the legume N supply. The data in Figure 15.23 illustrate that the corn yield

TABLE 15.13   Variation of N Fixation Capacity with Legume Species, Legume Productivity, and Initial Soil N Concentration

Species	N from Symbiosis by Harvest (%)				Dry Matter Yield (pounds/acre)
	1	2	3	Mean	
Hay and pasture legumes					
Alfalfa*	49	81	58	63	6,809
Red clover*	51	79	65	65	6,230
Birdsfoot trefoil*	27	67	25	40	4,880
Harvest at Grain Maturity					
Grain legumes					
Soybean[†]			76		2,494
Soybean[‡]			52		7,837

*Established in soil with 3.7% organic matter and the initial NO_3-N concentration of 12 ppm (0- to 6-inch depth).

[†]Established in soil with 1.8% organic matter and an initial NO_3-N concentration of 12 ppm (0- to 8-inch depth).

[‡]Established in soil with 4.8% organic matter and an initial NO_3-N concentration of 31 ppm (0- to 8-inch depth).

SOURCE: Heichel et al. *Crop Sel.* **21**:330–335 (1981).

TABLE 15.14   Nitrogen Budget of Soybeans Illustrating the Allocation of Soil and Symbiotic N Among Plant Components and the Net Return of N to the Soil with 40% and 90% of Plant Nitrogen from Symbiosis

Crop Component	Dry Matter Content	Total Reduced N Content	Content of Symbiotic N		Soil N Export in Grain		Symbiotic N Return in Residue		Loss (−) or Gain (+) or N	
			40% N Symbiosis	90% N Symbiosis	40% N	90% N	40% N	90% N	40% N	90% N
					lb/a					
Grain	2,100	151	61	136	90	15	—	—	—	—
Residue*	3,424	40	16	37	—	—	16	37	—	—
Total plant	5,524	191	77	173	—	—	—	—	−74	+22

*Pod walls, leaves, stems, roots, and nodules: incomplete grain harvest would increase this value.
SOURCE: Heichel and Barnes, *ASA Spec. Publ.* 46, 1984, pp. 46–59.

response to fertilizer N was similar in a corn crop following either soybeans or wheat. In this case, the rotation response compared to that of continuous corn is commonly referred to as the *rotation effect* and will be discussed in the next section.

With forage legumes, only part of the N fixed is returned to the soil because most of the forage would be harvested. Forages grown for green manure or as winter cover crops most likely return more fixed N to the soil, depending on species, yield, and management. For example, legume N availability can be greater in a one-cut system compared to a three-cut system because of the increased amount of N incorporated with less frequent harvests (Fig. 15.23).

Whether the nonlegume yield response following a legume is due to N or to a rotation effect, the benefit can be observed for several years (Fig. 5.2).

The decision on whether to use legumes or fertilizer N becomes a matter of economics, with selection based on the greatest net return on investment. Like other input costs, the cost of fertilizer N increases with time due to increased manufacturing and transportation costs (Fig. 14.1). As a consequence, interest in legumes to substitute partially for the fertilizer N requirements of nonlegume crops has increased.

In some developing countries, commercial N may not be available or is too

TABLE 15.15   Maximum N Transfer from Legumes to Grass. The Legume/Grass Ratio Was 2:3 for the Reed Canary-Grass-Alfalfa Community and 3:2 for the Reed Canary-Grass-Trefoil Community

Community	N in Grass from Legume (%)*		
	Harvest 1	Harvest 2	Harvest 3
Grass-alfalfa	64	68	68
Grass-trefoil	68	66	79

*Percent (proportion × 100, mass basis) of accumulated grass N received from legume via N transfer.
SOURCE: Brophy et al., *Crop Sci.* **27**:372–380 (1987).

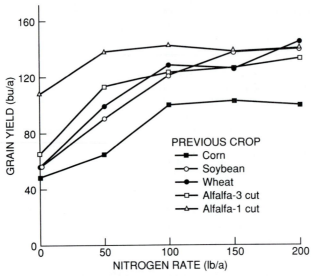

FIGURE 15.23   Corn grain yields as influenced by
previous crop and fertilizer N.   *Heichel,* Role of Legumes
in Cons. Tillage Systems, Soil Cons. Serv. Am., *p. 33,
1987.*

expensive. Therefore, a cropping system that includes legumes is essential to
supply some of the N needed for nonlegumes.

Adequacy of Legume N for Nonlegume Crops

Generally, fertilizer N must supplement the amount fixed by legumes, although the quantity depends on the N available from the legume and the N
requirement of the nonlegume crop. Good growth of a forage legume will
furnish sufficient N for average but not top yields of nonlegume grain crops.
Variable quantities of N_2 fixation can be caused by poor legume stands, improper inoculation, or inadequate fertility.

Rotations Versus Continuous Cropping

Continuous cropping, or monoculture, is practiced throughout the world.
Although monoculture was once considered a sign of poor farming, the increased supply of fertilizer, especially that of N in the 1950s, encouraged
continuous cropping on soils on which erosion was not a serious problem.

Most continuous cropping systems are currently used because of economics.
In some regions, the crop options may be limited because of the environment
or the market; thus, growers continuously produce a "most adapted" crop
that maximizes the profit potential. Conventionally tilled, continuous dryland
wheat, cotton, and irrigated corn are common examples.

Numerous long-term experiments have demonstrated that, in general, rotations increase long-term crop productivity compared to continuous cropping.
The data from the Morrill plots shown in Figure 15.24 provide an excellent

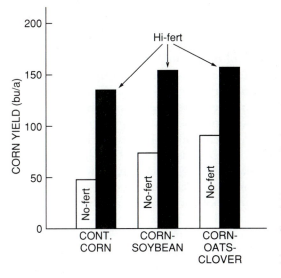

FIGURE 15.24   Influence of rotation and fertility on corn grain production on the 'Morrow' plots, University of Illinois. Grain yield is the average from 1967 to 1988. NO-FERT represents no nutrients added since the experiments began in 1876. HI-FERT represents application of manure, lime, N, P, and K.   Univ. of Illinois Agric. Exp. Sta. Bull. *775*.

example. The reasons for the rotation effect or production advantage with rotation cropping compared to continuous cropping are not fully understood; however, the following discussion provides some insights.

A crop may have a harmful effect on the following crop, be it another crop or the same crop. There is some evidence that substances released from roots or formed during the decomposition of residues are responsible for the toxicity. The comparison of continuous corn versus a corn-soybean rotation is an example. Seeding alfalfa following alfalfa is often unsatisfactory for unknown reasons. *Allelopathy* is the term used to describe the antagonistic action of one plant on another.

Control of Disease, Weeds, and Insects

In most cases, crop rotation or initiation of other cropping practices will help control certain diseases, weeds, or insects. For example, reducing root rot diseases in wheat and other cereals requires crop rotation, together with resistant varieties, clean seed, and field sanitation practices. Legumes, other dicotyledons, and even cereals such as oats, barley, or corn are often suitable alternate crops in place of wheat when take-all occurs. However, in some instances, this disease can be severe, even in wheat following alfalfa, soybeans, and grass crops. Corn root rots have been reduced by rotations, and the severity of several seedling diseases has been reduced by rotations combined with field sanitation. Susceptible crops should be grown on the same field only once in every 3 to 4 years.

Crop rotation is an important approach for the control of nematodes feeding on the roots of annual crops. Grass crops are commonly used in rotation to control root knot nematodes. Acceptable yields of irrigated cotton can be obtained following 2 or more years of root knot–resistant alfalfa. Two years of clean fallow also effectively controlled root knot. Few important bacterial or viral diseases are controlled by crop rotation.

The role of crop rotation for weed control depends on the particular weed and the ability to control it with available methods. If all the weeds can be conveniently and economically controlled with herbicides, then crop rotation is not a vital part of a weed control program. However, there are situations where rotations are necessary for control of a troublesome weed.

For example, downy brome and jointed goat grass can severely reduce yields in a wheat-fallow-wheat system. Use of atrazine in a reduced tillage wheat-sorghum/corn-fallow rotation will eliminate these weed problems.

Rotation was once a common practice for insect management, but its use declined with the development of economically effective insecticides. Interest in rotations has increased because of insect resistance to the chemicals and increased costs. Rotation can be helpful where the insects have few generations a year or where more than one season is needed for the development of a generation. For example, northern corn rootworm can be a serious problem in continuous corn. Rotation of soybean and corn replaces the need for insecticide control of this insect. Rotation is only partially successful in reducing damage by cotton bollworm. Sorghum, when planted at the proper time, will protect cotton from worms, while not harming the sorghum.

Effect on Soil Tilth

Most recommended crop production practices provide good plant cover and return large amounts of crop residues to the soil. In addition, there is less tillage, which reduces the detrimental effects of compaction and deterioration of soil structure. The important issue is not one of monoculture versus rotation but, rather, involves two factors: the amount of residues returned to the soil and the nature of the soil tillage necessary in the rotation.

Rotation can greatly improve the soil structure and tilth of many medium- and fine-textured soils. Pasture grasses and legumes in rotation exert significant beneficial effects on physical properties soil. When soils previously in sod are plowed, they crumble easily and readily shear into a desirably mellow seedbed. Plow draft is often reduced in fine-textured soils when less intensively tilled crops are grown in the rotation. Internal drainage can be improved so that ponding and the time needed for the soil to drain excess water are reduced. More information is required concerning the soil conditions under which deep-rooted legumes are needed in the rotation.

Corn in monoculture is unique, since on many soils it maintains reasonably acceptable soil physical conditions. The compensating factors are (1) the return of several tons of crop residues to the soil when corn is harvested for grain and (2) the corn crop is well adapted to reduced tillage and decreased damage from traffic on the soil.

Double cropping, such as small grain-soybeans or small grain-corn; triple cropping; or even quadruple cropping of rice in areas with long growing seasons and the possibilities of irrigation are becoming more common. With four crops a year, 27 tons/ha of rice are possibile. This necessitates utilizing soil, solar, and water resources to the maximum. If adequate fertility and pest control are provided and varieties are improved, soil productivity should gradually increase. Thus more attention will be directed to measuring yield per unit area per year.

Advantages of Both Systems

ROTATIONS

1. Deep-rooted legumes may be grown periodically over all fields.
2. There is more continuous vegetative cover, with less erosion and water loss.
3. Tilth of the soil may be superior.
4. Crops vary in extracting range of roots and nutrient requirements: deep-rooted versus shallow-rooted, strong feeder versus weak feeder, and N fixer versus nonlegume fixer.
5. Weed and insect control are favored.
6. Disease control is favored. Changing the crop residues fosters competition among soil organisms and may help reduce the pathogens.
7. Broader distribution of labor and diversification of income are effected.
8. Erosion is reduced.

CONTINUOUS CROPPING OR MONOCULTURE

1. Profits may be greater, but depends on the crops involved.
2. A soil may be especially adapted to one crop.
3. The climate may favor one crop.
4. Machinery costs may be lower.
5. The grower may prefer a single crop and become a specialist; however, monoculture demands greater skills, including pest, erosion, and fertilization control.
6. The grower may not wish to be fully occupied with farming year round.

Crop Removal of Nutrients in the Root Zone

Crop plants vary considerably in their content of primary, secondary, and micronutrients. In addition, crops may absorb nutrients from different soil zones, thus making the choice of cropping sequences important to plant nutrition. Deep-rooted crops absorb certain nutrients from the subsoil. As their residues decompose in the surface soil, shallow-rooted crops may benefit from the remaining nutrients. On a soil marginal in a particular micronutrient, it is possible that the preceding crop will have a considerable effect on the supply of this element to the current crop.

The net effect of cropping practices on P and K levels depends on the removal of nutrients by the harvested portion of the crop, nutrients supplied by the soil, and supplemental fertilization.

Effect of Rotation on Soil and Water Losses

Generally, increasing the crop yield and the residue produced and left on the soil surface will increase water infiltration and reduce runoff and soil loss. Some of the characteristics of cropping systems and/or fertility management practices related to soil losses are the following:

1. The denseness of the cover or canopy. This affects the amount of protection from the impact of rain and evaporation. Residues and stems reduce the velocity of water and evaporation. Residues, when turned back to the soil, make it more permeable to water.
2. The proportion of time that the soil is in a cultivated crop versus the amount of time in a close-growing crop such as small grains or forage.
3. The time that the crop grows in relation to the distribution and intensity of rainfall. The period from May to September is most vulnerable.
4. The type and amount of root system.
5. The amount of residues returned.

Adoption of improved soil and crop management practices, can increase yields and reduce runoff and erosion. As Table 15.16 shows, the improved management effects on reducing runoff and erosion were due in part to better surface protection through a quicker cover in the spring, a denser cover throughout the season, and a more extensive root system of the growing crop. The influence of maintaining a surface residue cover and crop rotation on reducing runoff and soil loss is demonstrated in Table 15.17.

Winter Cover/Green Manure Crops

Winter cover crops are planted in the fall and plowed down in the spring. These may be a nonlegume, a legume, or a combination grown together. There are several advantages to the last practice. A greater amount of OM is produced, the nonlegume can benefit from the N fixation, and because the nonlegume is usually more easily established, a stand of at least one crop is ensured.

Decomposition of green manure crops is rapid, but the residual effects are well recognized. The smallest residual effects generally are expected in areas in which the mean annual temperature is high and the soil is sandy.

Small grains or other crops can be grazed in late fall and winter when the amount of growth and soil conditions permit. Adequate fertility, either residual or added, is a necessity, and extra N may be needed. Grazing allows additional return from cover crop inputs.

TABLE 15.16   Effect of Management Level on Crop
Yields, Runoff, and Erosion (1945–1968)

	Prevailing Practices	Improved Practices
Corn (metric ton/ha)	5.1	7.3
Wheat (metric ton/ha)	1.5	2.3
Hay (metric ton/ha)	4.3	7.8
Runoff, growing season (cm)	1.9	1.0
Peak runoff rate (cm/hr)	2.3	1.5
Erosion from corn (metric ton/ha/yr)	10.6	3.1

SOURCE: Edwards et al., *SSSA Proc.*, **37**:927 (1973).

TABLE 15.17   Hydrologic Data from a 6- to 7-a Watershed Comparing Various
Cropping/Tillage Systems

Time Period	Cropping System*	Tillage	Annual Rainfall (in.)	Annual Runoff (in.)	Annual Soil Loss (tons/a)
1972–1974	Fallow/soybeans	Conventional	54.0	8.7	11.6
1974–1976	Barley/grain sorghum	No-till	52.0	3.5	0.2
1976–1979	Barley/soybeans	No-till + in-row chisel	46.5	0.8	0.06
1979–1983	Crimson clover/grain sorghum	No-till + in-row chisel	43.7	0.2	0.002

*Soybeans = *Glycine max.* L. Merr.; barley = *Hordeum vulgare* L.
SOURCE: Hargrove and Frye, *Role of Legumes in Cons. Tillage Systems.* Soil Cons. Soc. Am., 1987, p. 20.

N Added

One important reason for using green-manure legume crops is that they supply additional N, depending on the yield and N content (Table 15.18). These data show that increasing the quantity of N produced in the legume cover crop increased the corn yield (unfertilized). The grain yield after fallowing was greater than following the wheat cover crop because of N mineralization during the fallow period. When a nonlegume is turned under, only the N from the soil or that supplied in fertilizer is returned.

The nutrient content in several legume cover crops is shown in Table 15.19. Increasing the cover crop yield or biomass will subsequently increase the quantity of N_2 fixed and the N returned to the soil (Fig. 15.25). Legume cover crops can contribute large quantities of N to subsequent nonlegume crops (Table 15.20).

OM Added

One of the benefits attributed to winter cover crops is the OM supplied to the soil (Table 15.21). Green manures will help maintain the soil OM and will sometimes even increase it.

In rotations in which the crops return little residue, maintenance of soil productivity may be particularly difficult. The lengthening of the rotation to

TABLE 15.18   Dry Matter and N Concentration of Various Cover Crops
and Influence of Cover Crops on Corn Grain Yield

Cover Crop	Dry Matter	N Conc.	N Content	Yield by N rate (lb N/a)	
				0	200
	lb/a	%	lb/a	------ bu/a ------	
Fallow	—	—	—	63	161
Wheat	1,178	2.01	35	32	121
Winter Pea	1,423	4.56	61	132	165
Hairy Vetch	2,526	4.62	113	156	168
Crimson Clover	2,883	3.67	102	143	172

SOURCE: Neely et al., *Role of Legumes in Cons. Tillage Systems.* Soil Cons. Soc. Am., 1987, p. 49.

TABLE 15.19   Biomass Yield and Nutrient Accruement by Selected Cover Crops

Cover Crop	Biomass*	N	K	Ca	P	Mg
	---------------------------------- lb/a ----------------------------------					
Hairy vetch	3,260	141	133	52	18	11
Crimson clover	4,243	115	143	62	16	11
Austrian winter peas	4,114	144	159	45	19	13
Rye	5,608	89	108	22	17	8

*Dry weight of aboveground plant material.
SOURCE: Hoyt, *Role of Legumes in Cons. Tillage Systems*, Soil Cons. Soc. Am., 1987, p. 96.

include green-manure crops could be beneficial. The acreage of corn and sorghum silage is increasing in some areas. This leaves the soil with almost no surface residues. Oats or rye seeded immediately after harvesting or seeded by airplane before harvesting will help to protect the soil and increase the residue returned.

Protection of the Soil Against Erosion

Protection against erosion is one of the most important reasons for winter cover crops. The benefits, however, should be related to the distribution of rain and the erosion potential during the year. The effect of cover crops on soil loss is generally small when winter cover crops are turned under in early spring.

Surface residues from summer crops, if left undisturbed, may provide more protection than cover crops seeded in the fall. The greater the percentage of the soil surface covered by mulch, the less the soil loss. Freshly tilled land is quite susceptible to erosion, and considerable time is required before the cover crop can provide enough protection to have much effect on reducing soil loss. For example, rye cover has been grown after corn, but in comparison to heavy corn residues, it is generally not as effective for erosion control.

Methods of handling cornstalks in the fall have a marked effect on soil loss Shredding reduced soil losses to about half that obtained from cornstalks left by corn harvesters (Fig. 15.26). The shredded cornstalks provide a cushion that protects the soil from the impact of raindrops. Disking the shredded cornstalks will increase soil erosion losses.

For perennial crops such as peaches and apples planted on steep slopes, continuous cover is helpful in reducing erosion. Since the trees and the cover crops occupy the land simultaneously, care must be taken, particularly in young orchards, to prevent competition for water and N. In some of the muck soils suitable for vegetables, a strip of small grain or a row of trees at intervals helps to reduce losses from erosion.

Animal Manure

Animal manures are by-products of the livestock industry, and greater attention is being given to effective disposal of animal manures because of (1) increased use of "confinement" production systems and associated manure-

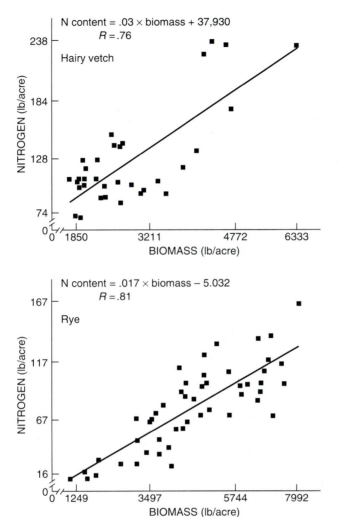

FIGURE 15.25   Effect of biomass on N uptake for hairy vetch and rye cover crops.   *Hoyt*, Role of Legumes in Conservation Tillage Systems, *Soil Cons. Serv. Am., 1987*, *p. 97.*

handling problems and (2) increased concern over contamination of groundwater and surface water by NO_3^- and other compounds originating from the manure.

   Methods for handling and storing manure will affect its nutrient content. Previously, the common method of disposal was to collect the manure or manure plus bedding and spread it on the fields. Liquid waste systems have since been developed in which the manure is diluted with water and stored in pits or lagoons.

   N losses from various systems are given in Table 15.22. P and K losses are only 5 to 15% under all but the open lot and lagoon waste systems. In an open lot, about 50% of these nutrients is lost. In a lagoon, much of the P settles out and is lost from the liquid applied on the land.

TABLE 15.20   Estimates of the N Contribution of
Winter Legumes to the N Requirements of No-till Corn,
Grain Sorghum, and Cotton

Location	Crop	Cover Crop	Fertilizer N (lb/a)
Kentucky	Corn	Hairy vetch	85
		Big flower vetch	45
Georgia	Grain	Crimson clover	75
	sorghum	Hairy vetch	81
		Common vetch	53
		Subterranean clover	51
Alabama	Cotton	Hairy vetch	61
		Crimson clover	61

SOURCE: Hargrove and Frye, *Role of Legumes in Cons. Tillage
Systems*, Soil Cons. Serv. Am., 1987, p. 2.

The three principal methods used for field application of manure are the
following:

1. Spreading of solid material when weather, soil, and crop permits.
2. Injecting the slurry of water and manure into the soil or spraying it on the
   surface.
3. Injecting the slurry into a sprinkler irrigation system.

N loss is greatly affected by the method of application (Table 15.23). Immediate incorporation will minimize N volatilization. The effectiveness of injected
liquid manure has been improved by maintaining the NH_4^+-N form by adding
nitrification inhibitors. In large operations where animals are confined to
feedlots, manure dried and bagged for the speciality turf and garden trade is
a valuable by-product.

The composition of animal manure varies according to the type and age of
animal, feed consumed, bedding used, and handling system (Tables 15.24 to
15.26). As might be expected, dry matter is highest in the solid waste system,
whereas N, P, and K are highest when manure is handled as a liquid.

TABLE 15.21   Influence of 5 Years of Various Cropping Sequences and Tillage
on Soil Organic C and N Concentrations in the Surface 7.5 cm of Soil

Cropping Sequence	Tillage Treatment	Fertilizer N Rate (pounds/acre/year)	Organic C	Organic N	C/N Ratio
			%		
Wheat/soybean	Conventional	70	1.4	0.12	11.7
Wheat/soybean	No-till	70	1.6	0.15	10.7
Clover/sorghum	No-till	0	2.2	0.17	13.0
Clover/sorghum	No-till	120	2.4	0.19	12.6

SOURCE: Hargrove and Frye, *Role of Legumes in Cons. Tillage Systems*. Soil Cons. Soc. Am., 1987,
p. 2.

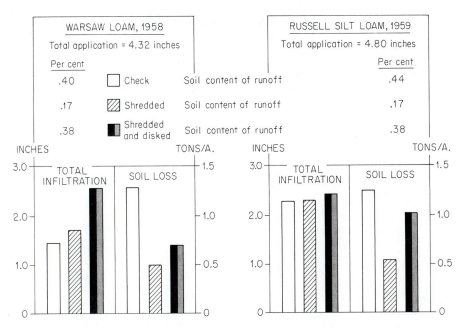

FIGURE 15.26   Effect of cornstalk residue management on infiltration and soil loss under simulated rainfall with an intensity of 2.4 in./hr. Warsaw, 4 to 4.5% slope; Russell, 3 to 3.5% slope. Check is cornstalks left by harvester.   *Mannering et al., SSSA Proc., 25:506, 1961.*

Fertilization with Manure

The fertility program on many livestock farms includes manure; however, unless an unusually large quantity of legumes is produced on the farm and considerable commercial feed is consumed, thereby capitalizing on fertility from elsewhere, a livestock program in itself will tend to gradually deplete soil fertility.

Many comparisons have been made of the effects of manure on crop production with those obtained from the application of equivalent amounts of N, P, and K in commercial fertilizers. On a fine sandy loam at Rothamsted in

TABLE 15.22   Effect of the Method of Handling and Storing on N Losses from Animal Manure

Handling and Storing Method	N Loss* (%)	Handling and Storing Method	N Loss* (%)
Solid systems		Liquid systems	
Daily scrape and haul	15–35	Anaerobic pit	15–30
Manure pack	20–40	Oxidation ditch	15–40
Open lot	40–60	Lagoon	70–80
Deep pit (poultry)	15–35		

*Based on composition of waste applied to the land vs. composition of freshly excreted waste, adjusted for dilution effects of the various systems.

SOURCE: Sutton et al., *Univ. of Minn. Ext. Bull. AG-FO-2613*, 1985.

TABLE 15.23   Effect of the Method of Application
of Manure on Volatilization Losses of N

Method of Application	Type of Waste	N Loss* (%)
Broadcast without cultivation	Solid	15–30
	Liquid	10–25
Broadcast with cultivation[†]	Solid	1–5
	Liquid	1–5
Knifing	Liquid	0–2
Irrigation	Liquid	30

*Percent of total N in waste applied that was lost within 4 days after
application.
[†]Cultivation immediately after application.
SOURCE: Sutton et al., *Univ. of Minn. Ext. Bull. AG-FO-2613*,
1985.

England, commercial fertilizers used for over 100 years have been just as
effective as manure for continuous wheat production. In Colorado, 27 tons/a
of manure increased corn yields an average of 20 bu/a over those obtained
with either equivalent or greater rates of fertilizers.

The favorable effect of manure on increasing the available moisture content
of soils is shown in Table 15.27. Another interesting finding in this study was
the capacity of manure to moderate soil acidification resulting from repeated
applications of N fertilizer.

TABLE 15.24   Approximate Dry Matter and Fertilizer Nutrient Composition
and Value of Various Types of Animal Manure at the Time Applied to the Land—
Solid Handling Systems

Type of Livestock	Waste-Handling System	Dry Matter (%)	Nutrient (lb/ton Raw Waste)				Value per Ton[‡]
			N		P_2O_5	K_2O	
			Available*	Total[†]			
Swine	Without bedding	18	6	10	9	8	$ 5.10
	With bedding	18	5	8	7	7	4.14
Beef cattle	Without bedding	15	4	11	7	10	4.26
	With bedding	50	8	21	18	26	10.44
Dairy cattle	Without bedding	18	4	9	4	10	3.36
	With bedding	21	5	9	4	10	3.60
Poultry	Without litter	45	26	33	48	34	24.72
	With litter	75	36	56	45	34	26.22
	Deep pit (compost)	76	44	68	64	45	35.16

*Primarily NH_4-N, which is available to the plant during the growing season.
[†]NH_4-N plus organic N, which is slow releasing.
[‡]Value per pound is 24 cents for available N, 30 cents for P_2O_5, and 12 cents for K_2O.
SOURCE: Sutton et al., *Univ. of Minn. Ext. Bull. AG-FO-2613*, 1985.

TABLE 15.25   Approximate Dry Matter and Fertilizer Nutrient Composition and Value of Various Types of Animal Manure at the Time Applied to the Land— Liquid Handling Systems*

Type of Livestock	Waste-Handling System	Dry Matter (%)	Nutrient (lb/ton Raw Waste)				Value per 1,000 Gallons§
			N		P_2O_5	K_2O	
			Available†	Total‡			
Swine	Liquid pit	4	20	36	27	19	$15.18
	Oxidation ditch	2.5	12	24	27	19	13.26
	Lagoon	1	3	4	2	0.4	1.80
Beef cattle	Liquid pit	11	24	40	27	34	17.94
	Oxidation ditch	3	16	28	18	29	12.72
	Lagoon	1	2	4	9	5	3.78
Dairy cattle	Liquid pit	8	12	24	18	29	11.76
	Lagoon	1	2.5	4	4	5	2.40
Poultry	Liquid pit	13	64	80	36	96	37.68

*Application conversion factors: 1000 gal = about 4 tons; 27,154 gal = 1 acre-in.

†Primarily NH_4-N, which is available to the plant during the growing season.

‡NH_4-N plus organic N, which is slow releasing.

§Value per pound is 24 cents for available N, 30 cents for P_2O_5, and 12 cents for K_2O.

SOURCE: Sutton et al., *Univ. of Minn. Ext. Bull. AG-FO-2613*, 1985.

TABLE 15.26   Nutrient Value of Manure per Animal Unit (lb/1,000 lb Live Weight) per Year

Handling and Disposal Method	Swine			Beef			Dairy			Broilers		
	N	P_2O_5	K_2O	N	P_2O_5	K_2O	N	P_2O_5	K_2O	N	P_2O_5	K_2O
Manure pack												
Broadcast												
Broadcast and	84	107	124	63	77	99	77	50	112	215	200	149
cultivation	102	107	124	77	77	99	91	50	112	263	200	149
Open lot												
Broadcast												
Broadcast and	58	61	80	44	45	64	51	30	59	—	—	—
cultivation	70	61	80	53	45	64	61	30	59	—	—	—
Manure pit												
Broadcast	95	111	119	69	82	95	87	54	107	—	—	—
Knifing	124	111	119	94	82	95	114	54	107	—	—	—
Irrigation	92	111	119	65	82	95	84	54	107	—	—	—
Lagoon												
Irrigation	24	25	89	18	18	71	23	14	80	—	—	—

SOURCE: Sutton et al., *Univ. of Minn. Ext. Bull. AG-FO-2613*, 1985.

Distribution of Manure by Grazing Animals

Distribution of manure by grazing animals presents a problem in the maintenance fertilization of pastures. For N, which does not remain in effective concentrations for more than a year, about 10% of a grazed area is effectively covered in 1 year. On the other hand, with P, which is not leached or removed

TABLE 15.27   Laboratory Measurements of Field
Capacity, Wilting Point, and Available Water of the
Topsoil from the Nil and Barnyard Manure Treatments
(Average of Six Years 1966–1971)

Soil	No Manure	Barnyard Manure
Field Capacity (% Moisture)		
Beryl FSL	19.4	22.6
Hazelmere L	21.6	24.3
Nampa CL	23.8	27.5
Wilting Point (% Moisture)		
Beryl FSL	4.8	5.8
Hazelmere L	7.4	8.8
Nampa CL	9.2	10.0
Available Water (% Moisture)		
Beryl FSL	14.6	16.7
Hazelmere L	14.1	15.6
Nampa CL	14.6	17.5

SOURCE: Hoyt and Rice, *Can. J. Soil Sci.*, **57**:425 (1977).

in large quantities, some effect might be obtained from a given application as much as 10 years later. In general, nearly all of a pasture area will receive deposits of manure in a 10-year period. K is intermediate between N and P in retention in the soil, and manure-deposited K is effective to some degree for at least 5 years. During this period, about 60% of a pasture will have been covered.

With low stocking rates, animal excreta will essentially have no effect on soil fertility. On highly productive pastures with a high carrying capacity, excreta may have a beneficial effect on soil fertility over a period of time. Grain feeding on pastures has a considerable effect on soil fertility, and each increase of 4.5 tons of grain fed per acre results in an increase of 53 steer-days of grazing.

Benefits of Manure

Reasons for the favorable action of manure are unclear, but they probably include one or more of the following:

1. An additional supply of NH_4-N.
2. Greater movement and availability of P and micronutrients due to complexation.
3. Increased moisture retention.
4. Improved soil structure, with a corresponding increase in infiltration rate and a decrease in soil bulk density.
5. Higher levels of CO_2 in the plant canopy, particularly in dense stands with restricted air circulation.
6. Increased pH and BC.
7. Complexation of Al^{3+} in acid soils.
8. Increased soil OM.

Sewage

Disposal of processed sewage materials from treatment plants is of increasing concern because of population pressures, more stringent laws, and increases in energy costs. About 75% of the sewage handled by municipal sewage treatment plants is of human origin, and the remaining 25% is from industrial sources. The end products of all sewage treatment processes are sewage sludge and sewage effluent (Fig. 15.27). Sewage sludge is the solids produced during sewage treatment. Sewage effluent is essentially clear water containing low concentrations of plant nutrients and traces of OM, which may be chlorinated and discharged into a stream or lake.

Sludge is a heterogeneous material, varying in composition from one city to another and even from one day to the next in the same city. Before developing plans for land application of sludge, it is essential to obtain representative samples of the sludge over a period of time and determine its typical chemical analysis (Table 15.28). Sewage sludge is disposed of by (1) application on cropland by approved methods; (2) incineration, with loss of OM and N; and (3) burial in landfill sites, where it will produce methane for many years.

Application of Sludge to Agricultural Land

Use of sludge on agricultural land has some benefits; at the same time, there can be problems. Sludge is a source of OM containing macro- and micronutrients, and is a cost-effective alternative to more costly methods of disposal, such as burning or burying.

It is essential that appropriate application and soil management techniques be used to protect the environment and the health of human beings and animals. Because of the possibility of applying excessive N and subsequent movement of NO_3-N into surface water and groundwater, careful monitoring is necessary.

Care must be taken to avoid cadmium (Cd) contamination of crops. Although there is no immediate human health problem, the U.S. Environmental Protection Agency has established recommendations concerning the maximum amount of Cd that can be applied.

There is a possibility of disease transmission due to the presence of bacteria, parasites, or viruses. Treatment of sludge by aerobic and anaerobic digestion, air drying, composting, and lime stabilization reduces the pathogen content.

Contamination of crops by stable organic compounds such as polychlorinated biphenyls (PCBs) can be minimized by soil incorporation of sludges containing less than 10 ppm PCBs.

Other difficulties include objectionable odors and impaired plant growth due to the antagonistic action of heavy metals such as Cu, Pb, Ni, and Zn.

APPLICATION RATES   Soil testing and fertilizer recommendations are used in conjunction with sewage sludge characteristics to determine application rates. The annual rate of sludge is based either on the lowest tonnage that will satisfy the N requirements of the crop or on the maximum quantity that can be used without exceeding permissible limits for Cd.

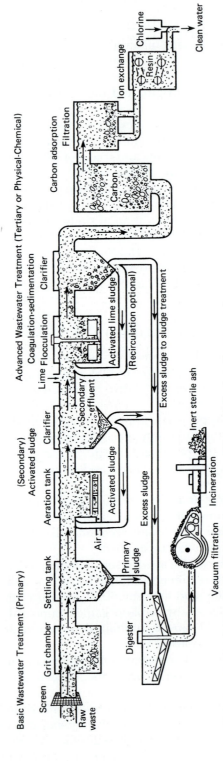

FIGURE 15.27 Stages in wastewater treatment. Water Pollution Causes and Cures, *Manufacturing Chemists Assn., Washington, D.C., 1972.*

TABLE 15.28   Typical Analysis of Sewage Sludge

Component	Concentration on Dry Weight Basis	Component	Concentration on Dry Weight Basis
	%		*ppm*
Organic carbon	50	Fe	40,000
N		Zn	5,000
Ammonium	2	Cu	1,000
Organic	3	Mn	500
Total	5	B	100
P_2O_5	6.8		
K_2O	0.5		
Ca	3	Cd	150
Mg	1	Pb	1,000
S	0.9	Ni	400

SOURCE: *Univ. Illinois Soil Manag. Conserv. Ser. Bull. SM-29, 1975.*

Sludges contain both inorganic and organic N (Table 15.28). Because of possible volatilization losses of up to 20 to 50% of the NH_4^+ from surface application of sludge and only partial mineralization (20 to 30%) of the organic N, exact application rates are difficult to determine. However, if the sludge is incorporated into the soil, very little NH_4^+ will be lost. After the first year, when 20% of the organic N is mineralized, about 3% of the remaining N will be released annually during the next few years.

N and P additions to soil from a single application of sludge can be as high as 800 lb/a (900 kg/ha) of total N in some areas, with about 50% as NH_4^+, and 1,025 lb/a (1,150 kg/ha) of P_2O_5. Use of fertilizer N and P on sludge-applied lands is generally not required for at least 2 years following sludge application. However, there is a possibility of having to correct imbalances of other nutrients.

When Cd is the main concern, sludge should be applied only to soils of pH 6.5 or above. The maximum allowable rate for crops must not exceed 4.4 lb/a of Cd on coarse-textured soils. The life of a sludge application site is determined by the cumulative addition of metals. Guidelines for the buildup of Cd and other heavy metals in soils are summarized in Table 15.29. Use of cation exchange capacity (CEC) as a controlling soil factor does not necessarily mean that all of these metals are retained on the exchange complex. Rather, CEC was chosen as a single soil property that can be easily measured and one that is positively related to soil components that may minimize plant availability of metals in sludges added to soils. On soils of little agricultural value, higher levels of Cd may be permitted.

Crop Response

Some regulatory agencies confine the application of sludges to crops such as forages, oilseed crops, small grains, commercial sod, and trees. Unacceptable crops may include root crops, vegetables and fruit, tobacco, and dairy pastures. Direct grazing of sludge-treated forage lands is not usually recommended for a period of 3 years immediately following application. Wheat is preferable to

TABLE 15.29   Maximum Amount of
Metal Suggested for Agricultural Soils
Treated wtih Sewage Sludge

Metal	Maximum Amount of Metal (lb/a) When CEC (meq/100 g) is:		
	<5	5–15	>15
Lead	440	880	1760
Zinc	220	440	880
Copper	110	220	440
Nickel	110	220	440
Cadmium	4.4	8.8	17.6

SOURCE: Sommers et al., *Purdue Univ. Bull.*
AY-240, 1980.

barley. Oats are not recommended in the first two growing seasons following
sludge treatment.

The response of crops to sewage sludge is at least equal to the response to
commercial fertilizer in the first year after application and may be somewhat
greater in subsequent years because of residual effects of the added plant
nutrients. Also, many of the favorable effects of the OM in sludges, probably
very similar to the ones associated with manure, may be long-lasting. Addition-
ally, it may take one or more years for the applied nutrients to become effectively distributed in the root zone.

Sewage Effluent

Sewage effluent can be either a valuable water and nutrient resource for crops
or a pollutant to land and waters. Large quantities of water are generally
involved; thus, it is essential that the soil be (1) internally well drained and
medium-textured, having a pH of between 6.5 and 8.2 and (2) be supporting
a dense stand of trees, shrubs, or grasses. The groundwater should be monitored periodically for NO_3^-.

Forage crops are commonly used for effluent application because of their
long growing season, with the resultant high seasonal evapotranspiration,
their high nutrient uptake, and their capacity to stabilize the soil and prevent
erosion. Because forages are not eaten directly by human beings, the transfer
of human diseases is unlikely.

Yields of five forage crops irrigated with sewage effluent are shown in Table
15.30. Alfalfa was the most suitable forage crop when the system was operated
for optimum utilization of the wastewater. The water did not supply sufficient
N for high production of the grasses.

The previous discussion indicates that there is much to be learned about
the use of sewage sludge or effluent on agricultural crops. Society will benefit
from wise application of this material and thus from recycling of a valuable
resource.

TABLE 15.30   Four-Year Annual Dry Matter Yield of Five Forage Species
as a Function of Wastewater and Fertilizer Levels

Wastewater Irrigation (cm/yr)	Fertilizer N-P (kg/ha/yr)	Forage (metric tons/ha)					
		Alfalfa	Reed Canary	Brome	Altai Wildrye	Tall Wheat	Mean
	0–0	8.7	5.8	6.0	5.8	5.3	6.3
62.5	56–48	9.4	8.8	8.9	8.4	7.7	8.6
	0–0	10.2	9.4	10.0	9.6	7.9	9.4
125	56–48	10.1	11.5	11.1	11.7	9.8	10.8
Mean		9.6	8.9	9.0	9.0	7.7	8.8

SOURCE: Bole and Bell, *J. Environ. Qual.*, **7**(2):222 (1978).

Summary

1. Marked increases in yields of crops have taken place in the past 10 years as a result of improved management practices, including more adequate fertilization. Before that time, yields increased slowly.

2. The aim of a crop- and soil-management program should be to realize a sustained maximum profit from the cropping program. Effects on tilth, water intake and erosion, plant nutrient supply, and pest control must be considered.

3. Cropping systems affect the amount of soil OM largely by the amounts of residue produced. As added plant nutrients increase yields, larger amounts of residues are obtained.

4. OM serves as a storehouse for nutrients, increases exchange capacity, provides energy for microorganisms, releases CO_2, improves tilth, increases moisture retention, and provides surface protection. All of these functions except the last depend on decomposition. Hence, OM accumulation in the soil is not an end in itself. The important point is the production of high quantities of OM and its subsequent decay.

5. One of the problems with soil erosion is related to the lower supply of nutrients, N in particular. On permeable subsoils, adequate fertilization and careful timing of tillage operations in regard to soil moisture content can go a long way toward producing satisfactory yields when combined with other good management practices.

6. The main function of legumes in a rotation, with the exception of legumes for grain, has been to furnish large quantities of high-quality forage. An additional benefit has been the N supplied on some soils. With the development of the fertilizer N industry, growing legumes for N alone is usually not economical.

7. It is recognized that supplemental N must be applied for highly profitable production of first-year corn after a legume.

8. Legumes or a nonlegume-legume mixture in forages reduces the fertilizer N requirement and generally results in superior animal performance.

9. The amount of N fixed by a legume is dependent on the amount of N the soil or residual fertilizer will supply. When the soil supplies a high amount of N the legume may fix less than 50% of its needs.

10. Where soil, solar, and water resources permit, two to four crops may be grown in a single year. Double cropping is common in the United States. In some tropical areas three or four crops may be possible.

11. Monoculture, or continuous cropping to one crop, is practiced in many parts of the world. Continuous corn cropping may yield somewhat less than corn in rotation, but economic factors may favor the former when production costs and the value of other crops in the rotation are considered.

12. Sod crops, including legumes, in rotation improve tilth and water intake, while continuous row cropping may favor compacting and breakdown of soil structure. Much information is needed on the long-time effects of high production.

13. Deep-rooted crops may bring nutrients to the surface. The net effect of cropping practices on fertility levels depends on removal of nutrients by the harvested portion, the amount supplied by the soil, and the amount of supplemental fertilization.

14. A primary cause of erosion is impoverishment of nutrients. The effects of cropping systems on soil losses can be drastically altered by the denseness of the cover, which in turn can be increased by more adequate fertilization and thicker planting.

15. Winter cover crops will supply residues and N if they are legumes. They are most useful in cropping systems in which few residues remain on the land over the winter. One of the uncertainties is the amount of growth produced. Early planting of these crops and late plowing to turn them under are helpful in increasing the amount of growth.

16. Farm manure is a valuable by-product of the livestock industry. Because of losses by volatilization and leaching, generally no more than one-third to one-half of the value is realized. Manure alone on a farm will result in a gradual depletion of soil fertility, but it is a valuable supplement to a well-designed lime, fertilization, and soil-management program.

17. Disposal of human waste or sludge from municipal treatment plants is receiving increasing attention. The sludge will supply large amounts of most essential elements. The possible toxic effects of heavy metals are still to be determined.

18. Conservation tillage is a principle, not a practice. More residues are left on the soil surface. Conservation tillage is based on fewer trips over the field. It results in better soil tilth, lower production costs, and generally higher yields, if other limiting factors are controlled. This results in more efficient use of fertilizers.

19. When maximum-profit-yield systems are used with conservation tillage, a deeper root zone is created, with greater water intake and greater water-holding capacity. An extra inch or two of water may be provided for critical growth stages.

20. No-till planting directly in sod or crop residues involves more herbicides but saves energy, water, and soil. It is particularly important on lighter soils and on sloping lands not suitable for other tillage because of erosion losses.

Questions

1. Why are yield trends over a period of years likely to be misleading as a measure of soil productivity? What might happen to yield trends if plant breeding studies ceased?
2. What is the aim of a crop- and soil-management program?
3. Explain why additions of N equal to crop removal help to reduce the loss of OM.
4. Under what soil condition may a corn-small grain-alfalfa rotation be preferable to corn plus commercial N as well as other nutrients each year? Under what conditions may the latter cropping system be preferable?
5. What influences how much of its total N a legume will fix?
6. On what soils in your area could OM be increased? Under what soil conditions in your area would additions of OM be beneficial other than for the nutrients supplied?
7. What nutrient may be most likely to give a marked yield response on corn grown on an eroded soil? Why?
8. Loss of surface soil is serious, but the problem varies considerably with the soil. In what soils in your area is the loss likely to be most serious?
9. Why is it important that plant residues decompose? What functions do they serve in the undecomposed state?
10. List the advantages of rotations and monoculture.
11. What cropping systems have depleted P and K in soils in your area? Explain. In what cropping systems have these elements been increased? Explain.
12. In what ways may N, P, and K be lost other than by crop removal? In what ways other than fertilization may the supplies be increased?
13. Are there places in your area in which the farmer is relying too heavily on a legume to furnish the N needs of a cropping system? Explain.
14. Explain the statement that a primary cause of erosion is depletion of plant nutrients.
15. In what cropping systems may winter cover crops fit? Why?
16. Why will the fertility level of a given farm gradually decrease if manure is the only carrier of plant nutrients used? Explain the fertility distribution problem in a pasture.
17. What measures can be taken to prevent losses of plant nutrients from manure?
18. What are the most popular manure storage and application systems used in your area? Are they effective?
19. What is sewage sludge? Are there problems in its disposal? Explain.
20. What is sewage effluent? Are there problems in its disposal? Explain.
21. In which of the major nutrients is sludge from sewage plants usually low? Is sludge being used in your area? If so, where?
22. Is sewage effluent being used in your area? If so, where, and what are its benefits?
23. What is conservation tillage? What are the advantages?
24. Would no-till fit in your area? If so, where? Why?
25. Is planting sometimes more difficult under no-till? Explain.

Selected References

POWER, J. F. (Ed.). 1987. *The Role of Legumes in Conservation Tillage Systems.* Soil Conservation Society of America.

SPRAGUE, M. A., and G. B. TRIPLETT (Eds.). 1986. *No-Tillage and Surface Tillage Agriculture: The Tillage Revolution.* John Wiley & Sons, New York.

TATE, R. L. 1987. *Soil Organic Matter: Biological and Ecological Effects.* John Wiley & Sons, New York.

TROEH, F. R., J. A. HOBBS, and R. L. DONAHUE. 1991. *Soil and Water Conservation.* Prentice-Hall, Englewood Cliffs, N.J.

Common Conversions and Constants

Conversion Factors for English and Metric Units

To Convert Column 1 Into Column 2, Multiply by:	Column 1	Column 2	To Convert Column 2 Into Column 1, Multiply by:
	Length		
0.621	kilometer, km	mile, mi	1.609
1.094	meter, m	yard, yd	0.914
0.394	centimeter, cm	inch, in.	2.54
	Area		
0.386	kilometer^2, km^2	mile^2, mi^2	2.590
247.1	kilometer^2, km^2	acre, a	0.00405
2.471	hectare, ha	acre, a	0.405
	Volume		
0.00973	meter^3, m^3	acre-inch, a-in.	102.8
3.532	hectoliter, hl	cubic foot, ft^3	0.2832
2.838	hectoliter, hl	bushel, bu	0.352
0.0284	liter, l	bushel, bu	35.24
1.057	liter, l	quart (liquid), qt	0.946
	Mass		
1.102	ton (metric)	ton (English)	0.9072
2.205	quintal, q	hundredweight, cwt (short)	0.454
2.205	kilogram, kg	pound, lb	0.454
0.035	gram, g	ounce (avdp.), oz	28.35
	Pressure		
14.50	bar	lb/in.^{2}, psi	0.06895
0.9869	bar	atmosphere, atm	1.013
0.9678	kg (weight)/cm^2	atmosphere, atm	1.033
14.22	kg (weight)/cm^2	lb/in.^{2}, psi	0.07031
14.70	atmosphere, atm	lb/in.^{2}, psi	0.06805
0.1450	kilopascal, kPa	lb/in.^{2}, psi	6.895
0.009869	kilopascal, kPa	atmosphere, atm	101.30

Conversion Factors for English and Metric Units (continued)

To Convert Column 1 Into Column 2, Multiply by:	Column 1	Column 2	To Convert Column 2 Into Column 1, Multiply by:
	Yield or Rate		
0.446	ton(metric)/hectare, ton/ha	ton (English)/acre, ton/a	2.240
0.891	kg/ha	lb/a	1.12
0.891	quintal/hectare, q/ha	hundredweight/acre, cwt/a	1.12
1.15	hectoliter/hectare, hl/ha	bushel/acre, bu/a	0.87

Convenient Conversion Factors

Multiply	By	To Get
acres	0.4048	hectare
acres	43,560	square feet
acres	160	square rods
acres	4,840	square yards
bushels	4	pecks
bushels	64	pints
bushels	32	quarts
centimeters	0.3937	inches
centimeters	0.01	meters
cubic feet	0.03382	ounces (liquid)
cubic feet	1,728	cubic inches
cubic feet	0.03704	cubic yards
cubic feet	7.4805	gallons
	29.92	quarts (liquid)
cubic yards	27	cubic feet
cubic yards	46,656	cubic inches
cubic yards	202	gallons
feet	30.48	centimeters
feet	12	inches
feet	0.3048	meters
feet	0.060606	rods
feet	⅓ or 0.33333	yards
feet	0.01136	miles per hour
gallons	0.1337	cubic feet
gallons	4	quarts (liquid)
gallons of water	8.3453	pounds of water
grams	15.43	grains
grams	0.001	kilograms
grams	1,000	milligrams
grams	0.0353	ounces
grams per liter	1,000	parts per million
hectares	2.471	acres
inches	2.54	centimeters
inches	0.08333	feet
kilograms	1,000	grams
kilograms	2.205	pounds
kilograms per hectare	0.892	pounds per acre
kilometers	3,281	feet
kilometers	0.6214	miles

Convenient Conversion Factors (continued)

Multiply	By	To Get
liters	1,000	cubic centimeters
liters	0.0353	cubic feet
liters	61.02	cubic inches
liters	0.2642	gallons
liters	1.057	quarts (liquid)
meters	100	centimeters
meters	3.2181	feet
meters	39.37	inches
miles	5,280	feet
miles	63,360	inches
miles	320	rods
miles	1,760	yards
miles per hour	88	feet per minute
miles per hour	1.467	feet per second
miles per minute	60	miles per hour
ounces (dry)	0.0625	pounds
ounces (liquid)	0.0625	pints (liquid)
ounces (liquid)	0.03125	quarts (liquid)
parts per million	8.345	pounds per million gallons H_2O
pecks	16	pints (dry)
pecks	8	quarts (dry)
pints (dry)	0.5	quarts (dry)
pints (liquid)	16	ounces (liquid)
pounds	453.5924	grams
pounds	16	ounces
pounds of water	0.1198	gallons
quarts (liquid)	0.9463	liters
quarts (liquid)	32	ounces (liquid)
quarts (liquid)	2	pints (liquid)
rods	16.5	feet
rods	5.5	yards
square feet	144	square inches
square feet	0.11111	square yards
square inches	0.00694	square feet
square miles	640	acres
square miles	27,878,400	square feet
square rods	0.00625	acres
square rods	272.25	square feet
square yards	0.0002066	acres
square yards	9	square feet
square yards	1,296	square inches
temperature (°C) + 17.98	$\frac{9}{5}$ or 1.8	temperature, °F
temperature (°F) − 32	$\frac{5}{9}$ or 0.5555	temperature, °C
ton	907.1849	kilograms
ton	2,000	pounds
ton, metric	2,240	pounds
yards	3	feet
yards	36	inches
yards	0.9144	meters

SOURCE: *Western Fertilizer Handbook,* 6th ed. Produced by the Soil Improvement Committee, Cal Fertilizer Association. Danville, Ill: Interstate Printers and Publishers, 1980.

Weight of Grain per Bushel

Crop	Weight (lb)
Barley	48
Canola/rapeseed	50
Corn	56
Flax	56
Oats	32
Rye	56
Sorghum	56
Soybeans	60
Wheat	60

Index

A

Acid, strong and weak, 365–366

Acidity (*see also* Acid; pH; Potential acidity; Soil acidity)
active, *defined*, 364
determination of, 374–375
potential, *defined*, 364
soil, reference to, in ancient agriculture, 5
subsoil, neutralization of by deep placement of lime, 388

Acidity (basicity) of nitrogen fertilizers, 371–372

Acidulation of soil
discussed, 392–395
materials used for, 393–395

Active ion uptake, 103–105

Activity index (AI) of urea-formaldehyde fertilizers, 170

Adenosine diphosphate and triphosphate role in energy transformations in plants, 51–52

Adsorption
of ions, 94–95
of phosphorus, 189, 193–204

Age
of crop, effect on nutrient requirements of corn, 256
of plant, effect on potassium uptake, 256

Air in soil, composition of as a factor in plant growth, 29–30

Aluminum
nature of compounds responsible for soil acidity, 368–370
reaction of with molybdenum, 348

Aluminum oxide, role in ion exchange, 93

Aluminum sulfate, use as a soil acidulant, 394

Aluminum toxicity of plants, effect of liming on, 384–385

Aminization in soil nitrification, 122–123

Ammonia, anhydrous, 154–162

Ammonia, aqua, 162

Ammonia, exchange by plants, 153

Ammonia, synthetic, manufacture of, 119

Ammonia, volatilization from soil, 148–153

Ammoniated superphosphate (*see* Phosphate fertilizers)

Ammoniation, effect on fertilizer phosphate availability, 222

Ammonification step in nitrification, 123

Ammonium bicarbonate fertilizer, 164

Ammonium bisulfite as sulfur-containing fertilizer, 285

Ammonium chloride fertilizer, 164

Ammonium fixation
   agricultural significance of, 138
   effect of
      clay minerals on, 138
      potassium on, 138
      type and amount of clay on,
         138–139
Ammonium nitrate fertilizer, 163
Ammonium nitrate-sulfate fertilizer,
   164
Ammonium nitrogen, effect on uptake
      of phosphorus, 492–493
   movement in soil, 136
   uptake by plants, 120–121
Ammonium phosphate fertilizer, 164,
   211
   composition of in fluid fertilizers, 213
Ammonium polyphosphate, 213
Ammonium polysulfide
   as sulfur-containing fertilizer,
      287–288
   use as soil acidulant, 395
Ammonium sulfate fertilizer, 164
Ammonium thiosulfate fertilizer, 172,
   287
Amorphous colloids in soil, effect on
      phosphate adsorption, 196
Animal manure
   benefits of, 600
   collection and disposal, 594–596
   composition, 597–598
   distribution by grazing animals, 600
   nitrogen losses from, 596
   use in cropping systems, 598–599
   value of, per animal unit, 597
Animal wastes, economics of use as
      fertilizer, 555
Anion exchange, 93
Anions, effect on soil phosphate
      adsorption, 199–200
Apatite
   phosphorus-containing mineral in
      phosphate rock, 207
   phosphorus minerals in soil, 190
Aspergillus niger, use in evaluating soil
      fertility levels, 428
Atmosphere, composition as a factor in
      plant growth, 29–30
Atomic weight, 87

Available phosphorus of fertilizers, 207
Azolla (Anabaena azolla), nitrogen
      fixation by, 118
Azotobacter, use for evaluating soil
      fertility levels, 428

                              B

Bacteria, nitrogen fixing, asymbiotic,
      117–118
Bacterial phosphate fertilization, 216
Band application of fertilizers, 477–481,
      483–484
Banding of acidulants in soil, 395
Base saturation of soils, 91–93
Basic slag
   as liming material, 379–380
   as source of fertilizer magnesium, 339
Biological soil tests, as diagnostic tool,
      426–428
Biotic factors as they affect plant
      growth, 31–32
Biotite as source of soil potassium, 239
Biuret, occurrence in urea, 165
Blast furnace slag as liming material,
      379–380
Blue green algae, role in nitrogen
      fixation, 117–118
Borates, boron-containing fertilizers,
      342
Boron (see also Borates)
   crop requirements for, 66
   deficiency symptoms of, 67
   effect of
      calcium on availability of, 340
      clay on availability of, 340
      liming on availability of, 340
      moisture on availability of, 341
      organic matter on availability of,
         340
      pH on availability of, 340
      plant factors on uptake of, 341
      soil texture on availability of, 339
   factors affecting availability of,
      339–341
   fertilizers, methods and rates of
      application, 342

forms of fertilizers containing,
    338–339
forms in soil, 66
geographical areas deficient in, 338
minerals containing, 339
plant growth processes requiring, 67
plant sensitivity to, 341
role in plant growth, 67
soil, adsorbed, 338
    amounts present, 337
soil solution, 338
specific plant functions requiring, 67
Boron cycle in soil, 339
Bray No. 1 Method, use in soil testing,
    204
Bray's nutrient mobility concept, 40
Broadcast application of fertilizers,
    477–481, 485–486
Buffer capacity, 94–95
Buffering of solutions against pH
    changes, 373–374, 366–367

C

Cadmium content of sewage, 603
Calcium
    effect of nitrogen form on adsorption
        of, 294
    forms utilized by plants, 289
    minerals containing, 289
    mobility of in plants, 61
    role of
        in cell division, 61
        in plant nutrition, 61
    soil, behavior of, 292–294
        effect of, on potassium availability,
            249
        factors determining availability to
            plants, 294
        leaching losses of, 294
    sources of, in soil, 289–292
Calcium carbonate
    coprecipitated with sulfate, 272
    as liming material, 379
    neutralizing value of, 380–381
    phosphorus adsorption on, 196
    soil, effect of, on phosphate
        adsorption, 196

Calcium deficiency, manifestations of,
    61
Calcium fertilizers, 294
Calcium hydroxide as liming material,
    379
Calcium magnesium carbonate, as a
    liming material, 379
Calcium and magnesium content of
    liming materials, 378
Calcium and magnesium oxide
        equivalent of liming materials,
        381
Calcium nitrate fertilizer, 295
Calcium oxide as liming material, 379
Calcium silicate as liming material,
    379–380
Calibration of soil test results, 437–453
Carbon-bonded organic soil sulfur, 277
Carbon dioxide
    atmospheric content of, 27
    effect on plant growth and
        supplemental use of, 27–28
Carbon:nitrogen ratio
    of organic materials, 124
    of soils, 124–126
Cation exchange, 86
    measurement in soils, 90–91
    soil fractions responsible for, 81–90
Cation exchange capacity
    effect on potassium availability, 242
    as influenced by nature of charge on
        soil colloids, 81–90
Cations
    determination of, in soil testing, 436
    effect of on soil phosphate
        adsorption, 199
CDTA as chelate, 310
Chelate(s), 309
    as fertilizers, 318
    stability, 310
    types of, 310
Chiseling, as a conservation tillage
        technique, 574
Chloride
    content of, in earth's crust, 342
    effect of
        on osmotic pressure of soil water,
            395
        on plant diseases, 344–345

Chloride (*cont.*)
    favorable effects of, on plant growth,
        395–396
    form absorbed by plants, 73
    functions of, in plant growth, 73–74
    osmotic effects of, in plant growth, 74
    plant content of, 73
    plant requirements for, 75
    plants benefiting from, 344
    plants requiring, 73
    role of, in plant disease control, 74
    soil
        behavior of, 343
        content of, 342
        environmental problems caused by,
            342
        interaction of, with nitrates and
            sulfates, 343
        mobility of, 343
Chloride fertilizers, listed, 346
Chlorite, 2:1:1 clay mineral, 86
Chlorophyll meter, diagnosis of leaf N
        status, 424
Citrate-insoluble phosphate of
        fertilizers, 207
Citrate-soluble phosphate of fertilizers,
        207
Clay fraction of soil, role in cation
        exchange, 82–86
Clay minerals, effect on potassium
        fixation, 238–241
Clay type in relation to soil phosphate
        adsorption, 196
CNC (*see* Critical nutrient levels)
C:N:P:S ratio in soils as related to
        organic soil phosphorus, 181
C:N:S ratio in soils, 276
Coal as a source of sulfur, 266
Cobalt
    content of, in earth's crust, 350
    crop responses to, 75
    fertilization with, 351
    fertilizers containing, 351
    functions of in plant growth, 75
    need for in ruminant nutrition, 351
    role of
        in nitrogen fixation, 75
        in vitamin B_{12} synthesis, 75

soil
    availability of, 350
    behavior of, 350
    content of in, 350
Cobalt bullet in animal nutrition, 351
Colemanite as boron-containing
        fertilizer, 342
Complementary ion effect, 90
Concentrated superphosphate (CSP) (*see*
        Phosphate fertilizers)
Conservation tillage
    advantages of, 573
    disadvantages of, 574
    effect of
        on fertilizer practices, 581
        on liming practices, 583
        on nitrogen fertilization practices,
            582
    as related to water use efficiency, 517
    types of, 574–576
Contact exchange, importance in plant
        nutrition, 95
Continuous cropping (monoculture)
    advantages of, 591
    value of, as compared to crop
        rotations, 588–592
Copper
    appearance of deficiency, 71
    concentration in plant tissue, 71
    content in earth's crust and in soils,
        326
    crops susceptible to deficiency of, 331
    deficient geographical areas, 327
    fertilizers containing, 332
    forms absorbed by plants, 71
    with organic matter, 328
    plants exhibiting deficiency of, 71
    role of, in plant nutrition, 71
    soil
        adsorbed, 328
        availability in, 330
        coprecipitated, 328
        forms of in, 327
        interaction of with nutrients, 331
        occluded, 328
        pH and availability, 330
        texture and availability, 330
    soil solution, 327

specific plant functions of, 71
toxicity of, 331
Copper deficiency of crops, 331
Copper fertilizers, 332
   foliar applications of, 332
   residual effects of, in soil, 332
Copper hydroxide as fertilizer, 332
Copper sulfate as fertilizer, 332
Corn production, changes in practices
      for, Minnesota, 564
Cost of plant nutrients, 541–554
Cover crops, 592–594
Critical nutrient levels, in plants, 45–46,
      414–416
Crop (see Plant species)
Crop logging as diagnostic tool, 424
Crop management (see also Cropping
      systems)
   effect of
      on crop yields, 563–568
      on runoff and soil erosion,
         563–568
      objectives of, 561
Crop production, economics of,
      540–558
Crop responses as related to soil test
      results, 438–444
   to sewage sludge, 604
Crop rotation
   advantages of, 591
   fertilization of, 499
   value of, compared to continuous
      cropping, 588–593
Crop yields as related to nutrient
      content of plant, 416–418
Cropping practices, changes in, Ohio,
      563
Cropping systems (see also Crop
      rotation; Cover crops; Winter
      cover crops; Crop management)
   characteristics of, related to soil
      erosion, 591–592
   continuous cropping or monoculture,
      advantages of, 591
   effect of
      on crop residue accumulation, 576
      on nutrient concentration in
         topsoil, 591

on phosphorus and potassium levels
      in soil, 495
   as related to plant nutrient
      additions, 581–583
   on soil tilth, 590
   on water losses, 591–592
   role of legumes in, 583–588
   rotation, advantages of, 591
Crotonylidene diurea (CDU) fertilizers,
      170

D

DCD, nitrification inhibitor, 171
Denitrification
   agricultural and environmental
      significance of, 145–146
   chemical pathway of, in soils, 139
   effect of
      soil aeration on, 143–144
      soil moisture on, 142–143
      soil nitrate level on, 144–145
      soil organic matter on, 141–142
      soil pH on, 144
      soil temperature on, 144
   organisms responsible for, 139
   in soil, 139–147
Desorption, of soil phosphorus, 196
Diammonium phosphate (DAP) (see
      Phosphate fertilizers)
Diammonium phosphate fertilizer,
      reactions in soil, 213
Dicalcium phosphate, 190
Diffusion
   as mechanism of ion movement to
      root surfaces, 99–102
   in micronutrient transport, 304, 310
   in potassium transport in soils, 233
   in relation to sulfate movement in
      soils, 267
Diffusion coefficient of ions in soil
      solution, 100
Diminishing returns, law of, 546
Dipotassium phosphate, fertilizer, 216
Dolomite (see also Calcium magnesium
      carbonate)
   as a source of soil calcium, 295

DRIS system
    advantages in making fertilizer
        recommendations, 420–423
    defined, 420
    as diagnostic tool, 422–423
    in making fertilizer
        recommendations, 420–421
DTPA as chelate, 310

E

Economics of crop production, 540–558
EDDHA as chelate, 310
EDTA as chelate, 310
Electric furnace slag as liming material,
    379
Elemental sulfur (*see* Sulfur, elemental)
Environment, effect of denitrification
    on, 145
Environmental factors affecting plant
    growth, 17–33
Epsomite as source of soil sulfur, 266
Equations, regression (*see* Regression
    equations)
Equivalent weight, 87
Erosion (*see also* Soil erosion; Soil loss)
    effect of tillage on, 576
    soil, characteristics of cropping system
        related to, 591–592
Exchangeable potassium (*see* Potassium,
    exchangeable)
Experiments, factorial (*see* Factorial
    experiments)

F

Factorial experiments, 38
Farm inputs, cost of, 541
Feeding power of crop, effect on
    fertilizer application, 467
Feldspars as sources of soil potassium,
    230–231
Ferric sulfate (*see* Iron sulfate)
Ferrous sulfate (*see* Iron sulfate)
Fertigation with fluid fertilizers,
    505–506

Fertility, soil, index of as related to soil
    test results, 428–454
Fertilization (*see also* Fertilizer
        application)
    effect of on winter killing of crops,
        475
    of subsoil (*see* Fertilizer application, to
        subsoil; Subsoil fertilization)
Fertilizer, application of (*see also*
        Fertilizer application; Fertilizer
        use)
    cost of (*see* Plant nutrients, cost of)
    movement of in soil (*see* Fertilizer
        movement in soils)
    placement of (*see* Fertilizer
        application; Fertilization)
Fertilizer application
    as affected by plant root
        characteristics, 462–468
    annual, based on soil test results,
        428
    band, 477–480, 483–484
    based on soil test results, 442–450
    broadcast (*see* Broadcast application of
        fertilizers)
    carry-over effects of, 500
    with conservation tillage, 495,
        581–583
    to cover crops, 593
    crop characteristics affecting,
        462–468
    in crop rotation, 499
    effect of
        crop price on rate of, 548
        fertility level of soil on, 442–450
        on production costs, 543–545
        soil characteristics on, 469–475
        uneven spreading, 485
    fluid, 505–507
    foliar (*see also* Foliar application of
        fertilizers), 506–507
    as influenced by
        crop species and variety, 464
        feeding power of crop, 467
        phosphorus, 486–491
    methods of, 477–480
    micronutrient,
        nitrogen, 492–493
        potassium, 494–495

most profitable rate of, 547
in rotation, 499
to subsoil, 502–505
time of, as affected by nutrient, 495–499
Fertilizer bands, acidification of soil in, 395
Fertilizer movement in soils, 481
Fertilizer placement (*see* Fertilizer application; Placement of nutrients)
Fertilizer prices, relative to other farm costs, 541
Fertilizer recommendations, based on soil test results, 442–453
Fertilizer use
  effect of
    on water absorption by plants, 516–517
    on water use efficiency, 517–523
  in relation to
    farm income, 541
    level of stored soil water, 524–528
  role of in increasing land values, 555
Fick's law, 99–100
Field tests for determining soil nutrient levels, 426–427
Fixed nitrogen (*see also* Nitrogen, fixation of)
  amounts returned to soil from atmosphere, 118–119
Fluid fertilizers (*see also* Fertilizer application, fluid), 505–507
Foliar application of fertilizers, 506
Freundlich equation in relation to soil phosphate adsorption, 194
Fuel oil as source of sulfur, 266
Fungi, mycorrhizal, effect of on nutrient absorption by higher plants, 467

G

Genetics as a factor in plant growth, 15–17
Grass tetany in ruminants, resulting from magnesium deficiency, 299–300

Greenhouse tests, for determining nutrient availability in soils, 427
Growth, 14
  plant, factors affecting, 14–42
Growth expressions, limited application of, 41
Growth inhibitors, as they affect plant growth, 32–33
Gypsum as source of
  soil calcium, 295
  soil sulfur, 285

H

Haber-Bosch process in manufacture of ammonia, 119
HI reducible soil sulfur, 278
Hidden hunger, of crops, 408–409
High crop yields, need for, 542
Humus, soil content of, 122
Hydrogen, use in ammonia synthesis, 119
Hydrogen sulfide (*see* Sulfides; Polysulfides)
Hydrous oxides of iron and aluminum, effect of phosphorus retention in soils, 196

I

Illite, 2:1 clay mineral, 84–85
  as source of soil potassium, 231
Immobilization of nitrogen in soils, 123–131
Infrared photography as diagnostic agriculture, 454–456
Inner space of root cells, role in ion absorption, 103–104
Inorganic soil phosphorus (*see* Phosphorus, inorganic soil)
Inosital phosphates in soils, 182
Interactions
  among nutrients and plant population, 531
  among plant nutrients, 528–531
  among plant nutrients and fertilizer placement, 532

Interactions (*cont.*)
  among plant nutrients and plant
        variety, 535–536
  among plant nutrients and planting
        date, 532
  among plant population and planting
        date, 531–532
  between fertilizer placement and
        tillage method, 533–535
  between variety and row width, 532
    discussed, 511–512
    explanation of, 39–40
Ion absorption by plant roots,
        mechanisms, 102–106
Ion exchange (*see also* Cation exchange;
        Anion exchange), 81
Ion transport, soil factors affecting,
        100–101
Ions, movement from soil to roots,
        96–102
Iron (*see also* Iron oxide)
  chelation in plants, 67
  content in earth's crust, 304
  factors affecting availability, 311–316
  fertilizers, stability of organic
        complexes in soil solution,
        309–311
    fertilizers containing, 318–319
    foliar applications, 319
    injections into plants, 319
  forms of absorbed by plants, 67
  forms of in solution, 305
  functions in plant growth, 68–69
  nature of compounds responsiblew
        for soil acidity, 368
  oxidation–reduction reactions
        affecting, 306
  plant
    effect of species on uptake, 316
    mechanisms of absorption, 316
  plant sensitivity to deficiency, 68
  role
    in enzyme activity, 69
    in oxidation–reduction reactions in
        plants, 69
  soil content of, 304
    aeration and availability, 314
    bicarbonate and carbonate ion and
        availability of, 314

    deficient geographical areas, 304
    effect of acidification on availability,
        305–306, 314
    ion imbalance and availability of,
        314
    moisture level and availability, 315
    organic matter and availability,
        309–310, 315
    other nutrients and availability, 315
    pH and availability, 305–306, 314
  soil tests for, 317
  soluble organic complexes of,
        309–310, 315
    solubility of, 306
  in solution, 305
  sufficiency range in crops, 67
Iron deficiency
  appearance in plants, 68
  plants sensitive to, 316–317
Iron fertilizers (*see also* Iron, fertilizers
        containing)
    complex organic forms, 318–319
    effect of polyphosphates on
        availability of, 319
Iron oxide, role in ion exchange, 93
Iron polymers in soil, role in soil acidity,
        368–370
Iron sulfate as fertilizer, 318–319
Irrigation, efficient use of, 524
Isobutylidene diurea (IBDU) fertilizers,
        170
Isomorphous substitution in clay
        minerals, 83

K

Kaolinite, 1:1 clay mineral, 84

L

Langmuir equation in relation to soil
        phosphate adsorption, 194–195
Law of diminishing returns, 546
Leaching requirement, 399
Legumes
  amounts of nitrogen fixed by, 111

nitrogen fixation in cropping systems, 586–588
role in cropping systems, 583–585
Light (*see* Radiant energy)
Lime (*see also* Liming materials)
  economics of use, 554–555
  use of, based on soil test
    recommendations, 376–377
Lime requirement
  of crop as factor in selecting liming
    program, 392
  of soil, determination of, 376–377
  SMP method for determining,
    376–377
Limestone (*see* Liming materials)
Liming
  as affected by conservation tillage,
    583
  direct benefits of, 384–385
  effect of
    on micronutrient availability,
      385
    on nitrification, 387
    on nitrogen fixation, 387
    on phosphate availability, 385
    on physical condition of soil, 387
    on plant diseases, 388
  equipment used for, 390–392
  program of, factors determining
    selection, 392
Liming materials (*see also* Lime)
  application of, 388–392
    time and frequency, 392
  calcium carbonate equivalent (CCE) of
    (*see* Neutralizing value of liming
    materials)
  deep placement in soils, 388
  effect of fineness on value of,
    381–384
  fluid lime, 389
  miscellaneous, 380
  neutralizing value of, 380–384
  placement in soil, 388
  tests for evaluating effectiveness of,
    380–384
  types of, 378–380
Luxury consumption of nutrients by
  crops, 45
Lyotropic series, 89–90

M

Magnesium
  deficiency of in relation to grass
    tetany in ruminants, 299
  fertilizer sources, 299
  form utilized by plants, 296
  minerals containing, 296
  mobility in plants, 61
  plant factors affecting uptake, 296
  role of
    in enzyme activation, 61
    in photosynthetic activity, 61
    in plant nutrition, 61
  soil
    behavior of, 298
    effect of high levels of potassium
      on, 298
    interaction of other cations with,
      298
    leaching losses of, 299
    potassium availability, 249
  sources of in soils, 296
Magnesium chloride as source of
    fertilizer magnesium, 299
Magnesium nitrate as source of fertilizer
    magnesium, 299
Magnesium sulfate as source of fertilizer
    magnesium, 299
Management, level of in efficient crop
    production, 540
Manganese
  content of
    in earth's crust, 332–333
    in plant tissue, 70
  fertilizers containing (*see* Manganese
    fertilizers)
  forms absorbed by plants, 70
  plant, effect of species on sensitivity
    to deficiency of, 336
  role of
    in enzyme systems in plants, 71
    in oxidation-reduction reactions in
      plants, 70
  soil
    availability of, factors affecting,
      335–336
    conditions associated with
      deficiency, 335–336

Manganese (*cont.*)
  content of, 333
  deficient geographical areas, 333
  effect of aeration on availability,
    335
  effect of carbonates on availability,
    335
  effect of forms of in, 333
  effect of interaction with other
    nutrients, 335
  effect of metal ions on availability,
    335
  effect of microorganisms on
    availability, 336
  effect of moisture, 335
  effect of organic matter on
    availability, 335
  effect of pH on availability, 335
  effect of redox potential on
    availability, 333, 335
  effect of seasonal changes on
    availability, 335
  in solution, 333
  toxicity from excess of, 71
Manganese deficiency, appearance in
    plants, 70
Manganese fertilizers, application of,
    336–337
Manganese oxide as fertilizer, 337
Manganese sulfate as fertilizer, 337
Manganese toxicity of plants, effect of
    liming, 386
Manure, animal (*see* Animal manure)
Marl, as liming material, 379
Mass flow
  as mechanism of movement of ions to
    root surfaces, 99
  phosphorus movement by, 119
  in relation to sulfate movement in
    soils, 267
  role in absorption of potassium, 233
  significance in absorption of ions, 99
Maximum crop yield as related to water
    use efficiency, 557
Maximum economic yield (*see also*
    Maximum profit yield), 542
Maximum profit yield, 542
Mehlich's method of soil testing,
    204–205

Metal ions as urease inhibitors, 171
Methyl urea, urease inhibitor, 171
Mica
  potassium content of, 230
  transformations in soils, 238–240
Microbial activity, inhibition of,
    170–172
Microbiological methods, for evaluating
    soil fertility level, 428
Micronutrient availability in soil as
    affected by liming, 386
Micronutrient fertilizers, approaches to
    application of, 501–502
Micronutrients (*see also* Micronutrient
    fertilizers; Fertilizer application,
    micronutrient)
  determination in soil testing, 317,
    436–437
  effect of soil moisture on absorption,
    523
Mineral nutrient supply as it affects
    plant growth, 32
Mineralization
  of nitrogen in soils, 122–131
    effect of tillage on, 129–130
  patterns of, 382
  of soil sulfur (*see* Sulfur, soil
    mineralization of)
Mitscherlich's laws and growth
    equations, 34–37
Modeling in soil fertilization, 41–42
Moisture, soil, effect on mineralization
    of soil sulfur, 280
Moisture supply as factor in plant
    growth, 21–24
Molybdenum
  amount in earth's crust, 346
  fertilizers containing (*see* Molybdenum
    fertilizers)
  forms absorbed by plants, 72, 346
  plant
    crops sensitive to deficiency, 349
    ruminant toxicities caused by, 349
  role in plant growth, 72–73
  soil
    amounts in, 346
    amounts in solution, 346
    deficient geographical areas, 346
    effect of forms in solution, 346

effect of interaction with nutrients, 348

effect of nitrate on uptake, 349

effect of pH and liming on availability of, 348

reaction with iron and aluminum, 348

Molybdenum fertilizer application, 350

Molybdenum toxicity, 349

Molybdenum trioxide fertilizer, 350

Monoammonium phosphate (MAP) (*see* Phosphate fertilizers)

Monopotassium phosphate fertilizer, reactions in soil, 213

Montmorillonite (smectite) 2:1 clay mineral, 83–85

Muriate of potash (*see* Potassium chloride fertilizers)

Muriatic acid (*see* Muriate of potash)

Muscovite as source of soil potassium, 231

Mycorrhizal fungi, effect on absorption of nutrients by higher plants, 97, 467

N

Natural gas, use in ammonia synthesis, 119

Neubauer seedling method for evaluating soil fertility level, 427

Neutralization of acids and bases, 364–365

Neutralizing value of liming materials, 380–384

Nickel
  content in crops, 77
  functions in plants, 77

Nickel compounds as urease inhibitors, 171

Nitrate fertilizer materials, 168–169

Nitrate nitrogen
  determination in soil testing, 435, 449
  mobility in soils, 137–138

Nitrates
  gaseous losses from soil, 139
  uptake by plants, 120

Nitric phosphate fertilizers, 214

Nitric phosphates (*see* Phosphate fertilizers)

Nitrification
  of soil nitrogen, 132
    effect of liming, 387
  in soils, factors affecting, 132–138

Nitrification inhibitors, 170–172

Nitrite, accumulation in soils, 146–147
  effect of fertilizers, 147

Nitrite oxidation, conditions suitable for, in soils, 147

Nitrogen
  amounts fixed by legumes, 111
  appearance of deficiency symptoms in plants, 50
  content in sewage sludge, 603
  effect of
    form of on absorption of calcium, 294
    soil moisture on absorption of, 517
  elemental, conversion to forms utilizable by plants, 109
  fertilization of legumes with, 115
  fixation of
    by asymbiotic (free-living) bacteria, 118
    by free-living organisms, 117–118
    by leguminous trees and shrubs, 117
    by symbiotic bacteria, 112–117
  fixed
    by legumes, 586–588
    presence in atmosphere, 118–119
    transfer from one crop to another, 113–114
  forms of
    absorbed by plants, 48, 120–121
    as related to plant diseases, 121–122
    in sewage sludge, 601–605
  functions in plant nutrition, 48–50
  gaseous losses from soils, 139
  immobilization in soils, 122–131, 133–134
  losses from animal manure, 596–597
  mineralization in soils, 122–131
  organic forms present, 119–120
  organic-mineral balance in soil, 122

Nitrogen (*cont.*)
  soil
    determination in soil testing, 434
    effect of winter cover crop, 592
    forms of, 119–120
    inorganic forms present in, 119
  soil and fertilizer (*see* Nitrates;
      Ammonia; Nitrogen fertilizers)
  transformations in soil, 122–131
Nitrogen content of organic substances
    as related to nitrogen release in
    soils, 124
Nitrogen cycle in soils, 109
Nitrogen fertilization as affected by
    conservation tillage, 582–583
Nitrogen fertilizers
  acidity and basicity, 371
  ammoniacal sources of, 154–168
  effect of
    temperature on application, 135
    time of application, 497–498
  materials involved, discussion of,
    153–172
  nitrate movement in soils, 137–138
  organic, 153
  placement, 492–493
  recommendations for use based on
    soil test results, 447–453
  requiring decomposition, 169–170
  residual effects, 500
  slowly available, 169–170
  soluble, treated to impede dissolution,
    170
Nitrogen fixation
  factors affecting, 115–117
  on leaf surfaces in tropics, 118
  by legumes, amounts fixed, 111
    in cropping systems, 586–588
  nonleguminous crops, 117–118
  organisms responsible for, 111
  in soil, effect of liming, 387
  by symbiotic bacteria, amounts fixed,
    111–113
Nitrogen solutions
  growth in consumption, reasons for,
    154
  nonpressure, 162
Nodulation of legumes, effectiveness of,
    112

No till, as conservation tillage technique,
    575–576
No-till farming, factors favoring
    adoption, 583
N-Serve, nitrification inhibitor,
    discussed, 17
Nucleic acids in soils, 182
Nutrient absorption, effect of soil
    moisture, 516–523
Nutrient availability in soil as affected
    by organic matter, 568–572
Nutrient-deficiency symptoms
  in plants, 405–408
  precautions in use as diagnostic tool,
    405–408
Nutrient interactions as factor in plant
    growth, 39–40
Nutrient level in plants (*see also* Plant
    nutrient levels)
  in relation to crop yield, 416
Nutrients, levels of, 45
Nutrient uptake, effect of plant age,
    416

O

Oceans as potential source of sulfur,
    267
Olsen's method, use in soil testing,
    204–205
Organic carbon in soils, 126–131
  effect of cultivation and fertilizers on,
    126–131, 568–572
Organic matter, soil (*see also* Soil organic
    matter)
  effect of
    on nutrient availability in cropping
    systems, 568–572
    on soil phosphorus adsorption,
    201
  role in cation exchange, 86
  as source of soil sulfur, 277–278
Organic nitrogen fertilizers, 153–154
Organic phosphorus, 180–189
Organic soil sulfur (*see* Sulfur, soil,
    organic forms)
Outer space of cells, role in ion
    adsorption by roots, 102–105

P

Partially acidulated rock phosphate as phosphatic fertilizer, 221
Passive ion uptake, 102–103
Pentlandite as source of soil sulfur, 292
pH (*see also* Acidity, Soil acidity, Reaction, soil)
  concept, *defined*, 364
  effect on mineralization of soil sulfur, 276
  as indicator of exchangeable aluminum, 373
  levels suitable for crop production, 377
  soil, effect of
    on crop production, 377
    on phosphate adsorption, 198
  soluble salts on determination of, 370
pH dependent charge on clay minerals and soil organic matter, 84–86
pH meter for determination of active acidity in soils, 374
Phosphatase enzymes, role in mineralization of organic soil phosphorus, 183
Phosphate absorption by plants as influenced by form of nitrogen, 222
Phosphate esters in soils, 182
Phosphate fertilization with conservation tillage, 581–582
Phosphate fertilizers
  acid treated, 209–210
  ammonium phosphates, 211–213
  ammonium polyphosphate, 213–214
  calcium orthophosphates, 210–211
  compounds contained in their saturated solutions, 218
  effect of
    granule size on efficiency, 222
    rate of application on efficiency, 223
    soil moisture on efficiency, 223
    soil residual phosphorus on efficiency, 223–226
    on uptake by ammonium ion, 492
  importance of placement of, 486–491

interactions with
  nitrogen fertilizers in soil, 222
  zinc, 386–387
nitric phosphates, 214
phosphoric acid, 209–210
terminology of, 207
Phosphate rock (*see also* Rock phosphate), deposits of, origin of, 207
  processing of, for fertilizer manufacture, 209–210
  as a source of phosphate fertilizers, 207
  types of, 207
  world reserves of, 207
Phosphatic fertilizers
  behavior in soil, 216–226
  composition, 205
  movement and redistribution in soil, 221–222
Phosphobacterins as phosphatic fertilizing agents, 216
Phospholipids, in soils, 182
Phosphoric acid (*see also* Phosphate fertilizers)
  wet process in fertilizer manufacture, 209
Phosphorus
  adsorption on calcium carbonate, 196–197
  adsorption in soil, 193–204
  content in soils, 176
  determination of in soil testing, 204–205
  dynamics in soil, 176
  effect of
    cations on adsorption, 199
    factors affecting retention, 196–204
    organic forms, 194–196
    organic matter on adsorption, 201
    reaction time on adsorption, 201–203
    solubility diagrams, 190–193
    temperature on adsorption, 201–203
  forms adsorbed by plants, 51, 178
  in soil solution, 178–180
  functions in plant growth, 51–52
  inorganic soil, 189–204

Phosphorus (*cont.*)
  manufacture of, 207–216
  movement in plants, 52
  organic (*see* Organic phosphorus)
    mineralization in soils, 180–189
  role in energy transfers in plants,
      51–52
  soil
    adsorption, as affected by clay
        content, 196
    as affected by pH, 198
    availabiilty as affected by liming,
        386
    soil–fertilizer reaction products,
        216–222
    in soil solution, 178–180
      movement of (*see* Soil solution
          phosphorus)
    turnover in soils, 183–189
Phosphorus cycle in soils, 176
Phosphorus fertilizers, placement in
      soil, 486–491
Photoperiodism as factor in plant
      growth, 27
Placement of nutrients, 477–481
Plant analyses
  as diagnostic tool, 410–425
  interpretation of, results of, 414
    using DRIS system, 420–423
  methods employed, 411
  plant parts used for, 411–412
  stage of maturity of plant, 412–414
  total, 414
Plant diseases
  effect of soil liming on, 388
  as influenced by form of soil
      nitrogen, 121–122
Plant growth (*see also* Growth, plant)
  as related to time, 33
Plant growth expressions, 33–34
Plant nutrient deficiencies,
      determination of, using plant
      analyses (*see also* Plant analyses)
  effect of
    season on, 409–410
    soil moisture and temperature on,
        409–410
Plant nutrient levels (*see* Critical nutrient
    levels, in plants)

Plant nutrient requirements as
      determined by DRIS system,
      420–423
Plant nutritient supply as related to
      water use efficiency, 516–524
Plant nutrients (*see also* Mineral
      nutrients, Nutrients)
  balance of
    as calculated by DRIS system (*see*
        DRIS system)
    as related to plant growth,
        416–418
  cost of, 553–554
  crop content of in relation to yield,
      416–418
  interactions among, 528–531
  supply of, in winter cover crops, 593
Plant nutrition, elements required in,
      46–47
Plant population, effect of on potassium
      availability, 254
Plant residues, effect of cropping system
      on, 576
Plant spacinig (*see* Spacing, plant)
Plant species, effect on potassium
      availability, 252
Polyphosphate fertilizers, reactions in
      soils, 213–214
Polysulfides, soil, 287–288
Potash fertilization with conservation
      tillage, 581–582
Potash fertilizers
  application of, 494
  starter responses to, 494
  time of application, 499
Potash resources, world, 258
Potassium (*see also* Potash entries)
  absorption by plants, 233–235
  activity ratio in soils, 233
  availability to plants, soil factors
      affecting, 242–251
  content of in secondary minerals,
      242
  content in soils, 230–231
  deficiency symptoms in plants, 56
  diffusion of as affected by
      temperature and moisture, 249
  effect of
    on ammonium fixation, 138

soil moisture level on absorption of, 520

on stalk strength of grain crops, 56

exchangeable, amounts of in soils, 235–238

fixation of factors affecting, 240–242

fixation in soils, 240–242

forms of in soils, 231–242

functions of in plant nutrition, 53–60

losses of by leaching, 250

minerals containing, 230–231

mobility of in plants, 56

nonexchangeable (fixed) amounts in soil, 230

origin of in soils, 230

plant factors affecting availability, 231–256

role of
  diffusion in transport of in soils, 233–234
  in disease resistance, 57
  in energy transfer in plants, 55
  in enzyme activation in plants, 55
  mass flow in absorption, 233
  in plant protein synthesis, 55
  in plant–water relations, 55
  in starch synthesis, 55
  in translocation of assimilates in plants, 55

soil availability of as affected by
  calcium and magnesium, 249
  level of exchangeable soil potassium, 242
  nutrient level of soil, 249
  rooting depth of plants, 243
  soil aeration, 247
  soil capacity to fix potassium, 243
  soil CEC, 242
  soil moisture, 244
  soil pH, 247
  soil temperature, 244
  soil tillage, 249

soil solution, amounts in soils, 231–235

subsoil, effect on potassium availability, 243

use of Q/I concept for predicting status in soils, 236–237

water-soluble (see Potassium, soil solution), amounts of, in soils, 231–233

Potassium availability (see Potassium, availability; Potassium, soil, availability)

Potassium carbonate and bicarbonate fertilizers, 260, 282

Potassium chloride fertilizers, 259

Potassium fertilizers, 258–262
  agronomic value of various forms, 261
  calculation of potassium and K_2O content, 258
  nutrient content of various forms, 261
  placement, 494
  production areas in North America, 258

Potassium fixation (see Potassium, fixation in soils)

Potassium hydroxide fertilizers, 260

Potassium magnesium sulfate
  as fertilizer, 260
  as source of fertilizer magnesium, 299

Potassium nitrate (saltpeter) as fertilizer, 168–169, 260

Potassium phosphate as phosphatic fertilizer, 216

Potassium phosphate fertilizers, 260

Potassium polyphosphate fertilizer, 260

Potassium sulfate fertilizer, 260

Potassium thiosulfate fertilizer, 261

Potassium uptake (see Potassium, absorption of)

Potential acidity, 374

Presidedress soil test, 451–453

Production, crop, unit cost of, 543–545

Production costs, crop, 541

Profit per acre and fertilizer use, 547–551

Pyrites as source of soil sulfur, 266

Pyrrhotite as source of soil sulfur, 266

R

Radiant energy as factor in plant growth, 24–27

Reaction, soil, as factor in plant growth, 30

Reaction time, effect on adsorption of soil phosphorus, 201
Reclamation sodic soils, 399–401
Reduced sulfur, inorganic forms in soils, 272
Regression equations, 38–39
Remote sensing as diagnostic tool, 455–456
Residual effect of fertilizers, value of, 223–226, 500, 552
Residues on soil as affected by tillage, 576
Rhizobia, symbiotic nitrogen-fixing bacteria, 110–113
Ridge tillage, 574
Rock phosphate
    acidulation with nitric acid, 209
    as phosphatic fertilizer, 207–209
Root cation exchange capacity, 95
    effect of on potassium availability, 251
Root characteristics of plants as affected by species, 464–466
Root growth as affected by fertilization, 473–475
Root interception as mechanism in ion absorption, 96–97
Root penetration as affected by soil properties, 469–472
    effect of chemical makeup of soils on, 473–475
Root system, plant, effect on potassium availability, 251–252

S

Salt index of fertilizers, 481–483
Salt injury from fertilizers, 481–483
Selenates in soils, 357
Selenides in soils, 357
Seleniferous soils (see Selenium, soils with high content of)
Selenites in soils, 357
Selenium
    accumulation of by plants, 358
    amounts in earth's crust, 356
    elemental, in soils, 357
    fertilizers containing (see Selenium fertilizers)

organic forms in soils, 357
soil
    amounts of in, 356
    beneficial effects of, 358
    chemistry of, 357
    effect of low levels on animal health, 358
    effect of nitrogen on plant uptake of, 358
    effect of pH on, 358
    effect of phosphate fertilizers on plant uptake of, 358
    forms of, 356–357
    toxic effects, 356
soils low in, geographical distribution, 356
soils with high content, geographical distribution, 356
Selenium fertilizers, 358
Sensing, remote (see Remote sensing)
Sequestering agents (see Chelate(s))
Sewage (see also Sewage effluent), 601
    application to agricultural land, 601–603
    cadmium content, 601
    composition, 601
    crop responses to, 603
    problems associated with, 601
    treatment of, 601
    use of, as fertilizer, 603
Sewage effluent, crop responses to application of, 603–604
Short-term method for evaluating soil fertility level, 427–428
Sideband as method of fertilizer application, 478–479
Side-dressed, as method of fertilizer application, 480
Silicon
    appearance of deficiency symptoms, 76
    content in earth's crust, 353
    crops requiring, 76
    fertilizers, sources of (see Silicon fertilizers)
    forms absorbed by plants, 76
    functions of in plant growth, 76
    geographical areas in which deficiencies occur, 76

plant contents of, 76
soil content of, 353–354
factors affecting availability of, 355
Silicon fertilizers, 355–356
Slags as liming materials, types of, 379
Slowly available nitrogen fertilizers (*see* Nitrogen fertilizers, slowly available)
SMP method for determining lime requirement of soils, 376–377
Sodic soils, reclamation of, 399–401
Sodium
content of in earth's crust, 351
deficiency symptoms, 76
effect of
on plant growth, 352
on soil properties, 352
essentiality for plants with C_4 photosynthetic pathway, 352
exchangeable amounts in soil, 352
fertilizers containing (*see* Sodium fertilizers)
functions in plant growth, 76
plant, effect of on forage palatability, 353
plants requiring, 76
role in C_4 photosynthetic pathway, 76
in water economy of plants, 76
soil (*see also* Sodic soils)
forms of, 351
specific roles in plant metabolism, 76
uptake of, by plants, 352
Sodium bicarbonate, use in soil testing, 204
Sodium fertilizers, 352
Sodium metasilicate fertilizer, 355
Sodium molybdate fertilizer, 350
Sodium nitrate fertilizer, 353
Sodium selenite fertilizer, 358
Soil
acidulation of (*see* Acidulation of soil)
buffer activity of (*see* Buffer activity, soil)
factors affecting, 370–373
lime requirement of (*see* Lime requirement, of soil)
Soil acidity (*see also* pH, potential acidity)
active, determination of, 374

determination in soil testing, 376–377
effect of
carbon dioxide on, 370
on denitrification, 144
fertilizers on, 371–372
on potassium availability, 247–248
on potassium fixation, 241
removal of cations on, 371
soluble salts on, 370
sulfur materials on, 383–385
factors affecting, 370–373
nature of, 486–494
potential, 374
role of
aluminosilicate clay minerals in production of, 367
aluminum and iron polymers in development of, 368–370
humus in production of, 367
Soil aeration, effect of
on denitrification, 143
on potassium availability, 247
Soil air (*see* Air in soil)
Soil analysis (*see also* Soil testing)
chemical, methods used for, 436–437
Soil compaction (*see* Soil structure)
Soil erosion, value of winter cover crops in reducing, 594
Soil fertility
building, 555–556
effect of level on fertilizer application, 437–442
evaluation of (*see* Soil fertility evaluation)
level of, as related to
efficient use of irrigation, 524
time of fertilizer application, 495–499
water use efficiency, 513
as related to season, 409
Soil fertility evaluation, approaches employed, 405
Soil fertility index, as related to soil test results, 437–442
Soil–fertilizer reaction products (*see* Phosphorus, soil–fertilizer reaction products)
Soil loss, as affected by cropping system, 591–592

Soil management (*see* Crop management)
Soil moisture (*see also* Moisture, soil)
  fertilizer placement, 523
  micronutrient absorption, 523
  nitrogen absorption, 517
  nutrient absorption, 516–524
  phosphorus absorption, 520
  potassium absorption, 520–523
  potassium availability; losses, 520–523
Soil nitrogen (*see* Nitrogen, soil)
Soil organic matter
  effect of
    cropping systems on, 568–572
    on denitrification, 141
    winter cover crops on level of, 593
  functions of in soil, 573
  and liming, 387
Soil pH (*see* pH; Soil acidity; Acidity, soil)
Soil phosphorus, beneficial effects in increasing level of, 488
Soil phosphorus level, as affected by cropping system, 591
Soil physical condition, effect of liming on, 387
Soil potassium level, as affected by cropping system, 591
Soil productivity, 563–568
Soil reaction (*see* Reaction, pH; Acidity, soil; Soil acidity)
Soil sampling (*see also* Soil testing)
  areas involved in, 430
  depth of, 434–435
  effect of soil variability on, 429
  number of subsamples required, 432–434
  time of, 435
  tools used for, 430
Soil solution phosphorus (*see also* Phosphorus, soil solution), 178
  movement by diffusion and mass flow, 179–180
Soil structure, effect on root penetration, 469–473
Soil temperature (*see also* Temperature, soil)
  effect of
    on denitrification, 144

on potassium availability, 244
  tillage on, 577–580
Soil test recommendations, for nitrogen, 447–453
  as related to crop responses on high-testing soils, 453
  types of, 443–451
  use of lime, based on, 376–377
Soil test results
  calibration with crop responses, 437–442
  interpretation of, 442–453
  recommendations based on, for different yield levels, 448
  summaries of, 454
Soil testing (*see also* Microbiological methods; Soil tests, biological; Greenhouse tests)
  chemical, 436–437
  determination of, lime requirement in, 376–377
  importance of to agricultural lime industry, farmer, and fertilizer industry, 454
  objectives of, 428
  soil sampling for, 429–435
Soil tests, biological (*see also* Biological soil tests)
  calibration of results of, 437–442
  recommendations base on (*see* Soil test recommendations)
  results of, 442–453
Soil texture, as factor in selection of liming program, 374
Soil tilth (*see also* Soil structure) as affected by continuous cropping and crop rotation, 590
Spacing, plant, effect on potassium availability, 254
Sphalerite as source of soil sulfur, 292
Starter fertilizers, 479, 483–484
Stored soil water as related to fertilizer use, 524–528
Strip tests, use on farmers' fields, 427
Stubble mulching as conservation tillage technique, 574
Subsoil, suitability for crop production, 502

Subsoil fertilization, specific effects of, 502–504
Sulfatase, activity in soils, 281
Sulfate
  adsorbed, 270–272
    at various depths in different soils, 271
    movement in soil by mass flow and diffusion, 267
    reduction of, in soils, 272
  soil
    adsorbed (*see* Sulfate adsorption)
    adsorption of, factors affecting, 271
    coprecipitated with calcium carbonate, 272
    fluctuations of, 269
  soluble
    behavior in soil, 267
    quantities in soil, 266
Sulfate adsorption, soil mechanisms, 270–272
Sulfate salts, effect on soil pH, 271
Sulfates
  concentrations in soils, 267
  reduction of in water-logged soils, 276
  soluble, soil, effect of cations, 267–271
Sulfides, forms in water-logged soils, 272–273
Sulfur (*see also* Sulfur dioxide; Sulfates; Sulfides)
  content in ocean waters, 267
  crop requirements for, 283
  cycling in soils, 268
  deficiency symptoms in plants, 63
  determination of in soil testing, 284
  effect of
    on nitrogen utilization by plants, 66
    oil synthesis in plants, 65
  efficiency of absorption of by different crops, 283
  elemental
    behavior in soils, 286
    effect of particle size on oxidation in soil, 286
    effect of properties on oxidation in soil, 287
    effect of rate and placement on oxidation of, in soil, 287
    as a fertilizer, 286

forms and content of in atmosphere, 267
functions of
  in plant metabolism, 65
  plant nutrition, 62–66
inorganic
  reduced forms, in soils, 272
  soil mineralization of, factors affecting, 279–282
minerals containing, 266
occurrence of elemental S
  in gypsum deposits, 266
  as oil acidulant, 393–395
  oxidation in soils, factors affecting, 273–276
  in salt domes, 266
  suspensions of, as sulfur-containing fertilizer, 287
  in volcanic deposits, 266
in organic mineral deposits, 266
organic transformation in soils, 279–282
oxidation by soil microorganisms, 273–274
in soils (*see* Sulfur oxidation), effect of
  plant, volatilization of, 282
  temperature on, 275
soil, adsorbed, 270
  easily soluble sulfate, 267
  formation of, under reducing conditions, 272
  forms of, 267–278
  HI reducible, 278
  organic, carbon-bonded, 278
  practical aspects of, 283
  volatilization of, 282
sources of in soils, 266–267
Sulfur-bentonite, as sulfur fertilizer, 287
Sulfur-coated urea (*see* Urea, sulfur-coated)
Sulfur-containing fertilizers, 286–289
Sulfur cycle, 268
Sulfur cycling in soils, organisms responsible for, 278
Sulfur dioxide
  atmospheric content of, 267
  as atmospheric pollutant, 267
  beneficial effects of atmospheric emissions of, 267

Sulfur oxidation
    effect of soil properties on, 276
    in soil, effect of
        aeration on, 276
        moisture on, 276
        organisms responsible for, 273
        pH on, 276
Sulfur-oxidizing microorganisms, 273
Sulfuric acid, as a soil acidulant, 394
Sustainable agriculture, definition and
        discussion, 561–562

T

Temperature
    effect of
        on mineralization of soil sulfur, 274
        on plant nutrient deficiencies,
            409–410
    as a factor in plant growth, 18–21
    soil, effect of on adsorption of soil
        phosphorus, 201
Thiobacillus, sulfur-oxidizing bacteria,
    273
Thiourea, urease inhibitor, 171
Thomas slag (*see* Basic slag)
Till plant, as conservation tillage
        technique, 574
Tillage, conservation (*see also*
        Conservation tillage; Stubble
        mulching, No till; Till plant),
        574–575
    depth of, effect of on selection of
        liming program, 388
    effect of
        on soil, properties, 577–580
        on soil erosion, 576–577
        on soil organic matter, 126–130,
            572–576
        in relation to soil organic matter,
            568–572
Tissue tests (*see* Plant analyses)
Top-dressed, as a method of fertilizer
        application, 479
Total carbonate equivalent of liming
        materials, 380–384
Total phosphate, of phosphate
        fertilizers, 207

Tourmaline, boron-containing mineral,
        337
Toxic atmospheric substances, 28–29
Toxicity of molybdenum in ruminants,
        349
Triple or concentrated superphosphate
        (TSP or CSP) (*see* Phosphate
        fertilizers)
Triple superphosphate fertilizer,
        manufacture of, 210
Tropical soils, determination of lime
        requirement of, 377

U

Urea
    advances in production of, 165
    behavior of in soil, 165–166
    biuret levels of, 165
    environmental factors influencing
        effectiveness as a fertilizer, 166
    formaldehyde treatment of, 169
    hydrolysis of to ammonia in soils, 165
    importance of proper placement of,
        166–167
    improving effectiveness of as
        fertilizer, 166–167
    losses from surface applications of,
        150, 166
    management in soils, 166–167
    placement of with respect to seed,
        166–167
    properties and fertilizer uses of, 165
    sulfur-coated, 170
Urea-ammonium nitrate solution,
        properties and use of, 162–163
Urea-formaldehyde, activity index of,
        170
Urea-formaldehyde fertilizers,
        169–170
Urea-phosphate fertilizers, 168
Urea-sulfate fertilizers, 168
Urea-sulfur as sulfur-containing
        fertilizer, 288
Urea-sulfuric acid as sulfur-containing
        fertilizer, 288
Urea-Z fertilizers (UZ), 170

Urease
    effect on hydrolysis of urea, 165
    factors affecting activity in soils, 166
Urease inhibitors, 171–172

V

Vanadium
    crops responding to, 75
    plant requirements and content of, 75
    role in plant and animal growth, 75
    specific functions in plant metabolism, 75
Vermiculite, 2:1 clay mineral, 85–86
    as a source of soil potassium, 231
Volatilization of nitrogen, 148–152
Volcanic sulfur, 267

W

Water (*see* Moisture; Soil moisture)
Water absorption by plants as affected by fertilization, 515–516
Water losses from soil as affected by cropping system, 513–515
Water relations, soil, as affected by tillage, 577–579
Water requirement of plants (*see also* Water use efficiency), effect of plant nutrient supply on, 516–523
Water-soluble phosphate of fertilizers, 207
Water use efficiency
    as affected by fertility level on unirrigated soils, 517
    effect of plant nutrient supply on, 516–523
Winter cover crops (*see also* Cropping systems, Cover crops), 592–594
    effect of
        on addition of nitrogen to soil, 593
        on soil organic matter level, 593
    use of for grazing, 592
    value of in reducing soil erosion, 594
Winter killing of crops, effect of fertilization on, 475

"With seed," as method of fertilizer application, 479

Y

Yield (*see also* Crop yields)
    effect of increasing, on production costs, 543
    maximum profit benefits of, 546–550
Yield level (*see* Crop yields)
Yield potential, 442–443
    as a goal for growers, 448, 542

Z

Zinc
    appearance of crop deficiency symptoms, 72
    concentration in plants, 71
    content in earth's crust, 319
    crop sensitivity to deficiency, 72, 325
    differential uptake by crops, 325
    effect of
        climate on availability, 325
        flooding on availability, 325
        liming on availability, 322
        nitrogen fertilizers on availability, 325
        pH on availability, 322
        pH on solubility, 320–321
    fertilizers containing (*see* Zinc fertilizers)
    forms absorbed by plants, 71
        function of in plant nutrition, 72
    forms of, 320–321
        in solution, 320–321
    interactions with
        other nutrients, 323
        phosphate fertilizers, 326
        mechanisms of adsorption, 322
    movement in soil, factors affecting, 321
    organic complexes of, 321
    plant, soil, adsorption
        amounts of, 319
        availability to plants, factors affecting, 322–325

Zinc (*cont.*)
    by carbonate minerals, 322
    by clay minerals, 322
    complex forms of, 321
    deficiency induced by phosphate
      fertilizers, 323–325
    deficient geographical areas of, 320
    by metal oxides, 322
    in solutions, 320

    reaction with organic matter,
      323
    specific functions of, in plant growth,
      72
    function of in plant nutrition, 72
Zinc chelate fertilizers, 326
Zinc fertilizers, application of,
    326
Zinc sulfate fertilizer, 326